T0293784

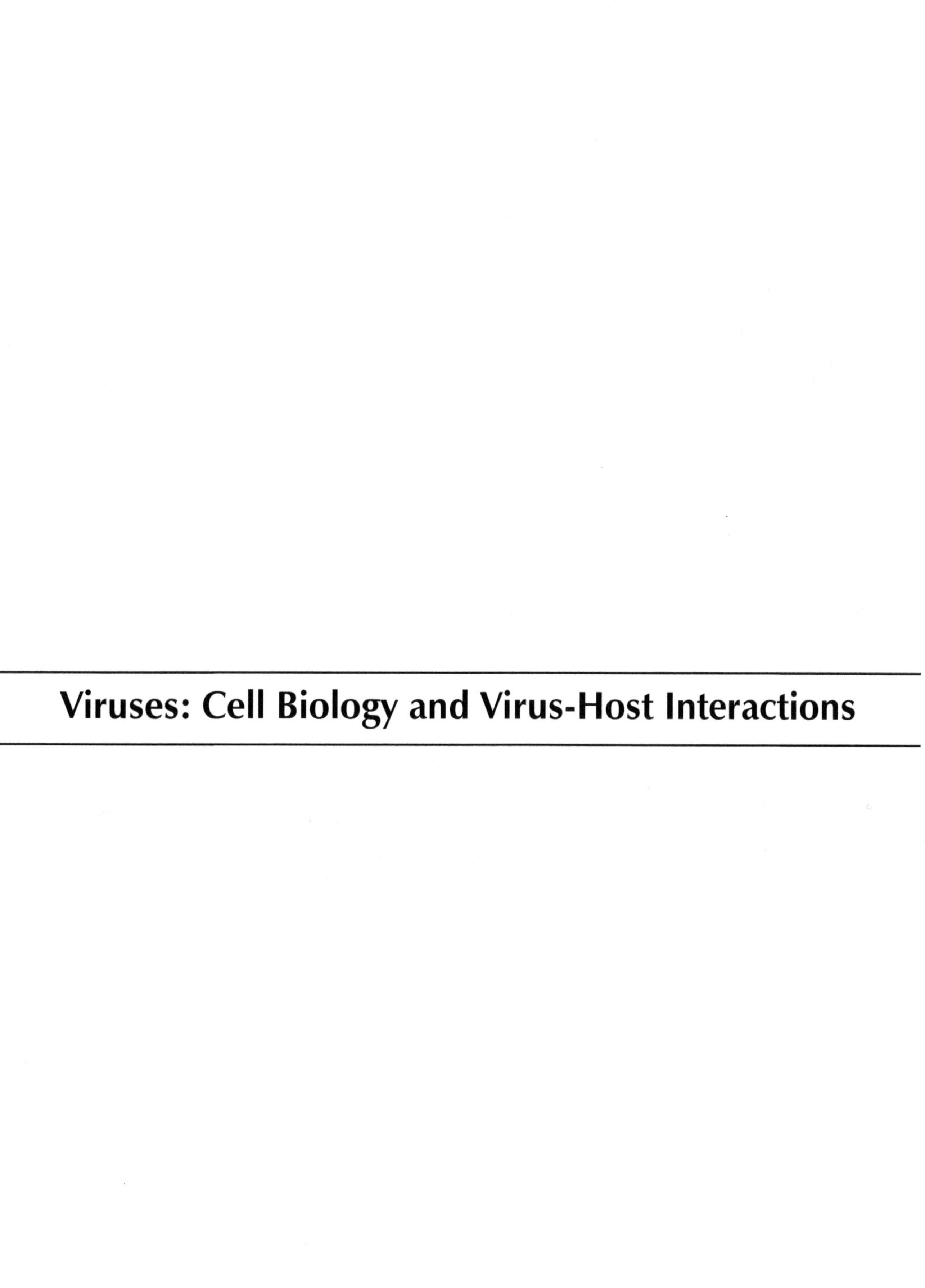

Viruses: Cell Biology and Virus-Host Interactions

Viruses: Cell Biology and Virus-Host Interactions

Editor: Kristin Allen

Cataloging-in-Publication Data

Viruses : cell biology and virus-host interactions / edited by Kristin Allen.
p. cm.
Includes bibliographical references and index.
ISBN 978-1-63927-822-0
1. Viruses. 2. Viruses--Receptors. 3. Cytology. 4. Host-virus relationships. I. Allen, Kristin.
QR360 .V57 2023
576.64--dc23

American Medical Publishers,
41 Flatbush Avenue,
1st Floor, New York,
NY 11217, USA

ISBN 978-1-63927-822-0 (Hardback)

Contents

Preface

The main aim of this book is to educate learners and enhance their research focus by presenting diverse topics covering this vast field. This is an advanced book which compiles significant studies by distinguished experts in the area of analysis. This book addresses successive solutions to the challenges arising in the area of application, along with it; the book provides scope for future developments.

Viruses are intracellular parasites that are found in all types of life forms, from plants and animals to prokaryotes and archaebacteria. These obligate infectious agents subvert various essential molecular and cellular processes of the host cell. Viruses exploit host cells to replicate and proliferate from cell to cell and from host to host which leads to an increase in their own population within the host. Infection begins when viruses attach to the surface of the host cell. This means that the viruses bind to one or more cellular receptors, including proteins, carbohydrates and lipids. Viruses not only show the ability to evade immune responses to infect and replicate host cells, but also show the ability to control cellular stress responses such as degradation pathways, autophagy and cell death. Vector-borne viruses usually persist within the vector, but not on other hosts. Viruses often assemble within virus factories, but they can also mature and fold their surface and fusion proteins, depending on the endoplasmic reticulum (ER), Golgi apparatus, and subsequent extracellular pathways. This book is a compilation of chapters that discuss the most vital concepts related to the cell biology of viruses as well as virus-host interactions. It aims to serve as a resource guide for students and experts alike and contribute to the growth of the research on viruses.

It was a great honour to edit this book, though there were challenges, as it involved a lot of communication and networking between me and the editorial team. However, the end result was this all-inclusive book covering diverse themes in the field.

Finally, it is important to acknowledge the efforts of the contributors for their excellent chapters, through which a wide variety of issues have been addressed. I would also like to thank my colleagues for their valuable feedback during the making of this book.

Editor

1

Paneth Cells during Viral Infection and Pathogenesis

Mayumi K. Holly and Jason G. Smith *

Department of Microbiology, University of Washington, Box 357735, 1705 NE Pacific St., Seattle, WA 98195, USA; mkh24@uw.edu

* Correspondence: jgsmith2@uw.edu

Abstract: Paneth cells are major secretory cells located in the crypts of Lieberkühn in the small intestine. Our understanding of the diverse roles that Paneth cells play in homeostasis and disease has grown substantially since their discovery over a hundred years ago. Classically, Paneth cells have been characterized as a significant source of antimicrobial peptides and proteins important in host defense and shaping the composition of the commensal microbiota. More recently, Paneth cells have been shown to supply key developmental and homeostatic signals to intestinal stem cells in the crypt base. Paneth cell dysfunction leading to dysbiosis and a compromised epithelial barrier have been implicated in the etiology of Crohn's disease and susceptibility to enteric bacterial infection. Our understanding of the impact of Paneth cells on viral infection is incomplete. Enteric α-defensins, produced by Paneth cells, can directly alter viral infection. In addition, α-defensins and other antimicrobial Paneth cell products may modulate viral infection indirectly by impacting the microbiome. Here, we discuss recent insights into Paneth cell biology, models to study their function, and the impact, both direct and indirect, of Paneth cells on enteric viral infection.

Keywords: Paneth cell; defensin; virus; regulated secretion; dense core vesicles

1. Location and Secretory Function of Paneth Cells

Paneth cells are located at the base of the crypts of Lieberkühn in the small intestine of various animals and are interspersed amongst the intestinal stem cells from which they differentiate [1–8]. Unlike other differentiated epithelial cell types (goblet cells, enteroendocrine cells, tuft cells, and enterocytes), which migrate out of the intestinal crypt, Paneth cells remain within the crypt base [9]. Also, unique to Paneth cells are their long lifespans (≈30 days) compared to the other differentiated epithelial cell types (≈3–5 days) [10–12]. Paneth cells are characterized by apically located, large, cytoplasmic, and electron-dense granules that contain a mixture of antimicrobial peptides and proteins, cytokines, scaffolding molecules, and proteases. Mature granules are surrounded by an acidic mucopolysaccharide complex, which is visible as a halo in electron micrographs [13]. Like many dedicated secretory cells that undergo regulated secretion, Paneth cells package only a subset of their secreted proteins into their cytoplasmic granules. They also utilize constitutive secretion mechanisms common to most epithelial cells. Because of a lack of long-term culture models for Paneth cells that have only been recently resolved (see Section 7 below), the mechanism by which Paneth cells specifically package contents into their granules remains poorly understood. However, general models of regulated vs constitutive secretion in other cell systems likely also apply to Paneth cells [14–16]. A series of studies have elucidated secreted products, both granule-dependent and independent, that are specific to Paneth cells through a variety of methods including laser capture microdissection and single cell transcriptomics [1,3,7,17–22]. The most well-known and characterized granule contents are the antimicrobial peptides and proteins; however, other proteins have also been localized to these structures. A subset of notable granule constituents is discussed below.

2. Granule Contents (Regulated Secretion)

Antimicrobials. Several types of antimicrobial peptides and proteins are packaged into and secreted via Paneth cell granules including enteric α-defensins, lysozyme, secretory phospholipase A2 ($sPLA_2$), angiogenin-4 (Ang4), RegIIIγ, and α1-antitrypsin. Collectively, these molecules have broad antimicrobial activity against a wide range of organisms. Enteric α-defensins are the most abundant secreted product [23,24]. Additionally, of the antimicrobial products packaged into Paneth cell granules, only α-defensins have known anti-viral activity, which has been recently reviewed [25,26]. α-defensins are small, cationic, and amphipathic peptides with two β-sheets stabilized by three disulfide bonds [25,27,28]. Humans encode genes for two α-defensins, human defensin 5 (HD5), and human defensin 6 (HD6), whereas mice encode over 25 different enteric α-defensin genes, although only a subset is expressed abundantly in any given mouse strain [29]. The broad antimicrobial activity of HD5, including potent antiviral activity, is well documented [25,26,30,31], while HD6 has a unique mode of action and functions by trapping bacteria and fungi [32].

The other antimicrobial constituents of the Paneth cell granules are not known to be active against viruses. Lysozyme is an enzyme that cleaves peptidoglycan in the cell walls of bacteria [33]. Mice encode two genes for lysozyme; one is expressed in Paneth cells and the other is expressed in macrophages [34]. In contrast, humans encode only one lysozyme gene that is expressed in both Paneth cells and macrophages. RegIIIβ and RegIIIγ (HIP/PAP in humans) are antibacterial C-type lectins that target peptidoglycan of Gram-positive bacteria [21,35]. Human and mouse $sPLA_2$ catalyzes the hydrolysis of phospholipids and is bactericidal for Gram-positive, but not Gram-negative, bacteria [36,37]. Mouse Ang4 is a bactericidal member of the RNase superfamily with activity against both Gram-positive and Gram-negative bacteria, although the human ortholog angiogenin is not localized to Paneth cells [20]. α1-antitrypsin is a serine protease inhibitor with some antimicrobial activity including inhibition of hemolysis by enteropathogenic *Escherichia coli* and *Cryptosporidium parvum* infection [38]. A naturally occurring peptide derived from this protein also has anti-HIV activity [39]. Of these antimicrobial peptides and proteins, Paneth cells are the sole epithelial source within the intestine of α-defensins [40], lysozyme [41], $sPLA_2$ [42], and Ang4 [20], while RegIIIγ, RegIIIβ, and α1-antitrypsin are also produced by other epithelial cell types [43,44].

Cytokines. Two cytokines have been identified in Paneth cell granules: interleukin (IL)-17A [19] and tumor necrosis factor-α (TNF-α) [45]. IL-17A, a molecule that was traditionally thought to be produced only by immune cells, was found specifically in Paneth cells using laser capture microdissection and has been localized to granules by immunoelectron microscopy [19]. IL-17A is an important factor during immune responses to invading pathogens through stimulation of antimicrobial peptide production and cytokine secretion [46]. Following hepatic ischemia and reperfusion injury, high amounts of IL-17A were detected in the serum and small intestines of treated mice [47]. Paneth cells were identified as the source of this IL-17A by a reduction in IL-17A levels in mice lacking Paneth cells due to deletion of *Sox9* in the intestinal epithelium.

TNF-α is typically produced by immune cells and functions to potentiate systemic inflammation. TNF-α mRNA was first identified in Paneth cells in 1990 [45] and was then localized specifically to the granules [48]. Endotoxin treatment of mice results in degranulation of Paneth cells and the subsequent release of TNF-α into the crypt lumen [48]. However, the exact role that Paneth cell-derived TNF-α plays in homeostasis and disease remains to be determined.

Proteases. A characteristic of regulated secretion is the need to package granule contents at high density and in forms that are not cytotoxic. To limit cytotoxicity, the granule contents of Paneth cells, like those of other cells that use regulated secretion, are often subject to post-translational activation. One common paradigm is the proteolytic cleavage of an inactive precursor into a mature, active form. This mechanism has been well-characterized for enteric α-defensins. In the mouse, matrix metalloproteinase 7 (MMP7) is co-packaged in Paneth cell granules and activates pro-defensins [49]. This cleavage can occur intracellularly, although packaged enteric α-defensins are not completely processed [50]. MMP7 also cleaves a related class of antimicrobial peptides, called cryptdin-related sequences peptides, that are unique to the mouse [51]. In humans, HD5 and HD6 are processed by Paneth cell trypsin [30]. Trypsin is packaged into granules as a zymogen and is activated through an unknown mechanism upon degranulation; thus, no intracellular processing of HD5 and HD6 occurs. There are three isoforms of trypsin that are produced in the pancreas, but Paneth cells appear to only produce two of the three isoforms. Remarkably, mouse Paneth cells do not appear to make their own trypsin, and *MMP7* is not expressed in the human intestinal mucosa.

3. Non-Granule Products (Constitutive Secretion)

In addition to their granule contents, Paneth cells also secrete proteins through constitutive secretion. These include innate immune molecules (e.g., interferon- β (IFN-β) and IL-1β) common to many epithelial cells as well as homeostatic cues essential for maintenance of the stem cell niche that are unique to Paneth cells: Wnt, epidermal growth factor (EGF), and Notch ligands [52].

Wnt proteins are membrane-bound ligands critical for development, cell migration, and cell polarization [53]. Within the intestine, Wnts are required for proper development and health of stem cells [54,55]. Several Wnts are expressed in the small intestine by Paneth cells and by non-epithelial stromal cells, including Wnt2b, Wnt3, Wnt4, Wnt5a, Wnt6, and Wnt9b [56]. However, only a subset of these (Wnt3, Wnt6, and Wnt9b) is expressed in isolated intestinal crypts, indicating they play a direct role in stem cell maintenance [56]. Wnt signaling in intestinal development and homeostasis has been studied most extensively in the mouse and is supported by in vitro studies in the recently developed enteroid model (see Section 7 below) [54,56,57]. Wnt signaling controls the expression of some Paneth cell-specific secretion products (e.g., MMP7, enteric α-defensins), transcription factors required for Paneth cell differentiation (SOX9), and signaling molecules that dictate proper positioning along the crypt/villus axis (EphB2 and EphB3) (see Section 6 below) [58].

EGF is important for crypt cell proliferation through activation of the ERK pathway. EGFR, the receptor for EGF, is expressed on intestinal stem cells [52,59]. Thus, close association of the EGF-producing Paneth cells and intestinal stem cells in the crypt base facilitates stem cell proliferation. However, the mitogenic potential of Paneth cell-derived EGF is checked by tight regulation of the EGF-ErbB pathway [59].

Paneth cells express Notch ligands, such as Dll1 and Dll4 on their membranes, which bind the Notch receptor on intestinal stem cells [52]. Notch receptor engagement results in activation of target genes important in intestinal stem cell homeostasis, such as *Hes1*. Inhibition of this signaling pathway either through small molecules or genetic knockout of *Hes1*, *Notch1*, or *Notch2* results in increased goblet cell differentiation in the mouse small intestine [60–64]. In contrast, activation of the Notch pathway results in increased proliferation in the stem cell compartment with a concomitant loss in secretory cell differentiation [60].

4. Mechanisms of Packaging

Two potential mechanisms that are not necessarily mutually exclusive have been proposed for sorting of specific cargo into dense core vesicles (DCVs) such as Paneth cell granules: sorting by entry and sorting by retention [65,66]. In the sorting by entry model, proteins must interact with a specific receptor to enter a budding DCV [65]. Sorting by entry can be further subdivided into two classes: sorting by aggregation and sorting by insertion of the protein into lipid rafts [67]. Sorting by aggregation occurs when proteins targeted to DCVs aggregate via a pH- and cation-dependent mechanism or are nucleated by scaffolding molecules [65–68]. The other subclass of sorting by entry entails insertion of a C-terminal domain directly into lipid rafts of the trans-Golgi network (TGN). In contrast, the sorting by retention model posits that all proteins are initially packaged into immature DCVs, but only specific proteins are retained [65,67]. Proteins not destined for DCVs are removed in constitutive-like vesicles; however, the basis for identifying which proteins are to be removed is not known for all proteins [67]. Studies using a variety of different cell types have found evidence for all of these mechanisms, suggesting that the method of sorting may be protein- or cell type-dependent [65,69].

Dense core vesicle formation begins with vesicle budding from the TGN [67,69]. Following release from the TGN, vesicles undergo a series of maturation steps to form mature DCVs as they traffic to the plasma membrane [69]. It has been proposed that two populations of DCVs exist in a secretory cell, a readily releasable pool (RRP), and a reserve pool (RP). The RRP is primed for release upon stimulation with a given ligand, whereas the RP requires several additional steps to be released (trafficking, docking, and priming). The applicability of these general models to Paneth cells has yet to be elucidated.

Lysozyme is the only Paneth cell granule constituent for which packaging has been worked out in detail. Two recent studies demonstrated a role for NOD2, RIP2, LRRK2, and RAB2a in this process [70,71]. NOD2 sensing of microbiome-derived peptidoglycan is necessary for lysozyme sorting into Paneth cell granules [71]. In the absence of either LRRK2 or RIP2 within the granule, lysozyme is initially packaged into immature granules in the subapical region but is not retained in the mature granules due to selective targeting for lysosomal degradation [70,71]. RAB2a, a protein involved in vesicle sorting in *Caenorhabditis elegans* [72], is recruited to Paneth cell granules in an LRRK2- and RIP2-dependent manner and is important for correct DCV trafficking [70,71]. Interestingly, enteric α-defensin packaging into Paneth cell DCVs was unaffected when this pathway was perturbed. In contrast, Paneth cell packaging and secretion can be altered more broadly during bacterial infection. Bel and colleagues found that infection of mice with *Salmonella enterica* serovar Typhimurium (STm) induces endoplasmic reticulum (ER) stress in Paneth cells, leading to accumulation of autophagosomes containing lysozyme [73]. Under these conditions, lysozyme was secreted through the secretory autophagy pathway. In addition, although not co-localized with lysozyme in vesicles, α-defensin packaging into Paneth cell granules was also perturbed. Therefore, further studies on the mechanisms of packaging of lysozyme and other granule contents, the maturation of DCVs in Paneth cells, and alternative secretion pathways under homeostatic and inflammatory conditions are needed.

5. Mechanisms of Secretion

Granule contents of Paneth cells appear to be regulated at the level of secretion rather than at the level of transcription or translation. However, there is some data for transcriptional regulation that contributes to differences in expression of specific paralogs of mouse α-defensins [74,75]. Quantitative analyses are complicated, because Paneth cell expansion, which has been shown to occur in response to infection and injury [76–78], would also manifest as apparent transcriptional upregulation. Additionally, an appropriate factor that is unique to Paneth cells has not been identified to normalize for cell numbers.

Two non-mutually exclusive modes of secretion, holocrine (cell extrusion from the intestinal epithelium) and merocrine (secretory granule release by exocytosis), have been put forth to explain Paneth cell degranulation [79–81]. Furthermore, there is conflicting data on the types of physiologic stimuli that trigger secretion. Merocrine secretion was initially posited as the sole mechanism of Paneth cell degranulation and is consistent with the long lifespan of Paneth cells in vivo [79]. In these experiments, crypts isolated from the small intestine of mice were exposed to bacteria and bacterial products, and antibacterial activity in the supernatant was measured to assess the stimulatory capacity of the ligands. From these experiments, it was concluded that Paneth cells degranulate in response to lipopolysaccharide (LPS), muramyl dipeptide, lipid A, lipoteichoic acid, and live bacteria. However, these experiments are complicated by the instability of purified intestinal crypts; once isolated, crypts rapidly undergo anoikis.

Paneth cells also undergo merocrine secretion in response to cholinergic agents and specific cytokines [2,80,82–86]. The enteric nervous system is in close proximity with intestinal crypts, and stimulation of the epithelium triggers secretion by Paneth and goblet cells [87]. In conjunction with this phenomenon is the idea that eating stimulates Paneth and goblet cell secretion in order to prepare the intestine for the arrival of food and associated micro-organisms, which is consistent with the observation that Paneth cell granules accumulate in fasting animals [3,8,88]. Nervous stimulation can be modeled in vitro through the addition of cholinergic agents, chemicals that function as or mimic neurotransmitters, to cell cultures containing Paneth cells [79,80,89]. Moreover, treatment of mice with cholinergic agents, such as carbamylcholine and aceclidine, stimulates Paneth cells to degranulate [82] (Figure 1). Both in vitro and in vivo, crypts exposed to cholinergic agents still maintain morphologically distinct Paneth cells, but they contain fewer granules and large cytoplasmic vacuoles, which are suggestive of merocrine secretion. IL-4 and IL-13 treatment of small intestinal tissue explants induces degranulation without death of the Paneth cells, which is consistent with merocrine secretion [90]. Additionally, the model for merocrine secretion is in agreement with the concept of two separate pools of granules. Paneth cells secrete the RRP immediately following stimulation, which is then repopulated from the RP.

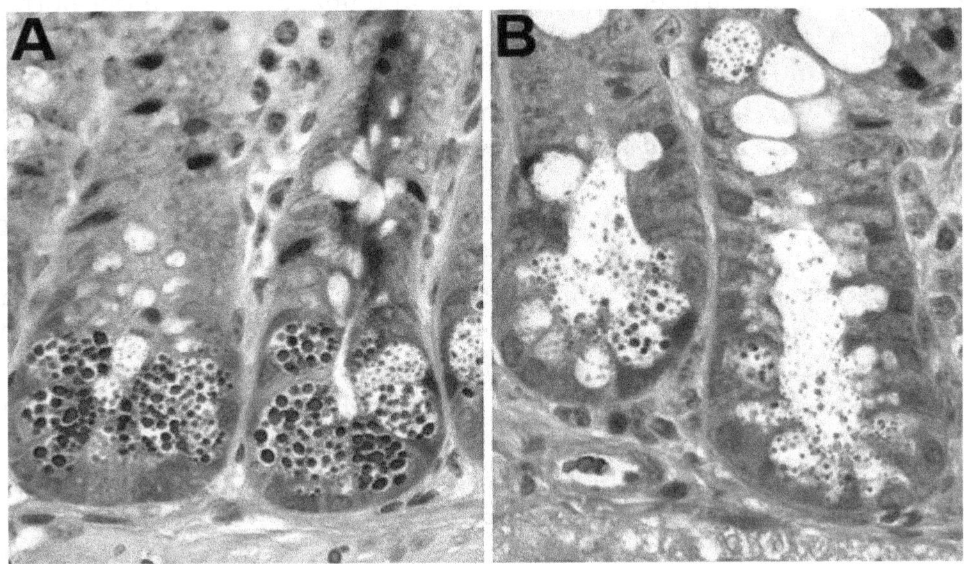

Figure 1. Mouse Paneth cell degranulation. (**A**) Trichrome staining of untreated mouse intestinal epithelium showing prominent Paneth cell granules (red); (**B**) mouse crypts 7 min after intraperitoneal injection with 10 μg/g aceclidine showing granule release into the crypt lumen.

Recently, a second model of Paneth cell secretion was proposed in which Paneth cells die and extrude into the lumen upon IFN-γ treatment [80]. These authors found no evidence for the previously described bacterial product-dependent Paneth cell degranulation or extrusion. While IFN-γ-dependent Paneth cell extrusion is likely a physiologically relevant event [91,92], extrusion is unlikely the only mechanism of Paneth cell degranulation, since it requires Paneth cell death followed by regeneration of Paneth cells from intestinal stem cells and would be inconsistent with a long lifespan in vivo. Since compelling evidence exists for both models of Paneth cell secretion, it is possible that one operates primarily under homeostatic conditions (merocrine), while the other functions during inflammation (holocrine).

Zinc has long been known to be concentrated in Paneth cells [93,94]; however, the functional importance of zinc in Paneth cells was unknown until recently. ZnT2, a zinc transporter, localized specifically to Paneth cell granules [95]. Mice lacking ZnT2 exhibited abnormal Paneth cell granules. Moreover, loss of ZnT2 resulted in reduced bacterial killing by crypt secretions compared to wild-type secretions, suggesting that Paneth cell secretion was impaired by the lack of zinc.

6. Paneth Cell Development

Paneth cell origins. All of the intestinal epithelial lineages are derived from intestinal stem cells. Intestinal crypts contain two classes of multipotent intestinal stem cells: $Lgr5^+$ crypt base columnar cells (CBCs) and +4 label retaining cells (LRCs) [52]. $Lgr5^+$ CBCs undergo symmetrical division and stochastically become another CBC or a transit amplifying cell [96]. Of the subset of CBC daughter cells that become transit amplifying cells, some will differentiate into Paneth cells and remain in the crypt base. The remainder of the transit amplifying cells will migrate out of the stem cell compartment and differentiate into one of the other intestinal epithelial cell types along the villi. Recent studies have shown that the +4 LRCs are transit amplifying cells that are precursors of Paneth cells and enteroendocrine cells [97]. This population is not well-defined by markers (*Bmi1*, *mTert*, *Hopx*, and *Lrig1*), because expression of these markers is not exclusive to the crypt base [9]. However, lineage tracing experiments, as well as studies on sensitivity to radiation treatment, have identified a population of cells at approximately the +4 position that retain label for a prolonged period of time, are actively cycling, and are sensitive to radiation damage. Thus, under normal conditions, Paneth cells and enteroendocrine cells differentiate from CBCs via the +4 LRCs; however, there is evidence for plasticity in the intestinal compartment whereby the +4 LRCs and even more committed lineages such as enterocytes can revert to a more stem cell state when the CBCs are lost due to damage [97,98]. Inflammation and physical damage can also stimulate an expansion of the Paneth cell compartment [76,78].

Signaling pathways involved in the expression of specific gene products. Many of the Paneth cell-specific gene products are controlled by Wnt signaling [99–101]. When Wnt is present, it binds to its receptor Frizzled, leading to an intracellular signaling cascade that culminates in translocation of β-catenin to the nucleus and association with specific members of the T cell factor (TCF) family to act in concert to turn on transcription of target genes [52,53]. The promoters of enteric α-defensin genes in mice, *DEFA5* (HD5) and *DEFA6* (HD6) in humans, and *Mmp7* in both species contain consensus sequences for β-catenin/TCF-binding sites [99,101–103]. Co-transfection of luciferase reporters driven by mouse enteric α-defensin, *DEFA5*, and *DEFA6* promoters with activated β-catenin and TCF4 constructs greatly increased luciferase activity over basal activity [99]. Additionally, mutation of the β-catenin/TCF-binding sites in the *DEFA6* promoter significantly reduced luciferase activity of these constructs. Furthermore, deletion of *Tcf4* in the embryonic mouse small intestine significantly reduced enteric α-defensin expression [58]. Thus, the Wnt-β-catenin/TCF axis is important in regulating expression of specific Paneth cell products.

Although lysozyme is also packaged into Paneth cell granules, its promoter does not contain β-catenin/TCF-binding sites [58,99]. The specific signaling pathways involved in lysozyme expression in Paneth cells are difficult to parse from the pathways involved in Paneth cell differentiation, since lysozyme staining is often used to quantify Paneth cell numbers. Inhibition of PI3K in mouse enteroids results in increased lysozyme expression; however, this effect is likely mediated through expansion of the Paneth cell population [104].

Similarly, fibroblast growth factor receptor-3 (FGFR-3) signaling has also been implicated in the expression of Paneth cell-specific secretory products but also appears to affect Paneth cell development. $Fgfr^{-/-}$ mice have fewer crypts, fewer Paneth cells per crypt, and reduced expression of lysozyme and mouse enteric α-defensin-5 [105]. Interestingly, although these mice had fewer Paneth cells, the Paneth cells that did develop looked morphologically normal. Moreover, treatment of a human colorectal cancer cell line (Caco-2) with FGFR-3 ligands resulted in increased expression of the Paneth cell specific markers HD5, HD6, and lysozyme, and increased TCF4/β-catenin activity [106]. Thus, FGFR-3 signaling plays an important role in Paneth cell differentiation and gene product expression.

Signaling pathways involved in morphology. In addition to unique gene expression, secretory morphology is a defining feature of Paneth cells. The genetic factors that contribute to this characteristic are not completely known, and there is evidence that signaling pathways that specify Paneth cell secretory morphology are distinct from those that control the expression of Paneth cell secretory products. For example, Wnt/β-catenin signaling is dysregulated in some tumors that express Paneth cell gene products without a secretory morphology [58]. Additionally, in mice, Paneth cell effectors are expressed prior to the development of mature Paneth cells, which begins 7 days after birth and reaches adult levels 30 days after birth [107–109].

Wnt signaling is critical for Paneth cell formation in addition to Paneth cell-specific gene expression [110]. Ectopic expression of Wnt3 in mouse enteroids (see Section 7 below) induces Paneth cell differentiation [56]. Conditional deletion of APC, a protein complex that prevents β-catenin translocation to the nucleus, in intestinal epithelial cells results in increased Paneth cell differentiation and mislocalization out of the crypt base in the small intestine and Paneth cell formation in the colon, which does not normally contain Paneth cells [100]. Conversely, expression of a hypomorphic allele of the β-catenin gene led to reduced numbers of granular cells in intestinal crypts and an associated decrease in production of lysozyme and Ang4. Abnormal Paneth cell localization in these models may be due to disruption of the EphB receptor/B-type ephrin gradient. EphB receptors are TCF/β-catenin responsive receptors required to correctly position Paneth cells along the crypt/villus axis [102]. *EphB3* is normally expressed in the crypt up to the +4 LRC position. Cells expressing this receptor are positioned inversely to a gradient of the ligand ephrin-B1, expression of which is decreased by TCF/β-catenin signaling [102,111]. Thus, deletion of *EphB3* also leads to mislocalization of Paneth cells outside of the crypt base [102]. Interestingly, these mislocalized Paneth cells lack nuclear β-catenin but express lysozyme, demonstrating that Paneth cell morphology and lysozyme expression can occur in the absence of β-catenin stabilization and translocation to the nucleus.

Several transcription factors are essential for the specification and differentiation of Paneth cells [110]. MATH1/ATOH1 is negatively regulated by Notch signaling [52] and is an essential transcription factor for differentiation of the secretory lineage [112]. Loss of *Math1* in vivo results in loss of Paneth cells, goblet cells, and enteroendocrine cells. Additionally, activation of PI3K by the neuregulin receptor ErbB3 negatively regulates *Math1*, controlling Paneth cell numbers in vivo [104]. Downstream of MATH1 is GFI1, a zinc finger transcriptional repressor [113]. Loss of *Gfi1* in mice results in a similar phenotype to *Math1*$^{-/-}$ mice with a loss of Paneth cells and fewer goblet and enteroendocrine cells [113]. Another transcription factor important for Paneth cell development is SPDEF. *Spdef*$^{-/-}$ mice do not have mature Paneth cells or goblet cells but do express markers of secretory lineage commitment in their intestines [114]. These mice also express higher levels of Dll1, a Notch ligand typically expressed on secretory progenitors, in their crypts, suggesting that SPDEF functions promote differentiation of Dll1^{+} cells into the secretory lineage. Thus, *Spdef* plays a role upstream of Paneth cell and goblet cell specification and prior to full Paneth cell differentiation. Finally, *Sox9*, a member of the Sox family of transcription factors, is a proximal factor required for Paneth cell development. Deletion of *Sox9* in intestinal epithelial cells results in loss of Paneth cells and enlargement of the intestinal crypts with no effect on other lineages [115,116]. In the absence of Paneth cells, proliferating cells occupied the base of the crypts [115].

Autophagy and the unfolded protein response. Due to their longevity, Paneth cells are sensitive to defects in the autophagy pathway, which is the regulated lysosomal degradation of organelles and cellular proteins that have been damaged by various cellular processes. Autophagy proteins ATG16L1, ATG5, and ATG12 form a complex that catalyzes microtubule-associated protein light chain 3 (LC3) lipidation [117]. ATG7 is critical for association of ATG5 and ATG12 [118]. Mice deficient in *Atg16l1*, *Atg5*, or *Atg7* exhibit reduced numbers of granules, increased numbers of cytoplasmic vacuoles, degenerating mitochondria, and diffuse cytoplasmic staining of lysozyme in their Paneth cells [22,119]. Interestingly, these phenotypes are dependent upon concomitant infection with murine norovirus (MNV), a member of the *Caliciviridae* family [120]. Mice lacking *Atg16l1* specifically in intestinal epithelial cells (*Atg16l1*$^{\Delta IEC}$) were also more susceptible to MNV-triggered epithelial damage with pronounced loss of Paneth cells upon dextran sulfate sodium (DSS) treatment [121]. This effect was likely mediated through a role of ATG16L1 in blocking necroptosis, a form of programmed cell death, in intestinal epithelial cells. Mice deficient for *Atg16l1* in their intestinal epithelial cells also show impaired responses to STm, notably decreased antimicrobial peptide expression, elevated inflammation, and increased bacterial translocation compared to wild-type mice [122]. However, mice expressing a hypomorphic allele of *Atg16l1* were not more susceptible to *Listeria monocytogenes* infection [22]. The discrepancy in these results could be due to the difference between Gram-negative and Gram-positive bacteria or the difference between expressing a hypomorphic allele and cleanly knocking out the gene in a specific cell lineage.

An additional protein, IRGM1 in mice (IRGM in humans), is potentially involved in autophagy and Paneth cell physiology. IRG proteins are similar to dynamins in their ability to control membrane fusion and vesicle trafficking, and IRGM1 has recently been linked to regulating autophagy [123]. *Irgm1*$^{-/-}$ mice exhibit Paneth cell abnormalities similar to *Atg16l1*$^{-/-}$ mice. The Paneth cell granules of these mice were abnormal in size with less dense granules. This reduction in density was associated with increased halos around the electron dense granules as visualized by transmission electron microscopy. Additionally, lysozyme-positive cells were found outside of the crypt, indicating dysregulation of Paneth cell localization. Notably, *Irgm1*$^{-/-}$ mice are more susceptible to ileal injury following DSS treatment compared to wild-type mice. This is unexpected, as DSS treatment typically causes disease in the colon and not the small intestine. An important caveat to these studies is the impact of the microbiome on the observed phenotypes. *Irgm1*$^{-/-}$ mice re-derived under specific pathogen free (SPF) conditions have only a modest increase in susceptibility to DSS treatment compared to SPF wild-type mice [124]. Additionally, goblet and Paneth cells did not exhibit any abnormal phenotypes. Since the bacterial communities differed significantly between the conventionally reared mice and the SPF mice, this is another example in which a combination of susceptibility genes and environmental factors determines the disease outcome.

The secretory function of Paneth cells demands significant protein synthesis and folding in the ER, making them vulnerable to ER stress. Several proteins involved in the unfolded protein response (UPR) have been implicated in Paneth cell function [117,125]. X box binding protein 1 (XBP1) is a transcription factor essential for the UPR. Crypts from *Xbp1*$^{-/-}$ mice lack cells with electron dense granules [126] and have reduced lysozyme and enteric α-defensin staining, indicative of Paneth cell loss [125]. These mice undergo spontaneous enteritis [125,126], implicating *Xbp1* in initiating intestinal inflammation. Interestingly, *Xbp1* deletion specifically in Paneth cells resulted in spontaneous enteritis similar to mice in which *Xbp1* was deleted in all intestinal epithelial cells. Moreover, *Xbp1*$^{-/-}$ mice were more susceptible to *L. monocytogenes* infection, which is likely due to the reduced bactericidal activity of Paneth cells, since crypt supernatants from *Xbp1*$^{-/-}$ mice were unable to kill *L. monocytogenes* [125]. The UPR and autophagy pathways both function to modulate intestinal inflammation. Loss of key components of both pathways (XBP1 and ATG16L1) resulted in ileitis more severe than with the loss of either protein alone [126]. Clearly, *Xbp1* and *Atg16l1* expression in intestinal epithelial cells, and specifically Paneth cells, plays an important role in mediating intestinal homeostasis. However, given the variable effects of the microbiome on inflammation in the gastrointestinal (GI) tract [120,124], it is important to consider the interplay of host susceptibility genes and environmental factors.

7. Models for Studying Paneth Cells

Mice, rats, chickens, equines, nonhuman primates, and humans have Paneth cells in their small intestines, although they are not found in all animals, such as sheep, cows, and seals [3,4,127–129]. The prevalence of Paneth cells in animals is not fully known due to both a lack of thorough investigation and the absence of uniform criteria for identifying Paneth cells in GI tracts. Moreover, there appears to be no clear evolutionary relationship that explains the presence or absence of Paneth cells among species. Most of what we know about Paneth cell development and function is derived from mouse studies. Until recently, with the exception of limited studies of short-lived intestinal explants or crypt preparations, Paneth cells could only be studied in vivo due to a lack of a culture system. In 2009, pioneering work by the Clevers group established a new model for culturing primary intestinal epithelial cells in vitro, termed enteroids [130]. Enteroids are three-dimensional tissue culture structures that contain the diversity of intestinal epithelial cell types found in the small intestine or colon [52]. They are untransformed and can be derived from adult intestinal epithelial, embryonic, or induced pluripotent stem cells, and their method of derivation determines the nomenclature used to describe them [131]. They can also be genetically manipulated, cryopreserved, and cultured continuously for extended periods of time. The robust formation of Paneth cells in mouse enteroids allows the manipulation and investigation of these cells in vitro for the first time, and enteroids are being used extensively in recent studies of intestinal development. Paneth cells in enteroids secrete mature α-defensins that maintain antimicrobial activity, and they can be used to model oral infection by microinjecting bacteria and viruses into the lumen of enteroids, which is topologically equivalent to the small intestinal lumen [89,132–134]. Enteroids provide a unique opportunity to study the interaction of enteric viruses with primary, intestinal epithelial cells. They have been shown to support the replication of rotaviruses, noroviruses, enteroviruses, and adenoviruses [132,135–138]. Additionally, the utility of enteroids in investigating host-pathogen interactions has become well established.

8. Paneth Cell Functions In Vivo

Mucus barrier augmentation and stem cell protection. Unlike the colon, the small intestinal epithelium is coated by a relatively porous mucus layer that is attached loosely to epithelial cells [139]. α-defensins and other antimicrobial proteins are concentrated in the mucus to enhance the mucosal barrier [140]. The combination of antimicrobial factors and mucus makes it difficult for luminal bacteria to interact directly with intestinal epithelial cells. Additionally, because they are secreted in close proximity to the crypt base stem cells, α-defensins (estimated at 4–24 mM) and other antimicrobial factors can reach very high concentrations within the crypt lumen [79]. Therefore, they function to protect the stem cells from the microbiome or invading pathogens.

α-defensins are potently antiviral against both enveloped and non-enveloped viruses [25,26], and the concentrations of α-defensins that are antiviral in cell culture are within the physiologic range estimated in the gut. A major mechanism whereby α-defensins inhibit enveloped viruses is by preventing viral glycoprotein interactions with their cellular receptors leading to inhibition of fusion [25,26]. In contrast, α-defensins inhibit non-enveloped viruses by binding to and stabilizing the capsid, thereby perturbing uncoating [25,26]. Direct neutralization of viral infection by α-defensins has not yet been demonstrated in vivo; however, recent studies have shown that for at least one enteric virus, mouse adenovirus 2 (MAdV-2), infection is not only resistant to neutralization by α-defensins but is actually increased or enhanced [132]. MAdV-2 is a natural pathogen of mice that infects that GI tract without causing overt disease [141]. MAdV-2 infection of traditional cell culture was increased ≈2-fold by mouse enteric α-defensins [132]. In an enteroid model, naturally secreted α-defensins increased MAdV-2 infection ≈2- to 4-fold by increasing the initial interaction of the virus with the host cell, allowing both receptor-dependent and -independent entry. Furthermore, wild-type mice infected orally with MAdV-2 shed more virus in feces than mice lacking functional enteric α-defensins. Thus, α-defensin-mediated enhancement of MAdV-2 infection occurs in two-dimensional cell culture, three-dimensional enteroid culture, and in vivo. Consistent with this finding, MAdV-2 infects Paneth cells in vivo [127]. Infection by the closely related MAdV-1 is inhibited by the same enteric α-defensins that enhance MAdV-2 infection [142], suggesting that resistance to α-defensin neutralization and the ability to utilize these host defense peptides to increase infection may be a consequence of the evolution of this fecal/orally transmitted virus under selective pressure from abundant α-defensin secretion in the mouse intestine.

Although there are only a few human adenovirus (HAdV) serotypes that are known to primarily cause gastroenteritis in humans (HAdV-12, HAdV-40, and HAdV-41), many respiratory serotypes also infect the GI tract [143–145]. HD5 is a potent inhibitor of only a subset of HAdVs, which cause disease outside of the GI tract [146]. The remaining serotypes of HAdV are either resistant or enhanced by HD5. These experiments were performed in a transformed lung epithelial cell line that is commonly used for studying HAdV. This effect was recently recapitulated in a more physiologically relevant system, human enteroids, which supports HAdV replication [135]. Thus, like MAdV-2 resistance to mouse enteric α-defensins, HD5 resistance of fecal/orally transmitted HAdVs may also reflect viral evolution.

Another case in which barrier integrity mediated by Paneth cells is compromised by viral infection occurs in intestinal dysbiosis following human immunodeficiency virus (HIV) infection. HIV exerts significant and detrimental effects on the GI immune system. Early studies of HIV-infected patients revealed severe enteropathy throughout the GI tract [147]. Moreover, HIV rapidly depletes $CCR5^{+}CD4^{+}$ T cells in the lamina propria following initial infection, and T cell numbers in the GI associated lymphoid tissue (GALT) do not fully recover even under effective anti-retroviral therapy [147]. This disruption may also impact epithelial integrity and function, which has been examined in a simian immunodeficiency virus (SIV) intestinal loop model. As early as 2.5 d post-infection, SIV disrupts intestinal epithelial barrier integrity [148]. Loss of epithelial integrity was accompanied by increased IL-1β expression in Paneth cells, suggesting a possible mechanism whereby Paneth cells respond to viral infection by secreting IL-1β, although the Paneth cells themselves were not infected. IL-1β in turn decreases expression of tight junction proteins, leading to epithelial permeability. In a previous study by the same group, rhesus macaques with simian AIDS (SAIDS) due to chronic SIV infection had increased numbers of Paneth cells per crypt, although the Paneth cells had reduced numbers of cytoplasmic granules [149]. Interestingly, this correlated with an increase in enteric α-defensin RNA levels but a decrease in α-defensin protein levels in Paneth cells, which was not observed in the ileal loop model [148]. The authors proposed that Paneth cells were undergoing frequent secretion in macaques with SAIDS, accounting for the absence of detectable α-defensins in Paneth cells [149]. The loss of α-defensin protein in Paneth cells correlated with an increase in bacterial and eukaryotic infections of the GI tract, suggesting that either the defensins are ineffective against pathogens in the context of SAIDS or that α-defensin protein synthesis is lost in SAIDS, resulting in reduced antimicrobial activity. Therefore, although SIV and HIV are not tropic for Paneth cells, infection by these viruses significantly impacts Paneth cell function.

Microbiome composition. Paneth cell antimicrobial products play a direct role in shaping the intestinal microbiome. Two mouse models have been critical in understanding the impact of α-defensins in particular on the composition of the host microbiome: *Mmp7*$^{-/-}$ mice and *DEFA5* transgenic mice [150]. MMP7 is produced by mouse Paneth cells and converts pro-defensins into mature enteric α-defensins [49]. Thus, the *Mmp7*$^{-/-}$ mouse is a functional α-defensin knockout in the ileum, although this is an imperfect model, because mature α-defensins that result from processing by other luminal proteases can be recovered in the caecum and colon [151]. *DEFA5* transgenic mice express HD5 at levels comparable to native mouse enteric α-defensins under the control of the *DEFA5* promoter, which restricts expression to Paneth cells [31]. *Mmp7*$^{-/-}$ mice have an altered microbiome relative to wild-type littermate control mice with an increase in Firmicutes species and a decrease in Bacteroidetes species in the ileum [150,152]. In contrast, *DEFA5* transgenic mice had a reciprocal change with a decrease in Firmicutes and an increase in Bacteroidetes. Interestingly, while *Mmp7*$^{-/-}$ mice were colonized by segmented filamentous bacteria (SFB), *DEFA5* transgenic mice lacked a detectable SFB population. Moreover, *DEFA5* transgenic mice and wild-type mice with low levels of SFB have fewer $CD4^{+}$ T cells expressing IL-17A than *Mmp7*$^{-/-}$ or wild-type mice with high levels of SFB.

The mechanism of SFB modulating T cell development in the gut has been partially elucidated. Upon SFB-intestinal epithelial cell contact, which only occurs in the ileum, type 3 innate lymphoid cells secrete IL-22, which stimulates production of epithelial serum amyloid A proteins 1 and 2 (SAA1/2) from intestinal epithelial cells [153]. It is important to note that SAA1/2 production could be due to SFB contact with intestinal epithelial cells or a combination of SFB contact and IL-22 signaling. SAA1/2 could then act directly upon Th17 cells. Thus, SFB colonization impacts Th17 effector functions in the GI tract, shaping not only the composition of the microbiome but also potentially the functionality of the GALT [150,153,154].

MNV has two strain-dependent disease profiles, causing either a persistent intestinal infection or an acute multi-organ infection [155]. Antibiotic treatment of wild-type mice infected with the persistent

MNV CR6 strain completely abolished MNV replication through an interferon-λ signaling-dependent mechanism [156]. The authors hypothesized that the bacterial components of the microbiome dampened the interferon-λ signaling pathway, and antibiotic treatment relieved this dampening allowing for clearing of MNV. Although no specific component of the microbiome (i.e., LPS) was identified as important for this phenotype, the dependence of MNV infection on the presence of the enteric microbiome underscores the potential importance of Paneth cells during MNV infection in vivo. Since antimicrobial peptides secreted by Paneth cells modulate the composition of the microbiome, Paneth cells may indirectly impact MNV infection by adjusting the population of bacteria in the gut.

Paneth cells may have a similar indirect effect on poliovirus. Oral infection of mice with poliovirus requires transgenic expression of the poliovirus receptor (PVRtg) and the absence of functional interferon signaling (*Ifnar*$^{-/-}$) [157]. Upon antibiotic treatment, fewer PVRtg/*Ifnar*$^{-/-}$ mice succumb to poliovirus infection compared to untreated mice [157], implicating the enteric microbiome in poliovirus infection and pathogenesis. Further investigations revealed that poliovirus was thermo-stabilized by feces from conventional but not germ-free or antibiotic-treated mice and by LPS [157,158]. Moreover, poliovirus infection in vitro was enhanced by purified LPS from several different enteric bacteria, but not by other components of the small intestinal milieu (i.e., peptidoglycan, mucin). Given that Paneth cells are a critical element in shaping the microbiome, it is possible that Paneth cells indirectly impact poliovirus infection by modulating the bacterial phyla present in the small intestine.

Innate immune sensing. Paneth cells also play a key role in host defense by sensing microorganisms. There are numerous innate sensing pathways including inflammasomes, RIG-I-like receptors, and toll-like receptors (TLRs). The importance of TLR-mediated sensing of bacteria by Paneth cells has been specifically addressed. MyD88 is a signaling adaptor protein involved in transducing signals from TLRs, IL-1 receptor, and IL-18 receptor. Deletion of *MyD88* or expression of a dominant negative allele of *MyD88* results in decreased production of RegIIIγ, RELMβ, and RegIIIβ by the intestinal epithelium and increased susceptibility to STm [43,159]. *MyD88* expression was selectively reconstituted in Paneth cells of *MyD88*$^{-/-}$ mice through use of the Paneth cell-specific cryptdin-2 (CR2) promoter [43]. CR2-MyD88 transgenic mice infected with STm had fewer bacteria in their mesenteric lymph node compared to infected *MyD88*$^{-/-}$ mice, suggesting that Paneth cell intrinsic sensing and function is sufficient to restore the mucosal barrier. Interestingly, STm infection did not increase expression of the *MyD88*-dependent gene program in conventional mice, indicating that the microbiome stimulates Paneth cells to express antimicrobial genes.

Beyond a direct role in microbial sensing, Paneth cells also alter the immune response through the activities of their secreted products. α-defensins in particular function as chemoattractants [25]. HD5 induces migration of macrophages and mast cells but not immature DCs and recruits both naïve and memory T cells. Although HD5 functions as a chemokine for a variety of immune cell types, the receptor for α-defensins is unidentified. Thus, HD5 is a potent chemotactic signal for immune cells, which may also be true for enteric α-defensins from other species.

α-defensins also function to modulate the adaptive immune response. Mouse adenovirus-1 (MAdV-1) causes disease in the central nervous system, manifesting as encephalitis [141]. Additionally, MAdV-1 infection is sensitive to neutralization by mouse and human enteric α-defensins [142]. *Mmp7*$^{-/-}$ mice are more susceptible to oral MAdV-1 challenge than WT mice; however, there was no evidence of direct enteric α-defensin antiviral activity. Rather, *Mmp7*$^{-/-}$ mice did not develop germinal centers and had delayed neutralizing antibody responses to MAdV-1 [142]. However, *Mmp7*$^{-/-}$ mice did not generate a sufficient antibody response to MAdV-1 infection; they were not universally impaired for humoral immunity. Intranasal challenge of wild-type and *Mmp7*$^{-/-}$ mice with ovalbumin (OVA) resulted in development of OVA specific antibodies with similar kinetics and similar titers for both genotypes. Moreover, parenteral challenge of both wild-type and *Mmp7*$^{-/-}$ mice with MAdV-1 abrogated any difference between the two genotypes, strongly implicating a role for α-defensins in the GI tract in generating an adaptive immune response to infection. It is possible that enteric α-defensins function as an adjuvant or in a paracrine signaling fashion to stimulate an adaptive immune response to other enteric pathogens.

9. Conclusions

Paneth cells play an integral part in stem cell maintenance, microbiome shaping, and host defense. Although Paneth cells were identified over a century ago, there still remains a lot to be learned about their basic biology and role in microbial infections, particularly viral infections. Although likely, it remains to be demonstrated that naturally produced enteric α-defensins either directly or indirectly impact the pathogenesis of viruses besides human and mouse adenoviruses. Other Paneth cell-derived antimicrobial peptides and proteins that shape the host intestinal microbiome also likely influence viral infection, an area that warrants further research. Enteroids have revolutionized the study of Paneth cell development and biology, allowing direct comparisons between species. Additionally, they are a reductionist system amenable to host-pathogen studies that bridge traditional cell culture and in vivo models. Studies that take advantage of the multiple platforms now available to study Paneth cells are likely to provide exciting new insights into the biology of these fascinating cells.

Author Contributions: M.K.H. and J.G.S. wrote the review.

Acknowledgments: This work was supported by R01 AI104920 from the National Institute for Allergy and Infectious Diseases to Jason G. Smith.

References

1. Porter, E.M.; Liu, L.; Oren, A.; Anton, P.A.; Ganz, T. Localization of human intestinal defensin 5 in Paneth cell granules. *Infect. Immun.* **1997**, *65*, 2389–2395. [PubMed]
2. Satoh, Y.; Yamano, M.; Matsuda, M.; Ono, K. Ultrastructure of Paneth cells in the intestine of various mammals. *J. Electron. Microsc. Tech.* **1990**, *16*, 69–80. [CrossRef] [PubMed]
3. Porter, E.M.; Bevins, C.L.; Ghosh, D.; Ganz, T. The multifaceted Paneth cell. *Cell. Mol. Life Sci.* **2002**, *59*, 156–170. [CrossRef] [PubMed]
4. Takehana, K.; Masty, J.; Yamaguchi, M.; Kobayashi, A.; Yamada, O.; Kuroda, M.; Park, Y.S.; Iwasa, K.; Abe, M. Fine structural and histochemical study of equine Paneth cells. *Anat. Histol. Embryol.* **1998**, *27*, 125–129. [CrossRef] [PubMed]
5. Clevers, H.C.; Bevins, C.L. Paneth cells: Maestros of the small intestinal crypts. *Annu. Rev. Physiol.* **2013**, *75*, 289–311. [CrossRef] [PubMed]
6. Bevins, C.L.; Salzman, N.H. Paneth cells, antimicrobial peptides and maintenance of intestinal homeostasis. *Nat. Rev. Microbiol.* **2011**, *9*, 356–368. [CrossRef] [PubMed]
7. Ouellette, A.J. Paneth cells and innate mucosal immunity. *Curr. Opin. Gastroenterol.* **2010**, *26*, 547–553. [CrossRef] [PubMed]
8. Klein, S. On the nature of the granule cells of Paneth in the intestinal glands of mammals. *Am. J. Anat.* **1906**, *5*, 315–330. [CrossRef]
9. Tan, D.W.; Barker, N. Intestinal stem cells and their defining niche. *Curr. Top. Dev. Biol.* **2014**, *107*, 77–107. [PubMed]
10. Troughton, W.D.; Trier, J.S. Paneth and goblet cell renewal in mouse duodenal crypts. *J. Cell Biol.* **1969**, *41*, 251–268. [CrossRef] [PubMed]
11. Cheng, H.; Merzel, J.; Leblond, C.P. Renewal of Paneth cells in the small intestine of the mouse. *Am. J. Anat.* **1969**, *126*, 507–525. [CrossRef] [PubMed]
12. Ireland, H.; Houghton, C.; Howard, L.; Winton, D.J. Cellular inheritance of a Cre-activated reporter gene to determine Paneth cell longevity in the murine small intestine. *Dev. Dyn.* **2005**, *233*, 1332–1336. [CrossRef] [PubMed]
13. Selzman, H.M.; Liebelt, R.A. A cytochemical analysis of Paneth cell secretion in the mouse. *Anat. Rec.* **1961**, *140*, 17–22. [CrossRef] [PubMed]

14. Sheshachalam, A.; Srivastava, N.; Mitchell, T.; Lacy, P.; Eitzen, G. Granule protein processing and regulated secretion in neutrophils. *Front. Immunol.* **2014**, *5*, 448. [CrossRef] [PubMed]
15. Adler, K.B.; Tuvim, M.J.; Dickey, B.F. Regulated mucin secretion from airway epithelial cells. *Front. Endocrinol. (Lausanne)* **2013**, *4*, 129. [CrossRef] [PubMed]
16. Hammel, I.; Meilijson, I. The econobiology of pancreatic acinar cells granule inventory and the stealthy nano-machine behind it. *Acta Histochem.* **2016**, *118*, 194–202. [CrossRef] [PubMed]
17. Stappenbeck, T.S.; Mills, J.C.; Gordon, J.I. Molecular features of adult mouse small intestinal epithelial progenitors. *Proc. Natl. Acad. Sci. USA* **2003**, *100*, 1004–1009. [CrossRef] [PubMed]
18. Haber, A.L.; Biton, M.; Rogel, N.; Herbst, R.H.; Shekhar, K.; Smillie, C.; Burgin, G.; Delorey, T.M.; Howitt, M.R.; Katz, Y.; et al. A single-cell survey of the small intestinal epithelium. *Nature* **2017**, *551*, 333–339. [CrossRef] [PubMed]
19. Takahashi, N.; Vanlaere, I.; de Rycke, R.; Cauwels, A.; Joosten, L.A.; Lubberts, E.; van den Berg, W.B.; Libert, C. IL-17 produced by Paneth cells drives TNF-induced shock. *J. Exp. Med.* **2008**, *205*, 1755–1761. [CrossRef] [PubMed]
20. Hooper, L.V.; Stappenbeck, T.S.; Hong, C.V.; Gordon, J.I. Angiogenins: A new class of microbicidal proteins involved in innate immunity. *Nat. Immunol.* **2003**, *4*, 269–273. [CrossRef] [PubMed]
21. Cash, H.L.; Whitham, C.V.; Behrendt, C.L.; Hooper, L.V. Symbiotic bacteria direct expression of an intestinal bactericidal lectin. *Science* **2006**, *313*, 1126–1130. [CrossRef] [PubMed]
22. Cadwell, K.; Liu, J.Y.; Brown, S.L.; Miyoshi, H.; Loh, J.; Lennerz, J.K.; Kishi, C.; Kc, W.; Carrero, J.A.; Hunt, S.; et al. A key role for autophagy and the autophagy gene *Atg16l1* in mouse and human intestinal Paneth cells. *Nature* **2008**, *456*, 259–263. [CrossRef] [PubMed]
23. Wehkamp, J.; Chu, H.; Shen, B.; Feathers, R.W.; Kays, R.J.; Lee, S.K.; Bevins, C.L. Paneth cell antimicrobial peptides: Topographical distribution and quantification in human gastrointestinal tissues. *FEBS Lett.* **2006**, *580*, 5344–5350. [CrossRef] [PubMed]
24. Bevins, C.L. Innate immune functions of α-defensins in the small intestine. *Dig. Dis.* **2013**, *31*, 299–304. [CrossRef] [PubMed]
25. Holly, M.K.; Diaz, K.; Smith, J.G. Defensins in viral infection and pathogenesis. *Annu. Rev. Virol.* **2017**, *4*, 369–391. [CrossRef] [PubMed]
26. Wilson, S.S.; Wiens, M.E.; Smith, J.G. Antiviral mechanisms of human defensins. *J. Mol. Biol.* **2013**, *425*, 4965–4980. [CrossRef] [PubMed]
27. Selsted, M.E.; Ouellette, A.J. Mammalian defensins in the antimicrobial immune response. *Nat. Immunol.* **2005**, *6*, 551–557. [CrossRef] [PubMed]
28. Lehrer, R.I.; Lu, W. α-Defensins in human innate immunity. *Immunol. Rev.* **2012**, *245*, 84–112. [CrossRef] [PubMed]
29. Shanahan, M.T.; Tanabe, H.; Ouellette, A.J. Strain-specific polymorphisms in Paneth cell α-defensins of C57BL/6 mice and evidence of vestigial myeloid α-defensin pseudogenes. *Infect. Immun.* **2011**, *79*, 459–473. [CrossRef] [PubMed]
30. Ghosh, D.; Porter, E.; Shen, B.; Lee, S.K.; Wilk, D.; Drazba, J.; Yadav, S.P.; Crabb, J.W.; Ganz, T.; Bevins, C.L. Paneth cell trypsin is the processing enzyme for human defensin-5. *Nat. Immunol.* **2002**, *3*, 583–590. [CrossRef] [PubMed]
31. Salzman, N.H.; Ghosh, D.; Huttner, K.M.; Paterson, Y.; Bevins, C.L. Protection against enteric salmonellosis in transgenic mice expressing a human intestinal defensin. *Nature* **2003**, *422*, 522–526. [CrossRef] [PubMed]
32. Chu, H.; Pazgier, M.; Jung, G.; Nuccio, S.P.; Castillo, P.A.; de Jong, M.F.; Winter, M.G.; Winter, S.E.; Wehkamp, J.; Shen, B.; et al. Human α-defensin 6 promotes mucosal innate immunity through self-assembled peptide nanonets. *Science* **2012**, *337*, 477–481. [CrossRef] [PubMed]
33. Callewaert, L.; Michiels, C.W. Lysozymes in the animal kingdom. *J. Biosci.* **2010**, *35*, 127–160. [CrossRef] [PubMed]
34. Hammer, M.F.; Schilling, J.W.; Prager, E.M.; Wilson, A.C. Recruitment of lysozyme as a major enzyme in the mouse gut: Duplication, divergence, and regulatory evolution. *J. Mol. Evol.* **1987**, *24*, 272–279. [CrossRef] [PubMed]
35. Vaishnava, S.; Yamamoto, M.; Severson, K.M.; Ruhn, K.A.; Yu, X.; Koren, O.; Ley, R.; Wakeland, E.K.; Hooper, L.V. The antibacterial lectin RegIIIγ promotes the spatial segregation of microbiota and host in the intestine. *Science* **2011**, *334*, 255–258. [CrossRef] [PubMed]

36. Koduri, R.S.; Gronroos, J.O.; Laine, V.J.; Le Calvez, C.; Lambeau, G.; Nevalainen, T.J.; Gelb, M.H. Bactericidal properties of human and murine groups I, II, V, X, and XII secreted phospholipases A(2). *J. Biol. Chem.* **2002**, *277*, 5849–5857. [CrossRef] [PubMed]
37. Qu, X.D.; Lehrer, R.I. Secretory phospholipase A2 is the principal bactericide for staphylococci and other gram-positive bacteria in human tears. *Infect. Immun.* **1998**, *66*, 2791–2797. [PubMed]
38. Janciauskiene, S.M.; Bals, R.; Koczulla, R.; Vogelmeier, C.; Kohnlein, T.; Welte, T. The discovery of α1-antitrypsin and its role in health and disease. *Respir. Med.* **2011**, *105*, 1129–1139. [CrossRef] [PubMed]
39. Munch, J.; Standker, L.; Adermann, K.; Schulz, A.; Schindler, M.; Chinnadurai, R.; Pohlmann, S.; Chaipan, C.; Biet, T.; Peters, T.; et al. Discovery and optimization of a natural HIV-1 entry inhibitor targeting the gp41 fusion peptide. *Cell* **2007**, *129*, 263–275. [CrossRef] [PubMed]
40. Ouellette, A.J.; Selsted, M.E. Paneth cell defensins: Endogenous peptide components of intestinal host defense. *FASEB J.* **1996**, *10*, 1280–1289. [CrossRef] [PubMed]
41. Deckx, R.J.; Vantrappen, G.R.; Parein, M.M. Localization of lysozyme activity in a Paneth cell granule fraction. *Biochim. Biophys. Acta* **1967**, *139*, 204–207. [CrossRef]
42. Kiyohara, H.; Egami, H.; Shibata, Y.; Murata, K.; Ohshima, S.; Ogawa, M. Light microscopic immunohistochemical analysis of the distribution of group I1 phospholipase A2 in human digestive organs. *J. Histochem. Cytochem.* **1992**, *40*, 1659–1664. [CrossRef] [PubMed]
43. Vaishnava, S.; Behrendt, C.L.; Ismail, A.S.; Eckmann, L.; Hooper, L.V. Paneth cells directly sense gut commensals and maintain homeostasis at the intestinal host-microbial interface. *Proc. Natl. Acad. Sci. USA* **2008**, *105*, 20858–20863. [CrossRef] [PubMed]
44. Molmenti, E.P.; Perlmutter, D.H.; Rubin, D.C. Cell-specific expression of α1-antitrypsin in human intestinal epithelium. *J. Clin. Investig.* **1993**, *92*, 2022–2034. [CrossRef] [PubMed]
45. Keshav, S.; Lawson, L.; Chung, L.P.; Stein, M.; Perry, V.H.; Gordon, S. Tumor necrosis factor mRNA localized to Paneth cells of normal murine intestinal epithelium by in situ hybridization. *J. Exp. Med.* **1990**, *171*, 327–332. [CrossRef] [PubMed]
46. Matsuzaki, G.; Umemura, M. Interleukin-17 family cytokines in protective immunity against infections: Role of hematopoietic cell-derived and non-hematopoietic cell-derived interleukin-17s. *Microbiol. Immunol.* **2018**, *62*, 1–13. [CrossRef] [PubMed]
47. Park, S.W.; Kim, M.; Brown, K.M.; D'Agati, V.D.; Lee, H.T. Paneth cell-derived interleukin-17A causes multiorgan dysfunction after hepatic ischemia and reperfusion injury. *Hepatology* **2011**, *53*, 1662–1675. [CrossRef] [PubMed]
48. Schmauder-Chock, E.A.; Chock, S.P.; Patchen, M.L. Ultrastructural localization of tumour necrosis factor-α. *Histochem. J.* **1994**, *26*, 142–151. [CrossRef] [PubMed]
49. Wilson, C.L.; Ouellette, A.J.; Satchell, D.P.; Ayabe, T.; Lopez-Boado, Y.S.; Stratman, J.L.; Hultgren, S.J.; Matrisian, L.M.; Parks, W.C. Regulation of intestinal α-defensin activation by the metalloproteinase matrilysin in innate host defense. *Science* **1999**, *286*, 113–117. [CrossRef] [PubMed]
50. Ayabe, T.; Satchell, D.P.; Pesendorfer, P.; Tanabe, H.; Wilson, C.L.; Hagen, S.J.; Ouellette, A.J. Activation of Paneth cell alpha-defensins in mouse small intestine. *J. Biol. Chem.* **2002**, *277*, 5219–5228. [CrossRef] [PubMed]
51. Hornef, M.W.; Putsep, K.; Karlsson, J.; Refai, E.; Andersson, M. Increased diversity of intestinal antimicrobial peptides by covalent dimer formation. *Nat. Immunol.* **2004**, *5*, 836–843. [CrossRef] [PubMed]
52. Date, S.; Sato, T. Mini-gut organoids: Reconstitution of the stem cell niche. *Annu. Rev. Cell Dev. Biol.* **2015**, *31*, 269–289. [CrossRef] [PubMed]
53. Nusse, R.; Clevers, H. Wnt/β-catenin signaling, disease, and emerging therapeutic modalities. *Cell* **2017**, *169*, 985–999. [CrossRef] [PubMed]
54. Farin, H.F.; Jordens, I.; Mosa, M.H.; Basak, O.; Korving, J.; Tauriello, D.V.; de Punder, K.; Angers, S.; Peters, P.J.; Maurice, M.M.; et al. Visualization of a short-range Wnt gradient in the intestinal stem-cell niche. *Nature* **2016**, *530*, 340–343. [CrossRef] [PubMed]
55. Schuijers, J.; Clevers, H. Adult mammalian stem cells: The role of Wnt, Lgr5 and R-spondins. *EMBO J.* **2012**, *31*, 2685–2696. [CrossRef] [PubMed]
56. Farin, H.F.; Van Es, J.H.; Clevers, H. Redundant sources of Wnt regulate intestinal stem cells and promote formation of Paneth cells. *Gastroenterology* **2012**, *143*. [CrossRef] [PubMed]

57. Sato, T.; van Es, J.H.; Snippert, H.J.; Stange, D.E.; Vries, R.G.; van den Born, M.; Barker, N.; Shroyer, N.F.; van de Wetering, M.; Clevers, H. Paneth cells constitute the niche for Lgr5 stem cells in intestinal crypts. *Nature* **2011**, *469*, 415–418. [CrossRef] [PubMed]
58. Van Es, J.H.; Jay, P.; Gregorieff, A.; van Gijn, M.E.; Jonkheer, S.; Hatzis, P.; Thiele, A.; van den Born, M.; Begthel, H.; Brabletz, T.; et al. Wnt signalling induces maturation of Paneth cells in intestinal crypts. *Nat. Cell Biol.* **2005**, *7*, 381–386. [CrossRef] [PubMed]
59. Wong, V.W.; Stange, D.E.; Page, M.E.; Buczacki, S.; Wabik, A.; Itami, S.; van de Wetering, M.; Poulsom, R.; Wright, N.A.; Trotter, M.W.; et al. Lrig1 controls intestinal stem-cell homeostasis by negative regulation of ErbB signalling. *Nat. Cell Biol.* **2012**, *14*, 401–408. [CrossRef] [PubMed]
60. Carulli, A.J.; Keeley, T.M.; Demitrack, E.S.; Chung, J.; Maillard, I.; Samuelson, L.C. Notch receptor regulation of intestinal stem cell homeostasis and crypt regeneration. *Dev. Biol.* **2015**, *402*, 98–108. [CrossRef] [PubMed]
61. Milano, J.; McKay, J.; Dagenais, C.; Foster-Brown, L.; Pognan, F.; Gadient, R.; Jacobs, R.T.; Zacco, A.; Greenberg, B.; Ciaccio, P.J. Modulation of notch processing by γ-secretase inhibitors causes intestinal goblet cell metaplasia and induction of genes known to specify gut secretory lineage differentiation. *Toxicol. Sci.* **2004**, *82*, 341–358. [CrossRef] [PubMed]
62. Suzuki, K.; Fukui, H.; Kayahara, T.; Sawada, M.; Seno, H.; Hiai, H.; Kageyama, R.; Okano, H.; Chiba, T. Hes1-deficient mice show precocious differentiation of Paneth cells in the small intestine. *Biochem. Biophys. Res. Commun.* **2005**, *328*, 348–352. [CrossRef] [PubMed]
63. Van Es, J.H.; van Gijn, M.E.; Riccio, O.; van den Born, M.; Vooijs, M.; Begthel, H.; Cozijnsen, M.; Robine, S.; Winton, D.J.; Radtke, F.; et al. Notch/γ-secretase inhibition turns proliferative cells in intestinal crypts and adenomas into goblet cells. *Nature* **2005**, *435*, 959–963. [CrossRef] [PubMed]
64. VanDussen, K.L.; Carulli, A.J.; Keeley, T.M.; Patel, S.R.; Puthoff, B.J.; Magness, S.T.; Tran, I.T.; Maillard, I.; Siebel, C.; Kolterud, A.; et al. Notch signaling modulates proliferation and differentiation of intestinal crypt base columnar stem cells. *Development* **2012**, *139*, 488–497. [CrossRef] [PubMed]
65. Dikeakos, J.D.; Reudelhuber, T.L. Sending proteins to dense core secretory granules: Still a lot to sort out. *J. Cell Biol.* **2007**, *177*, 191–196. [CrossRef] [PubMed]
66. Kim, T.; Gondre-Lewis, M.C.; Arnaoutova, I.; Loh, Y.P. Dense-core secretory granule biogenesis. *Physiology (Bethesda)* **2006**, *21*, 124–133. [CrossRef] [PubMed]
67. Park, J.J.; Loh, Y.P. How peptide hormone vesicles are transported to the secretion site for exocytosis. *Mol. Endocrinol.* **2008**, *22*, 2583–2595. [CrossRef] [PubMed]
68. Pejler, G.; Abrink, M.; Wernersson, S. Serglycin proteoglycan: Regulating the storage and activities of hematopoietic proteases. *Biofactors* **2009**, *35*, 61–68. [CrossRef] [PubMed]
69. Brunner, Y.; Schvartz, D.; Coute, Y.; Sanchez, J.C. Proteomics of regulated secretory organelles. *Mass Spectrom. Rev.* **2009**, *28*, 844–867. [CrossRef] [PubMed]
70. Wang, H.; Zhang, X.; Zuo, Z.; Zhang, Q.; Pan, Y.; Zeng, B.; Li, W.; Wei, H.; Liu, Z. Rip2 is required for Nod2-mediated lysozyme sorting in Paneth cells. *J. Immunol.* **2017**, *198*, 3729–3736. [CrossRef] [PubMed]
71. Zhang, Q.; Pan, Y.; Yan, R.; Zeng, B.; Wang, H.; Zhang, X.; Li, W.; Wei, H.; Liu, Z. Commensal bacteria direct selective cargo sorting to promote symbiosis. *Nat. Immunol.* **2015**, *16*, 918–926. [CrossRef] [PubMed]
72. Edwards, S.L.; Charlie, N.K.; Richmond, J.E.; Hegermann, J.; Eimer, S.; Miller, K.G. Impaired dense core vesicle maturation in *Caenorhabditis elegans* mutants lacking Rab2. *J. Cell Biol.* **2009**, *186*, 881–895. [CrossRef] [PubMed]
73. Bel, S.; Pendse, M.; Wang, Y.; Li, Y.; Ruhn, K.A.; Hassell, B.; Leal, T.; Winter, S.E.; Xavier, R.J.; Hooper, L.V. Paneth cells secrete lysozyme via secretory autophagy during bacterial infection of the intestine. *Science* **2017**, *357*, 1047–1052. [CrossRef] [PubMed]
74. Stockinger, S.; Duerr, C.U.; Fulde, M.; Dolowschiak, T.; Pott, J.; Yang, I.; Eibach, D.; Backhed, F.; Akira, S.; Suerbaum, S.; et al. TRIF signaling drives homeostatic intestinal epithelial antimicrobial peptide expression. *J. Immunol.* **2014**, *193*, 4223–4234. [CrossRef] [PubMed]
75. Menendez, A.; Willing, B.P.; Montero, M.; Wlodarska, M.; So, C.C.; Bhinder, G.; Vallance, B.A.; Finlay, B.B. Bacterial stimulation of the TLR-MyD88 pathway modulates the homeostatic expression of ileal Paneth cell α-defensins. *J. Innate Immun.* **2013**, *5*, 39–49. [CrossRef] [PubMed]
76. Martinez Rodriguez, N.R.; Eloi, M.D.; Huynh, A.; Dominguez, T.; Lam, A.H.; Carcamo-Molina, D.; Naser, Z.; Desharnais, R.; Salzman, N.H.; Porter, E. Expansion of Paneth cell population in response to enteric *Salmonella enterica* serovar Typhimurium infection. *Infect. Immun.* **2012**, *80*, 266–275. [CrossRef] [PubMed]

77. Roth, S.; Franken, P.; Sacchetti, A.; Kremer, A.; Anderson, K.; Sansom, O.; Fodde, R. Paneth cells in intestinal homeostasis and tissue injury. *PLoS ONE* **2012**, *7*, e38965. [CrossRef] [PubMed]
78. King, S.L.; Mohiuddin, J.J.; Dekaney, C.M. Paneth cells expand from newly created and preexisting cells during repair after doxorubicin-induced damage. *Am. J. Physiol. Gastrointest. Liver Physiol.* **2013**, *305*, G151–G162. [CrossRef] [PubMed]
79. Ayabe, T.; Satchell, D.P.; Wilson, C.L.; Parks, W.C.; Selsted, M.E.; Ouellette, A.J. Secretion of microbicidal α-defensins by intestinal Paneth cells in response to bacteria. *Nat. Immunol.* **2000**, *1*, 113–118. [CrossRef] [PubMed]
80. Farin, H.F.; Karthaus, W.R.; Kujala, P.; Rakhshandehroo, M.; Schwank, G.; Vries, R.G.; Kalkhoven, E.; Nieuwenhuis, E.E.; Clevers, H. Paneth cell extrusion and release of antimicrobial products is directly controlled by immune cell-derived IFN-γ. *J. Exp. Med.* **2014**, *211*, 1393–1405. [CrossRef] [PubMed]
81. Stappenbeck, T.S. Paneth cell development, differentiation, and function: New molecular cues. *Gastroenterology* **2009**, *137*, 30–33. [CrossRef] [PubMed]
82. Satoh, Y.; Ishikawa, K.; Oomori, Y.; Yamano, M.; Ono, K. Effects of cholecystokinin and carbamylcholine on Paneth cell secretion in mice: A comparison with pancreatic acinar cells. *Anat. Rec.* **1989**, *225*, 124–132. [CrossRef] [PubMed]
83. Satoh, Y. Effect of live and heat-killed bacteria on the secretory activity of Paneth cells in germ-free mice. *Cell Tissue Res.* **1988**, *251*, 87–93. [CrossRef] [PubMed]
84. Satoh, Y.; Ishikawa, K.; Oomori, Y.; Takeda, S.; Ono, K. Bethanechol and a G-protein activator, NaF/$AlCl_3$, induce secretory response in Paneth cells of mouse intestine. *Cell Tissue Res.* **1992**, *269*, 213–220. [CrossRef] [PubMed]
85. Satoh, Y. Atropine inhibits the degranulation of Paneth cells in ex-germ-free mice. *Cell Tissue Res.* **1988**, *253*, 397–402. [CrossRef] [PubMed]
86. Satoh, Y.; Habara, Y.; Ono, K.; Kanno, T. Carbamylcholine- and catecholamine-induced intracellular calcium dynamics of epithelial cells in mouse ileal crypts. *Gastroenterology* **1995**, *108*, 1345–1356. [CrossRef]
87. Yoo, B.B.; Mazmanian, S.K. The enteric network: Interactions between the immune and nervous systems of the gut. *Immunity* **2017**, *46*, 910–926. [CrossRef] [PubMed]
88. Ahonen, A.; Penttilä. Effects of fasting and feeding and pilocarpine on paneth cells of the mouse. *Scand. J. Gastroenterol.* **1975**, *10*, 347–352. [PubMed]
89. Wilson, S.S.; Tocchi, A.; Holly, M.K.; Parks, W.C.; Smith, J.G. A small intestinal organoid model of non-invasive enteric pathogen-epithelial cell interactions. *Mucosal. Immunol.* **2015**, *8*, 352–361. [CrossRef] [PubMed]
90. Stockinger, S.; Albers, T.; Duerr, C.U.; Menard, S.; Putsep, K.; Andersson, M.; Hornef, M.W. Interleukin-13-mediated Paneth cell degranulation and antimicrobial peptide release. *J. Innate Immun.* **2014**, *6*, 530–541. [CrossRef] [PubMed]
91. Burger, E.; Araujo, A.; Lopez-Yglesias, A.; Rajala, M.W.; Geng, L.; Levine, B.; Hooper, L.V.; Burstein, E.; Yarovinsky, F. Loss of Paneth cell autophagy causes acute susceptibility to *Toxoplasma gondii*-mediated inflammation. *Cell Host Microbe* **2018**, *23*. [CrossRef] [PubMed]
92. Raetz, M.; Hwang, S.H.; Wilhelm, C.L.; Kirkland, D.; Benson, A.; Sturge, C.R.; Mirpuri, J.; Vaishnava, S.; Hou, B.; Defranco, A.L.; et al. Parasite-induced Th1 cells and intestinal dysbiosis cooperate in IFN-γ-dependent elimination of Paneth cells. *Nat. Immunol.* **2013**, *14*, 136–142. [CrossRef] [PubMed]
93. Elmes, M.E.; Jones, J.G. Ultrastructural studies on Paneth cell apoptosis in zinc deficient rats. *Cell Tissue Res.* **1980**, *208*, 57–63. [CrossRef] [PubMed]
94. Dinsdale, D. Ultrastructural localization of zinc and calcium within the granules of rat Paneth cells. *J. Histochem. Cytochem.* **1984**, *32*, 139–145. [CrossRef] [PubMed]
95. Podany, A.B.; Wright, J.; Lamendella, R.; Soybel, D.I.; Kelleher, S.L. ZnT2-Mediated Zinc Import into Paneth cell granules is necessary for coordinated secretion and Paneth cell function in mice. *Cell. Mol. Gastroenterol. Hepatol.* **2016**, *2*, 369–383. [CrossRef] [PubMed]
96. Snippert, H.J.; van der Flier, L.G.; Sato, T.; van Es, J.H.; van den Born, M.; Kroon-Veenboer, C.; Barker, N.; Klein, A.M.; van Rheenen, J.; Simons, B.D.; et al. Intestinal crypt homeostasis results from neutral competition between symmetrically dividing Lgr5 stem cells. *Cell* **2010**, *143*, 134–144. [CrossRef] [PubMed]
97. Buczacki, S.J.; Zecchini, H.I.; Nicholson, A.M.; Russell, R.; Vermeulen, L.; Kemp, R.; Winton, D.J. Intestinal label-retaining cells are secretory precursors expressing Lgr5. *Nature* **2013**, *495*, 65–69. [CrossRef] [PubMed]

98. Tetteh, P.W.; Basak, O.; Farin, H.F.; Wiebrands, K.; Kretzschmar, K.; Begthel, H.; van den Born, M.; Korving, J.; de Sauvage, F.; van Es, J.H.; et al. Replacement of lost Lgr5-positive stem cells through plasticity of their enterocyte-lineage daughters. *Cell Stem Cell* **2016**, *18*, 203–213. [CrossRef] [PubMed]
99. Andreu, P.; Colnot, S.; Godard, C.; Gad, S.; Chafey, P.; Niwa-Kawakita, M.; Laurent-Puig, P.; Kahn, A.; Robine, S.; Perret, C.; et al. Crypt-restricted proliferation and commitment to the Paneth cell lineage following Apc loss in the mouse intestine. *Development* **2005**, *132*, 1443–1451. [CrossRef] [PubMed]
100. Andreu, P.; Peignon, G.; Slomianny, C.; Taketo, M.M.; Colnot, S.; Robine, S.; Lamarque, D.; Laurent-Puig, P.; Perret, C.; Romagnolo, B. A genetic study of the role of the Wnt/β-catenin signalling in Paneth cell differentiation. *Dev. Biol.* **2008**, *324*, 288–296. [CrossRef] [PubMed]
101. Crawford, H.C.; Fingleton, B.M.; Rudolph-Owen, L.A.; Goss, K.J.; Rubinfeld, B.; Polakis, P.; Matrisian, L.M. The metalloproteinase matrilysin is a target of β-catenin transactivation in intestinal tumors. *Oncogene* **1999**, *18*, 2883–2891. [CrossRef] [PubMed]
102. Batlle, E.; Henderson, J.T.; Beghtel, H.; van den Born, M.M.; Sancho, E.; Huls, G.; Meeldijk, J.; Robertson, J.; van de Wetering, M.; Pawson, T.; et al. β-catenin and TCF mediate cell positioning in the intestinal epithelium by controlling the expression of EphB/ephrinB. *Cell* **2002**, *111*, 251–263. [CrossRef]
103. Brabletz, T.; Jung, A.; Dag, S.; Hlubek, F.; Kirchner, T. β-catenin regulates the expression of the matrix metalloproteinase-7 in human colorectal cancer. *Am. J. Pathol.* **1999**, *155*, 1033–1038. [CrossRef]
104. Almohazey, D.; Lo, Y.H.; Vossler, C.V.; Simmons, A.J.; Hsieh, J.J.; Bucar, E.B.; Schumacher, M.A.; Hamilton, K.E.; Lau, K.S.; Shroyer, N.F.; et al. The ErbB3 receptor tyrosine kinase negatively regulates Paneth cells by PI3K-dependent suppression of Atoh1. *Cell Death Differ.* **2017**, *24*, 855–865. [CrossRef] [PubMed]
105. Vidrich, A.; Buzan, J.M.; Brodrick, B.; Ilo, C.; Bradley, L.; Fendig, K.S.; Sturgill, T.; Cohn, S.M. Fibroblast growth factor receptor-3 regulates Paneth cell lineage allocation and accrual of epithelial stem cells during murine intestinal development. *Am. J. Physiol. Gastrointest. Liver Physiol.* **2009**, *297*, G168–G178. [CrossRef] [PubMed]
106. Brodrick, B.; Vidrich, A.; Porter, E.; Bradley, L.; Buzan, J.M.; Cohn, S.M. Fibroblast growth factor receptor-3 (FGFR-3) regulates expression of Paneth cell lineage-specific genes in intestinal epithelial cells through both TCF4/β-catenin-dependent and -independent signaling pathways. *J. Biol. Chem.* **2011**, *286*, 18515–18525. [CrossRef] [PubMed]
107. Bry, L.; Falk, P.; Huttner, K.; Ouellette, A.; Midtvedt, T.; Gordon, J.I. Paneth cell differentiation in the developing intestine of normal and transgenic mice. *Proc. Natl. Acad. Sci. USA* **1994**, *91*, 10335–10339. [CrossRef] [PubMed]
108. Darmoul, D.; Brown, D.; Selsted, M.E.; Ouellette, A.J. Cryptdin gene expression in developing mouse small intestine. *Am. J. Physiol.* **1997**, *272*, G197–G206. [CrossRef] [PubMed]
109. Inoue, R.; Tsuruta, T.; Nojima, I.; Nakayama, K.; Tsukahara, T.; Yajima, T. Postnatal changes in the expression of genes for cryptdins 1-6 and the role of luminal bacteria in cryptdin gene expression in mouse small intestine. *FEMS Immunol. Med. Microbiol.* **2008**, *52*, 407–416. [CrossRef] [PubMed]
110. Van der Flier, L.G.; Clevers, H. Stem cells, self-renewal, and differentiation in the intestinal epithelium. *Annu. Rev. Physiol.* **2009**, *71*, 241–260. [CrossRef] [PubMed]
111. Clevers, H. The intestinal crypt, a prototype stem cell compartment. *Cell* **2013**, *154*, 274–284. [CrossRef] [PubMed]
112. Yang, Q.; Bermingham, N.A.; Finegold, M.J.; Zoghbi, H.Y. Requirement of Math1 for secretory cell lineage commitment in the mouse intestine. *Science* **2001**, *294*, 2155–2158. [CrossRef] [PubMed]
113. Shroyer, N.F.; Wallis, D.; Venken, K.J.; Bellen, H.J.; Zoghbi, H.Y. Gfi1 functions downstream of Math1 to control intestinal secretory cell subtype allocation and differentiation. *Genes Dev.* **2005**, *19*, 2412–2417. [CrossRef] [PubMed]
114. Gregorieff, A.; Stange, D.E.; Kujala, P.; Begthel, H.; van den Born, M.; Korving, J.; Peters, P.J.; Clevers, H. The Ets-domain transcription factor Spdef promotes maturation of goblet and Paneth cells in the intestinal epithelium. *Gastroenterology* **2009**, *137*. [CrossRef] [PubMed]
115. Mori-Akiyama, Y.; van den Born, M.; van Es, J.H.; Hamilton, S.R.; Adams, H.P.; Zhang, J.; Clevers, H.; de Crombrugghe, B. SOX9 is required for the differentiation of Paneth cells in the intestinal epithelium. *Gastroenterology* **2007**, *133*, 539–546. [CrossRef] [PubMed]

116. Bastide, P.; Darido, C.; Pannequin, J.; Kist, R.; Robine, S.; Marty-Double, C.; Bibeau, F.; Scherer, G.; Joubert, D.; Hollande, F.; et al. Sox9 regulates cell proliferation and is required for Paneth cell differentiation in the intestinal epithelium. *J. Cell Biol.* **2007**, *178*, 635–648. [CrossRef] [PubMed]
117. Hubbard, V.M.; Cadwell, K. Viruses, autophagy genes, and Crohn's disease. *Viruses* **2011**, *3*, 1281–1311. [CrossRef] [PubMed]
118. Klionsky, D.J.; Codogno, P. The mechanism and physiological function of macroautophagy. *J. Innate Immun.* **2013**, *5*, 427–433. [CrossRef] [PubMed]
119. Cadwell, K.; Patel, K.K.; Komatsu, M.; Virgin, H.W.; Stappenbeck, T.S. A common role for Atg16L1, Atg5 and Atg7 in small intestinal Paneth cells and Crohn disease. *Autophagy* **2009**, *5*, 250–252. [CrossRef] [PubMed]
120. Cadwell, K.; Patel, K.K.; Maloney, N.S.; Liu, T.C.; Ng, A.C.; Storer, C.E.; Head, R.D.; Xavier, R.; Stappenbeck, T.S.; Virgin, H.W. Virus-plus-susceptibility gene interaction determines Crohn's disease gene Atg16L1 phenotypes in intestine. *Cell* **2010**, *141*, 1135–1145. [CrossRef] [PubMed]
121. Matsuzawa-Ishimoto, Y.; Shono, Y.; Gomez, L.E.; Hubbard-Lucey, V.M.; Cammer, M.; Neil, J.; Dewan, M.Z.; Lieberman, S.R.; Lazrak, A.; Marinis, J.M.; et al. Autophagy protein ATG16L1 prevents necroptosis in the intestinal epithelium. *J. Exp. Med.* **2017**, *214*, 3687–3705. [CrossRef] [PubMed]
122. Conway, K.L.; Kuballa, P.; Song, J.H.; Patel, K.K.; Castoreno, A.B.; Yilmaz, O.H.; Jijon, H.B.; Zhang, M.; Aldrich, L.N.; Villablanca, E.J.; et al. Atg16l1 is required for autophagy in intestinal epithelial cells and protection of mice from *Salmonella* infection. *Gastroenterology* **2013**, *145*, 1347–1357. [CrossRef] [PubMed]
123. Liu, B.; Gulati, A.S.; Cantillana, V.; Henry, S.C.; Schmidt, E.A.; Daniell, X.; Grossniklaus, E.; Schoenborn, A.A.; Sartor, R.B.; Taylor, G.A. Irgm1-deficient mice exhibit Paneth cell abnormalities and increased susceptibility to acute intestinal inflammation. *Am. J. Physiol. Gastrointest. Liver Physiol.* **2013**, *305*, G573–G584. [CrossRef] [PubMed]
124. Rogala, A.R.; Schoenborn, A.A.; Fee, B.E.; Cantillana, V.A.; Joyce, M.J.; Gharaibeh, R.Z.; Roy, S.; Fodor, A.A.; Sartor, R.B.; Taylor, G.A.; et al. Environmental factors regulate Paneth cell phenotype and host susceptibility to intestinal inflammation in Irgm1-deficient mice. *Dis. Model Mech.* **2018**. [CrossRef] [PubMed]
125. Kaser, A.; Lee, A.H.; Franke, A.; Glickman, J.N.; Zeissig, S.; Tilg, H.; Nieuwenhuis, E.E.; Higgins, D.E.; Schreiber, S.; Glimcher, L.H.; et al. XBP1 links ER stress to intestinal inflammation and confers genetic risk for human inflammatory bowel disease. *Cell* **2008**, *134*, 743–756. [CrossRef] [PubMed]
126. Adolph, T.E.; Tomczak, M.F.; Niederreiter, L.; Ko, H.J.; Bock, J.; Martinez-Naves, E.; Glickman, J.N.; Tschurtschenthaler, M.; Hartwig, J.; Hosomi, S.; et al. Paneth cells as a site of origin for intestinal inflammation. *Nature* **2013**, *503*, 272–276. [CrossRef] [PubMed]
127. Takeuchi, A.; Hashimoto, K. Electron microscope study of experimental enteric adenovirus infection in mice. *Infect. Immun.* **1976**, *13*, 569–580. [PubMed]
128. Satoh, Y.; Vollrath, L. Quantitative electron microscopic observations on Paneth cells of germfree and ex-germfree Wistar rats. *Anat. Embryol.* **1986**, *173*, 317–322. [CrossRef] [PubMed]
129. Wang, L.; Li, J.; Li, J., Jr.; Li, R.X.; Lv, C.F.; Li, S.; Mi, Y.L.; Zhang, C.Q. Identification of the Paneth cells in chicken small intestine. *Poult. Sci.* **2016**, *95*, 1631–1635. [CrossRef] [PubMed]
130. Sato, T.; Vries, R.G.; Snippert, H.J.; van de Wetering, M.; Barker, N.; Stange, D.E.; van Es, J.H.; Abo, A.; Kujala, P.; Peters, P.J.; et al. Single Lgr5 stem cells build crypt-villus structures in vitro without a mesenchymal niche. *Nature* **2009**, *459*, 262–265. [CrossRef] [PubMed]
131. Stelzner, M.; Helmrath, M.; Dunn, J.C.; Henning, S.J.; Houchen, C.W.; Kuo, C.; Lynch, J.; Li, L.; Magness, S.T.; Martin, M.G.; et al. A nomenclature for intestinal in vitro cultures. *Am. J. Physiol. Gastrointest. Liver Physiol.* **2012**, *302*, G1359–G1363. [CrossRef] [PubMed]
132. Wilson, S.S.; Bromme, B.A.; Holly, M.K.; Wiens, M.E.; Gounder, A.P.; Sul, Y.; Smith, J.G. α-defensin-dependent enhancement of enteric viral infection. *PLoS Pathog.* **2017**, *13*, e1006446. [CrossRef] [PubMed]
133. Bartfeld, S.; Bayram, T.; van de Wetering, M.; Huch, M.; Begthel, H.; Kujala, P.; Vries, R.; Peters, P.J.; Clevers, H. In vitro expansion of human gastric epithelial stem cells and their responses to bacterial infection. *Gastroenterology* **2015**, *148*. [CrossRef] [PubMed]
134. Leslie, J.L.; Huang, S.; Opp, J.S.; Nagy, M.S.; Kobayashi, M.; Young, V.B.; Spence, J.R. Persistence and toxin production by *Clostridium difficile* within human intestinal organoids result in disruption of epithelial paracellular barrier function. *Infect. Immun.* **2015**, *83*, 138–145. [CrossRef] [PubMed]
135. Holly, M.K.; Smith, J.G. Adenovirus infection of human enteroids reveals interferon sensitivity and preferential infection of goblet cells. *J. Virol.* **2018**. [CrossRef] [PubMed]

136. Saxena, K.; Blutt, S.E.; Ettayebi, K.; Zeng, X.L.; Broughman, J.R.; Crawford, S.E.; Karandikar, U.C.; Sastri, N.P.; Conner, M.E.; Opekun, A.R.; et al. Human intestinal enteroids: A new model to study human rotavirus infection, host restriction, and pathophysiology. *J. Virol.* **2015**, *90*, 43–56. [CrossRef] [PubMed]
137. Ettayebi, K.; Crawford, S.E.; Murakami, K.; Broughman, J.R.; Karandikar, U.; Tenge, V.R.; Neill, F.H.; Blutt, S.E.; Zeng, X.L.; Qu, L.; et al. Replication of human noroviruses in stem cell-derived human enteroids. *Science* **2016**, *353*, 1387–1393. [CrossRef] [PubMed]
138. Drummond, C.G.; Bolock, A.M.; Ma, C.; Luke, C.J.; Good, M.; Coyne, C.B. Enteroviruses infect human enteroids and induce antiviral signaling in a cell lineage-specific manner. *Proc. Natl. Acad. Sci. USA* **2017**, *114*, 1672–1677. [CrossRef] [PubMed]
139. Johansson, M.E.; Hansson, G.C. Immunological aspects of intestinal mucus and mucins. *Nat. Rev. Immunol.* **2016**, *16*, 639–649. [CrossRef] [PubMed]
140. Meyer-Hoffert, U.; Hornef, M.W.; Henriques-Normark, B.; Axelsson, L.G.; Midtvedt, T.; Putsep, K.; Andersson, M. Secreted enteric antimicrobial activity localises to the mucus surface layer. *Gut* **2008**, *57*, 764–771. [CrossRef] [PubMed]
141. Spindler, K.R.; Moore, M.L.; Cauthen, A.N. Mouse Adenoviruses. In *The Mouse in Biomedical Research*, 2nd ed.; Fox, J.G., Davisson, M.T., Quimby, F.W., Barthold, S.W., Newcomer, C.E., Smith, A.L., Eds.; Academic Press: Burlington, MA, USA, 2007; Volume 2, pp. 49–65.
142. Gounder, A.P.; Myers, N.D.; Treuting, P.M.; Bromme, B.A.; Wilson, S.S.; Wiens, M.E.; Lu, W.; Ouellette, A.J.; Spindler, K.R.; Parks, W.C.; et al. Defensins potentiate a neutralizing antibody response to enteric viral infection. *PLoS Pathog.* **2016**, *12*, e1005474. [CrossRef] [PubMed]
143. Kosulin, K.; Geiger, E.; Vecsei, A.; Huber, W.D.; Rauch, M.; Brenner, E.; Wrba, F.; Hammer, K.; Innerhofer, A.; Potschger, U.; et al. Persistence and reactivation of human adenoviruses in the gastrointestinal tract. *Clin. Microbiol. Infect.* **2016**, *22*. [CrossRef] [PubMed]
144. Fox, J.P.; Hall, C.E.; Cooney, M.K. The Seattle Virus Watch. VII. Observations of adenovirus infections. *Am. J. Epidemiol.* **1977**, *105*, 362–386. [CrossRef] [PubMed]
145. Fox, J.P.; Brandt, C.D.; Wassermann, F.E.; Hall, C.E.; Spigland, I.; Kogon, A.; Elveback, L.R. The virus watch program: A continuing surveillance of viral infections in metropolitan New York families. VI. Observations of adenovirus infections: Virus excretion patterns, antibody response, efficiency of surveillance, patterns of infections, and relation to illness. *Am. J. Epidemiol.* **1969**, *89*, 25–50. [PubMed]
146. Smith, J.G.; Silvestry, M.; Lindert, S.; Lu, W.; Nemerow, G.R.; Stewart, P.L. Insight into the mechanisms of adenovirus capsid disassembly from studies of defensin neutralization. *PLoS Pathog.* **2010**, *6*, e1000959. [CrossRef] [PubMed]
147. Brenchley, J.M.; Douek, D.C. HIV infection and the gastrointestinal immune system. *Mucosal. Immunol.* **2008**, *1*, 23–30. [CrossRef] [PubMed]
148. Hirao, L.A.; Grishina, I.; Bourry, O.; Hu, W.K.; Somrit, M.; Sankaran-Walters, S.; Gaulke, C.A.; Fenton, A.N.; Li, J.A.; Crawford, R.W.; et al. Early mucosal sensing of SIV infection by Paneth cells induces IL-1β production and initiates gut epithelial disruption. *PLoS Pathog.* **2014**, *10*, e1004311. [CrossRef] [PubMed]
149. Zaragoza, M.M.; Sankaran-Walters, S.; Canfield, D.R.; Hung, J.K.; Martinez, E.; Ouellette, A.J.; Dandekar, S. Persistence of gut mucosal innate immune defenses by enteric α-defensin expression in the simian immunodeficiency virus model of AIDS. *J. Immunol.* **2011**, *186*, 1589–1597. [CrossRef] [PubMed]
150. Salzman, N.H.; Hung, K.; Haribhai, D.; Chu, H.; Karlsson-Sjoberg, J.; Amir, E.; Teggatz, P.; Barman, M.; Hayward, M.; Eastwood, D.; et al. Enteric defensins are essential regulators of intestinal microbial ecology. *Nat. Immunol.* **2010**, *11*, 76–83. [CrossRef] [PubMed]
151. Mastroianni, J.R.; Costales, J.K.; Zaksheske, J.; Selsted, M.E.; Salzman, N.H.; Ouellette, A.J. Alternative luminal activation mechanisms for Paneth cell α-defensins. *J. Biol. Chem.* **2012**, *287*, 11205–11212. [CrossRef] [PubMed]
152. Salzman, N.H. Paneth cell defensins and the regulation of the microbiome: Detente at mucosal surfaces. *Gut Microbes* **2010**, *1*, 401–406. [CrossRef] [PubMed]
153. Sano, T.; Huang, W.; Hall, J.A.; Yang, Y.; Chen, A.; Gavzy, S.J.; Lee, J.Y.; Ziel, J.W.; Miraldi, E.R.; Domingos, A.I.; et al. An IL-23R/IL-22 Circuit regulates epithelial serum amyloid a to promote local effector Th17 responses. *Cell* **2015**, *163*, 381–393. [CrossRef] [PubMed]
154. Schnupf, P.; Gaboriau-Routhiau, V.; Sansonetti, P.J.; Cerf-Bensussan, N. Segmented filamentous bacteria, Th17 inducers and helpers in a hostile world. *Curr. Opin. Microbiol.* **2017**, *35*, 100–109. [CrossRef] [PubMed]

155. Baldridge, M.T.; Turula, H.; Wobus, C.E. Norovirus regulation by host and microbe. *Trends Mol. Med.* **2016**, *22*, 1047–1059. [CrossRef] [PubMed]
156. Baldridge, M.T.; Nice, T.J.; McCune, B.T.; Yokoyama, C.C.; Kambal, A.; Wheadon, M.; Diamond, M.S.; Ivanova, Y.; Artyomov, M.; Virgin, H.W. Commensal microbes and interferon-λ determine persistence of enteric murine norovirus infection. *Science* **2015**, *347*, 266–269. [CrossRef] [PubMed]
157. Kuss, S.K.; Best, G.T.; Etheredge, C.A.; Pruijssers, A.J.; Frierson, J.M.; Hooper, L.V.; Dermody, T.S.; Pfeiffer, J.K. Intestinal microbiota promote enteric virus replication and systemic pathogenesis. *Science* **2011**, *334*, 249–252. [CrossRef] [PubMed]
158. Robinson, C.M.; Jesudhasan, P.R.; Pfeiffer, J.K. Bacterial lipopolysaccharide binding enhances virion stability and promotes environmental fitness of an enteric virus. *Cell Host Microbe* **2014**, *15*, 36–46. [CrossRef] [PubMed]
159. Gong, J.; Xu, J.; Zhu, W.; Gao, X.; Li, N.; Li, J. Epithelial-specific blockade of MyD88-dependent pathway causes spontaneous small intestinal inflammation. *Clin. Immunol.* **2010**, *136*, 245–256. [CrossRef] [PubMed]

2

Myeloid Cells during Viral Infections and Inflammation

Ashley A. Stegelmeier, Jacob P. van Vloten, Robert C. Mould †, Elaine M. Klafuric †, Jessica A. Minott †, Sarah K. Wootton ‡, Byram W. Bridle ‡ and Khalil Karimi *

Department of Pathobiology, Ontario Veterinary College, University of Guelph, Guelph, ON N1G 2W1, Canada; aross14@uoguelph.ca (A.A.S.); jvanvlot@uoguelph.ca (J.P.v.V.); rmould@uoguelph.ca (R.C.M.); eklafuri@uoguelph.ca (E.M.K.); minott@uoguelph.ca (J.A.M.); kwootton@uoguelph.ca (S.K.W.); bbridle@uoguelph.ca (B.W.B.)

* Correspondence: kkarimi@uoguelph.ca

† These authors contributed equally to this work.

‡ These authors contributed equally to this work.

Abstract: Myeloid cells represent a diverse range of innate leukocytes that are crucial for mounting successful immune responses against viruses. These cells are responsible for detecting pathogen-associated molecular patterns, thereby initiating a signaling cascade that results in the production of cytokines such as interferons to mitigate infections. The aim of this review is to outline recent advances in our knowledge of the roles that neutrophils and inflammatory monocytes play in initiating and coordinating host responses against viral infections. A focus is placed on myeloid cell development, trafficking and antiviral mechanisms. Although known for promoting inflammation, there is a growing body of literature which demonstrates that myeloid cells can also play critical regulatory or immunosuppressive roles, especially following the elimination of viruses. Additionally, the ability of myeloid cells to control other innate and adaptive leukocytes during viral infections situates these cells as key, yet under-appreciated mediators of pathogenic inflammation that can sometimes trigger cytokine storms. The information presented here should assist researchers in integrating myeloid cell biology into the design of novel and more effective virus-targeted therapies.

Keywords: neutrophils; inflammatory monocytes; inflammation; viral infection; myeloid cells; type I interferon

---

1. Introduction

The ability of the immune system to recognize invading pathogens and tissue damage, and subsequently respond in a targeted and reproducible manner bestows longevity to our existence. Within the diverse cellular network of the immune system, recent research has shown that myeloid cells deserve new-found attention due to their ability to detect and mitigate viral infections and promote inflammation. Upon viral infection, there are a number of myeloid cell subsets that play various roles in the subsequent inflammatory, cellular, and humoral responses. Myeloid cells are granulocytic and phagocytic leukocytes that traverse blood and solid tissues. When they recognize virus-infected cells or tissues damaged by viruses, these sentinels rapidly initiate an innate immune response [1]. This multifaceted response involves cellular activation [2], signaling cascades [3], and the release of cytokines [4] to guide leukocytes to mount an effective response. Evidence is accumulating that two myeloid cell subsets, in particular, are playing a larger role in recognizing and halting viral infections than was previously thought. Researchers are discovering that both neutrophils and inflammatory monocytes are intertwined in the immune system's anti-viral response. Moreover, they play unique immuno-regulatory roles post-infection, and are critical for restoring homeostasis.

Neutrophils are the most abundant leukocyte subset in mammals, ranging from 40–70% of white blood cell counts [5]. They are responsible for both pro-inflammatory and anti-viral responses, and, therefore, constitute a first line of defense against invading pathogens and cell damage [6]. Neutrophils are effector innate cells that live for a relatively brief five days [7] and exist in one of three states: quiescent, primed, or active. Although they are predominately considered cells that target extracellular organisms such as bacteria via phagocytic uptake, their control of other cell subsets enables them to play important indirect roles in clearing viral infections and modulating inflammation.

Monocytes are large mononuclear leukocytes that are involved with the inflammation and clearance of pathogens. These non-dividing cells are able to further differentiate into other myeloid subsets such as dendritic cells (DCs) and macrophages. Monocytes constitute a heterogeneous population that is endowed with a high degree of plasticity, allowing them to respond to environmental cues in tissues. Current research is uncovering the role that inflammatory monocytes play during inflammation and viral infections. This subset preferentially traffics to inflamed regions, where they secrete inflammatory cytokines [8]. However, they can also function as regulatory cells [9]. For example, alveolar macrophages have been shown to recruit inflammatory monocytes through a type I interferon (IFN)-mediated mechanism [8]. These monocytes can then provide protection against virus-induced pathology.

Evidence exists that both neutrophils and monocytes can contribute to viral clearance or exacerbate pathological damage depending on the context of the infection (Figure 1). In terms of myeloid cells contributing to virus-induced pathologies, a linkage can be made between the induction of cytokine storms and dysregulated type I IFN responses. In cases where these cells are beneficial, they can be therapeutically boosted, whereas they can also be depleted when viruses have commandeered them towards destructive fates. Exploring the pronounced involvement that myeloid subsets have in mitigating viral replication and pathology, therefore, has the potential to create novel therapeutics that are more efficacious against viral infections.

The aim of this review is to explore recent advances in our understanding of the roles that neutrophils and inflammatory monocytes play during viral infections. Although previous reviews have provided comprehensive coverage on the impact that these myeloid subsets have during bacterial infections [1,10,11], there is no current review with an extensive focus on their contributions to mitigating viral infections. Further, this review has a novel focus on the expanding literature discussing the regulatory roles of these cell types during viral infections, as well as a possible link between the virus-mediated blockade of type I interferon signaling and virus-induced cytokine storms.

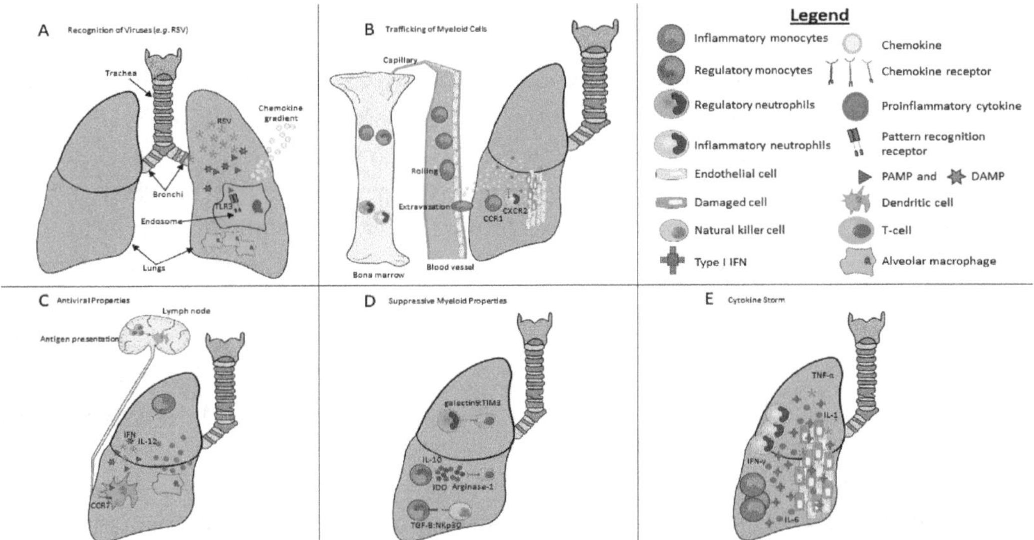

Figure 1. Schematic of myeloid cells highlighting their ability to respond to pulmonary viral infections via the initiation and modulation of anti-viral inflammatory activity. Lung-resident myeloid cells, such as alveolar macrophages, utilize a complex sensory system to integrate disturbances of pulmonary tissues by viruses such as respiratory syncytial virus (RSV) into the activation of local effector leukocytes. (**A**) RSV enters and infects the lungs. Viral pathogen-associated molecular patterns (PAMPs), such as double-stranded RNA or danger associated molecular patterns (DAMPs), are detected by pattern recognition receptors (PRRs) in or on sentinel cells in the lungs, such as TLR3 in the endosomes of lung-resident macrophages. TLR stimulation activates the NF-κß signaling cascade, resulting in the release of chemokines and inflammatory cytokines. A chemokine gradient forms between the lungs and bone marrow. (**B**) Homeostatic bone marrow tends to retain $CXCR4^+$ neutrophils and monocytes through endogenous expression of high levels of CXCL12. However, the release of PAMPs, as well as the secretion of cytokines and chemokines as a consequence of pulmonary RSV infections, is sensed by cells in the bone marrow, which in turn allow recruitment of new neutrophils and monocytes from the bone marrow into the lungs. Specifically, G-CSF downregulates CXCR4 on neutrophils, triggering their release. Similarly, CCL2 is produced in the bone marrow by endothelial cells following TLR signaling in infected lungs, which is crucial for inflammatory monocyte release into the bloodstream. Once in the bloodstream, these cells sense disrupted endothelium from the viral infection, which triggers a complex adhesion cascade. Activated $Ly6C^{hi}$ inflammatory monocytes are recruited to the site of infection by a variety of chemokine receptors including CCR1, 5 and 6, as well as CXCR2 binding to their respective ligands. (**C**) Once at the site of infection, they differentiate into dendritic cells and macrophages that initiate an inflammatory cascade that includes copious amounts of inflammatory cytokines, in particular IL-12 and IFN-γ, which are potent inducers of Th1-biased immune responses. Once these dendritic cells and macrophages acquire viral antigens, they home to lymph nodes via chemokine receptors, including CCR7. Monocyte-derived dendritic cells that home to lymph nodes present viral antigens to naïve $CD4^+$ and $CD8^+$ T-cells that are required to kill infected cells. (**D**) The basic neutrophil function of clearing an inflamed area by removing killed pathogens and host cells contributes to reduced inflammation and wound debridement. Neutrophils are also capable of promoting tissue repair and increased angiogenesis. Further, monocytes can suppress lymphocytes in various clinical scenarios. In lungs, myeloid cells are able to inhibit pro-inflammatory tissue-resident leukocytes through direct cell-to-cell contact through galectin9/TIM3 and the effect of TGF-β on NKp30 in order to regulate T-cells and NK cells, respectively. Myeloid cells can also exert suppressive functions through secretion of soluble factors such as IL-10, arginase-1 and indoleamine 2,3-dioxygenase. (**E**) We speculate that disruption of the cellular sensing of type I IFN responses can result in excessive production of pro-inflammatory cytokines, including IFN-γ, IL-1, IL-6, and TNF-α, leading to a toxic cytokine storm. The fatal outcome of severe lung infections is shown to be correlated with the early persistent production of inflammatory cytokines and chemokines that recruit neutrophils and monocytes. While inflammatory cytokines and chemokines are essential for effective control of viral infections, they can also contribute to the severity of disease and tissue damage.

2. Development of Myeloid Cells

Multiple progenitor cell types arise from self-renewing multi-potent hematopoietic stem cells that become committed in the bone marrow to lineage-specific myeloid cells of the immune system [12]. Clonogenic myeloid-primed precursors (CMPs) give rise to myeloid cells, which then further differentiate into granulocyte/monocyte progenitors (GMPs) [13]. Subsequently, GMPs undergo multiple stages of differentiation before they are terminally differentiated into neutrophils or monocytes in the bone marrow [14]. Monocytes require the growth factor colony-stimulating factor-1 to develop [15] and high levels of the transcription factor PU.1 to steer GMPs to commit to a monocyte lineage [16]. Monocyte-DC progenitors (MDPs) create descendants that are destined to become either DCs or monocytes [17]. Recent discoveries have expanded our understanding of neutrophil development. Advances in isolation techniques have elucidated that neutrophils are derived from unique $CD11b^{+}Ly6G^{lo}Ly6B^{int}CD115^{-}$ precursors that possess proliferation capabilities [18]. Indeed, transcriptional profiling coupled with mass cytometry has provided additional information on the process required for GMPs to differentiate into neutrophils [19]. Researchers determined the bone marrow possesses three distinct subsets: the aforementioned proliferative precursor cells, as well as non-proliferative immature and non-proliferative mature neutrophils. Precursors required the transcription factor C/EBPε to differentiate from GMPs. As precursors further shift into non-proliferative populations, they exchange proliferation capacity for increases in effector function and migration [19]. Further experiments on neutrophil precursors demonstrated their ability to expand in the presence of cancers such as melanoma and suppress regulatory T cells [20]. The role of precursor neutrophils during viral infections has not been determined and represents a novel avenue of research.

Once viruses such as influenza virus or respiratory syncytial virus (RSV) manage to infect a tissue (Figure 1A), type I IFNs are released from the infected cells and stimulate the expression of hundreds of genes [21], appropriately known as IFN-stimulated genes (ISGs), in neighboring cells. This induces an antiviral state within minutes to hours that is characterized by reduced transcription and translation [22], the induction of enzymes that degrade viral RNAs and proteins, and even the sensitization of cells to apoptosis [23]. Products of ISGs, including cytokines and chemokines, also recruit leukocytes, including neutrophils and monocytes to the virally infected tissue [24]. The induction of the IFN response following viral infections fundamentally changes the bone marrow microenvironment (Figure 1B), leading to the enhanced differentiation of myeloid cells [24] and emigration of neutrophils and monocytes to the site of infection, which is facilitated by chemokine gradients interacting with their cognate receptors (Figure 1A) [25].

Murine $Ly6C^{hi}$ monocytes originate from the bone marrow and travel to sites such as skin, lungs, and lymph nodes [26], whereas $Ly6C^{low}$ monocytes typically scan vasculature and the endothelial cells lining the lumen for damage. The human counterparts to these subsets are $CD14^{+}CD16^{-}$ and $CD14^{low}CD16^{+}$ monocytes, respectively [27]. Monocytes require Kruppel-like factor-4 to differentiate into inflammatory monocytes in vivo [28]. A recent advance by Yáñez and colleagues demonstrated that GMPs and MDPs can independently generate functionally distinct monocytes [29]. GMPs and MDPs are both derived from CMP-$Flt3^{+}$ progenitors but differentiate into the above subsets when either the Toll-like receptor (TLR)-4 agonist lipopolysaccharide (LPS) or the TLR9 agonist CpG DNA, respectively, are injected into mice. $Ly6C^{hi}$ monocytes can be derived from either subset [29]. Therefore, the innate pathogen-associated molecular patterns (PAMPs) and their cognate receptors, known as pattern recognition receptors (PRRs), will dictate which monocyte subset is preferentially generated.

Infections can affect hematopoiesis and influence the proportions of cell subsets. Human immunodeficiency virus (HIV) is capable of infecting bone marrow microvascular endothelial cells and provoking hematopoietic dysfunction [30]. A plethora of regulatory signals required to differentiate and release myeloid cells, including granulocyte-colony-stimulating factor and interleukin (IL)-6, can be suppressed by HIV. Human T-cell leukemia virus (HTLV)-1 has recently been shown to infect several lineages of hematopoietic stem cells in addition to T-cells [31]. Both neutrophil and monocyte lineages were permissive to infection, as evidenced by the viral Tax protein in neutrophils and the ability of monocyte progenitors to become infected. Indeed, these infected monocytes were capable of differentiating into DCs and spreading the infection to T-cells. Moreover, four subsets of neutrophils have been characterized in infants with viral respiratory infections [32]. These subsets include suppressive, progenitor, mature, and immature neutrophils, which are present in the blood of infected individuals. However, CD16^{high}CD62L^{low} suppressive neutrophils were only observed in patients with bacterial co-infections.

Strikingly, the viral dysregulation of hematopoiesis can lead to numerous diseases [24]. For example, the Epstein–Barr Virus (EBV) causes infectious mononucleosis, characterized by a dramatic increase in white blood cells in the bloodstream. In rare instances, this virus can cause pancytopenia, which is a severe reduction in the number of platelets, red, and white blood cells [33]. Pancytopenia has also been found in patients who have contracted hepatitis C virus (HCV) [34]. Reducing the number of progenitor cells available to differentiate into lymphoid and myeloid cells may be a reasonably common viral strategy to avoid clearance by the immune system. The functional capacity of myeloid cells to respond to viral particles is influenced by the origin of their precursors. Defining the molecular and cellular mechanisms underlying myeloid cell precursor development in viral illnesses will provide a better understanding of the susceptibility of patients to different viruses and the immunological events that may ultimately be exploited for therapeutic benefit. Indeed, progenitor cells constitute a promising gene therapy target to treat HIV infections because they can differentiate into multiple cell lineages, all possessing a therapeutic transgene such as an anti-HIV ribozyme [35]. Other viral infections that, in theory, may be successfully treated by targeting HSCs with gene therapies include viruses that dysregulate hematopoiesis, such as HCV or EBV.

3. Recognition of Danger Signals and Pathogens by Myeloid Cells

The human body relies on a robust innate sensory system to quickly eliminate many viruses. PRRs are present in and on a variety of cells including neutrophils and monocytes to recognize PAMPs (Figure 1A). TLRs are a subset of PRRs that recognize PAMPs. There are multiple TLRs in and on neutrophils and monocytes that specifically recognize viral PAMPs or danger associated molecular patterns (DAMPs) released from virus-damaged cells. The nucleic acid from RNA and DNA viruses constitutes a predominant source of viral PAMPs that can be recognized either via phagocytosis of cellular debris such as epithelial cells, or in cases where viruses infect myeloid cells. Within endosomes, TLR3 recognizes dsRNA from viruses (dsRNA constitutes the genome of one family of viruses, but is also generated during the life cycle of many viruses) [36], ssRNA is recognized by TLR7 and TLR8 [37], whereas TLR9 recognizes DNA viruses while distinguishing from host DNA [38]. Monocytes are activated via signaling through surface-bound TLR2 during Varicella–zoster virus [4], measles [39], and type 1 and 2 Herpes simplex virus (HSV) infections [40]. TLR2 can recognize a wide range of viral PAMPs including the glycoproteins gB and gH/gL from HSV [41] and hemagglutinin from measles [39]. TLR stimulation after phagocytosis activates the NF-κB signaling cascade, resulting in the release of inflammatory cytokines such as TNF-α, IL-1, and IL-6 from monocytes [4] to control virus infections by direct antiviral mechanisms and the recruitment of other leukocytes. Direct antiviral mechanisms of monocytes and neutrophils, including phagocytosis and oxidative burst, were reduced in patients who had contracted HCV and were taking IFN-based therapies [42]. Neutrophils also use TLRs to conduct anti-viral surveillance, and express ten out of eleven known human TLRs (they lack TLR3) [43]. The endosomal TLR7 is essential for recognition of influenza viruses by neutrophils via sensing viral single-stranded RNA when they phagocytose cell debris [2]. Lack of TLRs is associated with increased mortality during viral infections. For example, blocking TLR4 leads to increased mortalities associated with influenza virus infections by disrupting phagocytosis of infected cells [44]. Although influenza viruses do not contain LPS, TLR4 activation is also involved with delaying fusion between lysosomes and phagosomes, thereby preventing virus entry, and thus has an additional role in innate immunity besides recognition of PAMPs [45]. The multifaceted functions of TLRs should, therefore, be studied in greater detail to determine whether additional TLRs have unappreciated mechanisms to mitigate viral infections.

Another method for host recognition of viruses involves retinoic acid inducible gene-I (RIG-I) and melanoma differentiation factor 5 (MDA5) [46]. To clear viral infections, RIG-I-like receptors and MDA5 recognize cytosolic viral RNAs via the helicase domain [47]. In contrast to TLRs that are predominately present in leukocyte subsets, these receptors are ubiquitous in human cells. Neutrophils and monocytes themselves can become infected by viruses [48] and therefore possess cytoplasmic and endosomal mechanisms to recognize them, including RIG-I and MDA5 signaling cascades in the cytoplasm and endosomal TLRs. In fact, the double-stranded RNA mimetic poly(I:C) stimulates neutrophils to increase many antiviral genes, including type I IFN mRNA transcripts, IFN-responsive genes, TNF-α, and IFN regulatory factor (IRF)7 [49]. When infected with encephalomyocarditis virus (EMCV), MDA5-deficient mice mount significantly reduced TNF-α and IFN-β responses [49]. Similar results were also observed after infections with Coxsackie B virus (CVB) [50] and West Nile virus (WNV) [51]. Notably, TNF-α and IFN-β have the capacity to upregulate the expression of major histocompatibility complex molecules on antigen-presenting cells, which would make viruses more susceptible to T-cell-mediated clearance.

DAMPs are endogenous molecules that are released in response to tissue damage from trauma, including cells killed by viruses, and, like PAMPs, trigger an immune response (Figure 1A). DAMPs can be derived from a variety of cellular components, including the nucleus, cytoplasm, exosomes, plasma, or the extracellular matrix [52]. DAMPs that promote inflammation and immunogenic cell death [53] include the chromatin protein high mobility group box 1 (HMGB1) and mitochondrial DAMPs such as mitochondrial DNA and formyl peptides [54]. HMGB1 interacts with neutrophils and monocytes [55] by binding to the inflammatory Receptor for Advanced Glycation End-products (RAGE). This DAMP causes monocytes to secrete pro-inflammatory cytokines, including IL-1 and TNF-α, reorganize their cytoskeleton, and increases migration across epithelial barriers. Monocytes are also capable of secreting HMGB1 themselves when lysosome exocytosis is induced by the inflammatory lipid lysophosphatidylcholine [56]. Neutrophils, in turn, have upregulated transcription of genes for pro-inflammatory molecules involving the NF-κB, p38 MAPK, and ERK1/2 pathways in response to recognition of HMGB1 [57]. Cell damage from viral infections leads to a release of DAMPs and subsequent detection by myeloid cells. For example, infection of epithelial cells with dengue viruses results in the release of HMGB1 from necrotic cells [58]. The interaction between viral PAMPs and PRRs in or on myeloid cells can play an essential survival role in the response to viral infections but may, simultaneously, be responsible for tissue injury associated with severe virus-induced inflammation. In theory, mechanisms involved in the recognition of danger signals by neutrophils and monocytes could be targeted selectively to enhance protection against detrimental viral infections while, simultaneously, preventing exaggerated, pathological innate immune responses.

4. Myeloid Cell Migration and Trafficking

Chemokines and their receptors play a critical role in dictating the migration and positioning of myeloid cells. An extensive list of chemokines, their receptors, and their various functions has been described [59]. Neutrophils and monocytes begin their journey to a site of infection by first leaving the bone marrow (Figure 1B). Neutrophil and monocyte retention in the bone marrow is dictated by steady signaling between the chemokine (C-X-C motif) receptor 4 (CXCR4) and its ligand CXCL12 expressed on bone marrow stromal cells. During maturation, these cells downregulate CXCR4 and become less sensitive to CXCL12, causing their release into the bloodstream [60]. CXCR1, and mainly CXCR2 expression, on neutrophils grants an additional form of chemotaxis away from the bone marrow via their respective ligands, CXCL1 and CXCL2 [25], which are produced by macrophages and mast cells at the site of infection [61]. However, retention is typically favored in the steady state, as CXCL12 appears to be constitutively expressed in the bone marrow. Inflammation mediated by viral infections that induce G-CSF enhances CXCL2 release and decreases CXCR4 expression on bone marrow-resident neutrophils, tipping the balance in favor of neutrophil release [25]. Ly6C^{hi} inflammatory monocytes appear to require chemokine receptor 2 (CCR2) signaling to efficiently exit the bone marrow and travel to sites of inflammation, whereas CCR2 signaling appears to be contextually dependent for monocyte emigration from circulation into virus-infected tissues [62]. More research needs to be conducted to ascertain if CCR2 signaling is required to respond to viral infections of various tissues. CCR2 signaling, via CCL2 binding to the receptors on monocytes, causes the downregulation of CXCR4 and renders the monocytes less sensitive to CXC12, causing their release from the bone marrow [63]. Interestingly, low concentrations of circulating TLRs cause rapid CCL2 release by mesenchymal stem cells and their progeny in the bone marrow, which triggers the release of monocytes [64].

The dissemination of neutrophils and monocytes to virally infected tissues involves many complex processes (Figure 1B) [59,65,66]. Generally speaking, myeloid cell migration to infected tissues relies on transmigration through vascular endothelium from the blood. This transmigration is dictated by a milieu of cytokines, and chemokines produced by tissue injury and resident sentinel cells in response to DAMPs and viral PAMPs. The disruption of homeostasis confers a change to the vascular endothelium near sites of infection [67]. The multitude of changes to the endothelium can happen rapidly, and have been reviewed extensively elsewhere [68]. In brief, endothelial changes that start and subside within minutes are known as type I activation and can be mediated by factors such as histamine [69]). Alternatively, type II activation can last hours to days with substantial changes in gene expression profiles mediated by tumor necrosis factor (TNF)-α [70]. Both forms of activation cause increased blood flow, vascular leakage of plasma proteins, and the recruitment of leukocytes [70]. These disruptions in endothelium homeostasis can trigger a leukocyte adhesion cascade [71] that, in harmony with various cytokines released by inflamed endothelium, such as IL-8 and monocyte chemoattractant protein (MCP)-1 [72], initiates the selectin-mediated rolling of leukocytes along the surface of endothelial cells. Trafficking of neutrophils and monocytes through the endothelium towards the site of infection is then facilitated by crawling via macrophage-antigen-1 (Mac-1/CD11b) expressed on monocytes and the intercellular adhesion molecule-1 (ICAM-1/CD54) expressed on endothelial cells [73]. Crawling appears to facilitate the paracellular (between cells) transmigration of neutrophils and monocytes, which is generally the preferred method of trafficking (occurring 70–90% of the time), as opposed to transcellular (through cells) transmigration [71].

The dissemination of neutrophils and monocytes from the vasculature into infected tissues is critical for viral clearance. Neutrophils are initially recruited to sites of infection by their ability to recognize tissue damage via sensing of H_2O_2, DNA, N-formyl peptides, adenosine triphosphate, uric acid, and other DAMPs [74]. Further guidance to sites of infection is provided by a family of CXCL8 chemokines originating from concentrated sites of PAMPs and DAMPs, including CXCL1, CXCL2, CXCL3, CXCL5, CXCL6, CXCL7, and CXCL8 (IL-8), which are sensed by CXCR1 and CXCR2 on neutrophils and monocytes [74]. Within the context of viral infections, experimental data from mice infected with Theiler murine encephalomyelitis virus have demonstrated that CXCL1 released from epithelial cells, macrophages, and neutrophils recruits both neutrophils and monocytes to sites of infection [75]. Macrophages infected with rotaviruses release CXCL2 to recruit neutrophils [76] and Nipah virus C protein is capable of inducing the release of numerous chemokines, including CXCL2, CXCL3, and CXCL6, from endothelial cells [77]. PAMPs from viruses tend to amplify neutrophil recruitment. The inflammatory Ly6C^{hi} subset and the "patrolling" Ly6C^{low}CX3CR1^{hi} subset migrate along luminal and endothelial cell surfaces, with the latter being able to respond rapidly to infections in a CX3CR1-dependent fashion [78]. The migration of inflammatory monocytes to tissues is CCR2-dependent. However, as mentioned above, this tends to only be required to exit the bone marrow. Nonetheless, CCR2 signaling appears to be critical for inflammatory monocyte recruitment in cases of West Nile virus-induced encephalopathies, and influenza virus infections [1,66]. In summary, a range of trafficking signals and endothelial barrier regulatory molecules shape myeloid cell recruitment to virally infected and inflamed tissues.

5. Anti-Viral/Pro-Inflammatory Properties of Myeloid Cells during Viral Infections

Neutrophils are able to lyse and phagocytose virus-infected cells [44], and are one of the first leukocyte subsets to enter inflamed tissues (Figure 1C). The magnitude of the neutrophil response is a predictor of the host's ability to clear an influenza virus infection with minimal damage [79]. Depleting neutrophils causes greater viral spread and host mortality [79], and neutrophils are also crucial to mitigate HSV type-1 corneal infections in a murine model [80]. Moreover, Tate and colleagues demonstrated that neutrophils are critical for limiting the replication of influenza viruses [81] and that a loss of neutrophils increases disease severity. Thus, the antiviral response of neutrophils contributes to clearing viral infections.

Neutrophils can directly mediate innate immune responses, activate adaptive immunity and recruit lymphoid cells to sites of viral infections [82,83]. A key mechanism of action that enables neutrophils to neutralize invading viruses is the production of neutrophil extracellular traps (NETs) [82]. NETs are strands of DNA and granule proteins secreted by neutrophils that form around viral particles, preventing their spread [84]. Poxvirus infections in mice were mitigated in liver microvasculature via this mechanism [82]. In addition to the physical containment of infections, NETs are coated with antiviral enzymes that enable neutrophils to concentrate lethal antimicrobial proteins such as histones at sites of infection [84]. Neutrophils are also capable of mediating antibody-dependent cellular cytotoxicity (ADCC) or antibody-dependent phagocytosis, which involve the release of cytolytic granules or phagocytosis, respectively, after binding antibodies via Fc receptors [85]. These antibody-dependent processes are critical in the clearance and neutralization of certain viruses such as HIV [85]. ADCC responses peak quickly (i.e., within four hours) and are controlled by the FCγR family of receptors and can also utilize the extracellular release of reactive oxygen intermediates [85]. Reactive oxygen intermediates are also involved in other pathological responses, including exocytosis. Exocytosis is a cellular active transport process whereby membrane-bound vesicles transport molecules to the cell surface. Neutrophils emit an array of compounds including myeloperoxidase to control sepsis [86], antiviral lysozyme with anti-HIV properties [87], and N-formyl-methionyl-leucyl phenylalanine (fMLF)-stimulated superoxide release in the presence of periodontitis pathogens [88]. Exocytosis, therefore, expands the neutrophil arsenal to neutralize the array of pathogens they encounter.

Neutrophils are incredibly diverse in their functions. In addition to trafficking to sites of infection to phagocytize viruses and form NETs, they also stimulate virus-specific adaptive immune responses [83]. Neutrophils that have detected viral antigens can home to draining lymph nodes dependent on IL-1R, where they can act as antigen-presenting cells [83,89]. Neutrophils present processed viral antigens to naïve CD8$^+$ T-cells via the major histocompatibility complex I and T-cell receptor interactions, along with the expression of CD80 and CD86 to provide co-stimulation, thereby providing the two signals required to activate T-cells [83]. Furthermore, neutrophils are responsible for the recruitment of effector CD8$^+$ T-cells to sites of viral infections. The mechanism by which they recruit T-cells during influenza virus infections has been linked to CXCL12 deposits left behind like a "trail of breadcrumbs". CD8$^+$ T-cells follow this chemoattractant trail left behind by neutrophil uropods to the sites of influenza virus infections [90].

RSV causes lung infections that are characterized by neutrophils contributing to host damage [91]. RSV is capable of delaying the apoptosis of neutrophils and eosinophils, which is hypothesized to delay antigen presentation and increase tissue damage. IL-6 and TLR7/8 binding was determined to contribute to this delay and depended on NF-κB and PI3K activation. The authors of this study did not directly examine whether this delay resulted in an increase in host tissue damage in their model, but hypothesized this was the case, constituting an area of future study. During an RSV infection, neutrophils migrate through infected airway epithelial cells [92]. These neutrophils are characterized by the increased expression of myeloperoxidase and CD1b, and their migration promotes epithelial shedding and airway tissue damage. Aside from delaying apoptosis, RSV infection has also been shown to increase eosinophil recruitment and degranulation based on the macrophage inflammatory protein (MIP)1-α and eosinophil cationic protein concentrations measured in lower respiratory airway secretions [93].

Ly6C^{hi} monocytes migrate to injured sites, induce inflammation, and eliminate the cause of tissue injury (Figure 1C) [94]. For instance, type I IFNs amplify the production of MCP-1, the primary chemokine responsible for recruiting inflammatory monocytes to the lungs during influenza virus infections [95]. These monocytes have been implicated in influenza virus-induced lung injury [96]. Importantly, elevated MCP-1 levels have been associated with severity of illness in pediatric influenza virus infections [97]. In mice, the recruitment of monocytes to lungs was shown to be accompanied with an increase in type I IFN production, NLRP3 inflammasome activation, and alveolar epithelial barrier dysfunction [98]. It has been identified that increased pro-inflammatory monocytes are a major immunological determinant of severity of disease in previously healthy adults with life-threatening influenza virus infections [99]. This provides a possible mechanistic cause for disease severity in these patients, a potential early identifier and a modifiable immune pathway for therapeutic targeting. However, there is no role for recruited monocytes in the lungs of mice infected with the natural rodent pathogen, pneumonia virus of mice [100], indicating the pathogen-specific functions of these cells. Interestingly, monocytes have been proposed to be educated in the bone marrow to promote their tissue-specific functions at sites of persistent challenge [101]. Long-lasting epigenetic alterations in monocyte precursors may account for the "trained immunity" phenomena [102]. Indeed, monocytes have an immunological memory of past insults. Thus, this evidence shows that neutrophils and inflammatory monocytes participate in inflammation that is needed for an effective immune response against viruses. A shared feature of neutrophils and monocytes is their ability to synthesize pro-inflammatory cytokines that help the host overcome viral diseases. However, these responses can also be overly robust, thereby contributing to virus-induced tissue damage. Future research directions should include a focus on furthering our understanding of the diverse antiviral arsenal of myeloid cells.

6. Regulatory/Suppressive Properties of Myeloid Cells during Viral Infections and Inflammation

Robust immune responses are critical for protecting hosts against lethal viral infections. It is equally important that immune responses are of adequate magnitude and duration. The capacity for a host to resolve inflammation and return to homeostasis has important consequences for health (Figure 1D). The induction of an immune response that is too severe or the failure to return to homeostasis can result in immunopathology [103], including tissue and organ damage [65], cytokine storms (Figure 1D) [104], chronic inflammation [105], and autoimmune diseases [106]. As innate immune responders, myeloid cells are key players in orchestrating appropriate inflammatory responses and the return to homeostasis following virus infections. The role of myeloid cells in the regulation of immune responses is complex and involves specialized cellular subsets, suppressive receptors, and cytokines. In addition, much of what we know about the regulatory and immunosuppressive effects of myeloid cells originates from research investigating bacterial, fungal, and sterile inflammation models, but has implications for virus infections.

Neutrophils possess multiple mechanisms to control inflammation, despite their predominately pro-inflammatory role (Figure 1D) [107]. One mechanism involves the formation of aforementioned NETs [108]. These NETs function via serine proteases to degrade excess cytokines and chemokines in areas with high densities of neutrophils [108]. Neutrophils are also capable of reducing lung injury during influenza virus infections [79]. A neutrophil depletion study in a H3N2 murine model demonstrated that their absence led to weight loss, viremic spread, and increased inflammation. The basic neutrophil function of clearing an inflamed area by removing killed pathogens and host cells contributes to reduced inflammation and wound debridement [107]. They are also capable of healing mucosal regions of the intestine [109], and increasing angiogenesis [109]. A recent advance in our knowledge of neutrophils concerns their ability to de-prime [110]. Originally considered an irreversible process, neutrophils are capable of returning to quiescence. Neutrophils can be spontaneously de-primed in the circulatory system via the degradation of a superoxide anion response [111], with a de-priming half-life of approximately forty minutes [112], or retained in the bone marrow [113] to limit the number of primed cells that can traverse the body and cause damaging effects such as lung injury [114].

Recent experimental data have demonstrated that inflammatory monocytes are capable of exhibiting suppressive properties. Inflammatory monocytes are recruited to sites of vaccine-mediated inflammation via MCP-1 [115]. Within the vaccine draining lymph node, monocytes sequester cysteine, resulting in T-cell suppression [115]. Blocking monocyte suppression in this context may prove to be an effective mechanism to improve vaccine effectiveness. Monocytes are also capable of suppressing B cells. In vitro studies have demonstrated that monocytes suppress B cell differentiation, proliferation, and Ig class distribution [116]. Monocytes, therefore, represent a prime example of a cell type that can be both pro-inflammatory and suppressive, depending on the context.

The resolution of immune response is an active regulatory process, which is initiated via the release of soluble mediators such as cytokines and chemokines, as well as through cell-to-cell interactions mediated by surface-expressed ligands and receptors [117]. Evidence has revealed that monocytes that are part of inflammation also can be reprogrammed to cells that are highly anti-inflammatory and contribute to resolution of inflammation [117]. Moreover, during sepsis, human monocytes have been shown to undergo a transition from a pro-inflammatory to an anti-inflammatory status [118], although it remains unclear whether the conversion of monocytes from pro-inflammatory to a regulatory phenotype occurs in viral diseases. Further studies are needed to understand the mechanisms to explain how monocytes can be switched into suppressor/anti-inflammatory cells during a viral infection, which in turn would allow intervention with targeted therapeutics to control and down-modulate excessive inflammation in viral diseases.

An additional myeloid subset of interest is the myeloid-derived suppressor cells (MDSCs). MDSCs can suppress immune responses in numerous anatomical locations, including tumor microenvironments, virally infected tissues, and sites of inflammation. The subset of MDSCs with neutrophil-like properties have been designated polymorphonuclear (PMN)-MDSCs or granulocytic (G)-MDSCs, while their myeloid counterparts have the nomenclature M-MDSCs. Viral infections can induce MDSCs, as is the case with HCV [119]. $CD33^+$ MDSCs were upregulated upon co-culture with HCV infected hepatocytes, resulting in T cell suppression mediated by reactive oxygen species. Moreover, NK cells are also suppressed by MDSCs during HCV infection [120]. The production of MDSC-derived arginase-1 resulted in a decrease in IFN-γ production by NK cells. The suppression of key effector cells contributes to viral persistence. HCV is not the only virus to control MDSCs to evade the immune system. Patients with HIV-1 have M-MDSC populations that suppress helper T cells [121], and elevated levels of these myeloid cells were correlated with increased viral loads. Future research should focus on determining whether other viruses engage MDSCs to prolong infections. Additionally, more research is required to fully determine the position these subclasses have in myeloid cell differentiation. Although a recent review concluded that MDSCs constitute *bona fide* alternate lineages [122], future studies will be required to cement their status within the field of immunology.

7. Modulation of Innate Lymphoid Cells by Myeloid Cells during Viral Infections and Inflammation

Myeloid cells are able to translate micro-environmental cues into an effector profile that initiates lymphocyte responses [123]. Innate lymphoid cells (ILCs) react to pathogens indirectly through myeloid or epithelial cell-derived cytokines and other inflammatory mediators including IL-12, IL-23, and IL-33 [124]. ILCs are derived from a lymphoid progenitor but do not contain either a B or T-cell receptor due to the absence of the recombination-activating gene [125]. There are three major subsets of ILCs: groups 1, 2, and 3. Group 1 includes cells that produce IFN-γ and TNF-α and is predominately composed of classical natural killer (NK) cells. ILCs that require GATA3 and RORα to develop and express the cytokines IL-5 and IL-13 are denoted as group 2, while intestinal ILCs that express NKp46 and depend on RORγ comprise group 3 [126]. Since evidence shows that ILCs are tissue-resident cell types with limited capacity to directly recognize PAMPs [123], myeloid cells may play a crucial role in controlling ILC homeostasis and function [127].

In the steady state, monocytes enter tissues and replenish macrophages and DCs [128]. However, during viral infections they are recruited to infected tissues and mediate direct antiviral activities [129]. For instance, in mice infected with murine cytomegalovirus, inflammatory monocytes are recruited to the liver and produce MIP-1a, which recruits NK cells [130]. NK cells are relevant to viral infections because they target infected cells for destruction. NK cells are cytotoxic ILCs that require IL-15 to develop, differentiate, and survive [131]. IL-15 is secreted by several cell types, including monocytes after viral recognition [132], which therefore places NK cells under the control of myeloid cells. Expression of the activating receptor NKG2D is upregulated on NK cells in response to IL-15. IL-15-activated NK cells show preferential expression of the TNF-related apoptosis-inducing ligand (TRAIL) as well as activation and phosphorylation of ERK1 and 2, and increases in perforin production [133]. The increased expression of these activating receptors and effector compounds increases the killing potential of NK cells. Many viruses down-regulate the expression of MHC on infected cells to escape detection by $CD8^+$ T-cells [134]. Therefore, IL-15 secretion by monocytes constitutes a mechanism to upregulate multiple cell receptors. Changes in granzyme regulation were not documented in these studies, but represent an area of future investigation due to the role of this compound in the apoptosis of virus-infected cells. Human monocytes express membrane-bound IL-15 constitutively, with its expression increased in the presence of IFN-γ [135]. The monocyte-mediated production of IL-15 was increased in the presence of the anti-inflammatory cytokine IL-10, but was unaffected by IL-4 or IL-13 [135]. IL-15 also influences monocytes and can transform them into DCs in airway epithelia [136], which has implications for improving the presentation of viral antigens, suggesting a cross-talk between NK cells and myeloid cells under viral inflammatory conditions. Recently, Ashkar and colleagues [137] showed that type I IFNs produced during a viral infection stimulated vaginal MCP-1 production, which is a chemoattractant that is responsible for inflammatory monocyte migration to inflamed sites. Once recruited, type I IFNs stimulate inflammatory monocytes to produce IL-18, which then signals through the IL-18 receptor expressed by NK cells to induce their production of IFN-γ. Interestingly, cytokine IL-12 also promotes the secretion of IFN-γ by NK cells [138] and neutrophils [139]. Neutrophils can also increase IFN-γ production by NK cells using multiple pathways. The first method is to interact with DCs via ICAM-1 to further upregulate IL-12p70 [140], creating a positive feedback loop. The direct co-stimulation of NK cells also occurs with CD18 and ICAM-3 binding on neutrophils and NK cells, respectively [140]. Our unpublished data (personal observation by Karimi K and Bridle B) have demonstrated that the induction of viremia in mice, which induces the release of high concentrations of inflammatory cytokines into the circulation, is accompanied by increased numbers of pulmonary ILC subsets and the accumulation of multiple myeloid cell subsets that, interestingly, were type I IFN-dependent (data not shown). Additionally, we demonstrated that the induction of inflammation by concanavalin A in mice, which occurs due to macrophage activation downstream of the rapid stimulation of T-cells, led to increased numbers of ILC2 populations in all organs examined, including the bone marrow, spleen, and liver [141] (unpublished data). Recently, Mortha and Burrows [123] discussed how the feedback communication between ILCs and myeloid cells contributes to stabilize immunological homeostasis. Further studies are needed to dissect cell-to-cell interactions between myeloid cells and ILCs other than NK cells in viral inflammatory conditions.

8. Modulation of Adaptive Immune Responses by Myeloid Cells during Viral Infections

The concept that neutrophils can initiate, amplify and/or suppress adaptive immune effector responses by establishing direct bidirectional cross-talk with T-cells has garnered attention in the past few years [142]. A Th1 response can be induced by neutrophils in a murine model [143], which increases the number of $CD8^+$ cytotoxic T-cells available to lyse virally infected cells. Indeed, in vivo murine studies have demonstrated that neutrophils can cross-present ovalbumin to $CD8^+$ T-cells in a TAP- and proteasome-dependent manner [144]. Neutrophils can further impact the adaptive immune response by inducing DC maturation, which in turn increases antigen presentation to adaptive cells [145]. Neutrophils have been observed to cluster with immature DCs and bind their Mac-1 to DC-specific intercellular adhesion molecule-3-grabbing non-integrin (DC-SIGN). DC-SIGN is also referred to as CD209 and is a PRR that recognizes and binds to mannose residues, a conserved PAMP associated with a variety of viral infections. However, neutrophil depletion studies have demonstrated an increase in antigen presentation to $CD8^+$ T-cells. The mechanism by which this phenomenon occurs is thought to be a reduction in competition for viral antigens between neutrophils and DCs [146].

There are extensive demonstrations that neutrophils in humans and mice can also suppress T-cell responses (Figure 1D). Suppressive neutrophils that express low levels of CD62L are induced after acute inflammation arising from either viral infections or tissue injury [147]. They have been shown to impair T-cells by releasing hydrogen peroxide into an immunological synapse, which impairs T-cell migration via the CXCL11 chemokine gradient. Ball and colleagues have shown that CXCL11-induced migration to sites of infection decreases as the concentration of hydrogen peroxide released into the immunological synapse is increased. Results demonstrate the impaired recruitment of Th1 and $CD8^+$ T-cells to the periphery. Ultimately, the mechanistic consequence pertains to defective migration mechanisms rather than TCR:MHC signal transduction. It is also important to note that this interaction required Mac-1 (CD11b). Additional research has demonstrated that Mac-1-expressing neutrophils are crucial in limiting pathology caused by T-cells in a murine model of infection with influenza virus, presumably by suppressing T-cell proliferation [148]. We have demonstrated that a subset of neutrophils function as negative regulators of excessive cytokine production in a mouse model of viremia, in which type I IFN signaling has been disrupted (Karimi K and Bridle B, unpublished data). Altogether, these findings allow us to envision the therapeutic potential of subsets of neutrophils. However, one of the major challenges would be the heterogeneity of immunosuppressive or regulatory neutrophils. Future studies taking advantage of flow cytometry technology and next-generation sequencing to phenotypically and functionally define neutrophil subsets will extend our knowledge about the immunoregulatory role neutrophils play in viral infections and inflammation.

Neutrophils also have an indirect mechanism to modulate T cells during a viral infection. The bacteria *Mycobacterium tuberculosis* is capable of delaying neutrophil apoptosis, which delays an adaptive $CD4^+$ T-cell response [149]. Although this has not been demonstrated via a viral infection, it nonetheless demonstrates a key effect neutrophils have on controlling a $CD4^+$ T helper cell response. This response may be delayed because DCs ingest whole infected neutrophils [150] to acquire antigens and present them to T-cells. Additionally, DCs that ingest neutrophils possessing pathogen-derived antigens can migrate to lymph nodes more efficiently [151]. The differentiation of inflammatory monocytes into $CD11b^+$ pulmonary DCs is triggered by the presence of respiratory viruses such as influenza virus [152]. Defects in this differentiation delay the clearance of influenza viruses and significantly reduce the activation of $CD8^+$ T-cells [1].

While inflammatory monocytes are key regulatory cells in maintaining macrophage and DC populations in healthy tissues, a function of homeostasis, they are quintessential in the clearance of infections due to their ability to induce adaptive immunity and prime a variety of lymphocytes, including T-cells (Figure 1C) [152]. Upon viral infection, inflammatory monocytes in the blood are recruited to the primary site of infection or the draining lymph node. Cells that traffic to the primary site of infection play a critical role in the recruitment of T-cells and, thereby, the activation of inflammatory responses and cellular immunity [153]. However, inflammatory monocytes that traffic to draining lymph nodes acquire a DC phenotype that enables them to present viral antigens to naïve T-cells [153]. In particular, studies have shown that inflammatory monocytes stimulate a Th1-biased immune response via production of IL-12 that promotes production of IFN-γ by T cells primed in lymph nodes [153]. This Th1 immunity is critical in the defense against intracellular pathogens, such as viruses [153].

Although memory is traditionally considered a hallmark of the adaptive immune response, recent advances have shed light on the contributions of innate memory. Innate memory, also referred to as trained immunity, is a multifaceted response. A recent component of trained immunity involves its modulation of hematopoiesis [154]. Although myeloid cells have a short lifespan in circulation, the administration of the agonist ß-glucan resulted in myeloid progenitor expansion and subsequent improved responses to a secondary challenge with the agonist LPS. Trained immunity was able to reduce myelosuppression from chemotherapy, and was associated with metabolic shifts in cholesterol biosynthesis and glucose metabolism [154]. Other benefits of innate myeloid memory have been elegantly reviewed by Netea and colleagues [155]. In brief, monocytes are influenced by vaccination and viral infections, and are more responsive upon re-challenge. This innate memory response helps mitigate pathogens via upregulated cytokine production and enhanced pathogen elimination response times. This exciting new field may allow vaccines to be optimized for viruses by targeting the innate memory response.

Clearly, the cross-talk that is occurring between monocytes, neutrophils, and T-cells constitutes a crucial bridge between innate and adaptive immunity. Future investigations are encouraged to examine the full extent of communication between these cells, further elucidate the mechanisms, and the anatomical locations of these interactions. Depletion assays will be beneficial to determine which cell subsets can mount effective anti-viral responses, not just by T-cell and APC interactions, but also by direct interactions with neutrophils and monocytes.

9. Type I IFNs, Myeloid Cells and Cytokine Storms during Viral Infections

Extensive studies have highlighted the role type I IFNs play in initiating an anti-viral state in cells through the inhibition of viral replication [156]. In some cases, the disruption of this response results in the excessive production of cytokines, leading to a so-called cytokine storm that can be very toxic (Figure 1E) [157]. This is a cause of mortality in cases of severe acute respiratory syndrome (SARS) [158], infection with some strains of influenza viruses [3], Ebola virus [159], and dengue virus [104]. During viral infections, the regulation of cytokine networks and the mechanisms by which the cytokines may interact with neutrophils and monocytes are poorly documented.

The fatal outcome of severe influenza infections is shown to be correlated with the early persistent production of inflammatory cytokines and chemokines that recruit neutrophils and monocytes [65,160]. Lethal outcomes of H5N1 influenza infections in humans correlated with early excessive innate immune response, involving type I IFNs followed by prolonged inflammatory responses, and were associated with high viral loads and hypercytokinemia [65,160]. While inflammatory cytokines and chemokines are absolutely essential for the effective control of viral infections, they can also contribute to the severity of disease [161,162]. Other fatal viral infections that are hallmarked by dysregulated type I IFN responses and cytokine storms are hantaviruses [163] and WNV [103,164]. Given the dynamic nature of cytokines, the complexity of signaling pathways they interact with, and the fact that their excessive production is often associated with some of the worst clinical outcomes of viral infections, there is a need for much more research into the mechanisms by which virus-induced cytokine storms are triggered or controlled.

Investigation into the mechanisms involved in host responses to viral infections demonstrates a complex and carefully balanced interaction between type I IFNs and inflammatory neutrophils and monocytes. Recent analysis of mRNAs in the blood of humans responding to infections with influenza viruses revealed that early gene expression patterns of anti-viral molecules, such as the genes encoding for myxovirus resistance protein-1 (*MX1*) and ISG-15, are correlated with the heightened production and activation of type I IFNs after viral infections [165]. Late gene expression patterns were also induced by type I IFNs, but in contrast to patterns of antiviral molecules being observed, the transcriptional profiles of patients in the late stages of infections were highly reflective of neutrophil and inflammatory molecule activation [165], suggesting an important interplay between the secretion of type I IFNs and the activation of neutrophils and inflammatory monocytes.

It is important to study the receptors mediating the neutrophil antiviral response to reduce aberrant host responses and damage. NLRP12 is a nucleotide-binding domain leucine-rich repeat protein that is expressed on blood-derived leukocytes, including monocytes, and modulates neutrophil recruitment by increasing the chemokine CXCL1 through the IL-17-NLRP12 axis and increasing vascular permeability [166]. Another activator and recruiter of neutrophils is produced by liver cells and is entitled serum amyloid A (SAA) [167]. Injections of SAA increased phagocytosis of influenza viruses by neutrophils, resulting in the release of IL-8. Modulating these protein concentrations might represent a promising therapeutic strategy to achieve ideal neutrophil responses to promote elimination of influenza viruses without excessive bystander damage to tissues. Neutrophil-mediated antiviral responses have varying effects on the outcome of influenza virus infections, depending on the strain of virus [168]. Neutrophils contributed to terminating infections with H3N2 influenza virus strains of intermediate virulence and H1N1 strains that were highly virulent, while they did not limit the severity of disease during infection with an H3N2 strain of low virulence.

The early production of virus-induced type I IFNs has been observed to upregulate genes in neutrophils that encode pro-apoptotic molecules, such as IFN-induced dsRNA-activated protein kinase, and the oligoadenylate synthase-like proteins and the RNase L system [165]. Experiments with IRF-$3^{-/-}$ x IRF-$7^{-/-}$ double-knockout mice and WNV [169] concluded that the viral induction of cellular IFN-β secretion depends on interferon-β promoter stimulator-1-mediated signaling without requiring the IFN transcription factors IRF3/7, suggesting the essentiality of the immediate and optimal activation of the type I IFN response. SARS-coronaviruses are highly pathogenic and cause alveolar damage, fibrin deposition, and tissue necrosis [170]. The delayed expression of the type I IFN response in mice infected with SARS-coronaviruses was implicated in the promotion of inappropriate and chronic inflammatory responses, such as excessive inflammatory monocyte, neutrophil and cytokine accumulation, and impaired virus-specific T-cell responses due to augmented T-cell apoptosis, leading to lung damage [171]. In contrast, an early type I IFN response reduced the immunopathological damage observed, linking the early activation of the type I IFN response to the control of overly robust inflammation. Additionally, type I IFNs have been implicated in the regulation of myeloid cell migration during initial exposure to viral infections, heightening inflammatory and virus-specific B and T-cell responses [8,90]. The production of type I IFNs by sentinel leukocytes, in particular that of plasmacytoid DCs that serve as a potent source of IFNs, upon viral infection initiates a type I IFN-dependent secretion of neutrophil and inflammatory monocyte chemoattractants such as IL-1α, CXCL1 and CXCL2 [61,172], highlighting the role of virus-induced type I IFNs in the regulation of neutrophil and monocyte trafficking. Pollara et al. [173] demonstrated that the secretion of type I IFNs by HSV-1-infected myeloid DCs results in the activation of uninfected DCs. This process enables the adaptive immune system to become activated even during a viral infection that targets myeloid cells and prevents their maturation, such as in the case of HSV.

The protective functions of type I IFNs have been associated not only with the recruitment of neutrophils and inflammatory Ly6C^{hi} monocytes to sites of viral infections, but also with the prevention of excessive monocyte and neutrophil activation, thereby controlling inflammation caused by type II IFNs, such as IFN-γ [172]. The interplay between type I and II IFNs was crucial for mitigating damage stemming from influenza A virus-induced inflammation in Rag$2^{-/-}$, Ifnar$1^{-/-}$, Ifngr$1^{-/-}$and Stat$1^{-/-}$ C57Bl/6 mice [172]. Both IFNs were required to prevent excessive numbers of neutrophils trafficking into lungs. STAT1 was experimentally determined to coordinate inflammation via type I and II IFN receptors. When type I IFNs were absent, Ly6C^{lo} monocytes transitioned to being more inflammatory than Ly6C^{hi} monocytes. In the absence of type I IFN signaling, Ly6C^{lo} monocytes traditionally associated with tissue re-modeling became phenotypically and functionally more pro-inflammatory during infection with influenza A viruses [172]. Notably, infection of trophoblasts with Zika virus induced a lower secretion of type I IFNs, and higher immunopathological inflammatory immune responses when compared to trophoblasts infected with Yellow fever virus and dengue virus [174]. Measurement of immune mediators in nasal fluids from RSV-infected infants indicated that severe disease caused by heightened inflammatory responses was also associated with diminished type I IFN responses [175], furthering the idea that a link between type I IFNs and the promotion versus suppression of virus-induced inflammation exists. Taken together, these findings suggest that type I IFN signaling drives a balance of pro- and anti-inflammatory effects on the functions of monocytes and neutrophils in response to viral infections; providing protective immunity while simultaneously limiting immunopathology. These results suggest that the administration of type I IFNs at optimized time points and doses could prove beneficial in the limitation of toxic cytokine storm onset and the control of excessive immunopathological damage. Indeed, in vitro evidence suggests that the administration of exogenous type I IFNs can mitigate excessive cytokine production induced by SARS-coronaviruses [176]. Determining the means by which type I IFNs control excessive inflammation while ensuring effective anti-viral responses is required.

10. Conclusions and Future Directions

Neutrophils, inflammatory monocytes, and their roles in mitigating bacterial infections have been extensively studied and well characterized. Exciting new research in immunology and virology has demonstrated that these first responders of the innate immune system are also crucial in limiting viral infections, replication, and associated off-target pathological damage. A multifaceted range of tactics is utilized to combat an equally diverse range of viruses, including phagocytosis, the formation of extracellular traps, the production of cytokines such as IFNs, and modulation of ILCs and lymphocytes.

Despite rapid advances in the field, many exciting unknown aspects of the involvement of neutrophils and inflammatory monocytes in combating viral infections remain to be clarified. Current research has documented the impact of neutrophil/monocyte retention in the bone marrow as it pertains to viral infections, but we still do not completely understand all mechanisms by which myeloid cells are recruited from the blood stream to the primary sites of infection. Future studies should aim to elucidate the specific signaling cascades that recruit myeloid cells into infected tissues and the mechanistic consequences of disruptions in these cascades via the chemokine gradient as well as depletions of specific ligands. If the scientific community can determine how different cell subsets can influence the production of chemokine populations and hone in on the essential ligands required for migration into the primary sites of infection, drugs could potentially be developed to exploit this localized production of chemokines. The discovery of pharmaceuticals that could fine-tune myeloid cell trafficking could prove beneficial to inducing rapid antiviral responses. Differential ligation versus the blockade of PRRs associated with protective versus pathological inflammation constitutes another strategy to balance rapid viral clearance and minimize host damage. Current knowledge from myeloid cell studies in bacterial diseases demonstrated that neutrophils are essential for monocyte recruitment and function. Additionally, it has been shown that the ratio of neutrophils to lymphocytes is higher in bacterial than viral infections among patients hospitalized for fevers [177]. It is clear that neutrophils and monocytes work in concert to enhance immune responses against bacterial pathogens. However, future studies are needed to explore the mechanisms by which these myeloid cells collaborate with each other to control viral infections, with the aim of gaining new insights into how they function in virus-infected microenvironments to regulate cell-to-cell communication within the innate and adaptive arms of the immune system. Gaining a better understanding of the role of myeloid cells in the pathogenesis of viral diseases will facilitate the design of better therapies.

Importantly, viruses and virus-mediated tissue damage stimulate both neutrophils and monocytes, triggering a cascade of cytokine/chemokine-mediated innate immune responses. This antiviral activity is not always beneficial for a host and, when improperly regulated, may contribute to immunopathologies such as cytokine storms that have been observed in many severe viral infections and could be related to type I IFN signaling. Mechanisms, including the potential relationship between type I IFN signaling and the regulation of excessive cytokine responses, should be further examined to develop strategies to minimize detrimental tissue damage by neutrophils and monocytes, while maximizing their beneficial anti-viral features.

References

1. Shi, C.; Pamer, E.G. Monocyte recruitment during infection and inflammation. *Nat. Rev. Immunol.* **2011**, *11*, 762–774. [CrossRef] [PubMed]
2. Wang, J.P.; Bowen, G.N.; Padden, C.; Cerny, A.; Finberg, R.W.; Newburger, P.E.; Kurt-Jones, E.A. Toll-like receptor-mediated activation of neutrophils by influenza A virus. *Blood* **2008**, *112*, 2028–2034. [CrossRef] [PubMed]
3. Teijaro, J.R.; Walsh, K.B.; Rice, S.; Rosen, H.; Oldstone, M.B. Mapping the innate signaling cascade essential for cytokine storm during influenza virus infection. *Proc. Natl. Acad. Sci. USA* **2014**, *111*, 3799–3804. [CrossRef] [PubMed]
4. Wang, J.P.; Kurt-Jones, E.A.; Shin, O.S.; Manchak, M.D.; Levin, M.J.; Finberg, R.W. Varicella-zoster virus activates inflammatory cytokines in human monocytes and macrophages via toll-like receptor 2. *J. Virol.* **2005**, *79*, 12658–12666. [CrossRef] [PubMed]
5. Fung, Y.L.; Minchinton, R.M. The fundamentals of neutrophil antigen and antibody investigations. *ISBT Sci. Ser.* **2011**, *6*, 381–386. [CrossRef]
6. Witko-Sarsat, V.; Rieu, P.; Descamps-Latscha, B.; Lesavre, P.; Halbwachs-Mecarelli, L. Neutrophils: Molecules, functions and pathophysiological aspects. *Lab. Investig.* **2000**, *80*, 617–653. [CrossRef] [PubMed]
7. Pillay, J.; den Braber, I.; Vrisekoop, N.; Kwast, L.M.; de Boer, R.J.; Borghans, J.A.; Tesselaar, K.; Koenderman, L. In vivo labeling with 2H2O reveals a human neutrophil lifespan of 5.4 days. *Blood* **2010**, *116*, 625–627. [CrossRef] [PubMed]
8. Goritzka, M.; Makris, S.; Kausar, F.; Durant, L.R.; Pereira, C.; Kumagai, Y.; Culley, F.J.; Mack, M.; Akira, S.; Johansson, C. Alveolar macrophage-derived type I interferons orchestrate innate immunity to RSV through recruitment of antiviral monocytes. *J. Exp. Med.* **2015**, *212*, 699–714. [CrossRef]
9. Cheung, T.S.; Dazzi, F. Mesenchymal-myeloid interaction in the regulation of immunity. *Semin. Immunol.* **2018**, *35*, 59–68. [CrossRef]
10. Appelberg, R. Neutrophils and intracellular pathogens: Beyond phagocytosis and killing. *Trends Microbiol.* **2007**, *15*, 87–92. [CrossRef]
11. Nathan, C. Neutrophils and immunity: Challenges and opportunities. *Nat. Rev. Immunol.* **2006**, *6*, 173–182. [CrossRef] [PubMed]
12. Ceredig, R.; Rolink, A.G.; Brown, G. Models of haematopoiesis- seeing the wood for the trees. *Nat. Rev. Immunol.* **2009**, *9*, 293–300. [CrossRef]
13. Dexter, M.T. Introduction to the haemopoietic system. *Cancer Surv.* **1990**, *9*, 1–5.
14. Hong, C.W. Current understanding in neutrophil differentiation and heterogeneity. *Immune Netw.* **2017**, *17*, 298–306. [CrossRef]
15. Dai, X.-M.; Ryan, G.R.; Hapel, A.J.; Dominguez, M.G.; Russell, R.G.; Kapp, S.; Sylvestre, V.; Stanley, E.R. Targeted disruption of the mouse colony-stimulating factor 1 receptor gene results in osteopetrosis, mononuclear phagocyte deficiency, increased primitive progenitor cell frequencies, and reproductive defects. *Blood* **2001**, *99*, 111–120. [CrossRef]
16. Scott, E.W.; Simon, M.C.; Anastasi, J.; Singh, H. Requirement of transcription factor PU.1 in the development of multiple hematopoietic lineages. *Science* **1994**, *265*, 1573–1577. [CrossRef] [PubMed]
17. Hettinger, J.; Richards, D.M.; Hansson, J.; Barra, M.M.; Joschko, A.C.; Krijgsveld, J.; Feuerer, M. Origin of monocytes and macrophages in a committed progenitor. *Nat. Immunol.* **2013**, *14*, 821–830. [CrossRef] [PubMed]
18. Kim, M.H.; Yang, D.; Kim, M.; Kim, S.Y.; Kim, D.; Kang, S.J. A late-lineage murine neutrophil precursor population exhibits dynamic changes during demand-adapted granulopoiesis. *Sci. Rep.* **2017**, *7*, 39804. [CrossRef]
19. Evrard, M.; Kwok, I.W.H.; Chong, S.Z.; Teng, K.W.W.; Becht, E.; Chen, J.; Sieow, J.L.; Penny, H.L.; Ching, G.C.; Devi, S.; et al. Developmental analysis of bone marrow neutrophils reveals populations specialized in expansion, trafficking, and effector functions. *Immunity* **2018**, *48*, 364–379.e8. [CrossRef]
20. Zhu, Y.P.; Padgett, L.; Dinh, H.Q.; Marcovecchio, P.; Blatchley, A.; Wu, R.; Ehinger, E.; Kim, C.; Mikulski, Z.; Seumois, G.; et al. Identification of an early unipotent neutrophil progenitor with pro-tumoral activity in mouse and human bone marrow. *Cell Rep.* **2018**, *24*, 2329–2341.e8. [CrossRef]

21. De Veer, M.J.; Holko, M.; Frevel, M.; Walker, E.; Der, S.; Paranjape, J.M.; Silverman, R.H.; Williams, B.R.G. Functional classification of interferon-stimulated genes identified using microarrays. *J. Leuk. Biol.* **2001**, *69*, 912–920.
22. Balachandran, S.; Roberts, P.C.; Brown, L.E.; Truong, H.; Pattnaik, A.K.; Archer, D.R.; Barber, G.N. Essential role for the dsRNA-dependent protein kinase PKR in innate immunity to viral infection. *Immunity* **2000**, *13*, 129–141. [CrossRef]
23. Balachandran, S.; Roberts, P.C.; Kipperman, T.; Bhalla, K.N.; Compans, R.W.; Archer, D.R.; Barber, G.N. Alpha:Beta interferons potentiate virus-induced apoptosis through activation of the FADD:Caspase-8 death signaling pathway. *J. Virol.* **2000**, *74*, 1513–1523. [CrossRef] [PubMed]
24. Pascutti, M.F.; Erkelens, M.N.; Nolte, M.A. Impact of viral infections on hematopoiesis: From beneficial to detrimental effects on bone marrow output. *Front. Immunol.* **2016**, *7*, 364. [CrossRef] [PubMed]
25. Eash, K.J.; Greenbaum, A.M.; Gopalan, P.K.; Link, D.C. CXCR2 and CXCR4 antagonistically regulate neutrophil trafficking from murine bone marrow. *J. Clin. Investig.* **2010**, *120*, 2423–2431. [CrossRef]
26. Jakubzick, C.; Gautier, E.L.; Gibbings, S.L.; Sojka, D.K.; Schlitzer, A.; Johnson, T.E.; Ivanov, S.; Duan, Q.; Bala, S.; Condon, T.; et al. Minimal differentiation of classical monocytes as they survey steady-state tissues and transport antigen to lymph nodes. *Immunity* **2013**, *39*, 599–610. [CrossRef] [PubMed]
27. Geissmann, F.; Jung, S.; Littman, D.R. Blood monocytes consist of two principal subsets with distinct migratory properties. *Immunity* **2003**, *19*, 71–82. [CrossRef]
28. Alder, J.K.; Georgantas, R.W.; Hildreth, R.L.; Kaplan, I.M.; Morisot, S.; Yu, X.; McDevitt, M.; Civin, C.I. Kruppel-like factor 4 is essential for inflammatory monocyte differentiation in vivo. *J. Immunol.* **2008**, *180*, 5645–5652. [CrossRef]
29. Yanez, A.; Coetzee, S.G.; Olsson, A.; Muench, D.E.; Berman, B.P.; Hazelett, D.J.; Salomonis, N.; Grimes, H.L.; Goodridge, H.S. Granulocyte-monocyte progenitors and monocyte-dendritic cell progenitors independently produce functionally distinct monocytes. *Immunity* **2017**, *47*, 890–902.e4. [CrossRef]
30. Moses, A.V.; Williams, S.; Heneveld, M.L.; Strussenberg, J.; Rarick, M.; Loveless, M.; Bagbye, G.; Nelson, J.A. Human immunodeficiency virus infection of bone marrow endothelium reduces induction of stromal hematopoietic growth factors. *Blood* **1996**, *87*, 919–925.
31. Furuta, R.; Yasunaga, J.I.; Miura, M.; Sugata, K.; Saito, A.; Akari, H.; Ueno, T.; Takenouchi, N.; Fujisawa, J.I.; Koh, K.R.; et al. Human T-cell leukemia virus type 1 infects multiple lineage hematopoietic cells in vivo. *PLoS Pathog.* **2017**, *13*, e1006722. [CrossRef] [PubMed]
32. Cortjens, B.; Ingelse, S.A.; Calis, J.C.; Vlaar, A.P.; Koenderman, L.; Bem, R.A.; van Woensel, J.B. Neutrophil subset responses in infants with severe viral respiratory infection. *Clin. Immunol.* **2017**, *176*, 100–106. [CrossRef] [PubMed]
33. Ok, C.Y.; Li, L.; Young, K.H. EBV-driven B-cell lymphoproliferative disorders: From biology, classification and differential diagnosis to clinical management. *Exp. Mol. Med.* **2015**, *47*, e132. [CrossRef]
34. Klco, J.M.; Geng, B.; Brunt, E.M.; Hassan, A.; Nguyen, T.D.; Kreisel, F.H.; Lisker-Melman, M.; Frater, J.L. Bone marrow biopsy in patients with hepatitis C virus infection: Spectrum of findings and diagnostic utility. *Am. J. Hematol.* **2010**, *85*, 106–110. [CrossRef] [PubMed]
35. Amaldo, R.G.; Mitsuyasu, R.T.; Rosenblatt, J.D.; Ngok, F.K.; Bakker, A.; Cole, S.; Chorn, N.; Lin, L.-S.; Bristol, G.; Boyd, M.P.; et al. Anti-human immunodeficiency virus hematopoietic progenitor cell-delivered ribozyme in a phase I study: Myeloid and lymphoid reconstitution in human immunodeficiency virus type-1-infected patients. *Hum. Gene Ther.* **2004**, *15*, 251–262. [CrossRef] [PubMed]
36. Le Goffic, R.; Pothlichet, J.; Vitour, D.; Fujita, T.; Meurs, E.; Chignard, M.; Si-Tahar, M. Cutting edge: Influenza A virus activates TLR3-dependent inflammatory and RIG-I-dependent antiviral responses in human lung epithelial cells. *J. Immunol.* **2007**, *178*, 3368–3372. [CrossRef]
37. Carignan, D.; Herblot, S.; Laliberte-Gagne, M.E.; Bolduc, M.; Duval, M.; Savard, P.; Leclerc, D. Activation of innate immunity in primary human cells using a plant virus derived nanoparticle TLR7/8 agonist. *Nanomedicine* **2018**, *14*, 2317–2327. [CrossRef]
38. Barton, G.M.; Kagan, J.C.; Medzhitov, R. Intracellular localization of toll-like receptor 9 prevents recognition of self DNA but facilitates access to viral DNA. *Nat. Immunol.* **2006**, *7*, 49–56. [CrossRef]

39. Bieback, K.; Lien, E.; Klagge, I.M.; Avota, E.; Schneider-Schaulies, J.; Duprex, W.P.; Wagner, H.; Kirschning, C.J.; ter Meulen, V.; Schneider-Schaulies, S. Hemagglutinin protein of wild-type measles virus activates toll-like receptor 2 signaling. *J. Virol.* **2002**, *76*, 8729–8736. [CrossRef]

40. Kurt-Jones, E.A.; Chan, M.; Zhou, S.; Wang, J.; Reed, G.; Bronson, R.; Arnold, M.M.; Knipe, D.M.; Finberg, R.W. Herpes simplex virus 1 interaction with toll-like receptor 2 contributes to lethal encephalitis. *Proc. Natl. Acad. Sci. USA* **2004**, *101*, 1315–1320. [CrossRef]

41. Leoni, V.; Gianni, T.; Salvioli, S.; Campadelli-Fiume, G. Herpes simplex virus glycoproteins gH/gL and gB bind toll-like receptor 2, and soluble gH/gL is sufficient to activate NF-kB. *J. Virol.* **2012**, *86*, 6555–6562. [CrossRef] [PubMed]

42. Ahlenstiel, G.; Gambato, M.; Caro-Pérez, N.; González, P.; Cañete, N.; Mariño, Z.; Lens, S.; Bonacci, M.; Bartres, C.; Sánchez-Tapias, J.-M.; et al. Neutrophil and monocyte function in patients with chronic hepatitis C undergoing antiviral therapy with regimens containing ppotease inhibitors with and without interferon. *PLoS ONE* **2016**, *11*, e0166631.

43. Hayashi, F.; Means, T.K.; Luster, A.D. Toll-like receptors stimulate human neutrophil function. *Blood* **2003**, *102*, 2660–2669. [CrossRef] [PubMed]

44. Hashimoto, Y.; Moki, T.; Takizawa, T.; Shiratsuchi, A.; Nakanishi, Y. Evidence for phagocytosis of influenza virus-infected, apoptotic cells by neutrophils and macrophages in mice. *J. Immunol.* **2007**, *178*, 2448–2457. [CrossRef] [PubMed]

45. Shiratsuchi, A.; Watanabe, I.; Takeuchi, O.; Akira, S.; Nakanishi, Y. Inhibitory effect of Toll-like receptor 4 on fusion between phagosomes and endosomes/lysosomes in macrophages. *J. Immunol.* **2004**, *172*, 2039–2047. [CrossRef] [PubMed]

46. Yoneyama, M.; Fujita, T. Structural mechanism of RNA recognition by the RIG-I-like receptors. *Immunity* **2008**, *29*, 178–181. [CrossRef] [PubMed]

47. Jensen, S.; Thomsen, A.R. Sensing of RNA viruses: A review of innate immune receptors involved in recognizing RNA virus invasion. *J. Virol.* **2012**, *86*, 2900–2910. [CrossRef] [PubMed]

48. Feldmann, H.; Bugany, H.; Mahner, F.; Klenk, H.-D.; Drenckhahn, D.; Schnittler, H.-J. Filovirus-induced endothelial leakage triggered by infected monocytes:Macrophages. *J. Virol.* **1996**, *70*, 2208–2214.

49. Tamassia, N.; Moigne, V.L.; Rossato, M.; Donini, M.; McCartney, S.; Calzetti, F.; Colonna, M.; Bazzoni, F.; Cassatella, M.A. Activation of an immunoregulatory and antiviral gene expression program in poly(I-C)-transfected human neutrophils. *J. Immunol.* **2008**, *181*, 6563–6573. [CrossRef]

50. Wang, J.P.; Cerny, A.; Asher, D.R.; Kurt-Jones, E.A.; Bronson, R.T.; Finberg, R.W. MDA5 and mavs mediate type I interferon responses to coxsackie B virus. *J. Virol.* **2010**, *84*, 254–260. [CrossRef]

51. Fredericksen, B.L.; Keller, B.C.; Fornek, J.; Katze, M.G.; Gale, M., Jr. Establishment and maintenance of the innate antiviral response to Wst Nile Virus involves both RIG-I and MDA5 signaling through IPS-1. *J. Virol.* **2008**, *82*, 609–616. [CrossRef] [PubMed]

52. Tang, D.; Kang, R.; Coyne, C.B.; Zeh, H.J.; Lotze, M.T. PAMPs and DAMPs- signal 0s that spur autophagy and immunity. *Immunol. Rev.* **2012**, *249*, 158–175. [CrossRef] [PubMed]

53. Van Vloten, J.P.; Workenhe, S.T.; Wootton, S.K.; Mossman, K.L.; Bridle, B.W. Critical interactions between immunogenic cancer cell death, oncolytic viruses, and the immune system define the rational design of combination immunotherapies. *J. Immunol.* **2018**, *200*, 450–458. [CrossRef] [PubMed]

54. Zhang, Q.; Raoof, M.; Chen, Y.; Sumi, Y.; Sursal, T.; Junger, W.; Brohi, K.; Itagaki, K.; Hauser, C.J. Circulating mitochondrial damps cause inflammatory responses to injury. *Nature* **2010**, *464*, 104–107. [CrossRef] [PubMed]

55. Dumitriu, I.E.; Baruah, P.; Manfredi, A.A.; Bianchi, M.E.; Rovere-Querini, P. HMGB1: Guiding immunity from within. *Trends Immunol.* **2005**, *26*, 381–387. [CrossRef] [PubMed]

56. Gardella, S.; Andrei, C.; Ferrera, D.; Lotti, L.V.; Torrisi, M.R.; Bianchi, M.E.; Rubartelli, A. The nuclear protein HMGB1 is secreted by monocytes via a non-classical, vesicle-mediated secretory pathway. *EMBO Rep.* **2002**, *3*, 995–1001. [CrossRef] [PubMed]
57. Park, J.S.; Arcaroli, J.; Yum, H.-K.; Yang, H.; Wang, H.; Yang, K.-Y.; Choe, K.-H.; Strassheim, D.; Pitts, T.M.; Tracey, K.J.; et al. Activation of gene expression in human neutrophils by high mobility group box 1 protein. *Am. J. Physiol. Cell Physiol.* **2003**, *284*, C870–C879. [CrossRef]
58. Chen, L.C.; Yeh, T.M.; Wu, H.N.; Lin, Y.Y.; Shyu, H.W. Dengue virus infection induces passive release of high mobility group box 1 protein by epithelial cells. *J. Infect.* **2008**, *56*, 143–150. [CrossRef]
59. Griffith, J.W.; Sokol, C.L.; Luster, A.D. Chemokines and chemokine receptors: Positioning cells for host defense and immunity. *Annu. Rev. Immunol.* **2014**, *32*, 659–702. [CrossRef]
60. Suratt, B.T.; Petty, J.M.; Young, S.K.; Malcolm, K.C.; Lieber, J.G.; Nick, J.A.; Gonzalo, J.-A.; Henson, P.M.; Worthen, G.S. Role of the CXCR4/SDF-1 chemokine axis in circulating neutrophil homeostasis. *Blood* **2004**, *104*, 565–571. [CrossRef]
61. De Filippo, K.; Dudeck, A.; Hasenberg, M.; Nye, E.; van Rooijen, N.; Hartmann, K.; Gunzer, M.; Roers, A.; Hogg, N. Mast cell and macrophage chemokines CXCL1/CXCL2 control the early stage of neutrophil recruitment during tissue inflammation. *Blood* **2013**, *121*, 4930–4937. [CrossRef] [PubMed]
62. Serbina, N.V.; Pamer, E.G. Monocyte emigration from bone marrow during bacterial infection requires signals mediated by chemokine receptor CCR2. *Nat. Immunol.* **2006**, *7*, 311–317. [CrossRef] [PubMed]
63. Jung, H.; Mithal, D.S.; Park, J.E.; Miller, R.J. Localized CCR2 activation in the bone marrow niche mobilizes monocytes by desensitizing CXCR4. *PLoS ONE* **2015**, *10*, e0128387. [CrossRef] [PubMed]
64. Shi, C.; Jia, T.; Mendez-Ferrer, S.; Hohl, T.M.; Serbina, N.V.; Lipuman, L.; Leiner, I.; Li, M.O.; Frenette, P.S.; Pamer, E.G. Bone marrow mesenchymal stem and progenitor cells induce monocyte emigration in response to circulating toll-like receptor ligands. *Immunity* **2011**, *34*, 590–601. [CrossRef] [PubMed]
65. Gao, R.; Bhatnagar, J.; Blau, D.M.; Greer, P.; Rollin, D.C.; Denison, A.M.; Deleon-Carnes, M.; Shieh, W.J.; Sambhara, S.; Tumpey, T.M.; et al. Cytokine and chemokine profiles in lung tissues from fatal cases of 2009 pandemic influenza a (H1N1): Role of the host immune response in pathogenesis. *Am. J. Pathol.* **2013**, *183*, 1258–1268. [CrossRef] [PubMed]
66. Lim, J.K.; O'bara, C.J.; Rivollier, A.; Pletnev, A.G.; Kelsall, B.L.; Murphy, P.M. Chemokine receptor CCR2 is critical for monocyte accumulation and survival in west nile virus encephalitis. *J. Immunol.* **2011**, *186*, 471–478. [CrossRef] [PubMed]
67. Moses, A.V.; Fish, K.N.; Ruhl, R.; Smith, P.P.; Strussenberg, J.G.; Zhu, L.; Chandran, B.; Nelson, J.A. Long-term infection and transformation of dermal microvascular endothelial cells by human herpesvirus 8. *J. Virol.* **1999**, *73*, 6892–6902.
68. Ley, K.; Laudanna, C.; Cybulsky, M.I.; Nourshargh, S. Getting to the site of inflammation: The leukocyte adhesion cascade updated. *Nat. Rev. Immunol.* **2007**, *7*, 678–689. [CrossRef]
69. Ferstl, R.; Akdis, C.A.; O'Mahony, L. Histamine regulation of innate and adaptive immunity. *Front. Biosci.* **2012**, *17*, 40–53. [CrossRef]
70. Pober, J.S.; Sessa, W.C. Evolving functions of endothelial cells in inflammation. *Nat. Rev. Immunol.* **2007**, *7*, 803–815. [CrossRef]
71. Nourshargh, S.; Alon, R. Leukocyte migration into inflamed tissues. *Immunity* **2014**, *41*, 694–707. [CrossRef] [PubMed]
72. Julkunen, I.; Melén, K.; Nyqvist, M.; Pirhonen, J.; Sareneva, T.; Matikainen, S. Inflammatory responses in influenza A virus infection. *Vaccine* **2000**, *8*, S32–S37.
73. Sumagin, R.; Prizant, H.; Lomakina, E.; Waugh, R.E.; Sarelius, I.H. LFA-1 and MAC-1 define characteristically different intralumenal crawling and emigration patterns for monocytes and neutrophils in situ. *J. Immunol.* **2010**, *185*, 7057–7066. [CrossRef]
74. De Oliveira, S.; Rosowski, E.E.; Huttenlocher, A. Neutrophil migration in infection and wound repair: Going forward in reverse. *Nat. Rev. Immunol.* **2016**, *16*, 378–391. [CrossRef] [PubMed]
75. Rubio, N.; Sanz-Rodriguez, F. Induction of the CXCL1 (KC) chemokine in mouse astrocytes by infection with the murine encephalomyelitis virus of theiler. *Virology* **2007**, *358*, 98–108. [CrossRef]
76. Mohanty, S.K.; Ivantes, C.A.P.; Mourya, R.; Pacheco, C.; Bezerra, J.A. Macrophages are targeted by rotavirus in experimental biliary atresia and induce neutrophil chemotaxis by MIP2:CXCL2. *Ped Res.* **2010**, *67*, 345–351. [CrossRef]

77. Mathieu, C.; Guillaume, V.; Volchkova, V.A.; Pohl, C.; Jacquot, F.; Looi, R.Y.; Wong, K.T.; Legras-Lachuer, C.; Volchkov, V.E.; Lachuer, J.; et al. Nonstructural nipah virus C protein regulates both the early host proinflammatory response and viral virulence. *J. Virol.* **2012**, *86*, 10766–10775. [CrossRef]
78. Auffray, C.; Fogg, D.; Garfa, M.; Elain, G.; Join-Lambert, O.; Kayal, S.; Sarnacki, S.; Cumano, A.; Lauvau, G.; Geissmann, F. Monitoring of blood vessels and tissues by a population of monocytes with patrolling behavior. *Science* **2007**, *317*, 666–670. [CrossRef]
79. Tate, M.D.; Deng, Y.M.; Jones, J.E.; Anderson, G.P.; Brooks, A.G.; Reading, P.C. Neutrophils ameliorate lung injury and the development of severe disease during influenza infection. *J. Immunol.* **2009**, *183*, 7441–7450. [CrossRef]
80. Tumpey, T.M.; Chen, S.-H.; Oaes, J.E.; Lausch, R.N. Neutrophil-mediated suppression of virus replication after herpes simplex virus type 1 infection of the murine cornea. *J. Virol.* **1996**, *70*, 898–904.
81. Tate, M.D.; Brooks, A.G.; Reading, P.C. The role of neutrophils in the upper and lower respiratory tract during influenza virus infection of mice. *Respir. Res.* **2008**, *9*, 57. [CrossRef] [PubMed]
82. Jenne, C.N.; Wong, C.H.; Zemp, F.J.; McDonald, B.; Rahman, M.M.; Forsyth, P.A.; McFadden, G.; Kubes, P. Neutrophils recruited to sites of infection protect from virus challenge by releasing neutrophil extracellular traps. *Cell Host Microbe* **2013**, *13*, 169–180. [CrossRef] [PubMed]
83. Hufford, M.M.; Richardson, G.; Zhou, H.; Manicassamy, B.; Garcia-Sastre, A.; Enelow, R.I.; Braciale, T.J. Influenza-infected neutrophils within the infected lungs act as antigen presenting cells for anti-viral CD8(+) T cells. *PLoS ONE* **2012**, *7*, e46581. [CrossRef] [PubMed]
84. Brinkmann, V.; Reichard, U.; Goosmann, C.; Fauler, B.; Uhlemann, Y.; Weiss, D.S.; Weinrauch, Y.; Zychlinsky, A. Neutrophil extracellular traps kill bacteria. *Science* **2004**, *202*, 532–1535. [CrossRef] [PubMed]
85. Worley, M.J.; Fei, K.; Lopez-Denman, A.J.; Kelleher, A.D.; Kent, S.J.; Chung, A.W. Neutrophils mediate HIV-specific antibody-dependent phagocytosis and ADCC. *J. Immunol. Methods* **2018**, *457*, 41–52. [CrossRef] [PubMed]
86. Kothari, N.; Keshari, R.S.; Bogra, J.; Kohli, M.; Abbas, H.; Malik, A.; Dikshit, M.; Barthwal, M.K. Increased myeloperoxidase enzyme activity in plasma is an indicator of inflammation and onset of sepsis. *J. Crit. Care* **2011**, *26*. [CrossRef] [PubMed]
87. Lee-Huang, S.; Huang, P.L.; Sun, Y.; Huang, P.L.; Kung, H.F.; Blithe, D.L.; Chen, H.C. Lysozyme and rnases as anti-hiv components in β-core preparations of human chorionic gonadotropin. *Proc. Natl. Acad. Sci. USA* **1999**, *96*, 2678–2681. [CrossRef] [PubMed]
88. Jimenez Flores, E.; Tian, S.; Sizova, M.; Epstein, S.S.; Lamont, R.J.; Uriarte, S.M. *Peptoanaerobacter stomatis* primes human neutrophils and induces granule exocytosis. *Infect. Immun.* **2017**, 85. [CrossRef] [PubMed]
89. Lukens, M.V.; van de Pol, A.C.; Coenjaerts, F.E.; Jansen, N.J.; Kamp, V.M.; Kimpen, J.L.; Rossen, J.W.; Ulfman, L.H.; Tacke, C.E.; Viveen, M.C.; et al. A systemic neutrophil response precedes robust CD8(+) T-cell activation during natural respiratory syncytial virus infection in infants. *J. Virol.* **2010**, *84*, 2374–2383. [CrossRef] [PubMed]
90. Lim, K.; Hyun, Y.M.; Lambert-Emo, K.; Capece, T.; Bae, S.; Miller, R.; Topham, D.J.; Kim, M. Neutrophil trails guide influenza-specific CD8(+) T cells in the airways. *Science* **2015**, *349*, aaa4352. [CrossRef]
91. Lindemans, C.A.; Coffer, P.J.; Schellens, I.M.M.; Graaff, P.M.A.d.; Kimpen, J.L.L.; Koenderman, L. Respiratory syncytial virus inhibits granulocyte apoptosis through a phosphatidylinositol 3-kinase and NF-κB-dependent mechanism. *J. Immunol.* **2006**, *176*, 5529–5537. [CrossRef] [PubMed]
92. Deng, Y.; Herbert, J.A.; Smith, C.M.; Smyth, R.L. An in vitro transepithelial migration assay to evaluate the role of neutrophils in respiratory syncytial virus (RSV) induced epithelial damage. *Sci. Rep.* **2018**, *8*, 6777. [CrossRef] [PubMed]
93. Harrison, A.M.; Bonville, C.A.; Rosenburg, H.F.; Domachowske, J. Respiratory syncytical virus–induced chemokine expression in the lower airways eosinophil recruitment and degranulation. *Am. J. Respir. Crit. Care Med.* **1999**, *159*, 1918–1924. [CrossRef] [PubMed]
94. Ikeda, N.; Asano, K.; Kikuchi, K.; Uchida, Y.; Ikegami, H.; Takagi, R.; Yotsumoto, S.; Shibuya, T.; Makino-Okamura, C.; Fukuyama, H.; et al. Emergence of immunoregulatory $Ym1^+Ly6C^{hi}$ monocytes during recovery phase of tissue injury. *Sci. Immunol.* **2018**, *3*, eaat0207. [CrossRef] [PubMed]
95. Wherry, E.J.; Seo, S.-U.; Kwon, H.-J.; Ko, H.-J.; Byun, Y.-H.; Seong, B.L.; Uematsu, S.; Akira, S.; Kweon, M.-N. Type I interferon signaling regulates $Ly6C^{hi}$ monocytes and neutrophils during acute viral pneumonia in mice. *PLoS Pathog.* **2011**, *7*, e1001304.

96. Herold, S.; Steinmueller, M.; von Wulffen, W.; Cakarova, L.; Pinto, R.; Pleschka, S.; Mack, M.; Kuziel, W.A.; Corazza, N.; Brunner, T.; et al. Lung epithelial apoptosis in influenza virus pneumonia: The role of macrophage-expressed TNF-related apoptosis-inducing ligand. *J. Exp. Med.* **2008**, *205*, 3065–3077. [CrossRef]
97. Hall, M.W.; Geyer, S.M.; Guo, C.Y.; Panoskaltsis-Mortari, A.; Jouvet, P.; Ferdinands, J.; Shay, D.K.; Nateri, J.; Greathouse, K.; Sullivan, R.; et al. Innate immune function and mortality in critically ill children with influenza: A. multicenter study. *Crit. Care Med.* **2013**, *41*, 224–236. [CrossRef] [PubMed]
98. Coates, B.M.; Staricha, K.L.; Koch, C.M.; Cheng, Y.; Shumaker, D.K.; Budinger, G.R.S.; Perlman, H.; Misharin, A.V.; Ridge, K.M. Inflammatory monocytes drive influenza a virus-mediated lung injury in juvenile mice. *J. Immunol.* **2018**, *200*, 2391–2404. [CrossRef] [PubMed]
99. Cole, S.L.; Dunning, J.; Kok, W.L.; Benam, K.H.; Benlahrech, A.; Repapi, E.; Martinez, F.O.; Drumright, L.; Powell, T.J.; Bennett, M.; et al. M1-like monocytes are a major immunological determinant of severity in previously healthy adults with life-threatening influenza. *JCI Insight* **2017**, *2*, e91868. [CrossRef] [PubMed]
100. Percopo, C.M.; Ma, M.; Brenner, T.A.; Krumholz, J.O.; Break, T.J.; Laky, K.; Rosenberg, H.F. Critical adverse impact of IL-6 in acute pneumovirus infection. *J. Immunol.* **2019**, *202*, 871–882. [CrossRef] [PubMed]
101. Mildner, A.; Marinkovic, G.; Jung, S. Murine monocytes: Origins, subsets, fates, and functions. *Microbiol. Spectr.* **2016**, *4*. [CrossRef]
102. Netea, M.G.; Quintin, J.; van der Meer, J.W. Trained immunity: A memory for innate host defense. *Cell Host Microbe* **2011**, *9*, 355–361. [CrossRef] [PubMed]
103. Rossini, G.; Landini, M.P.; Gelsomino, F.; Sambri, V.; Varani, S. Innate host responses to west nile virus: Implications for central nervous system immunopathology. *World J. Virol.* **2013**, *2*, 49–56. [CrossRef] [PubMed]
104. Srikiatkhachorn, A.; Mathew, A.; Rothman, A.L. Immune-mediated cytokine storm and its role in severe dengue. *Semin. Immunopathol.* **2017**, *39*, 563–574. [CrossRef] [PubMed]
105. Griseri, T.; McKenzie, B.S.; Schiering, C.; Powrie, F. Dysregulated hematopoietic stem and progenitor cell activity promotes interleukin-23-driven chronic intestinal inflammation. *Immunity* **2012**, *37*, 1116–1129. [CrossRef] [PubMed]
106. O'Shea, J.J.; Ma, A.; Lipsky, P. Cytokines and autoimmunity. *Nat. Rev. Immunol.* **2002**, *2*, 37–45. [CrossRef] [PubMed]
107. Kolaczkowska, E.; Kubes, P. Neutrophil recruitment and function in health and inflammation. *Nat. Rev. Immunol.* **2013**, *13*, 159–175. [CrossRef] [PubMed]
108. Schauer, C.; Janko, C.; Munoz, L.E.; Zhao, Y.; Kienhofer, D.; Frey, B.; Lell, M.; Manger, B.; Rech, J.; Naschberger, E.; et al. Aggregated neutrophil extracellular traps limit inflammation by degrading cytokines and chemokines. *Nat. Med.* **2014**, *20*, 511–517. [CrossRef] [PubMed]
109. Fournier, B.M.; Parkos, C.A. The role of neutrophils during intestinal inflammation. *Mucosal. Immunol.* **2012**, *5*, 354–366. [CrossRef]
110. Vogt, K.L.; Summers, C.; Chilvers, E.R.; Condliffe, A.M. Priming and de-priming of neutrophil responses in vitro and in vivo. *Eur. J. Clin. Investig.* **2018**, *48* (Suppl. 2), e12967. [CrossRef]
111. Kitchen, E.; Rossi, A.G.; Condliffe, A.M.; Haslett, C.; Chilvers, E.R. Demonstration of reversible priming of human neutrophils using platelet-activating factor. *Blood* **1996**, *88*, 4330–4337. [PubMed]
112. Summers, C.; Chilvers, E.R.; Peters, A.M. Mathematical modeling supports the presence of neutrophil depriming in vivo. *Physiol. Rep.* **2014**, *2*, e00241. [CrossRef]
113. Summers, C.; Singh, N.R.; White, J.F.; Mackenzie, I.M.; Johnston, A.; Solanki, C.; Balan, K.K.; Peters, A.M.; Chilvers, E.R. Pulmonary retention of primed neutrophils: A novel protective host response, which is impaired in the acute respiratory distress syndrome. *Thorax* **2014**, *69*, 623–629. [CrossRef] [PubMed]
114. Singh, N.R.; Johnson, A.; Peters, A.M.; Babar, J.; Chilvers, E.R.; Summers, C. Acute lung injury results from failure of neutrophil de-priming: A. new hypothesis. *Eur. J. Clin. Investig.* **2012**, *42*, 1342–1349. [CrossRef] [PubMed]
115. Mitchell, L.A.; Henderson, A.J.; Dow, S.W. Suppression of vaccine immunity by inflammatory monocytes. *J. Immunol.* **2012**, *189*, 5612–5621. [CrossRef] [PubMed]
116. Knapp, W.; Baumgartner, G. Monocyte-mediated suppression of human B lymphocyte differentiation in vitro. *J. Immunol.* **1978**, *121*, 1177–1183. [PubMed]
117. Ortega-Gomez, A.; Perretti, M.; Soehnlein, O. Resolution of inflammation: An integrated view. *EMBO Mol. Med.* **2013**, *5*, 661–674. [CrossRef]

118. Shalova, I.N.; Lim, J.Y.; Chittezhath, M.; Zinkernagel, A.S.; Beasley, F.; Hernandez-Jimenez, E.; Toledano, V.; Cubillos-Zapata, C.; Rapisarda, A.; Chen, J.; et al. Human monocytes undergo functional re-programming during sepsis mediated by hypoxia-inducible factor-1alpha. *Immunity* **2015**, *42*, 484–498. [CrossRef]
119. Tacke, R.S.; Lee, H.C.; Goh, C.; Courtney, J.; Polyak, S.J.; Rosen, H.R.; Hahn, Y.S. Myeloid suppressor cells induced by hepatitis c virus suppress T-cell responses through the production of reactive oxygen species. *Hepatology* **2012**, *55*, 343–353. [CrossRef]
120. Goh, C.C.; Roggerson, K.M.; Lee, H.-C.; Golden-Mason, L.; Rosen, H.R.; Hahn, Y.S. Hepatitis C virus–induced myeloid-derived suppressor cells suppress NK cell IFN-γ production by altering cellular metabolism via arginase-1. *J. Immunol.* **2016**, *196*, 2283–2292. [CrossRef]
121. Qin, A.; Cai, W.; Pan, T.; Wu, K.; Yang, Q.; Wang, N.; Liu, Y.; Yan, D.; Hu, F.; Guo, P.; et al. Expansion of monocytic myeloid-derived suppressor cells dampens T cell function in HIV-1-seropositive individuals. *J. Virol.* **2013**, *87*, 1477–1490. [CrossRef] [PubMed]
122. Escors, D. Differentiation of murine myeloid-derived suppressor cells. In *Myeloid-Derived Suppressor Cells and Cancer*; Springer: Cham, Switzerland, 2016; pp. 25–37. [CrossRef]
123. Mortha, A.; Burrows, K. Cytokine networks between innate lymphoid cells and myeloid cells. *Front. Immunol.* **2018**, *9*, 191. [CrossRef] [PubMed]
124. Klose, C.S.; Artis, D. Innate lymphoid cells as regulators of immunity, inflammation and tissue homeostasis. *Nat. Immunol.* **2016**, *17*, 765–774. [CrossRef] [PubMed]
125. Spits, H.; Cupedo, T. Innate lymphoid cells: Emerging insights in development, lineage relationships, and function. *Annu. Rev. Immunol.* **2012**, *30*, 647–675. [CrossRef] [PubMed]
126. Walker, J.A.; Barlow, J.L.; McKenzie, A.N. Innate lymphoid cells-how did we miss them? *Nat. Rev. Immunol.* **2013**, *13*, 75–87. [CrossRef] [PubMed]
127. Nabatanzi, R.; Cose, S.; Joloba, M.; Jones, S.R.; Nakanjako, D. Effects of hiv infection and art on phenotype and function of circulating monocytes, natural killer, and innate lymphoid cells. *AIDS Res. Ther.* **2018**, *15*, 7. [CrossRef] [PubMed]
128. Lavin, Y.; Winter, D.; Blecher-Gonen, R.; David, E.; Keren-Shaul, H.; Merad, M.; Jung, S.; Amit, I. Tissue-resident macrophage enhancer landscapes are shaped by the local microenvironment. *Cell* **2014**, *159*, 1312–1326. [CrossRef] [PubMed]
129. Serbina, N.V.; Jia, T.; Hohl, T.M.; Pamer, E.G. Monocyte-mediated defense against microbial pathogens. *Annu. Rev. Immunol.* **2008**, *26*, 421–452. [CrossRef] [PubMed]
130. Thais, P.; Salazar-Mather, J.S.O.; Biron, C.A. Early murine cytomegalovirus (MCMV) infection induces liver natural killer (NK) cell inflammation and protection through macrophage inflammatory protein 1α (MIP-1α)–dependent pathways. *J. Exp. Med.* **1998**, *187*, 1–14.
131. Grabstein, K.; Eisenman, J.; Shanebeck, K.; Rauch, C.; Srinivasan, S.; Fung, V.; Beers, C.; Richardson, J.; Schoenborn, M.; Ahdieh, M. Cloning of a T cell growth factor that interacts with the beta chain of the interleukin-2 receptor. *Science* **1994**, *13*, 965–968. [CrossRef]
132. Perera, P.Y.; Lichy, J.H.; Waldmann, T.A.; Perera, L.P. The role of interleukin-15 in inflammation and immune responses to infection: Implications for its therapeutic use. *Microbes Infect.* **2012**, *14*, 247–261. [CrossRef] [PubMed]
133. Zhang, C.; Zhang, J.; Niu, J.; Zhang, J.; Tian, Z. Interleukin-15 improves cytotoxicity of natural killer cells via up-regulating nkg2d and cytotoxic effector molecule expression as well as STAT1 and ERK1/2 phosphorylation. *Cytokine* **2008**, *42*, 128–136. [CrossRef] [PubMed]
134. Hengel, H.; Flohr, T.; Hammerling, G.J.; Koszinowski, U.H.; Momburg, F. Human cytomegalovirus inhibits peptide translocation into the endoplasmic reticulum for mhc class I assembly. *J. Gen. Virol.* **1996**, *77*, 2287–2296. [CrossRef] [PubMed]
135. Musso, T.; Calosso, L.; Zucca, M.; Millesimo, M.; Ravarino, D.; Giovarelli, M.; Malavasi, F.; Ponzi, A.N.; Paus, R.; Bulfone-Paus, S. Human monocytes constitutively express membrane-bound, biologically active, and interferon-γ–upregulated interleukin-15. *Blood* **1999**, *93*, 3531–3539. [PubMed]
136. Regamey, N.; Obregon, C.; Ferrari-Lacraz, S.; van Leer, C.; Chanson, M.; Nicod, L.P.; Geiser, T. Airway epithelial IL-15 transforms monocytes into dendritic cells. *Am. J. Respir. Cell Mol. Biol.* **2007**, *37*, 75–84. [CrossRef] [PubMed]

137. Lee, A.J.; Chen, B.; Chew, M.V.; Barra, N.G.; Shenouda, M.M.; Nham, T.; van Rooijen, N.; Jordana, M.; Mossman, K.L.; Schreiber, R.D.; et al. Inflammatory monocytes require type i interferon receptor signaling to activate NK cells via IL-18 during a mucosal viral infection. *J. Exp. Med.* **2017**, *214*, 1153–1167. [CrossRef] [PubMed]
138. Ferlazzo, G.; Pack, M.; Thomas, D.; Paludan, C.; Schmid, D.; Strowig, T.; Bougras, G.; Muller, W.A.; Moretta, L.; Munz, C. Distinct roles of IL-12 and IL-15 in human natural killer cell activation by dendritic cells from secondary lymphoid organs. *Proc. Natl. Acad. Sci. USA* **2004**, *101*, 16606–16611. [CrossRef] [PubMed]
139. Ethuin, F.; Gerard, B.; Benna, J.E.; Boutten, A.; Gougereot-Pocidalo, M.A.; Jacob, L.; Chollet-Martin, S. Human neutrophils produce interferon gamma upon stimulation by interleukin-12. *Lab. Investig.* **2004**, *84*, 1363–1371. [CrossRef] [PubMed]
140. Costantini, C.; Calzetti, F.; Perbellini, O.; Micheletti, A.; Scarponi, C.; Lonardi, S.; Pelletier, M.; Schakel, K.; Pizzolo, G.; Facchetti, F.; et al. Human neutrophils interact with both 6-sulfo LacNAc+ DC and NK cells to amplify NK-derived IFN{gamma}: role of CD18, ICAM-1, and ICAM-3. *Blood* **2011**, *117*, 1677–1686. [CrossRef] [PubMed]
141. Neumann, K.; Karimi, K.; Meiners, J.; Voetlause, R.; Steinmann, S.; Dammermann, W.; Luth, S.; Asghari, F.; Wegscheid, C.; Horst, A.K. A proinflammatory role of type 2 innate lymphoid cells in murine immune-mediated hepatitis. *J. Immunol.* **2017**, *198*, 128–137. [CrossRef] [PubMed]
142. Costa, S.; Bevilacqua, D.; Cassatella, M.A.; Scapini, P. Recent advances on the crosstalk between neutrophils and B or T lymphocytes. *Immunology* **2019**, *156*, 23–32. [CrossRef] [PubMed]
143. Abi Abdallah, D.S.; Egan, C.E.; Butcher, B.A.; Denkers, E.Y. Mouse neutrophils are professional antigen-presenting cells programmed to instruct TH1 and TH17 T-cell differentiation. *Int. Immunol.* **2011**, *23*, 317–326. [CrossRef] [PubMed]
144. Beauvillain, C.; Delneste, Y.; Scotet, M.; Peres, A.; Gascan, H.; Guermonprez, P.; Barnaba, V.; Jeannin, P. Neutrophils efficiently cross-prime naive T cells in vivo. *Blood* **2007**, *110*, 2965–2973. [CrossRef] [PubMed]
145. Van Gisbergen, K.P.; Sanchez-Hernandez, M.; Geijtenbeek, T.B.; van Kooyk, Y. Neutrophils mediate immune modulation of dendritic cells through glycosylation-dependent interactions between Mac-1 and DC-SIGN. *J. Exp. Med.* **2005**, *201*, 1281–1292. [CrossRef] [PubMed]
146. Yang, C.W.; Strong, B.S.; Miller, M.J.; Unanue, E.R. Neutrophils influence the level of antigen presentation during the immune response to protein antigens in adjuvants. *J. Immunol.* **2010**, *185*, 2927–2934. [CrossRef] [PubMed]
147. Pillay, J.; Kamp, V.M.; van Hoffen, E.; Visser, T.; Tak, T.; Lammers, J.W.; Ulfman, L.H.; Leenen, L.P.; Pickkers, P.; Koenderman, L. A subset of neutrophils in human systemic inflammation inhibits T cell responses through MAC-1. *J. Clin. Investig.* **2012**, *122*, 327–336. [CrossRef] [PubMed]
148. Tak, T.; Rygiel, T.P.; Karnam, G.; Bastian, O.W.; Boon, L.; Viveen, M.; Coenjaerts, F.E.; Meyaard, L.; Koenderman, L.; Pillay, J. Neutrophil-mediated suppression of influenza-induced pathology requires CD11B/CD18 (MAC-1). *Am. J. Respir. Cell Mol. Biol.* **2018**, *58*, 492–499. [CrossRef]
149. Blomgran, R.; Desvignes, L.; Briken, V.; Ernst, J.D. Mycobacterium tuberculosis inhibits neutrophil apoptosis, leading to delayed activation of naive CD4 T cells. *Cell Host Microbe* **2012**, *11*, 81–90. [CrossRef]
150. Clayton, A.R.; Prue, R.L.; Harper, L.; Drayson, M.T.; Savage, C.O. Dendritic cell uptake of human apoptotic and necrotic neutrophils inhibits CD40, CD80, and CD86 expression and reduces allogeneic T cell responses: Relevance to systemic vasculitis. *Arthritis Rheum.* **2003**, *48*, 2362–2374. [CrossRef]
151. Blomgran, R.; Ernst, J.D. Lung neutrophils facilitate activation of naive antigen-specific CD4+ T cells during mycobacterium tuberculosis infection. *J. Immunol.* **2011**, *186*, 7110–7119. [CrossRef]
152. Hohl, T.M.; Rivera, A.; Lipuma, L.; Gallegos, A.; Shi, C.; Mack, M.; Pamer, E.G. Inflammatory monocytes facilitate adaptive CD4 T cell responses during respiratory fungal infection. *Cell Host Microbe* **2009**, *6*, 470–481. [CrossRef] [PubMed]
153. Nakano, H.; Lin, K.L.; Yanagita, M.; Charbonneau, C.; Cook, D.N.; Kakiuchi, T.; Gunn, M.D. Blood-derived inflammatory dendritic cells in lymph nodes stimulate acute T helper type 1 immune responses. *Nat. Immunol.* **2009**, *10*, 394–402. [CrossRef] [PubMed]
154. Mitroulis, I.; Ruppova, K.; Wang, B.; Chen, L.S.; Grzybek, M.; Grinenko, T.; Eugster, A.; Troullinaki, M.; Palladini, A.; Kourtzelis, I.; et al. Modulation of myelopoiesis progenitors is an integral component of trained immunity. *Cell* **2018**, *172*, 147–161.e12. [CrossRef] [PubMed]

155. Netea, M.G.; Joosten, L.A.; Latz, E.; Mills, K.H.; Natoli, G.; Stunnenberg, H.G.; O'Neill, L.A.; Xavier, R.J. Trained immunity: A program of innate immune memory in health and disease. *Science* **2016**, *352*, aaf1098. [CrossRef] [PubMed]
156. Samuel, C.E. Antiviral actions of interferons. *Clinc. Microbiol. Rev.* **2001**, *14*, 778–809. [CrossRef] [PubMed]
157. Chousterman, B.G.; Swirski, F.K.; Weber, G.F. Cytokine storm and sepsis disease pathogenesis. *Semin. Immunopathol.* **2017**, *39*, 517–528. [CrossRef] [PubMed]
158. Fehr, A.R.; Channappanavar, R.; Jankevicius, G.; Fett, C.; Zhao, J.; Athmer, J.; Meyerholz, D.K.; Ahel, I.; Perlman, S. The conserved coronavirus macrodomain promotes virulence and suppresses the innate immune response during severe acute respiratory syndrome coronavirus infection. *mBio* **2016**, *7*. [CrossRef] [PubMed]
159. Younan, P.; Iampietro, M.; Nishida, A.; Ramanathan, P.; Santos, R.I.; Dutta, M.; Lubaki, N.M.; Koup, R.A.; Katze, M.G.; Bukreyev, A. Ebola virus binding to TIM-1 on T lymphocytes induces a cytokine storm. *MBio* **2017**, *8*. [CrossRef]
160. De Jong, M.D.; Simmons, C.P.; Thanh, T.T.; Hien, V.M.; Smith, G.J.; Chau, T.N.; Hoang, D.M.; Chau, N.V.; Khanh, T.H.; Dong, V.C.; et al. Fatal outcome of human influenza A (H5N1) is associated with high viral load and hypercytokinemia. *Nat. Med.* **2006**, *12*, 1203–1207. [CrossRef]
161. Borges, A.A.; Campos, G.M.; Moreli, M.L.; Souza, R.L.; Aquino, V.H.; Saggioro, F.P.; Figueiredo, L.T. Hantavirus cardiopulmonary syndrome: Immune response and pathogenesis. *Microbes Infect.* **2006**, *8*, 2324–2330. [CrossRef]
162. Kawane, K.; Tanaka, H.; Kitahara, Y.; Shimaoka, S.; Nagata, S. Cytokine-dependent but acquired immunity-independent arthritis caused by DNA escaped from degradation. *Proc. Natl. Acad. Sci. USA* **2010**, *107*, 19432–19437. [CrossRef] [PubMed]
163. Macneil, A.; Nichol, S.T.; Spiropoulou, C.F. Hantavirus pulmonary syndrome. *Virus Res.* **2011**, *162*, 138–147. [CrossRef] [PubMed]
164. Kumar, M.; Belcaid, M.; Nerurkar, V.R. Identification of host genes leading to West Nile virus encephalitis in mice brain using RNA-seq analysis. *Sci. Rep.* **2016**, *6*, 26350. [CrossRef] [PubMed]
165. Dunning, J.; Blankley, S.; Hoang, L.T.; Cox, M.; Graham, C.M.; James, P.L.; Bloom, C.I.; Chaussabel, D.; Banchereau, J.; Brett, S.J.; et al. Progression of whole-blood transcriptional signatures from interferon-induced to neutrophil-associated patterns in severe influenza. *Nat. Immunol.* **2018**, *19*, 625–635. [CrossRef] [PubMed]
166. Hornick, E.E.; Banoth, B.; Miller, A.M.; Zacharias, Z.R.; Jain, N.; Wilson, M.E.; Gibson-Corley, K.N.; Legge, K.L.; Bishop, G.A.; Sutterwala, F.S.; et al. Nlrp12 mediates adverse neutrophil recruitment during influenza virus infection. *J. Immunol.* **2018**, *200*, 1188–1197. [CrossRef] [PubMed]
167. White, M.R.; Hsieh, I.-N.; De Luna, X.; Hartshorn, K.L. Effects of serum amyloid protein a on influenza a virus replication and viral interactions with neutrophils. *J. Immunol.* **2018**, *200*, 168.111.
168. Tate, M.D.; Ioannidis, L.J.; Croker, B.; Brown, L.E.; Brooks, A.G.; Reading, P.C. The role of neutrophils during mild and severe influenza virus infections of mice. *PLoS ONE* **2011**, *6*, e17618. [CrossRef]
169. Daffis, S.; Suthar, M.S.; Szretter, K.J.; Gale, M., Jr.; Diamond, M.S. Induction of ifn-beta and the innate antiviral response in myeloid cells occurs through an IPS-1-dependent signal that does not require IRF-3 and IRF-7. *PLoS Pathog.* **2009**, *5*, e1000607. [CrossRef]
170. Nicholls, J.M.; Poon, L.L.M.; Lee, K.C.; Ng, W.F.; Lai, S.T.; Leung, C.Y.; Chu, C.M.; Hui, P.K.; Mak, K.L.; Lim, W.; et al. Lung pathology of fatal severe acute respiratory syndrome. *Lancet* **2003**, *361*, 1773–1778. [CrossRef]
171. Channappanavar, R.; Fehr, A.R.; Vijay, R.; Mack, M.; Zhao, J.; Meyerholz, D.K.; Perlman, S. Dysregulated type I interferon and inflammatory monocyte-macrophage responses cause lethal pneumonia in SARS-CoV-infected mice. *Cell Host Microbe* **2016**, *19*, 181–193. [CrossRef]
172. Stifter, S.A.; Bhattacharyya, N.; Pillay, R.; Florido, M.; Triccas, J.A.; Britton, W.J.; Feng, C.G. Functional interplay between type I and II interferons is essential to limit influenza A virus-induced tissue inflammation. *PLoS Pathog.* **2016**, *12*, e1005378. [CrossRef]
173. Pollara, G.; Jones, M.; Handley, M.E.; Rajpopat, M.; Kwan, A.; Coffin, R.S.; Foster, G.; Chain, B.; Katz, D.R. Herpes simplex virus type-1-induced activation of myeloid dendritic cells: The roles of virus cell interaction and paracrine type I IFN secretion. *J. Immunol.* **2004**, *173*, 4108–4119. [CrossRef]
174. Luo, H.; Winkelmann, E.R.; Fernandez-Salas, I.; Li, L.; Mayer, S.V.; Danis-Lozano, R.; Sanchez-Casas, R.M.; Vasilakis, N.; Tesh, R.; Barrett, A.D.; et al. Zika, dengue and yellow fever viruses induce differential anti-viral immune responses in human monocytic and first trimester trophoblast cells. *Antiviral Res.* **2018**, *151*, 55–62. [CrossRef]

175. Malachowa, N.; Freedman, B.; Sturdevant, D.E.; Kobayashi, S.D.; Nair, V.; Feldmann, F.; Starr, T.; Steele-Mortimer, O.; Kash, J.C.; Taubenberger, J.K.; et al. Differential ability of pandemic and seasonal H1N1 influenzaa viruses to alter the function of human neutrophils. *mSphere* **2018**, *3*, e00567-17. [CrossRef] [PubMed]

176. Zorzitto, J.; Galligan, C.L.; Ueng, J.J.; Fish, E.N. Characterization of the antiviral effects of interferon-alpha against a SARS-like coronoavirus infection in vitro. *Cell Res.* **2006**, *16*, 220–229. [CrossRef] [PubMed]

177. Naess, A.; Nilssen, S.S.; Mo, R.; Eide, G.E.; Sjursen, H. Role of neutrophil to lymphocyte and monocyte to lymphocyte ratios in the diagnosis of bacterial infection in patients with fever. *Infection* **2017**, *45*, 299–307. [CrossRef] [PubMed]

3

Astrocyte Infection during Rabies Encephalitis Depends on the Virus Strain and Infection Route as Demonstrated by Novel Quantitative 3D Analysis of Cell Tropism

Madlin Potratz [1], Luca Zaeck [1], Michael Christen [1], Verena te Kamp [2], Antonia Klein [1], Tobias Nolden [3], Conrad M. Freuling [1], Thomas Müller [1] and Stefan Finke [1,*]

1 Friedrich-Loeffler-Institut (FLI), Federal Research Institute for Animal Health, Institute of Molecular Virology and Cell Biology, 17493 Greifswald-Insel Riems, Germany; Madlin.Potratz@fli.de (M.P.); Luca.Zaeck@fli.de (L.Z.); m.christen1994@gmail.com (M.C.); Antonia.Klein@fli.de (A.K.); Conrad.Freuling@fli.de (C.M.F.); Thomas.Mueller@fli.de (T.M.)

2 Thescon GmbH, 48653 Coesfeld, Germany; verena_tekamp@gmx.de

3 ViraTherapeutics GmbH, 6020 Innsbruck, Austria; tobias.nolden@boehringer-ingelheim.com

* Correspondence: stefan.finke@fli.de

Abstract: Although conventional immunohistochemistry for neurotropic rabies virus (RABV) usually shows high preference for neurons, non-neuronal cells are also potential targets, and abortive astrocyte infection is considered a main trigger of innate immunity in the CNS. While in vitro studies indicated differences between field and less virulent lab-adapted RABVs, a systematic, quantitative comparison of astrocyte tropism in vivo is lacking. Here, solvent-based tissue clearing was used to measure RABV cell tropism in infected brains. Immunofluorescence analysis of 1 mm-thick tissue slices enabled 3D-segmentation and quantification of astrocyte and neuron infection frequencies. Comparison of three highly virulent field virus clones from fox, dog, and raccoon with three lab-adapted strains revealed remarkable differences in the ability to infect astrocytes in vivo. While all viruses and infection routes led to neuron infection frequencies between 7–19%, striking differences appeared for astrocytes. Whereas astrocyte infection by field viruses was detected independent of the inoculation route (8–27%), only one lab-adapted strain infected astrocytes route-dependently [0% after intramuscular (i.m.) and 13% after intracerebral (i.c.) inoculation]. Two lab-adapted vaccine viruses lacked astrocyte infection altogether (0%, i.c. and i.m.). This suggests a model in which the ability to establish productive astrocyte infection in vivo functionally distinguishes field and attenuated lab RABV strains.

Keywords: rabies; uDISCO; 3D imaging; rabies pathogenicity; astrocyte infection

---

1. Introduction

The rabies virus (RABV) is a highly neurotropic virus, which inevitably causes lethal disease in mammals after onset of neurological signs [1]. As a non-segmented, single-stranded RNA virus of negative RNA polarity, RABV belongs to the *Rhabdoviridae* family in the order *Mononegavirales* [2]. With nucleoprotein N, phosphoprotein P, matrix protein M, glycoprotein G, and the large polymerase L, the 12 kb genome of RABV encodes five virus proteins, all of which are essential for virus replication and spread [3]. In addition to essential roles of the virus proteins in genome replication and virus assembly, multiple accessory functions of the RABV proteins have been identified. RABV pathogenicity has mainly been attributed to a potent interference with the innate immune system by N, P, and M [4–10], and neuronal survival regulation by G [11–14]. Most pathogenicity studies, however, were performed on already attenuated virus backbones. Thus, differences in their ability to cause disease between highly virulent field virus isolates and lab-adapted, less pathogenic RABV strains are poorly understood. Moreover, it is unclear how molecular differences identified in virulent and attenuated viruses affect virus replication and spread in the infected animal and how the complex virus–host interplay eventually results in either disease or an abortive infection.

In vivo, after infection of neurons, RABV spreads trans-synaptically from infected to connected neurons [15]. Retrograde axonal transport of RABV over long distances [16,17] along microtubules [18,19] is a key step in RABV neuroinvasion and is essential for infection of the central nervous system (CNS) through the peripheral nervous system. Co-internalization together with the neuronal p75NTR (tumor necrosis factor receptor superfamily member 16; TNFRSF16) receptor, subsequent retrograde axonal transport of RABV particles in endocytic vesicles, and post-replicative anterograde axonal transport of newly formed RABV have been visualized by live virus particle tracking in sensory neurons [20,21], emphasizing the capacity of hijacking neuron-specific machineries for long distance transport to synaptic membranes. However, internalization and axonal transport of lab-adapted viruses [20,21] together with the use of vaccine virus vectors for trans-synaptic tracing [22,23] demonstrate that the general capacity of axonal transport and trans-synaptic spread cannot explain mechanistic differences between highly virulent RABV and more attenuated lab strains. Differences between RABV lab strains in the efficiency of trans-synaptic spread [24] indicate that the efficacy of the involved processes, more than the capability itself, may contribute to RABV spread in vivo.

With nAChR (nicotinic acetylcholine receptor), NCAM (neuronal cell adhesion molecule), p75NTR, and mGluR2 (metabotropic glutamate receptor subtype 2) supporting RABV entry [25–28], several RABV receptors have been discussed. However, none of these receptors are essential for CNS infection by RABV, and a broad panel of non-neuronal cell types can be infected in vitro [29], indicating that cell tropism of RABV is not restricted to neurons by receptor specificity.

Most in vivo studies report a strict neurotropic infection. Directly after exposure by bite, however, muscle cells are infected (reviewed in [30]), and infection of non-neuronal cells in the CNS can occur [29,31,32]. Use of recombinant Cre recombinase-expressing RABV led to the identification of abortively infected glial cells in infected mouse brains, strongly suggesting infection of and virus elimination from these cells by a potent type I interferon response [33]. Accordingly, abortive infection of non-neuronal cells and induction of innate immune responses may play an important role in the infection process itself and in regulating downstream adaptive immune pathways. Indeed, a model based on in vitro-infected astrocytes suggests that, in contrast to wild-type RABV, attenuated RABV activates inflammatory responses in astrocytes through increased double-strand RNA (dsRNA) synthesis and recognition by retinoic acid-inducible gene I (RIG-I)/melanoma differentiation-associated protein 5 (MDA5) [34]. Highly virulent field RABV isolates are able to evade or at least delay host immune reactions [35], which may allow virus replication to reach pathogenic levels, whereas early innate immune induction via astrocytes or other glial cells by attenuated viruses does not. Nevertheless, all productive RABV infections eventually cause rabies as a disease, which is always associated with an encephalitis. However, differences appear to exist as field viruses cause less tissue damage and neuroinflammation than attenuated lab RABV [34,36,37].

To compare the cell tropism of highly virulent field and lab-adapted RABVs in the CNS, we employed the novel immunostaining-compatible tissue clearing technique uDISCO (ultimate 3D imaging of solvent-cleared organs) [38–40] for detection and quantification of RABV infection in neurons and astrocytes in solvent-cleared brain tissue. After confocal laser scan acquisition of large confocal z-stacks in thick tissue slices, three-dimensional (3D) reconstructions were performed to visualize the cellular context of RABV infection. Frequencies of RABV infection in neurons and astrocytes were determined after intramuscular (i.m.) and intracerebral (i.c.) inoculation, leading to novel insights regarding the ability to establish RABV infection in non-neuronal astrocytes, its correlation with RABV virulence, and its dependence on the infection route. These data strongly support a model in which, contrary to in vitro conditions where all viruses are able to infect non-neuronal astrocytes, there are differences between field and lab-adapted viruses in their ability to replicate to detectable levels in astrocytes in vivo. The infection of astrocytes with virulent field RABV and the accumulation of interferon antagonistic virus proteins may block rapid innate immune induction, thereby contributing to immune evasion, even in the context of glial cell infection.

2. Materials and Methods

2.1. Cell Lines, Primary Brain Cell Preparation, and Cultivation

Mouse neuroblastoma cells (Na42/13) were used for virus amplification. All cells were provided by the Collection of Cell Lines in Veterinary Medicine (CCLV), Friedrich-Loeffler-Institut, Insel Riems, Germany. Primary neurons and astrocytes were prepared from 1-day-old neonatal Sprague Dawley rats (P0–P1) [41]. The rats were decapitated, the heads were disinfected in 70% ethanol, the brains were removed, and the hippocampi were separated and mechanically minced. The hippocampal cells were transferred into ice-cold Hank´s Balanced Salt Solution (HBSS) and stored on ice. After preparation, the brain cells were dissociated by adding 0.25% trypsin + EDTA (Gibco; Thermo Fisher Scientific, Darmstadt, Germany) and DNase I (Applichem, Darmstadt, Germany) and incubated for 15 min at 37 °C. Afterwards, further dissociation in Neurobasal-A Medium (NB-A, Gibco; Thermo Fisher Scientific) was performed by using a glass Pasteur pipette with a fire-polished tip. The cells were purified with an Optiprep™ gradient (with concentrations from bottom to the top: 17.3%, 12.4%, 9.9%, and 7.4%) and centrifuged for 15 min at 800× *g* without brake. Next, the cells were washed with NB-A medium and counted with the LUNA-II Automated Cell Counter (Logos Biosystems, Villeneuve d'Ascq, France). The neuronal cells were seeded on coverslips in a 24-well plate and cultured for two weeks at 37 °C and 5% CO_2 in serum-free NB-A supplemented with 2% B27 (Gibco; Thermo Fisher Scientific), 1% GlutaMAX (Gibco; Thermo Fisher Scientific), 1% Penicillin-Streptomycin (stock: 10,000 U/mL penicillin and 10 mg/mL streptomycin; Sigma-Aldrich, Taufkirchen, Germany), and 0.2% Gentamicin (stock: 50 mg/mL; Gibco; Thermo Fisher Scientific) until they were used for infection experiments.

2.2. Viruses

RABV isolates used in this study comprised five recombinant virus clones and one non-recombinant lab virus. The recombinant field virus clones rRABV Dog and rRABV Fox have been described before [42]. SAD L16 is a recombinant virus clone of attenuated vaccine virus SAD B19 live vaccine virus [43]. The Evelyn Rokitnicki Abelseth (ERA) strain [44] is a progenitor virus strain of live vaccine virus SAD B19 [45] and was obtained from the FLI virus archive (FLI ID N 12829).

A cDNA full length clone (pRABV Rac) from a raccoon RABV isolate (Alabama, USA 1991; FLI archive ID N 13205) [46] was generated by full length RT-PCR amplification of the 12 kb cDNA genome with the primer pair 5′-TCGATCCCGGGTCACGCTTAACAACAAAA-3′/ 5′-TAATACACCTGCCCATGCCGACCCACGCTTAACAAAAAAACAA-3′. After PCR amplification of a 2.7 kb vector DNA fragment from pCMV HaHd ampR Br 322 ori with the primers 5′-TCTGTTTGCTTGATGGTTTTTTTGTCTTTGTTGTTTTTTTGTTAAGCGTGGGTCGGCATGGC ATCTCCAC-3′ and 5′-TTTTTGTAGATGATACTGTCTACTTCTTCTCTGATTTTGTTGTTAAGCGTGA CCCGGGACTCCGGGTTTCGTC-3′, the fragments were combined by linear to linear recombination, and the resultant recombinant rRABV Rac virus was rescued and amplified according to previously described protocols [42]. The sequence of the full-length cDNA clone was deposited at GenBank (accession no. MN862283).

A CVS-11 (Challenge Virus Street-11; sequence accession no. LT839616) cDNA full length clone (pCVS-11) was generated by RT-PCR amplification of 7.5kb and 3.7 kb cDNA fragments from CVS-11 RNA with the primers pairs 5′-AGTTTCAGACGTCTCAGTC-3′/5′-CTAGTAGGGATGATCTAGATC-3′ and 5′-GACTGAGACGTCTGAAACT-3′/5′ CATTGCAGATAGGATAGAG-3′, respectively. The resultant fragments were assembled in combination with EcoRI-linearized 3.4 kb vector plasmid pCVS-11-termini by Hot Fusion reaction [47]. Plasmid pCVS-11-termini was generated by PCR amplification of a 2.7 kb DNA fragment from pCMV HaHd ampR Br 322 ori [42] with the primers 5′-GACCCGGGACTCCGGGTTTC-3′ and 5′-GGGTCGGCATGGCATCTCCA-3′, and Hot Fusion assembly with a synthetic DNA fragment containing the CVS-11 (accession no. LT839616) regions from nucleotide positions 1–580 (3′-genome end) and 11759–11927 (5′-genome end) separated by a non-viral EcoRI restriction site. The full-length cDNA clone was 99.99% identical to CVS-11 GenBank sequence no. LT839616 (1 nt mismatch in the N gene at nucleotide position 1263, which was already present in #LT839616 as a minor SNP (single nucleotide polymorphism), resulting in the amino acid exchange E398V). Recombinant rCVS-11 was rescued in Na42/13 neuroblastoma cells by co-transfection of pCVS-11, pCAGGS-based expression plasmids for RABV N, P, and L [48], and a pCAGGS-T7Pol vector comprising a codon-optimized bacteriophage T7 RNA polymerase gene. Three to six days after transfection, the supernatants were transferred to new Na42/13 cells. Two days after the transfer, infectious virus was identified by N and G protein-specific indirect immunofluorescence.

rRABV Dog, rRABV Fox, rRABV Rac, and rCVS-11 were amplified on Na42/13 cells. BSR T7/5 cells [49] were used for amplification of SAD L16 and ERA viruses. Infectious virus titers in cell culture supernatants were determined by end point dilution and titration on Na42/13 cells.

2.3. Antibodies

To verify the presence of RABV, a polyclonal rabbit serum against recombinant RABV P protein (P160-5, immunofluorescence 1:5000, uDISCO 1:3000) and a polyclonal goat serum against RABV N (goat anti-RV N, immunofluorescence 1:4000), which have been described previously [40,50], were used. The polyclonal chicken anti-glial fibrillary acidic protein (GFAP) antibody (Thermo Fisher Scientific, Darmstadt, Germany; #PA1-10004, uDISCO 1:1500), the polyclonal rabbit anti-GFAP (Dako, #Z0334, immunofluorescence 1:500), the polyclonal guinea pig anti-NeuN antibody (Synaptic Systems, Goettingen, Germany; #266004, uDISCO 1:800), and the rabbit anti-MAP2 antibody (Abcam, Cambridge, UK; #ab32454, immunofluorescence 1:250) were purchased from their respective suppliers.

2.4. Infection of in Vitro Cell Cultures and Immunofluorescence Staining

Two-week-old primary rat hippocampal cell cultures were infected with 1×10^3 infectious units of rRABV Dog, rRABV Fox, rCVS-11, and SAD L16, and cultivated for 24 h at 37 °C and 5% CO_2. Indirect immunofluorescence was performed by standard techniques after fixation with 4% paraformaldehyde (PFA) in phosphate-buffered saline (PBS) for 30 min and 15 min permeabilization with 0.5% Triton X-100 in PBS. Afterwards, samples were blocked with 0.025% skim milk powder in PBS for 15 min. Immunostainings were executed by 1.5 h incubation with primary antibodies, three wash steps with PBS, followed by 1 h incubation with secondary antibodies and additional Hoechst33342 (1 μg/mL) for staining of the nuclear chromatin. Specimens were mounted on coverslips and were analyzed by confocal laser scanning microscopy.

2.5. Brain Samples and Mouse Infections

Three- to four-week-old BALB/c mice (Charles-River, Germany) were infected with rRABV Rac, rCVS-11, ERA, and SAD L16 viruses using two different inoculation routes and two different viral doses, essentially as described before [51]. Two groups of six animals each were anesthetized and infected i.m. with 10^2 or 10^5 $TCID_{50}$/30 μL, and an additional group of three mice was infected i.c. with 10^2 $TCID_{50}$/30 μL. The weight and the clinical score, ranging from zero to four, of all mice were observed for 21 days post infection (dpi). When reaching a clinical score of two or three (ruffled fur, slowed movement, weight loss >15%), the animals were anaesthetized with isoflurane and euthanized through cervical dislocation. Samples were taken, fixed with 4% paraformaldehyde (PFA) for one week, and stored for further processing. All remaining animals were euthanized at 21 dpi. Mouse experimental studies on the characterization of lyssaviruses were evaluated by the responsible animal care, use, and ethics committee of the State Office for Agriculture, Food Safety, and Fishery in Mecklenburg-Western Pomerania (LALFF M-V) and gained approval with permissions 7221.3–2–001/18.

2.6. *Archived Mouse Brains Infected with rRABV Dog and rRABV Fox*

To minimize animal experiments, archived PFA-fixed brains from previous pathogenicity trials [42] were used for the analysis of rRABV Dog and rRABV Fox virus infections in mice. Similar to the mouse experiments with rRABV Rac, rCVS-11, ERA, and SAD L16 described above, mice were inoculated via the i.c. or the i.m. route.

2.7. *Ultimate 3D Imaging of Solvent-Cleared Organs (uDISCO)*

The clearing of brain tissue slices was performed as described previously [40] in modification of earlier publications [38,39]. Briefly, the PFA-fixed tissues were sectioned into 1 mm-thick slices using a vibratome (Leica VT1200S; Leica Biosystems, Wetzlar, Germany). All subsequent incubations steps were performed with gentle oscillation. To increase antibody diffusion and reduce tissue autofluorescence [38], the sections were pretreated with increasing concentrations of methanol (20%, 40%, 60%, 80%, and twice in 100%; dilutions with distilled water, incubation for 1 h each) and bleached by overnight incubation at 4 °C with 5% H_2O_2 in 100% methanol. After removal of the bleaching solution, the samples were rehydrated with decreasing concentrations of methanol (80%, 60%, 40%, and 20%; dilutions with distilled water, incubation for 1 h each) and a subsequent wash with PBS for 1 h. To permeabilize the samples, they were washed twice for 1 h each with 0.2% Triton X-100 in PBS and subsequently incubated for 48 h at 37 °C in 0.2% Triton X-100/20% DMSO/0.3 M glycine in PBS. The samples were then blocked by incubation with 0.2% Triton X-100/10% DMSO/6% donkey serum in PBS for 48 h at 37 °C. Primary antibodies were diluted in 3% donkey serum/5% DMSO in PTwH (0.2% Tween-20 in PBS with 10 μg/mL heparin), and incubation of samples was performed at 37 °C for 5 days. The antibody solution was refreshed after 2.5 days. The samples were washed with PTwH by exchanging the solution four times during the course of the day and subsequently incubated overnight in PTwH. Incubation with secondary antibodies was performed in 3% donkey serum in PTwH for 5 days at 37 °C, refreshing the secondary antibody solution once after 2.5 days. Subsequent washing was performed as described above following primary antibody incubation.

For tissue clearing, the samples were dehydrated with a series of *tert*-butanol (TBA) solutions (30%, 50%, 70%, 80%, 90%, and 96%; dilutions with distilled water, incubation for 2 h each), leaving 96% TBA on overnight. Following further dehydration in 100% TBA for 2 h, the samples were cleared in BABB-D15 [39] [1:2 mixture of benzyl alcohol (BA) and benzyl benzoate (BB), which is mixed with diphenyl ether (DPE) at a ratio of 15:1 and supplemented with 0.4 vol% DL-α-tocopherol] until they were optically transparent (2–6 h).

For confocal laser scanning microscopy, the samples were mounted in 3D-printed imaging chambers (printer: Ultimaker 2 + [Ultimaker, Utrecht, Netherlands], material: co-polyester, nozzle: 0.25 mm, layer height: 0.06 mm, wall thickness: 0.88 mm, wall count: 4, infill: 100%, no support structure; the corresponding .STL file is provided in the Supplementary Materials of Zaeck et al. [40]).

2.8. *Confocal Laser Scanning Microscopy and Image Processing*

The immunofluorescent staining of primary brain cells and infected tissues was visualized with a confocal laser scanning microscope (Leica DMI 6000 TCS SP5; Leica Microsystems, Wetzlar, Germany) equipped with a long free working distance 40× water immersion objective (NA = 1.1; Leica, #15506360). For image processing, a Dell Precision 7920 workstation was used (CPU: Intel Xeon Gold 5118, GPU: Nvidia Quadro P5000, RAM: 128 GB 2666 MHz DDR4, SSD: 2 TB; Dell, Frankfurt am Main, Germany).

For quantification of infected cells in thick tissue sections, the image was split into individual channels using Fiji, an ImageJ (v1.52h) distribution package [52]. A bleach correction was performed (simple ratio; background intensity: 5.0), and brightness and contrast were adjusted for each channel. Objects were identified and counted with the 3D Objects Counter plugin [53]. The resulting objects map was overlaid with the RABV P channel to quantify infected objects. For each sample, at least six regions were imaged and analyzed. The 3D projections in Figure 2d,e were generated with Icy [54]. All other maximum z- and 3D projections, including the Supplementary Videos S1–S4, were generated with Fiji.

2.9. Statistical Analysis

Statistical significance was determined using two-way ANOVA followed by Tukey's multiple comparison test using GraphPad Prism 7.05.

3. Results

3.1. Astrocytes Are Readily Infected by both Field Viruses and Lab Strains in Mixed Primary Brain Cell Cultures

To investigate whether field and lab-adapted RABVs differ in their ability to infect primary neurons and non-neuronal astrocytes, hippocampal brain cells were prepared from neonatal rats and were cultivated as mixed cultures containing neurons and glial cells. After 13 days of cultivation, the cultures were infected with 10^3 infectious units of the recombinant field viruses rRABV Dog and rRABV Fox, and the lab-adapted strains rCVS-11 and SAD L16. After 24 h of infection, the cells were fixed and analyzed by confocal laser scanning microscopy using indirect immunofluorescence stainings against RABV nucleoprotein N, neuron marker MAP2 (microtubule-associated protein 2) and astrocyte marker GFAP (glial fibrillary acidic protein). All viruses infected both MAP2-positive neurons and GFAP-positive astrocytes, as demonstrated by the formation of RABV-positive cytoplasmic inclusion bodies (Figure 1). These data indicated that there was no obvious difference between the tested field and the lab-adapted viruses in their ability to infect neurons and non-neuronal astrocytes.

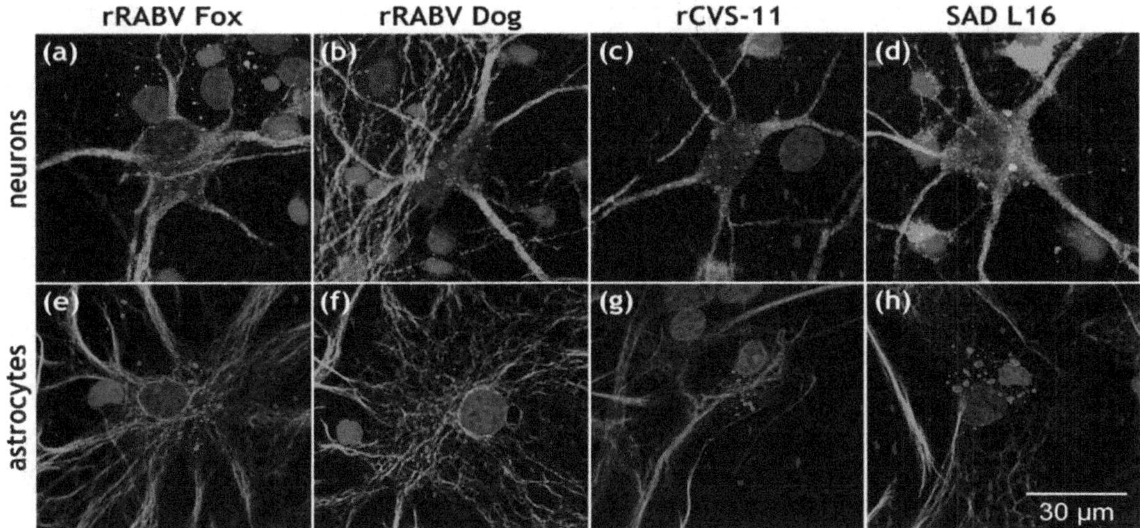

Figure 1. In vitro infection of primary hippocampus neurons and astrocytes by field and lab rabies virus (RABV). (**a–d**) Indirect immunofluorescence detection of RABV nucleoprotein N (red) and microtubule-associated protein 2 (MAP2) (green). Nuclei were counterstained with Hoechst33342 (blue). (**e–f**) Indirect immunofluorescence detection of RABV nucleoprotein N (red) and glial fibrillary acidic protein (GFAP) (green). Nuclei were counterstained with Hoechst33342 (blue). Negative controls for RABV detection are provided in Supplementary Figure S1.

Although all four viruses were able to infect GFAP-positive cells in the hippocampal cell cultures, the majority of the infected cells were MAP2-positive neurons. To quantitatively compare astrocyte infection between viruses, the number of RABV N-positive cells was counted for an area of 25 mm^2, and the percentage of GFAP-positive and -negative RABV-infected cells was determined (Table 1). All four viruses resulted in infection of GFAP-positive cells at levels ranging from 4.9% to 6.7%. Although a higher percentage of GFAP-positive cells could not be excluded because a high number of cells could not be classified into either positive or negative for GFAP, these results indicated that all four viruses could readily establish an infection in cultivated astrocytes and did not substantially differ in their ability to do so.

Table 1. Infection of GFAP- and MAP2-expressing cells in hippocampus cell cultures. RABV-infected cells were counted after indirect immunofluorescence detection of RABV N protein and subdivided in cells that were either negative (−) or positive (+) for GFAP/MAP2. Because of overlaying positive and negative cells in GFAP and MAP2 stainings, a clear assignment was not always possible. Such cells were classified as uncertain. CVS: Challenge Virus Street.

Virus	GFAP (−)		GFAP (+)		Uncertain	
	Count	%	Count	%	Count	%
SAD L16	256	48.3	30	5.7	244	46
rCVS-11	58	64.4	6	6.7	26	28.9
rRABV Dog	252	46.8	29	5.4	258	47.9
rRABV Fox	1002	55.2	89	4.9	712	39.5
Virus	**MAP2 (−)**		**MAP2 (+)**		**Uncertain**	
	Count	**%**	**Count**	**%**	**Count**	**%**
SAD L16	18	3.5	387	74.2	116	22.3
rCVS-11	1	1.0	44	43.2	57	55.9
rRABV Dog	27	5.3	347	73.3	109	21.4
rRABV Fox	190	7.3	2096	80.3	324	12.4

3.2. Field RABV Infects Neurons and Astrocytes In Vivo as Demonstrated by High-Resolution 3D Analysis of Infected Brain Tissue

In order to test whether the primary cell culture infection experiments are comparable to the more complex in vivo situation, a sample preparation and imaging pipeline was established, which allows three-dimensional, high-resolution confocal laser scanning image acquisition from RABV-infected tissue for 3D reconstruction and quantification of infected cells. To this end, 1 mm-thick brain slices (Figure 2a) from clinically diseased rRABV Fox field virus-infected mice (i.m. infection route) were immunostained for RABV P, GFAP, and NeuN, optically cleared, and imaged. Acquisition of z-stacks by confocal laser scanning microscopy resulted in z-volumes of 400 μm × 400 μm × 50–100 μm (x, y, and z) from different areas of the infected brain. RABV-infected neurons of variable morphologies were detected by presence of NeuN and RABV P protein accumulation in neuronal cell bodies and neurites (Figure 2b–e; details in Supplementary Figure S2). Astrocytes were detected by filamentous GFAP-positive structures (Figure 2b–g).

Notably, besides accumulation of RABV P in NeuN-positive neurons, P also accumulated at GFAP-positive filaments (Figure 2b,c,f,g; white arrows), indicating a robust infection of astrocytes by field virus rRABV Fox. Evaluation of brain areas with differing localization patterns of the highlighted cellular subpopulations (Figure 2b,c) revealed that astrocyte infections were not restricted to particular regions of the brain but appeared in all imaged neuron layers or brain areas.

Because of the complex 3D morphology of both neurons (long axons in Figure 2b,c) and astrocytes (filamentous structure of GFAP signals in Figure 2f,g), 3D reconstruction of the imaged tissues (Figure 2d,e,g; Supplementary Videos S1 and S2) was performed in order to achieve reliable visualization and quantification of cells in downstream analyses.

(a) 1 mm brain slices

pretreatment
immunostaining
clearing

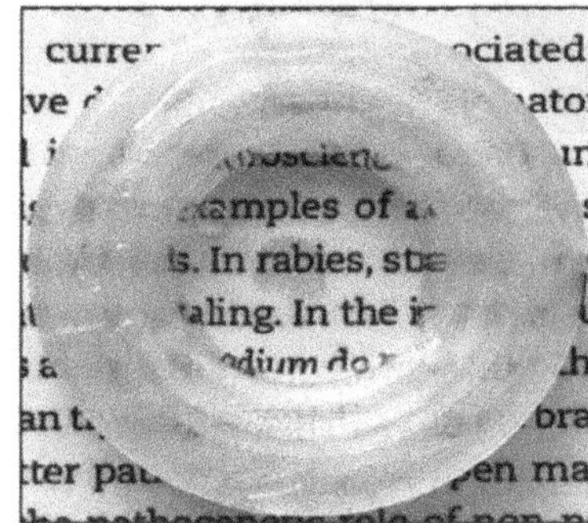

z-stack acquisition

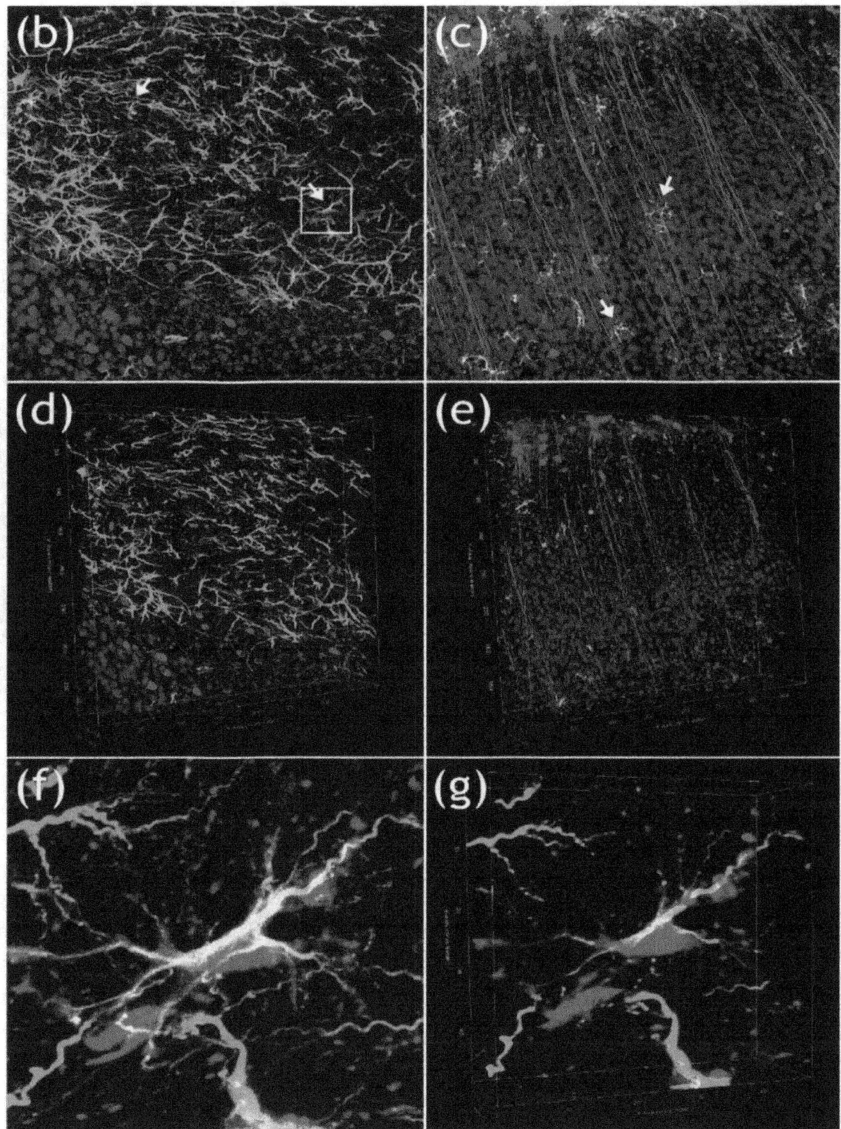

Figure 2. 3D immunofluorescence imaging of field RABV-infected neurons and astrocytes in a mouse brain. (**a**) Workflow for 3D immunofluorescence imaging, including vibratome sectioning into 1 mm slices, pretreatment, immunostaining, and subsequent optical clearing with organic solvents. Confocal imaging and acquisition of cleared tissue slices was done in custom-made imaging containers (see lower image). (**b**,**c**) Maximum z- and (**d**–**e**) 3D projections of z-stack (x, y, z = 400 μm, 400 μm, 59 μm for D and 400 μm, 400 μm, 103 μm for E) after indirect immunofluorescence for RABV phosphoprotein P (red), GFAP (green), and NeuN (blue). White arrows in Figure 2b indicate RABV P and GFAP-positive astrocytes. (**f**) Maximum projection of detail from Figure 2b (see white box) with GFAP-positive cell (green) and associated RABV P fluorescence (red). (**g**) 3D projection of detail view from Figure 2f.

3.3. Similar Levels of Field Virus Infection in both Neurons and Astrocytes in the Infected Mouse Brain

Whereas immunofluorescence analysis in Figure 2 clearly demonstrated infection of astrocytes and neurons, they also revealed that only a fraction of both cell types was positive for RABV P. To quantify the ratio of infected and non-infected neurons and astrocytes, 3D object segmentation and counting was performed for NeuN- (Figure 3a,b) or GFAP (Figure 3d,e)-positive cells. The object map was merged with their respective RABV fluorescence (Figure 3c,f; Supplementary Video S3), and the number of RABV-positive neurons and astrocytes was determined by manual counting of RABV P- and cell marker-positive cells. For the z-stack shown in Figures 2b and 3, total numbers of 762 neurons and 272 astrocytes were counted (Table 2, region 2), of which 16 and 18 were RABV positive, respectively.

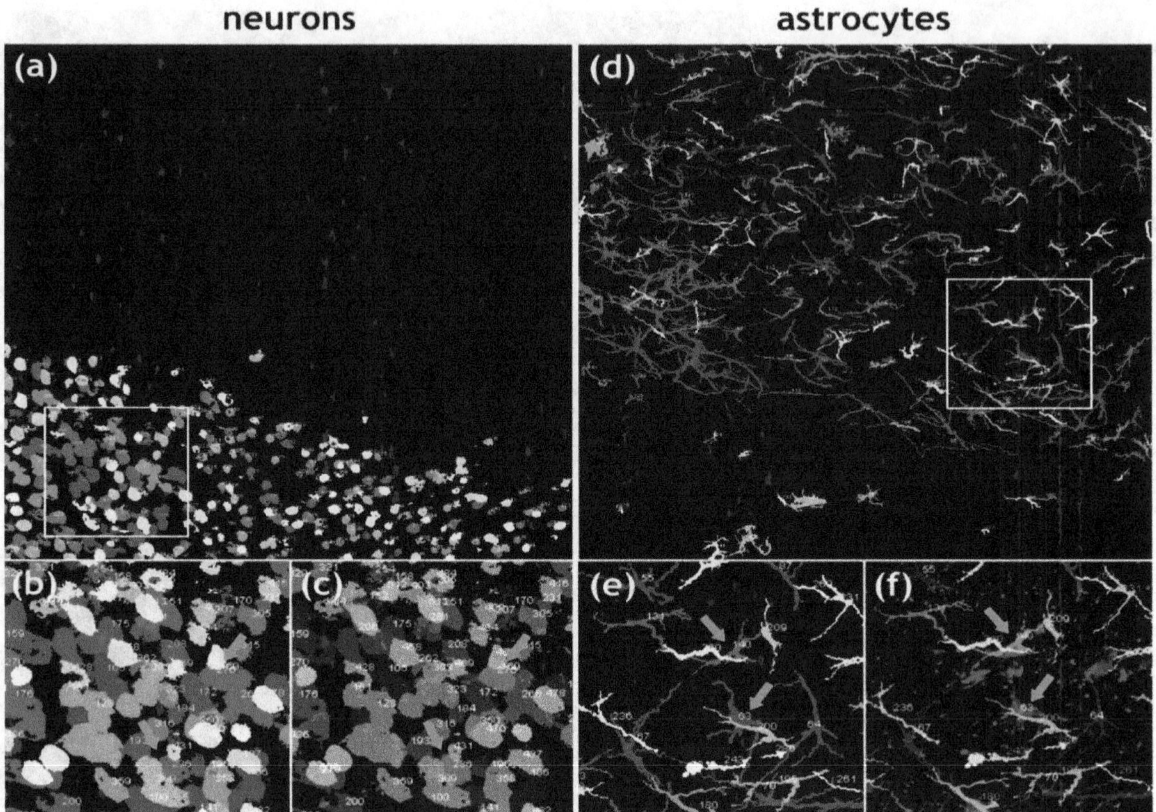

Figure 3. Quantification of RABV-infected neurons and astrocytes. (**a**) Maximum z-projection of objects map for NeuN-positive neurons generated from a confocal z-stack (see Figure 2b). Individual neurons in the z-stack were identified by NeuN-specific fluorescence and converted to objects. Numbers indicate individual cell counts (for improved legibility of the individual numbers, refer to the enlarged details in Supplementary Video S3). n = 762 neurons in a volume of 400 μm × 400 μm × 59 μm. The object colors indicate different z-positions (darker colors in the back and brighter colors in the front). (**b**) Detail of area indicated by white box in Figure 3a. Green arrows indicate RABV-positive neuron cell bodies. (**c**) Overlay of objects map (greyscale) with RABV P signals allows identification and counting of rRABV Fox-infected neurons. (**d**) Maximum z-projection of objects map for GFAP-positive astrocytes generated from a confocal z-stack (see Figure 2b). Individual astrocytes in the z-stack were identified by GFAP-specific fluorescence and converted to objects. Numbers indicate individual cell counts. n = 272 astrocytes in a volume of 400 μm × 400 μm × 59 μm. The object colors indicate different z-positions (darker colors in the back and brighter colors in the front). (**e**) Detail of area indicated by white box in Figure 3d. Green arrows indicate RABV-positive astrocytes. (**f**) Overlay of objects map (greyscale) with RABV P signals allows identification and counting of rRABV Fox-infected astrocytes.

Table 2. Quantification of rRABV Fox-infected neurons and astrocytes. Results of the analysis of six different areas of one infected mouse brain. The quantification pipeline of region 2 is depicted in Figure 3. Neurons and astrocytes were segmented, automatically counted, and overlaid with the respective RABV P signals. Infected cells were counted manually.

	Neurons					Astrocytes				
Region	**Counted**	**Infected**	**%**	**Mean**	**SD**	**Counted**	**Infected**	**%**	**Mean**	**SD**
1	485	21	4.3			362	11	3.0		
2	762	16	2.1			272	18	6.6		
3	448	13	2.9			106	18	17.0		
4	2518	132	5.2	3.9	1.4	488	9	1.8	7.0	4.9
5	3029	91	3.0			273	20	7.3		
6	3847	234	6.1			541	32	5.9		
Σ	11098	507				2042	108			

Analysis of six z-stacks from different RABV-infected brain areas of the same sample led to the detection of 11,089 neurons and 2042 astrocytes, of which 3.9% and 7.0% were RABV positive, respectively (Table 2). The fraction of RABV-positive cells ranged from 2.1% to 6.1% (SD: ± 1.4) for neurons and 1.8% to 17.0% for astrocytes (SD: ± 4.9) (Table 2). These data indicated that rRABV Fox infects astrocytes and neurons in the mouse brain to comparable levels.

Observation of three different rRABV Fox-infected mice revealed that the infection level for astrocytes differed between single animals from 1.2% to 15.6% (SD: ± 6.7) (Supplementary Table S1). Nevertheless, in spite of some variance between individual mice and/or between different analyzed brain regions, detection of infected non-neuronal astrocytes in all animals at surprisingly high levels indicates that astrocyte infection by RABV has been underestimated thus far.

3.4. *Astrocyte Infection by RABV Depends on the Type of Virus (Field vs. Lab-Adapted).*

To compare the astrocyte tropism of field and lab-adapted viruses after i.m. inoculation, brains of mice infected with three different field viruses (rRABV Fox, rRABV Dog, and rRABV Rac) and two lab-adapted viruses (rCVS-11 and ERA) were analyzed. SAD L16 was excluded from these analyses since it is not able to induce clinical signs after i.m. inoculation (Supplementary Figure S3d).

Whereas rRABV Fox and rRABV Dog caused disease even after infection with a very low virus dose [42], infections with rRABV Rac, rCVS-11, and ERA did not show any clinical signs at a dose of 10^2 $TCID_{50}$ (Supplementary Figure S3). High dose infection with the ERA strain (10^5 $TCID_{50}$) led to 100% disease development (Supplementary Figure S3c), whereas high dose infections with rRABV Rac and rCVS-11 only caused disease in 50% and 16.7% of the infected mice, respectively (Supplementary Figure S3a,b). Accordingly, available tissue samples for the latter two viruses were limited to two and one infected brain. Imaging and quantification of at least six confocal z-stacks from different RABV-infected regions per infected brain was performed.

RABV infections were readily detectable in the brain for all five viruses (Figure 4; Supplementary Video S4). Whereas RABV P antigen accumulated at GFAP-positive structures in rRABV Fox, rRABV Dog, and rRABV Rac-infected brains (Figure 4a,c,e), similar accumulations were not observed in rCVS-11 or ERA-infected brains, in which only infection of NeuN-positive neurons was detected (Figure 4b,d).

Infections with rRABV Fox, rRABV Dog, rRABV Rac, rCVS-11, and ERA led to a mean of 8.3%, 7.3%, 18.9%, 6.9%, and 15.1% RABV-positive neurons (Figure 5; Supplementary Table S1), respectively, demonstrating a comparable level of neuron infection by all five viruses in clinically diseased mice. However, significant differences were observed for the astrocyte infections. Whereas rRABV Fox, rRABV Dog, and rRABV Rac infected 7.6% (SD: ± 6.7), 10.1% (SD: ± 7.7), and 16.5% (SD: ± 15.0) of the astrocytes, no RABV-positive astrocytes were detected in rCVS-11 and ERA-infected samples (Figure 5; Supplementary Table S1). These data indicated that, in contrast to the tested field viruses, the lab-adapted RABVs rCVS-11 and ERA did not infect astrocytes to detectable levels in vivo after i.m. inoculation.

3.5. *Confirmation of the Specific Astrocyte Tropism of Field RABV, ERA, and SAD L16 after i.c. Inoculation and Route-Dependent Astrocyte Infection by rCVS-11*

To test whether the inoculation route affects astrocyte infection and whether the highly attenuated SAD L16 virus is comparable to the SAD vaccine progenitor strain ERA, brain samples from two animals i.c.-infected with rRABV Fox, rRABV Dog, rCVS-11, or SAD L16 and one infected with the ERA strain were analyzed.

With astrocyte infections at frequencies of 10.9% (SD: ± 10.3), 11.6% (SD: ± 5.6), and 27.2% (SD: ± 12.8) (Figure 6; Supplementary Table S2), rRABV Fox, rRABV Dog, and rRABV Rac led to robust astrocyte infection via the i.c. route and thus confirmed their ability to establish astrocyte infection in vivo (Figure 7a,c,e). Lack of detectable astrocyte infection for the ERA and SAD L16 further confirmed that these viruses are not able to infect astrocytes to a detectable level (Figure 7d,f and Figure 6), even after direct virus administration into the brain.

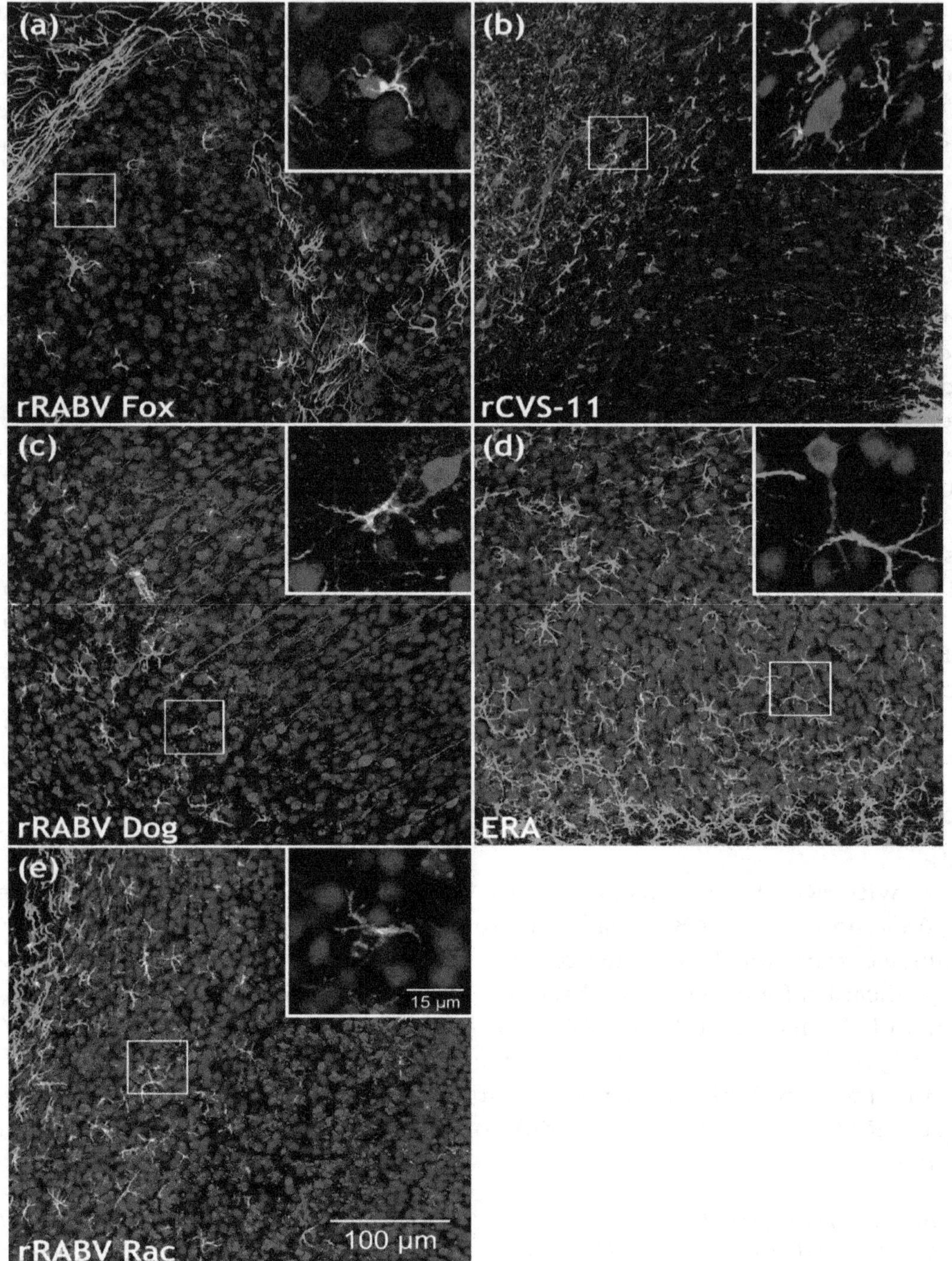

Figure 4. Comparison of field and lab RABV-infected brains after intramuscular (i.m.) infection with rRABV Fox, rRABV Dog, rRABV Rac, rCVS-11, and Evelyn Rokitnicki Abelseth (ERA). (**a–e**) Maximum z-projections of z-stacks [x, y = 400 µm, 400 µm (**a–e**); z = 59 µm (**a**, rRABV Fox), 66 µm (**b**, rCVS-11), 49 µm (**c**, rRABV Dog), 100 µm (**d**, ERA) and 89 µm (**e**, rRABV Rac)] after indirect immunofluorescence for RABV phosphoprotein P (red), GFAP (green), and NeuN (blue). To improve visualization of the maximum z-projections, some z-stacks were reduced in thickness. For the full z-stacks, refer to Supplementary Video S4. Insets in Figure 4a,c,e show RABV P accumulation (red) at GFAP-positive cells (green). Insets in Figure 4b,d show NeuN- (blue) and RABV P (red)-positive neurons. For the individual channels of the detail images, see Supplementary Figure S4.

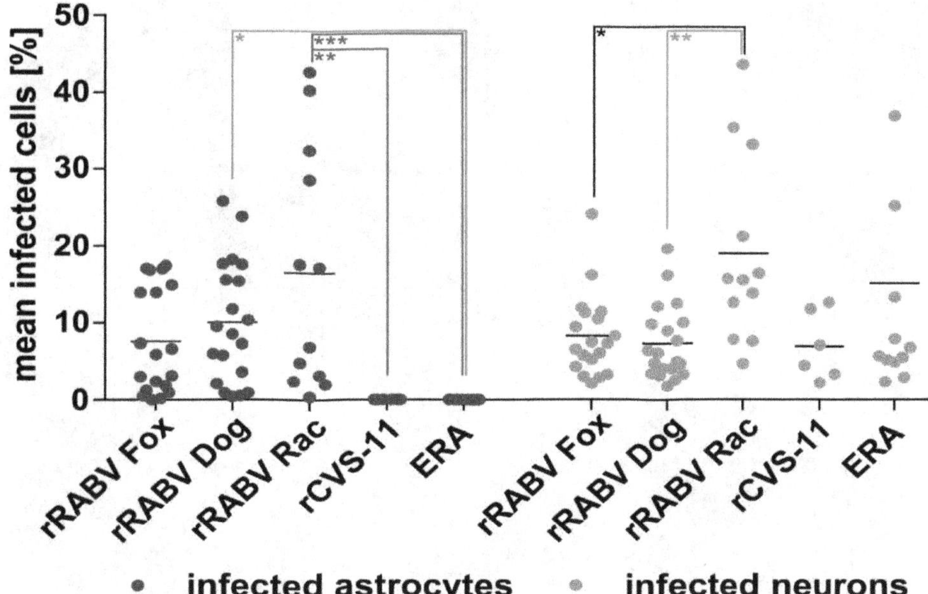

Figure 5. Percentage of field and lab RABV-infected neurons and astrocytes after i.m. inoculation. Per virus, 3 to 9 × 10^3 astrocytes and 1.5 to 4 × 10^4 neurons were counted in 19 (rRABV Fox), 20 (rRABV Dog), 12 (rRABV Rac and ERA), and six (rCVS-11) independent confocal z-stacks in three (rRABV Fox and rRABV Dog), two (rRABV RAC and ERA), and one (rCVS-11) animal (see Supplementary Table S1). Each dot represents the frequency of infected astrocytes or neurons in an analyzed z-stack. Mean values are provided as horizontal lines. * $p \leq .05$; ** $p \leq .01$; *** $p \leq .001$ (two-way ANOVA with Tukey's multiple comparison test).

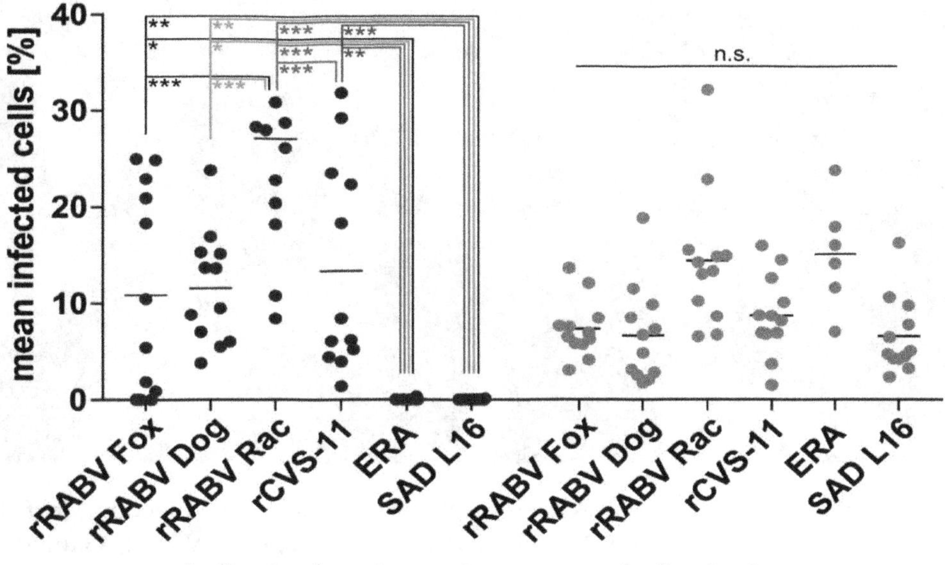

Figure 6. Percentage of field and lab RABV-infected neurons and astrocytes after i.c. inoculation. Per virus, 5 to 9 × 10^3 astrocytes and 1.3 to 3 × 10^4 neurons were counted in 12 (rRABV Dog, rRABV Fox, rRABV Rac, rCVS-11, and SAD L16) and six (ERA) independent confocal z-stacks in two (rRABV Fox, rRABV Dog, rRABV Rac, rCVS-11, and SAD L16) and one (ERA) animal (see Supplementary Table S2). Each dot represents the frequency of infected astrocytes or neurons in an analyzed z-stack. Mean values are provided as horizontal lines. * $p \leq .05$; ** $p \leq .01$; *** $p \leq .001$; n.s. = not significant (two-way ANOVA with Tukey's multiple comparison test).

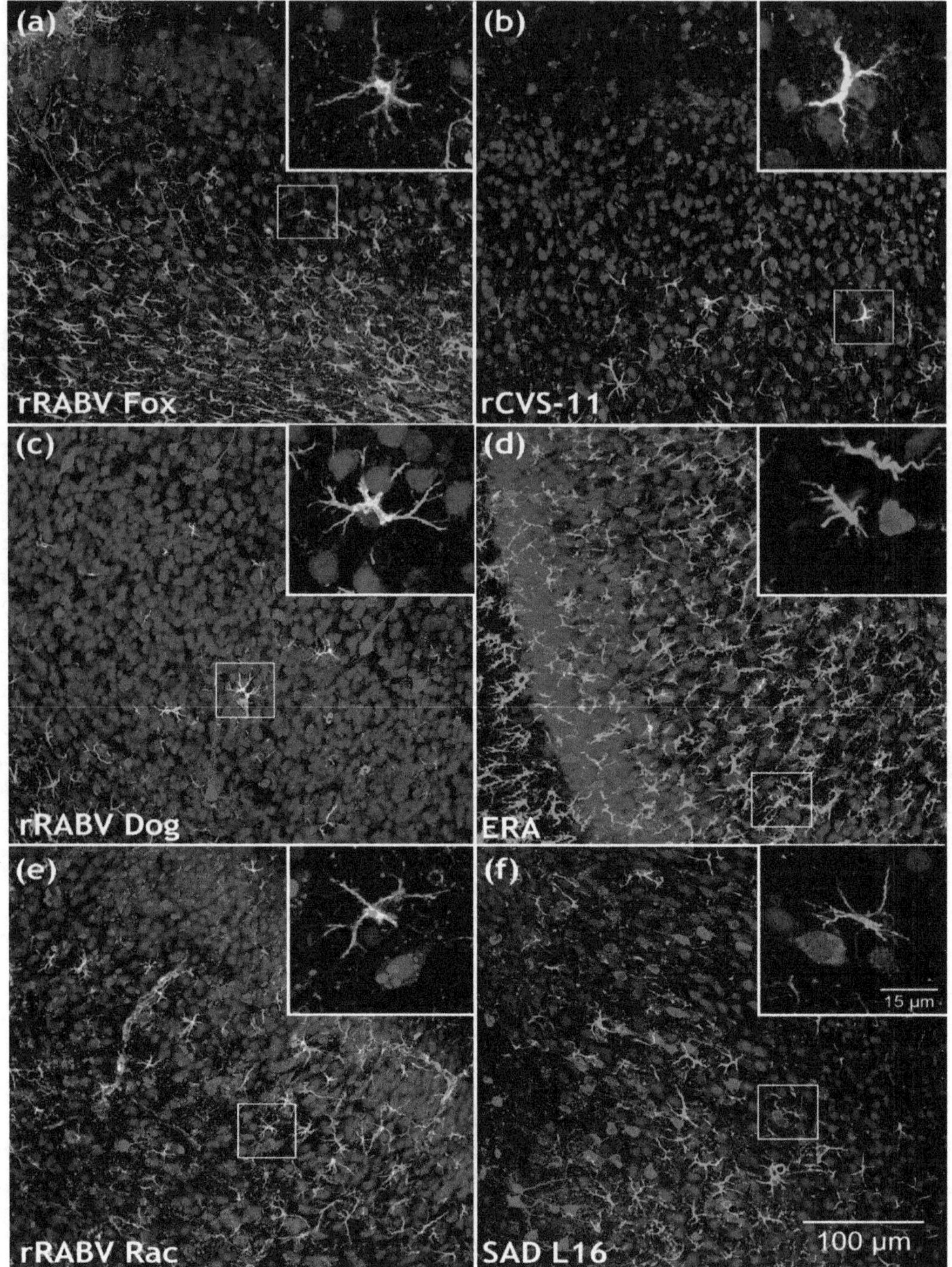

Figure 7. Comparison of field and lab RABV-infected brains after i.c. infection. (**a**–**f**) Maximum z-projections of z-stacks [x, y = 400 µm, 400 µm (**a**–**f**); z = 21 µm (**a**, rRABV Fox), 48 µm (**b**, rCVS-11), 74 µm (**c**, rRABV Dog), 98 µm (**d**, ERA), 75 µm (**e**, rRABV Rac) and 67 µm (**f**, SAD L16)] after indirect immunofluorescence for RABV phosphoprotein P (red), GFAP (green), and NeuN (blue). Insets in Figure 7a,b,c,e show RABV P accumulation (red) at GFAP-positive cells (green). To improve visualization of the maximum z-projections, some z-stacks were reduced in thickness. Insets in Figure 7d,f show NeuN- (blue) and RABV P (red)-positive neurons. For the individual channels of the detail images, see Supplementary Figure S5.

Notably, in contrast to the i.m. inoculation route and to the other i.c.-inoculated lab-adapted RABVs SAD L16 and ERA, 13.4% (SD: ± 10.4) of the astrocytes were RABV-positive in rCVS-11-infected animals, indicating that, in the case of rCVS-11, the infection route has a substantial influence on astrocyte infection in the brain. However, the ratio of rCVS-11-infected neurons (8.7%; SD: ± 4.0) remained comparable to i.m. infections.

4. Discussion

By use of a novel immunofluorescence-compatible technique for 3D immunofluorescence imaging of solvent-cleared brain tissue slices [38,39] adapted to the imaging of RABV infections in brain tissue [40], we investigated the cell tropism of six different RABV isolates and lab-adapted strains. Both neurons and astrocytes display heterogeneous morphologies with pronounced three-dimensional projections. Compared to conventional thin sections or even in silico 3D reconstructions of serial thin sections, optical slicing of 1 mm-thick tissue samples during image acquisition allowed fast and seamless 3D reconstruction of immunostained tissues and high-resolution dissection of the detected antigens [40]. For the first time, this allowed systematic analysis and comparison of the tropism of the different viruses in large 3D volumes of infected mouse brains. One example for the superiority of the employed technique is the unambiguous detection of rCVS-11-infected astrocytes after i.c. inoculation (Figures 6 and 7), whereas conventional thin layer immunohistochemistry analyses led to the assumption that infection of glial cells by CVS does not occur, even after i.c. inoculation [55].

Importantly, more than 10^4 neurons and 10^3 astrocytes (Supplementary Table S1 and S2) were quantitatively investigated for RABV infection in their authentic environment per virus and inoculation route. This provided reliable datasets about the frequencies of RABV detection in the two cellular subpopulations in vivo and statistically significant differences in astrocyte infections, which depended on the degree of virulence and the inoculation route of the tested RABVs (Figures 5 and 6).

Because of the clonal and the defined nucleotide sequences, the recombinant viruses rRABV Fox and rRABV Dog [42], rRABV Rac (this work), SAD L16 [43], and rCVS-11 (this work) were selected for the analyses (full genome nucleotide and G protein amino acid sequence comparisons are provided in Supplementary Tables S3 and S4, and Supplementary Figure S6). Since SAD L16 was directly derived from the live vaccine SAD B19 vaccine strain [56], which is equally apathogenic after intramuscular inoculation, the non-recombinant SAD B19 progenitor vaccine virus strain ERA was included to allow for the investigation of attenuated live vaccine virus astrocyte tropism after peripheral i.m. inoculation. Compared to SAD B19, the ERA strain has a less intense cell culture passage history [45] and is still pathogenic in mice after peripheral inoculation [57]. Indeed, whereas SAD L16 was not able to cause disease after i.m. inoculation, ERA was even more pathogenic than rCVS-11 at the high i.m. inoculation dose of 10^5 $TCID_{50}$ with 100% and 16.7% of mice developing clinical signs, respectively (Supplementary Figure S3).

Both the progenitor field virus isolates and the respective recombinant virus clones rRABV Fox and rRABV Dog have been shown to share comparable pathogenicity with efficient disease development after i.m. infection of mice [42]. For the two aforementioned field virus isolates as well as for the raccoon RABV isolate used here for the generation of rRABV Rac, virulence in raccoons has been demonstrated [46]. Notably, although the raccoon virus isolate was highly virulent in its natural reservoir host, the recombinant clone rRABV Rac generated here was less virulent in mice, showing a pathogenicity of 50% only at the high i.m. inoculation dose (Supplementary Figure S3).

In striking contrast to rRABV Fox and rRABV Dog, the lab virus clone rCVS-11 was less efficient via the i.m. route but still able to cause disease in one out of six mice at a high virus dose of 10^5 infectious units per mouse (Supplementary Figure S3). This perfectly matched previous studies, where the non-recombinant progenitor CVS-11 strain caused disease in only 16.7% of the mice at high i.m. dose inoculation, and, as observed here (Supplementary Figure S3), was apathogenic at low dose i.m. infection [51].

In contrast to the in vivo results, in vitro infection of mixed rat hippocampal neuron/astrocyte cultures with rRABV Fox, rRABV Dog, SAD L16, and rCVS-11 led to comparable levels of astrocyte infections, as demonstrated by GFAP-positive, RABV-infected cells (Figure 1) in a range from 4.9% to 6.7% for all viruses (Table 1). These data were in accordance with the general susceptibility of cultivated astrocytes for RABV infection repeatedly shown for field and lab-adapted viruses [29,34], with preferential replication in neurons [32]. Whereas Tsiang et al. reported 90–99% of the primary astrocytes free of RABV antigen, our results indicate that there is less variation within the tested viruses, with virus antigen detection in about 5% of the cultivated astrocytes independent of whether they were of field virus or lab strain origin (Table 1). Overall, comparison of the in vitro and the in vivo data showed that a general susceptibility of a primary CNS cell subpopulation in vitro (Figure 1) does not necessarily reflect the situation in vivo (Figures 2, 4 and 6).

Reasons for this could be altered antiviral response profiles in the more disordered cell culture conditions, different replication and spreading kinetics of the viruses in vivo, or differences in the non-synaptic release of infected neurons to allow infection of non-synaptically connected CNS cells. Whereas no information is available about the latter thus far, it is known that astrocytes are abortively infected by a chimeric SAD L16 virus expressing CVS-11 glycoprotein in mouse brains [33]. This supports the idea that replication of rCVS-11, ERA, and SAD L16 was similarly blocked in astrocytes by potent innate immune responses after infection via the i.m. or the i.c. inoculation routes, respectively. However, since neither viral genome copies nor virus mRNA levels were measured here, it remains to be clarified whether replication kinetics of the lab strains indeed differ from those of the field strains in the infected brains.

Nevertheless, it is conceivable that a more efficient replication of rRABV Fox, rRABV Dog, and rRABV Rac, and therefore the accumulation of the major interferon antagonist phosphoprotein P [8,9] in astrocytes, led to inhibition of antiviral responses. Indeed, lab-attenuated RABV has been shown to differ from wild-type RABV by the induction of an increased type I interferon (IFN) production and expression of inflammatory cytokines via the mitochondrial antiviral-signaling protein (MAVS) pathway [34]. Accordingly, the immunofluorescence detection of RABV P for the identification of infected cells (Figures 2, 4 and 6) confirmed the presence of abundant levels of the major interferon antagonist in the field virus-infected astrocytes. It is highly unlikely that the lack of P detection in the astrocytes of lab RABV-infected animals was due to a major difference in P gene expression of field viruses, since P was readily detectable in lab RABV-infected neurons.

Besides supporting virus replication in astrocytes by decreasing release of inflammatory cytokines, robust field virus replication in these cells may result in a less pronounced or delayed general antiviral response to field viruses. This would be in accordance with the observed immune escape of field viruses in infected animals [35] and may represent a major immunological difference to lab-adapted strains. However, further experimentation will be needed to test this hypothesis.

Notably, and in contrast to the ERA strain not being detected in astrocytes after i.c. inoculation, the astrocyte infection by rCVS-11 was at a frequency of 13.4% after i.c. inoculation (Figure 6 and Supplementary Table S2) and also at higher levels than observed for the field viruses rRABV Fox and rRABV Dog after both i.c. and i.m. infection. This revealed that the ability for rCVS-11 to establish an infection in these cell types depended on the infection route. Since infection from the periphery relies on trans-synaptic spread of the viruses and the virus may not become visible for immune and non-neuronal target cells, astrocyte infection could represent a late phase phenotype of brain infection, where abundant infection of neurons and non-synaptic release of virus particles may facilitate astrocyte infection. Consequently, astrocyte-mediated immune reactions would be delayed and virus elimination prior to disease onset not possible. Compared to the field viruses, slower virus replication and/or spreading kinetics of rCVS-11 could lead to lower levels or the absence of detectable astrocyte infection after i.m. inoculation (Figure 5), although the virus is principally able to infect these cell types (Figure 7b). Indeed, the efficacy of trans-synaptic retrograde spread can differ between a highly neurotropic CVS-24 virus variant and an SAD L16-like vaccine virus [24]. On a higher level, this may also distinguish highly virulent field viruses such as rRABV Fox and rRABV Dog from rCVS-11. Furthermore, it is conceivable that i.c. inoculations represent a shortcut to brain infection with simultaneous and multiple infection of neurons and astrocytes in a non-transsynaptic manner. Higher numbers of infected cells at the beginning of CNS infection compared to trans-synaptic invasion after i.m. infection may lead to faster virus spread and infection of multiple regions of the brain with abundant late phase virus release and astrocyte infection, similar to that speculated above for i.m. infections with highly virulent field viruses.

Indeed, astrocyte activation by other viruses have been shown to occur earlier after i.c. than after peripheral infections [58]. Both virus and cell response kinetics may differ between the two infection routes. Increased induction of neuronal cell death after i.c. RABV infection compared to no detectable apoptosis in i.m.-infected animals [59] further indicates qualitative differences in host reaction to the virus. The special role of astrocytes in the CNS as a main source of IFN-β expression and virus control through TLR (Toll-like receptor) and RLR (RIG-I-like receptor) activation pathways [33,60,61] in combination with the infection route-dependent differences described here, as observed for rCVS-11, may contribute to such differences in apoptosis and other host reaction patterns. However, further studies must clarify whether different infectious route-dependent and -independent virus kinetics can determine astrocyte tropism in vivo and how this affects downstream host reactions. Since differences in innate immune induction through dsRNA between field and attenuated viruses were demonstrated [34], the innate immune induction potential of rRABV Fox, rRABV Dog, rRABV Rac, rCVS-11, ERA, and SAD L16 in astrocytes has to be investigated in order to assess whether the route dependency of rCVS-11 is also affected by different levels of innate immune induction.

Most likely, lack of astrocyte infection by SAD L16 and ERA was the outcome of strong virus inhibition and elimination, as abortive astrocyte infection by a comparable virus has previously been suggested [33]. Whether less antiviral response induction may allow the more virulent rCVS-11 or the highly virulent field viruses to overcome a threshold of virus replication and antagonist expression—and thus may support further replication—will be addressed in future studies. Even though the underlying mechanisms cannot be clarified here, comparable results for SAD L16 and the still pathogenic ERA strain strongly suggest that a general block of detectable astrocyte replication of these vaccine strains represents a major difference to the other tested viruses.

Only by using novel 3D immunofluorescence techniques, we were able to provide comprehensive analyses of the cell tropism of highly virulent field RABVs and less virulent or attenuated lab strains. In particular, high-resolution 3D images and quantitative downstream analysis provided novel insights in the infection processes at the clinical phase of this deadly disease. Although independent of detectable astrocyte infection, symptoms and lethal progression of the disease occur once the virus efficiently spreads in the brain after infection with all six tested viruses (Figure 7 and Supplementary Figure S3). However, different virus kinetics and astrocyte-related innate immune reactions may affect the progression kinetics, immune pathogenicity, and further spread of the virus to peripheral salivary glands. The latter may represent a key issue in terms of field virus transmission and maintenance in host populations.

Whereas these aspects must be addressed in separate trials, this study provides a novel and quantitative basis for a new, dynamic view on RABV host interactions in vivo on the cellular level. Also, this approach showcases the potential of immunostaining-compatible tissue clearing/3D imaging techniques to specifically investigate virus–host interactions at high-resolution on cellular and subcellular levels. While this study focused on standard cellular markers for neurons and astrocytes, future approaches including immune and host pathway markers will pave the way for direct high-resolution imaging-based analysis of infection processes in complex and morphologically preserved tissues.

Supplementary Materials:
Captions_SupplementaryFiles: Captions and legends to supplementary figures, tables, and videos. Figure S1: Immunofluorescence of non-infected astrocytes and neurons in vitro. Figure S2: Details of rRABV Fox-infected, NeuN-positive neurons. Figure S3: Kaplan-Meyer survival plots for rRABV Rac, rCVS-11, ERA, and SAD L16. Figure S4: Details of field virus (rRABV Fox, rRABV Dog, and rRABV Rac) and lab RABV (rCVS-11 and ERA) infections after i.m. inoculation. Figure S5: Details of field virus (rRABV Fox, rRABV Dog, and rRABV Rac) and lab RABV (rCVS-11, ERA, and SAD L16) infections after i.c. inoculation. Figure S6. Amino acid alignment of glycoprotein G of rRABV Rac, rCVS-11, SAD L16, rRABV Fox, and rRABV Dog. Table S1: Quantification of RABV-infected neurons and astrocytes after i.m. infection. Table S2: Quantification of RABV-infected neurons

and astrocytes after i.c. infection. Table S3. Nucleotide sequence alignment of full genome sequences of rRABV Rac, rCVS-11, SAD L16, rRABV Dog, and rRABV Fox. Table S4. Alignment of G protein amino acid sequence of rRABV Rac, rCVS-11, SAD L16, rRABV Dog, and rRABV Fox. Video S1: 3D projections of rRABV Fox-infected astrocytes and neurons in two different areas of a mouse brain. Video S2: 3D projection of an rRABV Fox-infected astrocyte in a mouse brain. Video S3: 3D projection of the quantification of RABV-infected neurons and astrocytes. Video S4: 3D projections of field and lab RABV-infected brains after i.m. infection with rRABV Fox, rRABV Dog, rRABV Rac, rCVS-11, and ERA, and SAD L16 after i.c. infection.

Author Contributions: Conceptualization, S.F., L.Z., T.M. and C.M.F.; methodology, L.Z., V.t.K., T.N. and M.P.; investigation, M.P., L.Z., M.C. and A.K.; writing—original draft preparation, M.P., L.Z. and S.F.; writing—review and editing, M.P., L.Z., M.C., V.t.K., A.K., T.N., C.M.F., T.M. and S.F.; visualization, M.P. and L.Z.; supervision, S.F., T.M. and C.M.F.; project administration, S.F.; funding acquisition, S.F. All authors have read and agreed to the published version of the manuscript.

Acknowledgments: We thank Dietlind Kretzschmar and Angela Hillner for technical assistance.

References

1. Fooks, A.R.; Cliquet, F.; Finke, S.; Freuling, C.; Hemachudha, T.; Mani, R.S.; Müller, T.; Nadin-Davis, S.; Picard-Meyer, E.; Wilde, H.; et al. Rabies. *Nat. Rev. Dis. Primers* **2017**, *3*, 17091. [CrossRef]
2. Walker, P.J.; Blasdell, K.R.; Calisher, C.H.; Dietzgen, R.G.; Kondo, H.; Kurath, G.; Longdon, B.; Stone, D.M.; Tesh, R.B.; Tordo, N.; et al. ICTV Virus Taxonomy Profile: Rhabdoviridae. *J. Gen. Virol.* **2018**, *99*, 447–448. [CrossRef] [PubMed]
3. Finke, S.; Conzelmann, K.-K. Replication strategies of rabies virus. *Virus Res.* **2005**, *111*, 120–131. [CrossRef] [PubMed]
4. Zhang, G.; Wang, H.; Mahmood, F.; Fu, Z.F. Rabies virus glycoprotein is an important determinant for the induction of innate immune responses and the pathogenic mechanisms. *Vet. Microbiol.* **2013**, *162*, 601–613. [CrossRef] [PubMed]
5. Besson, B.; Sonthonnax, F.; Duchateau, M.; Ben Khalifa, Y.; Larrous, F.; Eun, H.; Hourdel, V.; Matondo, M.; Chamot-Rooke, J.; Grailhe, R.; et al. Regulation of NF-κB by the p105-ABIN2-TPL2 complex and RelAp43 during rabies virus infection. *PLoS Pathog.* **2017**, *13*, e1006697. [CrossRef] [PubMed]
6. Masatani, T.; Ito, N.; Shimizu, K.; Ito, Y.; Nakagawa, K.; Sawaki, Y.; Koyama, H.; Sugiyama, M. Rabies virus nucleoprotein functions to evade activation of the RIG-I-mediated antiviral response. *J. Virol.* **2010**, *84*, 4002–4012. [CrossRef] [PubMed]
7. Ben Khalifa, Y.; Luco, S.; Besson, B.; Sonthonnax, F.; Archambaud, M.; Grimes, J.M.; Larrous, F.; Bourhy, H. The matrix protein of rabies virus binds to RelAp43 to modulate NF-κB-dependent gene expression related to innate immunity. *Sci. Rep.* **2016**, *6*, 39420. [CrossRef]
8. Brzózka, K.; Finke, S.; Conzelmann, K.-K. Identification of the rabies virus alpha/beta interferon antagonist: Phosphoprotein P interferes with phosphorylation of interferon regulatory factor 3. *J. Virol.* **2005**, *79*, 7673–7681. [CrossRef]
9. Brzózka, K.; Finke, S.; Conzelmann, K.-K. Inhibition of interferon signaling by rabies virus phosphoprotein P: Activation-dependent binding of STAT1 and STAT2. *J. Virol.* **2006**, *80*, 2675–2683. [CrossRef]
10. Ito, N.; Moseley, G.W.; Blondel, D.; Shimizu, K.; Rowe, C.L.; Ito, Y.; Masatani, T.; Nakagawa, K.; Jans, D.A.; Sugiyama, M. Role of interferon antagonist activity of rabies virus phosphoprotein in viral pathogenicity. *J. Virol.* **2010**, *84*, 6699–6710. [CrossRef]
11. Faber, M.; Pulmanausahakul, R.; Hodawadekar, S.S.; Spitsin, S.; McGettigan, J.P.; Schnell, M.J.; Dietzschold, B. Overexpression of the rabies virus glycoprotein results in enhancement of apoptosis and antiviral immune response. *J. Virol.* **2002**, *76*, 3374–3381. [CrossRef] [PubMed]
12. Jackson, A.C.; Rasalingam, P.; Weli, S.C. Comparative pathogenesis of recombinant rabies vaccine strain SAD-L16 and SAD-D29 with replacement of Arg333 in the glycoprotein after peripheral inoculation of neonatal mice: Less neurovirulent strain is a stronger inducer of neuronal apoptosis. *Acta Neuropathol.* **2006**, *111*, 372–378. [CrossRef] [PubMed]

13. Morimoto, K.; Hooper, D.C.; Spitsin, S.; Koprowski, H.; Dietzschold, B. Pathogenicity of different rabies virus variants inversely correlates with apoptosis and rabies virus glycoprotein expression in infected primary neuron cultures. *J. Virol.* **1999**, *73*, 510–518. [CrossRef] [PubMed]
14. Sarmento, L.; Li, X.-q.; Howerth, E.; Jackson, A.C.; Fu, Z.F. Glycoprotein-mediated induction of apoptosis limits the spread of attenuated rabies viruses in the central nervous system of mice. *J. Neurovirol.* **2005**, *11*, 571–581. [CrossRef] [PubMed]
15. Dietzschold, B.; Li, J.; Faber, M.; Schnell, M. Concepts in the pathogenesis of rabies. *Future Virol.* **2008**, *3*, 481–490. [CrossRef] [PubMed]
16. Tsiang, H.; Lycke, E.; Ceccaldi, P.E.; Ermine, A.; Hirardot, X. The anterograde transport of rabies virus in rat sensory dorsal root ganglia neurons. *J. Gen. Virol.* **1989**, *70*, 2075–2085. [CrossRef]
17. Astic, L.; Saucier, D.; Coulon, P.; Lafay, F.; Flamand, A. The CVS strain of rabies virus as transneuronal tracer in the olfactory system of mice. *Brain Res.* **1993**, *619*, 146–156. [CrossRef]
18. Lycke, E.; Tsiang, H. Rabies virus infection of cultured rat sensory neurons. *J. Virol.* **1987**, *61*, 2733–2741. [CrossRef]
19. Ceccaldi, P.E.; Gillet, J.P.; Tsiang, H. Inhibition of the transport of rabies virus in the central nervous system. *J. Neuropathol. Exp. Neurol.* **1989**, *48*, 620–630. [CrossRef]
20. Gluska, S.; Zahavi, E.E.; Chein, M.; Gradus, T.; Bauer, A.; Finke, S.; Perlson, E. Rabies Virus Hijacks and accelerates the p75NTR retrograde axonal transport machinery. *PLoS Pathog.* **2014**, *10*, e1004348. [CrossRef]
21. Bauer, A.; Nolden, T.; Schröter, J.; Römer-Oberdörfer, A.; Gluska, S.; Perlson, E.; Finke, S. Anterograde glycoprotein-dependent transport of newly generated rabies virus in dorsal root ganglion neurons. *J. Virol.* **2014**, *88*, 14172–14183. [CrossRef] [PubMed]
22. Wickersham, I.R.; Finke, S.; Conzelmann, K.-K.; Callaway, E.M. Retrograde neuronal tracing with a deletion-mutant rabies virus. *Nat. Methods* **2007**, *4*, 47–49. [CrossRef] [PubMed]
23. Wickersham, I.R.; Lyon, D.C.; Barnard, R.J.O.; Mori, T.; Finke, S.; Conzelmann, K.-K.; Young, J.A.T.; Callaway, E.M. Monosynaptic restriction of transsynaptic tracing from single, genetically targeted neurons. *Neuron* **2007**, *53*, 639–647. [CrossRef] [PubMed]
24. Reardon, T.R.; Murray, A.J.; Turi, G.F.; Wirblich, C.; Croce, K.R.; Schnell, M.J.; Jessell, T.M.; Losonczy, A. Rabies Virus CVS-N2c(ΔG) Strain Enhances Retrograde Synaptic Transfer and Neuronal Viability. *Neuron* **2016**, *89*, 711–724. [CrossRef] [PubMed]
25. Thoulouze, M.I.; Lafage, M.; Schachner, M.; Hartmann, U.; Cremer, H.; Lafon, M. The neural cell adhesion molecule is a receptor for rabies virus. *J. Virol.* **1998**, *72*, 7181–7190. [CrossRef] [PubMed]
26. Tuffereau, C.; Bénéjean, J.; Blondel, D.; Kieffer, B.; Flamand, A. Low-affinity nerve-growth factor receptor (P75NTR) can serve as a receptor for rabies virus. *EMBO J.* **1998**, *17*, 7250–7259. [CrossRef]
27. Lentz, T.; Burrage, T.; Smith, A.; Crick, J.; Tignor, G. Is the acetylcholine receptor a rabies virus receptor? *Science* **1982**, *215*, 182–184. [CrossRef]
28. Wang, J.; Wang, Z.; Liu, R.; Shuai, L.; Wang, X.; Luo, J.; Wang, C.; Chen, W.; Wang, X.; Ge, J.; et al. Metabotropic glutamate receptor subtype 2 is a cellular receptor for rabies virus. *PLoS Pathog.* **2018**, *14*, e1007189. [CrossRef]
29. Ray, N.B.; Power, C.; Lynch, W.P.; Ewalt, L.C.; Lodmell, D.L. Rabies viruses infect primary cultures of murine, feline, and human microglia and astrocytes. *Arch. Virol.* **1997**, *142*, 1011–1019. [CrossRef]
30. Davis, B.M.; Rall, G.F.; Schnell, M.J. Everything You Always Wanted to Know About Rabies Virus (But Were Afraid to Ask). *Annu. Rev. Virol.* **2015**, *2*, 451–471. [CrossRef]
31. Jackson, A.C.; Phelan, C.C.; Rossiter, J.P. Infection of Bergmann glia in the cerebellum of a skunk experimentally infected with street rabies virus. *Can. J. Vet. Res.* **2000**, *64*, 226–228. [PubMed]
32. Tsiang, H.; Koulakoff, A.; Bizzini, B.; Berwald-Netter, Y. Neurotropism of rabies virus. An in vitro study. *J. Neuropathol. Exp. Neurol.* **1983**, *42*, 439–452. [CrossRef] [PubMed]
33. Pfefferkorn, C.; Kallfass, C.; Lienenklaus, S.; Spanier, J.; Kalinke, U.; Rieder, M.; Conzelmann, K.-K.; Michiels, T.; Staeheli, P. Abortively Infected Astrocytes Appear to Represent the Main Source of Interferon Beta in the Virus-Infected Brain. *J. Virol.* **2016**, *90*, 2031–2038. [CrossRef] [PubMed]
34. Tian, B.; Zhou, M.; Yang, Y.; Yu, L.; Luo, Z.; Tian, D.; Wang, K.; Cui, M.; Chen, H.; Fu, Z.F.; et al. Lab-Attenuated Rabies Virus Causes Abortive Infection and Induces Cytokine Expression in Astrocytes by Activating Mitochondrial Antiviral-Signaling Protein Signaling Pathway. *Front. Immunol.* **2018**, *8*, 2011. [CrossRef] [PubMed]

35. Ito, N.; Moseley, G.W.; Sugiyama, M. The importance of immune evasion in the pathogenesis of rabies virus. *J. Vet. Med. Sci.* **2016**, *78*, 1089–1098. [CrossRef]
36. Suja, M.S.; Mahadevan, A.; Madhusudana, S.N.; Shankar, S.K. Role of apoptosis in rabies viral encephalitis: A comparative study in mice, canine, and human brain with a review of literature. *Patholog. Res. Int.* **2011**, *2011*, 374286. [CrossRef]
37. Wang, Z.W.; Sarmento, L.; Wang, Y.; Li, X.-q.; Dhingra, V.; Tseggai, T.; Jiang, B.; Fu, Z.F. Attenuated Rabies Virus Activates, while Pathogenic Rabies Virus Evades, the Host Innate Immune Responses in the Central Nervous System. *J. Virol.* **2005**, *79*, 12554–12565. [CrossRef]
38. Renier, N.; Wu, Z.; Simon, D.J.; Yang, J.; Ariel, P.; Tessier-Lavigne, M. iDISCO: A simple, rapid method to immunolabel large tissue samples for volume imaging. *Cell* **2014**, *159*, 896–910. [CrossRef]
39. Pan, C.; Cai, R.; Quacquarelli, F.P.; Ghasemigharagoz, A.; Lourbopoulos, A.; Matryba, P.; Plesnila, N.; Dichgans, M.; Hellal, F.; Ertürk, A. Shrinkage-mediated imaging of entire organs and organisms using uDISCO. *Nat. Methods* **2016**, *13*, 859–867. [CrossRef]
40. Zaeck, L.; Potratz, M.; Freuling, C.M.; Müller, T.; Finke, S. High-Resolution 3D Imaging of Rabies Virus Infection in Solvent-Cleared Brain Tissue. *J. Vis. Exp.* **2019**, *30*. [CrossRef]
41. Brewer, G.J.; Torricelli, J.R. Isolation and culture of adult neurons and neurospheres. *Nat. Protoc.* **2007**, *2*, 1490–1498. [CrossRef] [PubMed]
42. Nolden, T.; Pfaff, F.; Nemitz, S.; Freuling, C.M.; Höper, D.; Müller, T.; Finke, S. Reverse genetics in high throughput: Rapid generation of complete negative strand RNA virus cDNA clones and recombinant viruses thereof. *Sci. Rep.* **2016**, *6*, 23887. [CrossRef] [PubMed]
43. Schnell, M.J.; Mebatsion, T.; Conzelmann, K.K. Infectious rabies viruses from cloned cDNA. *EMBO J.* **1994**, 4195–4203. [CrossRef]
44. Abelseth, M.K. An attenuated rabies vaccine for domestic animals produced in tissue culture. *Can. Vet. J.* **1964**, *5*, 279–286.
45. Höper, D.; Freuling, C.M.; Müller, T.; Hanke, D.; von Messling, V.; Duchow, K.; Beer, M.; Mettenleiter, T.C. High definition viral vaccine strain identity and stability testing using full-genome population data–The next generation of vaccine quality control. *Vaccine* **2015**, *33*, 5829–5837. [CrossRef]
46. Vos, A.; Nolden, T.; Habla, C.; Finke, S.; Freuling, C.M.; Teifke, J.; Müller, T. Raccoons (Procyon lotor) in Germany as potential reservoir species for Lyssaviruses. *Eur J. Wildl Res.* **2013**, *59*, 637–643. [CrossRef]
47. Fu, C.; Donovan, W.P.; Shikapwashya-Hasser, O.; Ye, X.; Cole, R.H. Hot Fusion: An efficient method to clone multiple DNA fragments as well as inverted repeats without ligase. *PLoS ONE* **2014**, *9*, e115318. [CrossRef]
48. Finke, S.; Granzow, H.; Hurst, J.; Pollin, R.; Mettenleiter, T.C. Intergenotypic replacement of lyssavirus matrix proteins demonstrates the role of lyssavirus M proteins in intracellular virus accumulation. *J. Virol.* **2010**, *84*, 1816–1827. [CrossRef]
49. Buchholz, U.J.; Finke, S.; Conzelmann, K.K. Generation of bovine respiratory syncytial virus (BRSV) from cDNA: BRSV NS2 is not essential for virus replication in tissue culture, and the human RSV leader region acts as a functional BRSV genome promoter. *J. Virol.* **1999**, *73*, 251–259. [CrossRef]
50. Orbanz, J.; Finke, S. Generation of recombinant European bat lyssavirus type 1 and inter-genotypic compatibility of lyssavirus genotype 1 and 5 antigenome promoters. *Arch. Virol.* **2010**, *155*, 1631–1641. [CrossRef]
51. Eggerbauer, E.; Pfaff, F.; Finke, S.; Höper, D.; Beer, M.; Mettenleiter, T.C.; Nolden, T.; Teifke, J.-P.; Müller, T.; Freuling, C.M. Comparative analysis of European bat lyssavirus 1 pathogenicity in the mouse model. *PLoS Negl. Trop. Dis.* **2017**, *11*, e0005668. [CrossRef] [PubMed]
52. Schindelin, J.; Arganda-Carreras, I.; Frise, E.; Kaynig, V.; Longair, M.; Pietzsch, T.; Preibisch, S.; Rueden, C.; Saalfeld, S.; Schmid, B.; et al. Fiji: An open-source platform for biological-image analysis. *Nat. Methods* **2012**, *9*, 676–682. [CrossRef] [PubMed]
53. Bolte, S.; Cordelières, F.P. A guided tour into subcellular colocalization analysis in light microscopy. *J. Microsc.* **2006**, *224*, 213–232. [CrossRef] [PubMed]
54. de Chaumont, F.; Dallongeville, S.; Chenouard, N.; Hervé, N.; Pop, S.; Provoost, T.; Meas-Yedid, V.; Pankajakshan, P.; Lecomte, T.; Le Montagner, Y.; et al. Icy: An open bioimage informatics platform for extended reproducible research. *Nat. Methods* **2012**, *9*, 690–696. [CrossRef]
55. Jackson, A.C.; Reimer, D.L. Pathogenesis of experimental rabies in mice: An immunohistochemical study. *Acta Neuropathol.* **1989**, *78*, 159–165. [CrossRef]

56. Conzelmann, K.K.; Cox, J.H.; Schneider, L.G.; Thiel, H.J. Molecular cloning and complete nucleotide sequence of the attenuated rabies virus SAD B19. *Virology* **1990**, *175*, 485–499. [CrossRef]
57. Weiland, F.; Cox, J.H.; Meyer, S.; Dahme, E.; Reddehase, M.J. Rabies Virus Neuritic Paralysis: Immunopathogenesis of Nonfatal Paralytic Rabies. *J. Virol.* **1992**, 5096–5099. [CrossRef]
58. Zlotnik, I. The reaction of astrocytes to acute virus infections of the central nervous system. *Br. J. Exp. Pathol* **1968**, *49*, 555–564.
59. Jackson, A.C. *Rabies. Scientific Basis of the Disease and Its Management*, 3rd ed.; Elsevier Science: San Diego, CA, USA, 2013; ISBN 9780123965479.
60. Detje, C.N.; Lienenklaus, S.; Chhatbar, C.; Spanier, J.; Prajeeth, C.K.; Soldner, C.; Tovey, M.G.; Schlüter, D.; Weiss, S.; Stangel, M.; et al. Upon intranasal vesicular stomatitis virus infection, astrocytes in the olfactory bulb are important interferon Beta producers that protect from lethal encephalitis. *J. Virol.* **2015**, *89*, 2731–2738. [CrossRef]
61. Kallfass, C.; Ackerman, A.; Lienenklaus, S.; Weiss, S.; Heimrich, B.; Staeheli, P. Visualizing production of beta interferon by astrocytes and microglia in brain of La Crosse virus-infected mice. *J. Virol.* **2012**, *86*, 11223–11230. [CrossRef]

4

A New ERAP2/Iso3 Isoform Expression is Triggered by Different Microbial Stimuli in Human Cells: Could it Play a Role in the Modulation of SARS-CoV-2 Infection?

Irma Saulle [1,2], Claudia Vanetti [1,2], Sara Goglia [1], Chiara Vicentini [1], Enrico Tombetti [1], Micaela Garziano [1], Mario Clerici [2,3] and Mara Biasin [1,*]

1 Department of Biomedical and Clinical Sciences-L. Sacco, University of Milan, 20157 Milan, Italy; irma.saulle@unimi.it (I.S.); claudia.vanetti@unimi.it (C.V.); sarag9623@gmail.com (S.G.); chiara.vicentini@studenti.unimi.it (C.V.); enrico.tombetti@unimi.it (E.T.); micaela.garziano@unimi.it (M.G.)
2 Department of Pathophysiology and Transplantation, University of Milan, 20122 Milan, Italy; mario.clerici@unimi.it
3 Don C. Gnocchi Foundation ONLUS, IRCCS, 20148 Milan, Italy
* Correspondence: mara.biasin@unimi.it

Abstract: Following influenza infection, rs2248374-G ERAP2 expressing cells may transcribe an alternative spliced isoform: ERAP2/Iso3. This variant, unlike ERAP2-wt, is unable to trim peptides to be loaded on MHC class I molecules, but it can still dimerize with both ERAP2-wt and ERAP1-wt, thus contributing to profiling an alternative cellular immune-peptidome. In order to verify if the expression of ERAP2/Iso3 may be induced by other pathogens, PBMCs and MDMs isolated from 20 healthy subjects were stimulated with flu, LPS, CMV, HIV-AT-2, SARS-CoV-2 antigens to analyze its mRNA and protein expression. In parallel, Calu3 cell lines and PBMCs were in vitro infected with growing doses of SARS-CoV-2 (0.5, 5, 1000 MOI) and HIV-1_{BAL} (0.1, 1, and 10 ng p24 HIV-1_{Bal}/1 $\times 10^6$ PBMCs) viruses, respectively. Results showed that: (1) ERAP2/Iso3 mRNA expression can be prompted by many pathogens and it is coupled with the modulation of several determinants (cytokines, interferon-stimulated genes, activation/inhibition markers, antigen-presentation elements) orchestrating the anti-microbial immune response (Quantigene); (2) ERAP2/Iso3 mRNA is translated into a protein (western blot); (3) ERAP2/Iso3 mRNA expression is sensitive to SARS-CoV-2 and HIV-1 concentration. Considering the key role played by ERAPs in antigen processing and presentation, it is conceivable that these enzymes may be potential targets and modulators of the pathogenicity of infectious diseases and further analyses are needed to define the role played by the different isoforms.

Keywords: ERAP2; ERAP2/Iso3; microbial infections; alternative splicing; SARS-CoV-2; host cell response

---

1. Introduction

ERAP1 and ERAP2 (endoplasmic reticulum aminopeptidases 1 and 2) are two IFNγ- and TNFα-inducible, ubiquitously-expressed human enzymes, which belong to the M1 family of zinc aminopeptidases [1]. In the endoplasmic reticulum (ER), ERAPs shape the antigenic repertoire by trimming the N-terminus of precursor peptides previously generated in the cytoplasm by the proteasome. In this way, ERAPs generate optimal-length peptides for loading onto MHC class I groove to be presented to CD8+ T lymphocytes [2,3]. Despite maintaining marked differences in their enzymatic specificity these two enzymes can act together in a concerted way, through the formation of homo- or heterodimers, thus allowing the generation of a variegated and more immunogenic antigenic repertoire [4]. In particular, ERAP1–ERAP2 heterodimer generation has been demonstrated to improve the shaping of peptides suitable for MHC class I molecule binding [5].

ERAPs are encoded by two genes, sharing ~49% sequence homology and situated on chromosome 5q15 in opposite directions, which are highly polymorphic [6]. Since their leading role in the antigen processing pathway, several studies have investigated any potential link between ERAP polymorphic variants and alterations in their functioning, which could result in MHC-I-associated disorder onset [2,7,8] as well as into variations in susceptibility/progression to microbial infections [9].

As for ERAP2, the most relevant single nucleotide polymorphism (SNP) is the non-coding rs2248374 (A/G) which identifies two haplotypes, hereafter referred to as HapA (A allele for rs2248374) and HapB (G allele for rs2248374). In HapB, the G allele for this SNP primes the transcription of a spliced ERAP2 variant (ERAP2/Iso2), presenting an extended exon 10 (56 extra nucleotides) and two in-frame TAG stop codons, which in turn lead to its nonsense-mediated decay (NMD) [10]. Conversely, HapA is translated into a 965-amino-acid protein and is associated with Crohn's disease [11], HLA-A29-associated birdshot uveitis [12], ankylosing spondylitis [13,14] and juvenile idiopathic arthritis [15], as well as natural resistance to HIV infection [9,16,17]. Since these variants are maintained by a balanced selection to a frequency of approximately 50% (HapB: 53% and HapA: 47%), nearly 25% of the population fails to express the ERAP2 protein [18]. This observation raises a logic question: in which peculiar setting does balancing selection operate to conserve the apparently loss-of-function HapB and the disease-causing HapA in the human population? Quite recently, Ye and co-workers provided an exhaustive explanation to this apparent paradox [19]. Indeed, for the first time, they documented the transcription of two novel short isoforms (ERAP2/Iso3, ERAP2/Iso4) from flu-infected monocyte-derived dendritic cells isolated from homozygous HapB-carrying subjects [19]. The two short isoforms are transcribed from HapB and differ from the full-length one—ERAP2/Iso1, transcribed from HapA—since their transcription begins in correspondence of exon 9 and undergoes alternative splicing of an extended exon 10. Besides, they diverge from each other by alternative splicing at a secondary splice site at exon 15 (Figure 1). Of note, while ERAP2/Iso4 is predicted to harbor a premature termination codon that could lead to NMD, ERAP2/Iso3, is expected to be translated into a protein [19]. Such protein misses the catalytic domain [19], but could still critically contribute to profile the cellular immune-peptidome as it preserves the capacity to dimerize with ERAP1 and possibly ERAP2-wild type (wt) [20].

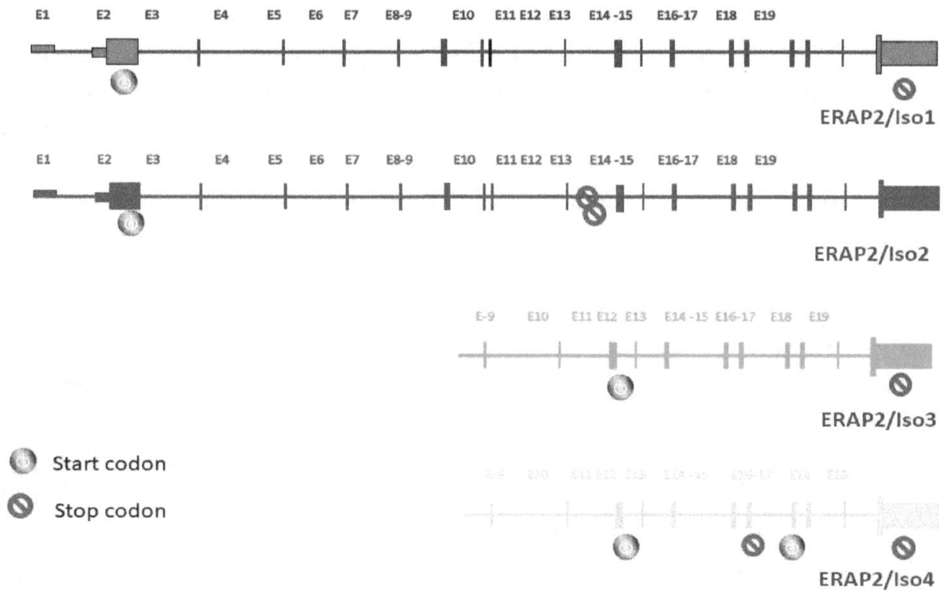

Figure 1. Genetics of ERAP2 isoforms regulation. Structures of transcripts derived from each ERAP2 isoform are represented. Start and stop codons for each isoform are reported.

Based on these premises, the aim of our study was to investigate if the transcription of ERAP2/Iso3 is either flu-specific or if it can be induced by other kind of stimuli, such as other viruses, bacteria, or inflammatory triggers. Given the foremost role played by ERAPs in the field of both acquired and innate immunity, the characterization of the different isoforms produced in a particular pathological setting, such as the one caused by microbial infections, could lead to the identification of new molecular targets to be exploited in the setting up of innovative therapeutically or vaccinal approaches.

2. Methods

2.1. Study Population

Twenty healthy controls (HC) were enrolled in the study in order to investigate the induction of ERAP isoform transcription in response to common recall antigens such as influenza-antigens (flu), Lipopolisaccaride (LPS), Citomegalovirus (CMV), in addition to Aldrithiol-2 (AT-2)-inactivated R5-tropic human immunodeficiency virus-$1_{\text{Ba-L}}$ (HIV-AT-2) and acute respiratory syndrome coronavirus 2 (SARS-CoV-2) inactivated virus (i-SARS-CoV-2). The Ethical Committee of the Fondazione IRCCS Ca' Granda Ospedale Maggiore Policlinico approved the study (Prot. N°0028257). All the donors signed an informed consent form, in agreement with the Declaration of Helsinki of 1975, revised in 2013.

2.2. Viruses

The laboratory-adapted HIV-1 strain used in the experiments was the R5 tropic HIV-1_{BaL} (courtesy of Drs. S. Gartner, M. Popovic, and R. Gallo; NIH AIDS Research and Reference Reagent Program) provided through the EU program EVA Centre for AIDS Reagents (NIBSC, Potter Bars, UK). The virus was inactivated with AT-2, which is able to change the zinc finger nucleocapsid proteins of HIV-1, therefore deactivating the viral infectivity as previously described [21].

SARS-CoV-2 (Virus Human 2019-nCoV strain 2019-nCoV/Italy-INMI1, Rome, Italy) was expanded on Calu-3 cells (ATCC® HTB-55™) and TCID_{50} was calculated as previously reported [22]. SARS-CoV-2 inactivation (i-SARS-CoV-2) was obtained by incubation at 65 °C for 30 min [23].

2.3. ERAP2 Genotyping Analyses

Whole blood was collected by all the subjects enrolled in the study by venipuncture in Vacutainer tubes containing EDTA (BD Vacutainer, San Diego, CA, USA). Total DNA was extracted by DNA purification Maxwell® RSC Instrument (Promega, Fitchburg, WI, USA) and quantified using the Nanodrop 2000 Instrument (Thermo Scientific, Waltham, MA, USA). Two-hundred ng of DNA were used to perform an SNP genotyping assay for ERAP2 rs2549782 (G/T) (TaqMan SNP Genotyping Assay; Applied Biosystems, Foster City, CA, USA), which is in linkage disequilibrium with the non-coding rs2248374 (A/G). Analyses were performed on Peripheral blood mononuclear cells (PBMCs) from all the subjects recruited in the study as well as on lung adenocarcinoma cells (Calu3; ATCC® HTB-55™). Allelic discrimination real-time PCR method was used to analyze the results.

2.4. Isolation of PBMCs and Monocyte-Derived Macrophages (MDMs) Differentiation

PBMCs, obtained from density gradient centrifugation on Ficoll (Cedarlane Laboratories Limited, Hornby, ON, Canada), were counted by automated cell counter ADAM-MC (NanoEnTek Inc., Seoul, Korea), which distinguishes viable from non-viable cells.

Flow cytometer analyses was used to quantify the percentage of CD14+ monocytes in PBMCs isolated from 3 HeteroAB HC. MDMs were obtained as previously described [24]. Briefly, 5×10^5 adherent monocytes were incubated for 5 days in RPMI with 20% of fetal bovine serum (FBS) (Euroclone, Milan, Italy) and 100 ng/mL macrophage-colony stimulating factor (M-CSF) (R&D Systems, Minneapolis, MN, USA). Optical microscope (ZOE™ Fluorescent Cell Imager, Bio-Rad, Hercules, CA, USA) observation allows to verify MDM differentiation.

2.5. *Cell Cultures for Microbial Antigen Stimulation*

PBMCs were resuspended at the concentration of 1×10^6 PBMCs/mL in RPMI 1640 medium (Euroclone, Milan, Italy) containing 10% fetal bovine serum (FBS), 1% levo-glutammin LG and 2% penstreptomicin. Subsequently, PBMCs and MDMs from HC were stimulated with antigens from different pathogens: 32 μg/mL of CMV grade 2 Antigen (Microbix Biosystem, Mississauga, ON, Canada), 1 μg/mL of LPS, 1 ng/mL of HIV-AT-2 equivalents and 5 multiplicity of infection (MOI) of i-SARS-CoV-2 inactivated virus. Two live UV-inactivated influenza viruses (flu) were used as well: an influenza A virus (A/RX73 and A/Puerto Rico/8/34 strains; 1:800) and the 1998–1999 formula of flu vaccine (1:5000; Wyeth Laboratories Inc., Marietta, PA, USA). Cells were stimulated even with non-microbial stimuli: 100U of IFNα and 1 μg/mL of IL-1β (Sigma, Saint Louis, MO, USA). Unstimulated PBMCs were cultured as control as well. Cells were harvested 10 (PBMCs) and 36 (MDMs) h post-treatment for RNA and protein analyses, respectively.

2.6. *In Vitro Infection of PBMCs and Calu3 Cells with SARS-CoV-2*

2.5×10^5 Calu3 cells (ATCC® HTB-55™) were cultured in DMEM medium (Euroclone, Milan, Italy) supplemented with 2% FBS in a 24-well plate. DMEM containing 100 U/mL penicillin and 100 μg/mL streptomycin was used as inoculum in the mock-infected cells. Cell cultures were incubated with serial dilutions of virus supernatant in duplicate, (1000 MOI, 5 MOI, 0.5 MOI) for three h at 37 °C and 5% CO_2. Cells were washed two times with lukewarm PBS and refilled with the proper growth medium (10%FBS). Optical microscope observation (ZOE™ Fluorescent Cell Imager, Bio-Rad, Hercules, CA, USA) was performed daily to investigate the cytopathic effect. The infected cells were harvested for SARS-CoV-2 detection in the supernatant and mRNA collection at 48 h. Each culture condition was run in triplicate. All the procedures were performed in agreement with the GLP guidelines adopted in our laboratory.

Maxwell® RSC Viral Total Nucleic Acid Purification Kit (Promega, Fitchburg, WI, USA) was used to extract RNA from Calu3 cell culture supernatants by the Maxwell® RSC Instrument (Promega, Fitchburg, WI, USA). Viral RNA was quantified, by single-step RT PCR -time PCR (GoTaq® 1-Step RT-qPCR) (Promega, Fitchburg, WI, USA) on a CFX96 (Bio-Rad, Hercules, CA, USA) by using TaqMan probes which target two portions of SARS-CoV-2 nucleocapsid (N) gene (N1 and N2). Specifically, we used the 2019-nCoV CDC qPCR Probe Assay emergency kit (IDT, Coralville, IA, USA), which allows also to amplify the human RNase P gene. Viral copy number quantification was performed by generating a standard curve from the quantified 2019-nCoV_N positive Plasmid Control (IDT, Coralville, IA, USA).

2.7. *In Vitro HIV-Infection Assay*

3×10^6 PBMCs from all the subjects included in the study were in vitro HIV-infected as previously described [25] with 10, 1, and 0.1 ng p24 HIV-1_{Bal}/1×10^6 PBMCs. After 5 days, supernatants were collected for p24 antigen ELISA (Cell Biolabs, San Diego, CA, USA), whereas PBMCs collected at 2 days post-infection were used for RNA extraction and gene expression analyses.

2.8. *Gene Expression Analysis*

RNA extracted from 1×10^6 PBMCs, and Calu3 cell lines were retrotranscribed as previously described [16]. cDNA quantification for ERAPs was performed on antigen-stimulated and HIV-infected PBMCs as well as on SARS-CoV-2 infected Calu3 cells through a real-time PCR (CFX96 connect, Bio-Rad, Hercules, CA, USA) and an SYBR Green PCR mix (Bio-Rad, Hercules, CA, USA); all the reactions were carried out in duplicate. Results are shown as the media of the relative expression units to the glyceraldehyde-3-phosphate dehydrogenase (GAPDH) and β-actin housekeeping genes calculated by the $2^{-\Delta\Delta Ct}$ equation. The following thermal protocol was used: initial denaturation (95 °C, 15 min) followed by 40 cycles of 15 s at 95 °C (denaturation), 20 s at 60 °C (annealing) and 20 s at 72 °C (extension).

Furthermore, a melting curve analysis was assessed for amplicon characterization. Ct values of 35 or higher were let off the analyses.

2.9. *Quantigene Plex Gene Expression Assay*

Gene expression of 8×10^5 PBMCs was performed by quantiGene Plex assay (Thermo Scientific, Waltham, MA, USA) which provides a fast and high-throughput solution for multiplexed gene expression quantitation, allowing the simultaneous measurement of 70 custom selected genes of interest in a single well of a 96-well plate. The QuantiGene Plex assay is hybridization-based and incorporates branched DNA (bDNA) technology, which uses signal amplification for direct measurement of RNA transcripts. The assay does not require RNA purification.

2.10. *Western Blot Analyses*

Cultured MDMs were removed by non-enzymatic cell dissociation solution (Sigma, Saint Louis, Missouri, USA), counted by the automated cell counter ADAM-MC (Digital Bio) and used for protein extraction by RIPA buffer (Sigma, Saint Louis, MO, USA). Extracted proteins were stored at −80 °C for further analyses. For WB analyses, samples from 3 HeteroAB subjects were sub-pooled (50 µg per pool). Equal amounts of proteins were separated by 4–20% SDS-polyacrylamide gel electrophoresis (Criterion TGX Stain-free precast gels and Criterion Cell system; Bio-Rad) and transferred onto nitrocellulose membrane using a Bio-Rad Trans-Blot Turbo System. Membranes were probed using a 1:1000 dilution of primary antibody goat anti-ERAP1 (AF2334; R&D Systems, Minneapolis, MN, USA) goat anti-ERAP2 polyclonal antibody (AF3830; R&D Systems, Minneapolis, MN, USA), rabbit anti-GAPDH polyclonal (VPA00187); Bio-Rad, Hercules, CA, USA] and a 1:10,000 dilution of secondary antibody conjugated with alkaline phosphatase anti-goat IgG (A4187; Sigma; goat anti-rabbit (STAR208P); Bio-Rad]. Membranes were incubated with the appropriate antibody and, after being excited using the Clarity Western ECL substrate, bands were visualized with a ChemiDoc MP imaging system (Bio-Rad) and quantified for densitometry with the Bio-Rad Image Lab software.

2.11. *Statistical Analyses*

Data are shown as mean and standard deviation. Analysis and figures were performed by GRAPHPAD PRISM version 5 (Graphpad software, La Jolla, CA, USA) and SPSS Statistics, version 25 (IBM software, Armonk, NY, USA). Gene expressions of ERAP2/iso1, ERAP2/iso1, and ERAP1 in PBMCs, monocyte-derived macrophages, and Calu-3 cells upon stimulation were compared to that of untreated cells by Wilcoxon test and Mann–Whitney test, as appropriate. P-values were corrected for false discovery rate (FDR) by using Microsoft R software, p-values ≤ 0.005 were considered to be significant.

3. Results

3.1. *ERAP2 Allelic Variants Analyses*

Analysis of ERAP2 SNP prevalence was aligned with European population distribution reported in the U.S. National Library of Medicine Database [https://www.ncbi.nlm.nih.gov/snp/rs2549782?fbclid=IwAR1ZdwC747PDWvtAzt6hZBV5j7oFZiPkLjY-JdSee1Plzvym7fhJVQc1Aks (data not shown). Among the 20 genotyped HC: 6 were HomoA, 8 heterozygous, 6 HomoB. Subsequent analyses were performed only on PBMCs and MDMs isolated from HomoB and heteroAB donors to exclude confounding results. Indeed, HomoA individuals express negligible levels of ERAP2/Iso2 and Iso3.

Calu3 cell lines were heterozygous for rs2248374 ERAP2 genotype.

3.2. mRNA Expression of ERAPs in PBMCs from Subjects Carrying Different ERAP2 Genotypes Following Microbial Stimulation

To verify whether the expression of ERAP2/Iso3 isoform is exclusively flu-specific or it may be triggered by other stimuli, we analyzed its expression on PBMCs isolated from HomoB and HeteroAB HC following HIV-AT-2, i-SARS-CoV-2, CMV, LPS, flu, IFNα, and IL-1β. As expected, the expression of the newly identified ERAP2/Iso3 was significantly augmented in PBMCs from all the subjects included in the study in response to flu ($p < 0.002$). However, even following CMV ($p < 0.004$), LPS ($p < 0.002$), HIV-AT-2 ($p < 0.003$), i-SARS-CoV-2 ($p < 0.002$) and IFNα ($p < 0.003$) stimulation, we observed a significant increase of its expression (Figure 2A). Conversely, IL-1β addition to cell culture did not result in ERAP2/Iso3 induction (Figure 2A). ERAP2/Iso1 expression was observed only in PBMCs from HeteroAB subjects and was induced following all the microbial-stimuli employed plus IFNα but not in response to IL-1β (Figure 2B). However, statistical significance was observed exclusively following HIV-AT-2 ($p < 0.025$), i-SARS-CoV-2 ($p < 0.03$) and CMV ($p < 0.02$). Likewise, ERAP1 expression was induced by all the stimuli employed excepting IL-1β and reached statistical significance following flu ($p < 0.002$), HIV-AT-2 ($p < 0.03$) CMV ($p < 0.002$), i-SARS-CoV-2 ($p < 0.008$) LPS ($p < 0.05$). We also observed a reduction following IL-1β ($p < 0,001$) stimulations (Figure 2C).

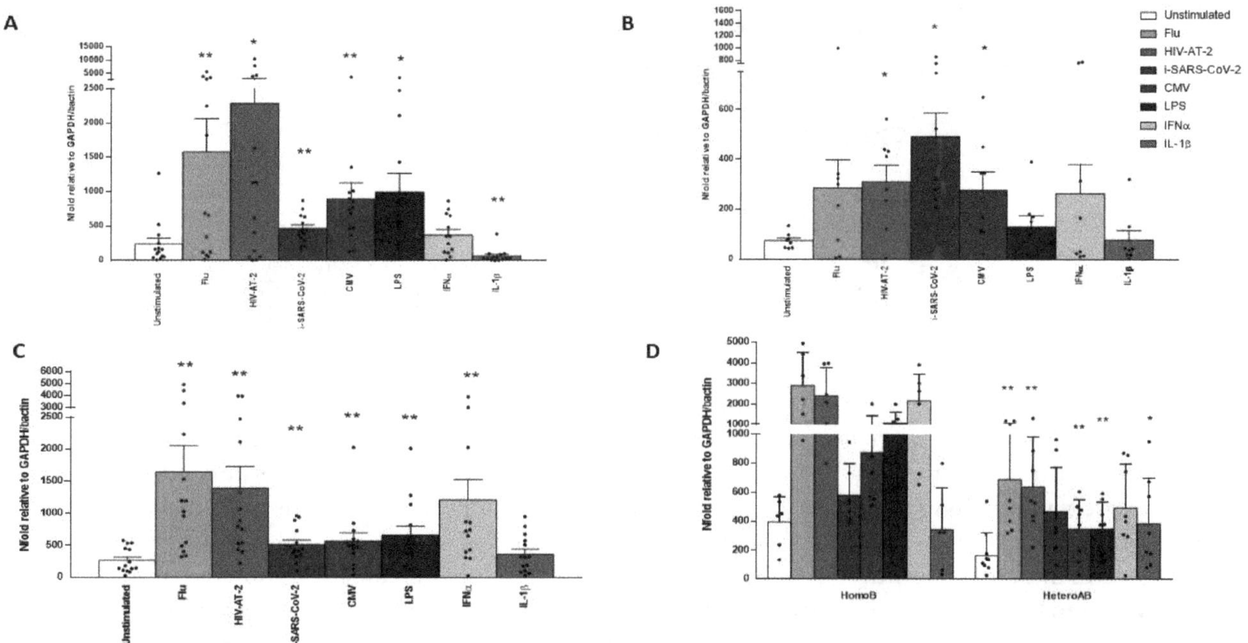

Figure 2. ERAP1, ERAP2/Iso1 and ERAP2/Iso3 mrna expression is increased following microbial stimulation. pbmcs isolated from 8 heteroab and 6 homob individuals were in vitro stimulated with microbial antigens (flu, CMV) inactivated viruses (i-SARS-CoV-2, HIV-AT-2) bacterial by-products (LPS) or inflammatory stimuli (IFNα, IL-1β) for ten h. mRNA expression for ERAP2/Iso3 (**A**), ERAP2/Iso1 (**B**) and ERAP1 (**C**) were assessed by RT-Real-Time PCR. Results are shown as the media of the relative expression units to the glyceraldehyde-3-phosphate dehydrogenase (GAPDH) and β-actin reference genes calculated by the $2^{-\Delta\Delta Ct}$ equation. (**D**) The microbial-dependent genetic control of ERAP2/Iso3 expression is underlined by the observation that its abundances are nearly doubled in HapB homozygotes compared to heterozygotes. Results are expressed as mean ± ES. * = $p < 0.05$; ** = $p < 0.01$.

Notably, the microbial-dependent genetic control of ERAP2 isoform usage is sustained by further evidence. Indeed, there was a significant correlation between whole microbial transcript quantity and ERAP2/Iso3 transcript abundances in heterozygotes compared to HapB homozygotes (Figure 2D).

3.3. Gene Expression of Immune Selected Effectors in PBMCs Following Microbial Stimulation

To verify if the increased expression of ERAPs in response to microbial stimulation could be extended to other factors involved in the orchestration of the immune response, mRNA expression of 70 selected effectors was investigated by the innovative Quantigene Plex Gene expression technology. The genes whose mRNA expression was upregulated are implicated in almost all phases of immune response, including: chemokines, cytokines and cytokine receptors, pathogen recognition receptor, inflammasome, cholesterol metabolism, interferon stimulated genes, adhesion molecules, activation/inhibition markers, antigen presentation factors (Figure 3). Notably, the gene expression pattern was partially shared by all the stimuli employed and partially pathogen-specific as summarized in Figure 3. In particular, following i-SARS-CoV-2-stimulation a significantly higher transcription rate was observed for: CCL2 ($p < 0.05$); CCL5 ($p < 0.035$), HMGCS1 ($p < 0.002$), PYCARD ($p < 0.002$), CASP1 ($p < 0.001$), CD44 ($p < 0.016$), CD274 ($p > 0.008$), IL-8 ($p < 0.003$); IL-1β ($p < 0.006$), ABCA1 ($p < 0.02$), IL-6R ($p < 0.004$), CCL3 ($p < 0.01$), IFNγ ($p < 0.05$); TAP1 ($p < 0.05$).

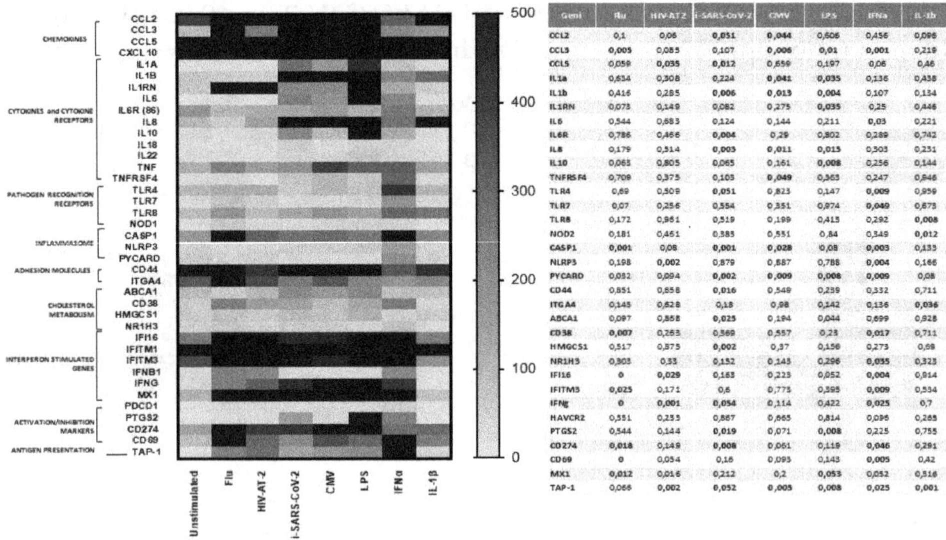

Geni	flu	HIV-AT2	i-SARS-CoV-2	CMV	LPS	IFNa	IL-1b
CCL2	0,1	0,06	0,051	0,044	0,606	0,456	0,096
CCL3	0,005	0,085	0,107	0,006	0,01	0,001	0,219
CCL5	0,059	0,035	0,012	0,659	0,197	0,06	0,46
IL1a	0,634	0,308	0,224	0,041	0,035	0,136	0,438
IL1b	0,416	0,285	0,006	0,013	0,004	0,107	0,134
IL1RN	0,073	0,149	0,082	0,275	0,035	0,25	0,446
IL6	0,544	0,683	0,124	0,144	0,211	0,03	0,221
IL6R	0,786	0,466	0,004	0,29	0,802	0,289	0,742
IL8	0,179	0,514	0,003	0,011	0,015	0,303	0,231
IL10	0,063	0,805	0,065	0,161	0,008	0,256	0,144
TNFRSF4	0,709	0,375	0,103	0,049	0,363	0,247	0,846
TLR4	0,69	0,509	0,051	0,823	0,147	0,009	0,959
TLR7	0,07	0,256	0,554	0,351	0,974	0,049	0,673
TLR8	0,172	0,961	0,519	0,199	0,413	0,292	0,008
NOD2	0,181	0,461	0,383	0,551	0,84	0,349	0,012
CASP1	0,001	0,08	0,001	0,028	0,08	0,003	0,133
NLRP3	0,198	0,002	0,879	0,887	0,788	0,004	0,166
PYCARD	0,032	0,094	0,002	0,009	0,008	0,009	0,08
CD44	0,851	0,858	0,016	0,549	0,239	0,332	0,711
ITGA4	0,145	0,628	0,18	0,98	0,142	0,136	0,036
ABCA1	0,097	0,858	0,025	0,194	0,044	0,699	0,928
CD38	0,007	0,263	0,369	0,557	0,23	0,017	0,711
HMGCS1	0,517	0,875	0,002	0,37	0,156	0,273	0,69
NR1H3	0,303	0,33	0,152	0,143	0,296	0,035	0,323
IFI16	0	0,029	0,163	0,223	0,052	0,004	0,914
IFITM3	0,025	0,171	0,6	0,773	0,395	0,009	0,534
IFNg	0	0,001	0,054	0,114	0,422	0,025	0,7
HAVCR2	0,551	0,233	0,867	0,665	0,814	0,096	0,263
PTGS2	0,544	0,144	0,019	0,071	0,008	0,225	0,755
CD274	0,018	0,149	0,008	0,061	0,119	0,046	0,291
CD69	0	0,054	0,16	0,093	0,143	0,005	0,42
MX1	0,012	0,016	0,212	0,2	0,053	0,052	0,516
TAP-1	0,066	0,002	0,052	0,005	0,008	0,025	0,001

Figure 3. mRNA expression of genes involved in the anti-microbial immune response was modulated in response to different pathogens. Quantigene Plex Gene expression technology was applied to quantify gene expression on PBMCs isolated from 8 HeteroAB and 6 HomoB individuals and stimulated with microbial antigens (flu, CMV) inactivated viruses (i-SARS-CoV-2, HIV-AT-2) bacterial by-products (LPS) or inflammatory stimuli (IFNα, IL-1β). Gene expression (mean values) is shown as a color scale from white to blue (Heatmap). Only statistically significant p values from T-test comparison between unstimulated and stimulated PBMCs are shown in table.

Unlike ERAP2 expression, no differences were observed in mRNA expression levels of all the analyzed genes in PBMCs from HomoB and HeteroAB subjects in response to all the stimuli engaged (data not shown).

3.4. mRNA Expression of ERAPs in In Vitro SARS-CoV-2 Infected Calu3 Cell Lines and HIV-Infected PBMCs

To verify whether ERAP2/Iso3 expression varies in response to growing viral concentrations we adopted two in vitro model of infection. Thus, Calu3 cell lines were infected with different SARS-CoV-2 viral input and after 48 hours' viral replication as well as ERAP mRNA expression were assessed. As expected, SARS-CoV-2 replication increased according to the rising viral input as assessed analyzing both N1 (MOI 0.5 vs. 5: $p < 0.01$; MOI 0.5 vs. 1000: $p < 0.001$) and N2 (MOI 0.5 vs. 5: $p < 0.01$; MOI 0.5 vs. 1000: $p < 0.001$) (Figure 4A). Images of cellular cytopathic effect on SARS-CoV-2 infected cells at 48 h showed that despite robust SARS-CoV-2 replication in Calu3 cells, substantial cell death was detected only in cells infected with 1000 MOI (Figure 4B). Notably, this increase was coupled with a progressive rise of ERAP2/Iso3 expression compared to the uninfected condition (MOI 0.5: $p < 0.04$; MOI 5: $p < 0.01$; MOI 1000: $p < 0.01$). Likewise, ERAP2/Iso1 and ERAP1 expression was induced in a viral-dose dependent manner, although statistical significance was observed only for ERAP2/Iso1 (MOI 0.5: $p < 0.05$) (Figure 4C).

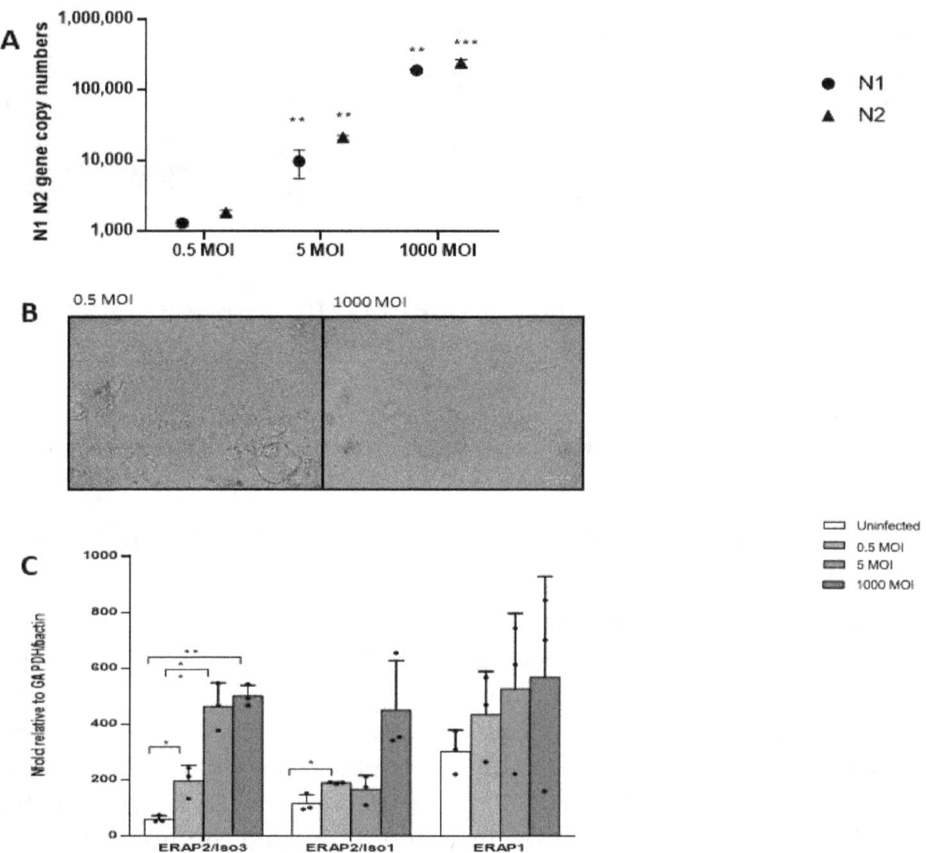

Figure 4. In vitro SARS-CoV-2 infection assay on Calu3 cells. (**A**) SARS-CoV-2 replication was assessed on Calu3 cells infected at 0.5, 5, and 1000 MOI 48 h post-infection. Viral copy number quantification was performed by generating a standard curve from the quantified 2019-nCoV_N positive plasmid control for Nucleocapsid (N) region 1 and 2, showing a significant increase in response to the increased viral input. (**B**) SARS-CoV-2-induced cytopathic effects were assessed in Calu3 infected cells. Representative images of SARS-CoV-2 infected-cells at 0.5 and 1000 MOI are reported. At 48 h post-infection, typical cytopathic effects, including cell rounding, detachment, degeneration, and syncytium formation were seen only in cells infected at 1000 MOI. Cells were imaged by optical microscope observation (ZOE™ Fluorescent Cell Imager, Bio-Rad, Hercules, CA, USA). (**C**) ERAP2/Iso3, Iso1, and ERAP1 mRNA expression by in vitro SARS-CoV-2 infected Calu3 cells increased according to the rising viral input. Mean values ± ES are reported. * = $p < 0.05$; ** = $p < 0.01$; *** = $p < 0.001$. MOI = multiplicity of infection.

In vitro HIV-1 infection of PBMCs isolated from HapB HC produced similar results. Indeed, as the viral input raised, viral replication quantified through p24 concentration analyses at 5 days post infection increased (0.1 vs. 1 ng p24 HIV-1_{Bal}/1 $\times 10^6$ PBMCs: $p < 0.02$; 0.1 vs. 10 ng p24 HIV-1_{Bal}/1 $\times 10^6$ PBMCs: $p < 0.001$) (Figure 5A). Likewise, ERAP2/Iso3 gene expression increased compared to the uninfected condition (0.1 ng p24 HIV-1_{Bal}/1 $\times 10^6$ PBMCs: $p < 0.003$; 0.1 ng p24 HIV-1_{Bal}/1 $\times 10^6$ PBMCs: $p < 0.04$; 10 ng p24 HIV-1_{Bal}/1 $\times 10^6$ PBMCs: $p < 0.03$) (Figure 5B). ERAP2/Iso1 (0.1 ng p24 HIV-1_{Bal}/1 $\times 10^6$ PBMCs: $p < 0.05$) and ERAP1 (0.1 ng p24 HIV-1_{Bal}/1 $\times 10^6$ PBMCs: $p < 0.05$) mRNA levels showed a similar trend (Figure 5B).

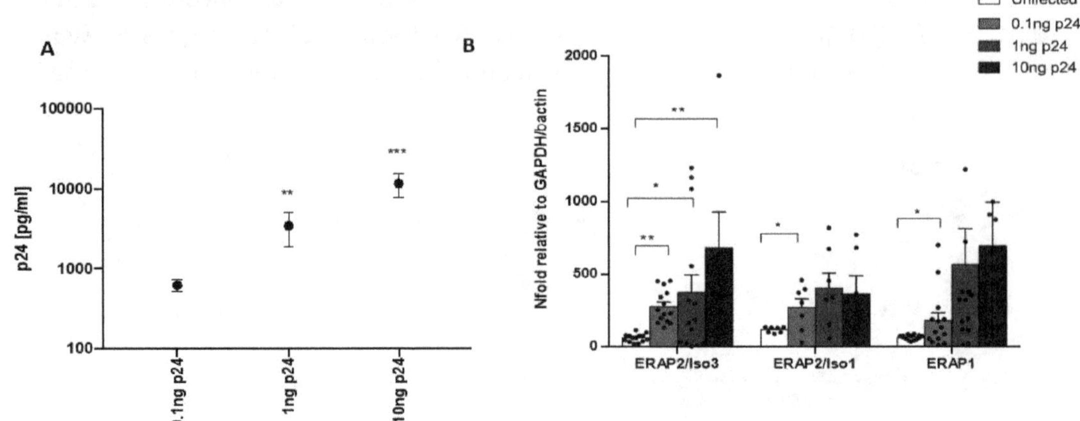

Figure 5. In vitro HIV-1 infection assay on PBMCs. (**A**) HIV-1 replication was assessed by p24 quantification on PBMCs isolated from 8 HeteroAB and 6 HomoB individuals infected with 0.1, 1, and 10 ng p24 HIV-1Bal/1 × 10^6 PBMCs 5 days post infection. Results showed a significant increase in viral replication according to the increased viral input. (**B**) ERAP2/Iso3, Iso1, and ERAP1 mRNA expression by in vitro HIV-1-infected PBMCs increased according to the rising viral input. Mean values ± ES are reported. * = $p < 0.05$; ** = $p < 0.01$; *** = $p < 0.001$.

3.5. ERAP2/Iso3 Protein Production by MDMs from HeteroAB Subjects Following Microbial Specific Stimulation

To verify if the short ERAP2/Iso3 isoform would function as an RNA or is translated into a protein product we performed a western blot assay on MDMs from 3 HeteroAB subjects triggered with different microbial antigens. Remarkably, the antibody which recognizes the full-length ERAP2 (Iso1) was able to detect also one short protein isoform (~50 kDa) in CMV, flu, HIV-AT-2, i-SARS-CoV-2, LPS, IFNα stimulated cells from HeteroAB subjects, suggesting the translation of the short microbial-specific ERAP2/Iso3 (Figure 6). Conversely, following IL-1β stimulation, only ERAP2/Iso1 isoform was detected. The production of ERAP1 protein was observed in all the stimulated conditions except IL-1β (Figure 6). As the proteins extracted from the 3 HeteroAB subjects were sub-pooled for WB analyses, statistical evaluation of the results was not possible.

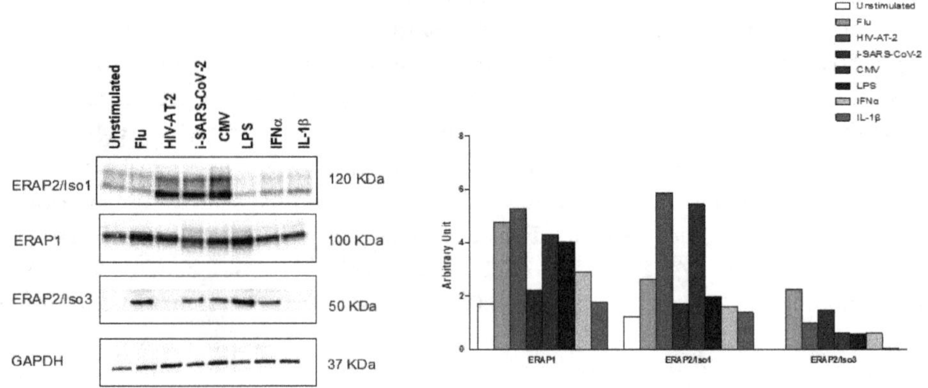

Figure 6. ERAP1, ERAP2/Iso1, and ERAP2/Iso3 production in pathogen stimulated monocyte-derived macrophages (MDMs). MDMs differentiated from 3 HeteroAB participants stimulated for 36 h with microbial antigens (flu, CMV) inactivated viruses (i-SARS-CoV-2, HIV-AT-2) bacterial by-products (LPS) or inflammatory stimuli (IFNα, IL-1β) were tested for protein using primary antibodies specific to a ERAP1 (goat polyclonal), ERAP2 and β-actin. Proteins extracted from the 3 HeteroAB subjects were sub-pooled for WB analyses. Histograms representing ERAP1, ERAP2/Iso1, and ERAP2/Iso3 densitometric quantification. Quantification was performed by Quantity One 4.6.6 software (Bio-Rad) and normalization was permed on GAPDH.

4. Discussion

Given the documented role of ERAP2 in antigen presentation [20] and viral infections [21], we examined the genetic control of ERAP2 transcripts in the human antimicrobial response. In particular, based on the results recently reported by Ye and co-workers [19], we investigated if the expression of the recently characterized ERAP2/Iso3 is flu-specific or if it can be triggered by other microbial stimuli. Our results suggest that: (1) ERAP2/Iso3 mRNA expression is not restricted to flu-infection but it can be prompted by other pathogens including HIV, SARS-CoV-2, CMV, Bacteria (LPS); (2) ERAP2/Iso3 mRNA is translated into a protein following microbial induction; (3) ERAP2/Iso3 mRNA expression is sensitive to viral concentration.

Remarkably, Ye et al. did not detect ERAP2/Iso3 expression in IFNB1 stimulated cells leading to the conclusion that the transcription of novel ERAP2 isoforms is likely initiated by viral sensing pathways upstream of type 1 interferon signaling [19]. Conversely, in our cell culture condition, IFNα-stimulation was able to induce ERAP2/Iso3 expression in a consistent way, suggesting the participation of type 1 interferon cascade to the induction of ERAP2/Iso3 transcription, as already documented for the wild type forms of ERAP1 and 2 [26]. The different cell kinds and IFN subtypes adopted in the two experimental settings could at least partially justify the discrepancies reported in the two studies, but further analyses are needed to clarify the pathways and molecules induced by microbial exposure, directly responsible for ERAP2/Iso3 synthesis. For example, in our study IL-1β stimulus was not able to trigger the expression of any ERAP variants, further strengthening the assumption of a specific response of ERAP gene transcription controlled by pathogen exposure. Furthermore, as IFNα is also strongly associated to various type of I IFN conditions such as Sjogren's disease [27], systemic lupus erythematosus [28] and Scleroderma [29] it would be valuable to verify if ERAP2/Iso3 expression varies in patients affected by these pathologies and/or following administration of IFNα therapy.

The discovery of these new ERAP2 isoforms is of great importance and gives a plausible explanation to the maintenance of one of the major gene expression quantitative trait loci (eQTL) and alternate isoform usage in most tissues and cell types [19,30]. Indeed, until last year, the preservation of HapB at intermediate frequency in human population was almost unexplained, as its transcript was believed to be addressed to NMD [9,10]. The identification of this previously uncharacterized short isoform, ERAP2/Iso3, transcribed from HapB, results in the partial rescue of ERAP2 expression, suggesting its involvement in the anti-microbial response.

The mechanism of action and functional impact of this new genetic variant to host defense; however, is still indefinite. The lack of the aminopeptidase domain proves that it does not directly participate in the shaping of antigenic peptides to be presented to CD8+ T cells. However, in 2005 Saveanu et al. conducted an immunoblot analysis detecting the presence of ERAP1-ERAP2 heterodimers and possibly homodimers of each enzyme [4]. This crystallographic dimer is described as mediated by domains I and II of the enzyme that is missing from ERAP2/Iso3 [31]. However, this observation has been reiterated by Ye et al. who hypothesized that ERAP2 isoforms could have a dominant-negative effect on either ERAP1 and possibly ERAP2 wt, through the formation of hetero- or homodimeric complexes [15]. Despite still speculative the fact that ERAP2/Iso3 may exert such an effect cannot be ruled out: this, in turn, could lead to an altered peptide processing, which could confer an advantage/disadvantage against infections by presenting a more/less immunogenic antigen repertoire. In relation to this aspect, previous studies have demonstrated that the functional skills of ERAP monomers, homo, or heterodimers may significantly differ in terms of both substrate specificity and trimming efficiency [32]. In particular, ERAP1/2 dimerization creates complexes with superior peptide-trimming efficacy and a higher affinity towards ERAP1 preferential substrates. This is allowed by the adoption of a modified physical conformation by ERAP1, caused by its interaction with ERAP2, which mainly works as an enhancer of ERAP1 role upon dimer assembling [33]. Further studies are needed to verify if also ERAP2/Iso3 physical interaction with the wt ERAP variants prompts an allosteric effect able to modify basic enzymatic parameters and to improve their substrate-binding affinity. However, the observation that ERAP2/Iso3 mRNA is translated into a protein allows to speculate on the generation of a new ERAP member which further contributes to enrich the non-redundant, yet a complete and potent system of aminopeptidases, warranting an efficient trimming of various kinds of precursors.

The results obtained in this study definitely establish a link between invading pathogens and ERAP2/Iso3 expression which further strengthens the significance of the results reported by Ye and collaborators. Supporting this hypothesis, we observed that ERAP2/Iso3 expression progressively increased in response to growing doses of viral input in both SARS-CoV-2 and HIV-1 in vitro infection assay, as if the production of this genetic variant, as well as one of the other elements within the ERAP family, were directly dependent on the viral dose of exposure. Additionally, the observation that the increased expression of ERAP2/Iso3 in response to pathogen exposure is accompanied by the modulation of many other determinants (chemokines, cytokines, pathogen recognition receptor, inflammasome, interferon-stimulated genes, adhesion molecules, activation/inhibition markers, antigen presentation elements) orchestrating the anti-microbial immune response, further supports its direct intervention in this defensive pathway. Notably, as its expression is increased by a wide range of microbial stimuli including viral antigens, inactivated virus as well as bacterial by-products, it is possible to assume that ERAP2/Iso3 expression is not pathogen-specific, but it's secondary to the activation of an antimicrobial cascade commonly shared by different pathogens. Meanwhile, we cannot exclude that ERAPs responses observed following microbial stimulations and in vitro viral infections result from an erroneous transcription and translation due to pathogen-induced cellular stress. Indeed, as ERAP1 and ERAP2 expression has been demonstrated to be prompted by IFNγ stimulation, ERAPs production could be secondary to the innate immune response of the cells to infections, rather than to an immune response or an immune evasion mechanism. Further analyses will be necessary to verify this hypothesis.

The involvement of ERAPs in modulating viral infections is widely recognized as recently reviewed in [34]. Several studies, indeed, have demonstrated the intervention of ERAP genetic variants in the life cycle of HCV, flu, CMV, HPV, HIV, and other pathogens at different levels. In particular, studies performed by our research group have established an association between ERAP2/Iso1 and HIV-infection in terms of both susceptibility [17,35] and progression [36]. However, to our knowledge ERAP expression and/or genetic variants have been correlated to the recent coronavirus disease 2019 (COVID-19), provoked by SARS-CoV-2, only by two recent studies [34,37,38]. In the first one by Stamatakis et al. ERAP2 trimming ability has been investigated in SARS-CoV-2 infection together with ERAP1 and IRAP, and it has been proved as the most stable of the enzymes generating optimal length antigenic peptides for HLA binding [38]. In the second one by Lu et al. by examining 193 deaths from 1,412 confirmed infections in a group of 5,871 UK Biobank participants tested for the virus, rs150892504 variant in ERAP2 gene came up as potentially being implicated in risk from SARS-CoV-2 infection. Although rs150892504 variant is not in linkage disequilibrium with rs2248374, this finding suggests the involvement of ERAP2 in the modulation of SARS-CoV-2 infection. Such an assumption is further supported by other intriguing observations. This virus enters the cells by spike protein binding to ACE2 (angiotensin-converting enzyme 2), which is responsible for the conversion of angiotensin I to angiotensin I-9 and of angiotensin II to angiotensin I-7, an effective vasodilator, thus working as a negative regulator of the renin-angiotensin system (RAS) Our results, for the first time demonstrate that SARS-CoV-2 exposure triggers the expression of ERAP1, ERAP2/Iso1 and also the recently detected ERAP2/Iso3 in a dose-dependent mode, suggesting its participation in the control of the anti-SARS-CoV-2 response. This observation is far more important, considering that besides their involvement in the antigen presentation pathway, ERAPs display several key anti-SARS-CoV-2 functions. Indeed, they intervene in the RAS, where ERAP1 efficiently cleaves angiotensin II to angiotensin III and IV, and ERAP2 cuts angiotensin III to angiotensin IV, thus influencing both ACE2 virus receptor bio-availability and blood pressure levels [39]. Furthermore, ERAPs modulate the proteolytic cleavage of IL-6 receptor (IL-6Rα) [40] a function which can improve the clinical conditions in COVID-19 patients, as recently documented following its pharmacological inhibition by Tocilizumab [41]. Last but not least, Ranjit and colleagues recently demonstrated that sex-specific differences in ERAP1 modulation influence blood pressure and RAS responses [42], and a male bias in mortality has emerged in the COVID-19 pandemic since the very beginning. As ERAPs genetic variants have been demonstrated to orchestrate and condition the result of infection of coronavirus in other animal species [43,44] detailed studies investigating SARS-CoV-2-host interplay is absolutely mandatory.

Another query which needs to be addressed in the near future concerns the cellular localization of ERAP2/Iso3 isoform. In particular, it would be interesting to verify if, as with the wt ERAP variants [17], even ERAP2/Iso3 may be secreted into the extracellular milieu following inflammatory stimulation, with which substrates it may interact, which functions may eventually exert in this environment and if its administration can interfere with viral replication.

Considering the key role played by ERAPs in antigen processing and presentation, it is plausible that these aminopeptidases may be potential targets and controllers of the pathogenicity of infectious diseases, shaping the susceptibility and response to microbial infections. The recent acquisition of ERAP intervention even in the modulation of innate immunity further reinforces this assumption.

Given the growing number of viral epidemics, the identification of molecular mechanisms driven by factors such as ERAPs that can interfere, control or modulate viral replication is unequivocally needed as they could be widely exploited for the inception of future, still unknown viral infections.

Author Contributions: Each author has approved the submitted version and agrees to be personally accountable for the author's own contributions and for ensuring that questions related to the accuracy or integrity of any part of the work. Conceptualization, M.B. and I.S.; Methodology, C.V. (Claudia Vanetti), S.G. and C.V. (Chiara Vicentini); Formal Analysis, I.S., M.G.; Statistical Analyses, E.T.; Investigation, S.G. and C.V. (Chiara Vicentini); Data Curation, C.V. (Claudia Vanetti), M.G.; Writing—Original Draft Preparation, I.S.; Writing—Review & Editing, M.B.; Supervision, M.B. and M.C.; Funding Acquisition, M.C. All authors have read and agreed to the published version of the manuscript.

References

1. Neefjes, J.; Jongsma, M.L.M.; Paul, P.; Bakke, O. Towards a systems understanding of MHC class I and MHC class II antigen presentation. *Nat. Rev. Immunol.* **2011**, *11*, 823–836. [CrossRef] [PubMed]
2. Cifaldi, L.; Romania, P.; Lorenzi, S.; Locatelli, F.; Fruci, D. Role of Endoplasmic Reticulum Aminopeptidases in Health and Disease: From Infection to Cancer. *Int. J. Mol. Sci.* **2012**, *13*, 8338–8352. [CrossRef] [PubMed]
3. Compagnone, M.; Fruci, D. Peptide Trimming for MHC Class I Presentation by Endoplasmic Reticulum Aminopeptidases. *Methods Mol. Biol. Clifton NJ* **2019**, *1988*, 45–57. [CrossRef]
4. Saveanu, L.; Carroll, O.; Lindo, V.; Del Val, M.; Lopez, D.; Lepelletier, Y.; Greer, F.; Schomburg, L.; Fruci, D.; Niedermann, G.; et al. Concerted peptide trimming by human ERAP1 and ERAP2 aminopeptidase complexes in the endoplasmic reticulum. *Nat. Immunol.* **2005**, *6*, 689–697. [CrossRef]
5. Evnouchidou, I.; van Endert, P. Peptide trimming by endoplasmic reticulum aminopeptidases: Role of MHC class I binding and ERAP dimerization. *Hum. Immunol.* **2019**, *80*, 290–295. [CrossRef]
6. Yao, Y.; Liu, N.; Zhou, Z.; Shi, L. Influence of ERAP1 and ERAP2 gene polymorphisms on disease susceptibility in different populations. *Hum. Immunol.* **2019**, *80*, 325–334. [CrossRef]
7. Stamogiannos, A.; Koumantou, D.; Papakyriakou, A.; Stratikos, E. Effects of polymorphic variation on the mechanism of Endoplasmic Reticulum Aminopeptidase 1. *Mol. Immunol.* **2015**, *67*, 426–435. [CrossRef]
8. López de Castro, J.A. How ERAP1 and ERAP2 Shape the Peptidomes of Disease-Associated MHC-I Proteins. *Front. Immunol.* **2018**, *9*. [CrossRef]
9. Cagliani, R.; Riva, S.; Biasin, M.; Fumagalli, M.; Pozzoli, U.; Lo Caputo, S.; Mazzotta, F.; Piacentini, L.; Bresolin, N.; Clerici, M.; et al. Genetic diversity at endoplasmic reticulum aminopeptidases is maintained by balancing selection and is associated with natural resistance to HIV-1 infection. *Hum. Mol. Genet.* **2010**, *19*, 4705–4714. [CrossRef]
10. Andrés, A.M.; Dennis, M.Y.; Kretzschmar, W.W.; Cannons, J.L.; Lee-Lin, S.-Q.; Hurle, B.; Schwartzberg, P.L.; Williamson, S.H.; Bustamante, C.D.; Nielsen, R.; et al. Balancing Selection Maintains a Form of ERAP2 that Undergoes Nonsense-Mediated Decay and Affects Antigen Presentation. *PLoS Genet.* **2010**, *6*. [CrossRef]

11. Jostins, L.; Ripke, S.; Weersma, R.K.; Duerr, R.H.; McGovern, D.P.; Hui, K.Y.; Lee, J.C.; Philip Schumm, L.; Sharma, Y.; Anderson, C.A.; et al. Host–microbe interactions have shaped the genetic architecture of inflammatory bowel disease. *Nature* **2012**, *491*, 119–124. [CrossRef] [PubMed]
12. Kuiper, J.J.W.; Van Setten, J.; Ripke, S.; Van 'T Slot, R.; Mulder, F.; Missotten, T.; Baarsma, G.S.; Francioli, L.C.; Pulit, S.L.; De Kovel, C.G.F.; et al. A genome-wide association study identifies a functional ERAP2 haplotype associated with birdshot chorioretinopathy. *Hum. Mol. Genet.* **2014**, *23*, 6081–6087. [CrossRef] [PubMed]
13. Wiśniewski, A.; Kasprzyk, S.; Majorczyk, E.; Nowak, I.; Wilczyńska, K.; Chlebicki, A.; Zoń-Giebel, A.; Kuśnierczyk, P. ERAP1-ERAP2 haplotypes are associated with ankylosing spondylitis in Polish patients. *Hum. Immunol.* **2019**, *80*, 339–343. [CrossRef] [PubMed]
14. Robinson, P.C.; Costello, M.E.; Leo, P.; Bradbury, L.A.; Hollis, K.; Cortes, A.; Lee, S.; Joo, K.B.; Shim, S.-C.; Weisman, M.; et al. ERAP2 is associated with ankylosing spondylitis in HLA-B27-positive and HLA-B27-negative patients. *Ann. Rheum. Dis.* **2015**, *74*, 1627–1629. [CrossRef] [PubMed]
15. Chiaroni-Clarke, R.C.; Munro, J.E.; Chavez, R.A.; Pezic, A.; Allen, R.C.; Akikusa, J.D.; Piper, S.E.; Saffery, R.; Ponsonby, A.-L.; Ellis, J.A. Independent confirmation of juvenile idiopathic arthritis genetic risk loci previously identified by immunochip array analysis. *Pediatr. Rheumatol.* **2014**, *12*, 53. [CrossRef]
16. Biasin, M.; Sironi, M.; Saulle, I.; de Luca, M.; la Rosa, F.; Cagliani, R.; Forni, D.; Agliardi, C.; lo Caputo, S.; Mazzotta, F.; et al. Endoplasmic reticulum aminopeptidase 2 haplotypes play a role in modulating susceptibility to HIV infection. *AIDS Lond. Engl.* **2013**, *27*, 1697–1706. [CrossRef]
17. Saulle, I.; Ibba, S.V.; Torretta, E.; Vittori, C.; Fenizia, C.; Piancone, F.; Minisci, D.; Lori, E.M.; Trabattoni, D.; Gelfi, C.; et al. Endoplasmic Reticulum Associated Aminopeptidase 2 (ERAP2) Is Released in the Secretome of Activated MDMs and Reduces in vitro HIV-1 Infection. *Front. Immunol.* **2019**, *10*. [CrossRef]
18. Evnouchidou, I.; Birtley, J.; Seregin, S.; Papakyriakou, A.; Zervoudi, E.; Samiotaki, M.; Panayotou, G.; Giastas, P.; Petrakis, O.; Georgiadis, D.; et al. A Common Single Nucleotide Polymorphism in Endoplasmic Reticulum Aminopeptidase 2 Induces a Specificity Switch That Leads to Altered Antigen Processing. *J. Immunol.* **2012**. [CrossRef]
19. Ye, C.J.; Chen, J.; Villani, A.-C.; Gate, R.E.; Subramaniam, M.; Bhangale, T.; Lee, M.N.; Raj, T.; Raychowdhury, R.; Li, W.; et al. Genetic analysis of isoform usage in the human anti-viral response reveals influenza-specific regulation of ERAP2 transcripts under balancing selection. *Genome Res.* **2018**, *28*, 1812–1825. [CrossRef]
20. Saveanu, L.; Carroll, O.; Hassainya, Y.; Endert, P.V. Complexity, contradictions, and conundrums: Studying post-proteasomal proteolysis in HLA class I antigen presentation. *Immunol. Rev.* **2005**, *207*, 42–59. [CrossRef]
21. Rossio, J.L.; Esser, M.T.; Suryanarayana, K.; Schneider, D.K.; Bess, J.W.; Vasquez, G.M.; Wiltrout, T.A.; Chertova, E.; Grimes, M.K.; Sattentau, Q.; et al. Inactivation of Human Immunodeficiency Virus Type 1 Infectivity with Preservation of Conformational and Functional Integrity of Virion Surface Proteins. *J. Virol.* **1998**, *72*, 7992–8001. [CrossRef] [PubMed]
22. Hui, K.P.Y.; Cheung, M.-C.; Perera, R.A.P.M.; Ng, K.-C.; Bui, C.H.T.; Ho, J.C.W.; Ng, M.M.T.; Kuok, D.I.T.; Shih, K.C.; Tsao, S.-W.; et al. Tropism, replication competence, and innate immune responses of the coronavirus SARS-CoV-2 in human respiratory tract and conjunctiva: An analysis in ex-vivo and in-vitro cultures. *Lancet Respir. Med.* **2020**, *8*, 687–695. [CrossRef]
23. Batéjat, C.; Grassin, Q.; Manuguerra, J.-C.; Leclercq, I. Heat inactivation of the Severe Acute Respiratory Syndrome Coronavirus 2. *bioRxiv* **2020**. [CrossRef]
24. Merlini, E.; Tincati, C.; Biasin, M.; Saulle, I.; Cazzaniga, F.A.; d'Arminio Monforte, A.; Cappione, A.J.I.; Snyder-Cappione, J.; Clerici, M.; Marchetti, G.C. Stimulation of PBMC and Monocyte-Derived Macrophages via Toll-Like Receptor Activates Innate Immune Pathways in HIV-Infected Patients on Virally Suppressive Combination Antiretroviral Therapy. *Front. Immunol.* **2016**, *7*. [CrossRef] [PubMed]
25. Saulle, I.; Ibba, S.V.; Vittori, C.; Fenizia, C.; Mercurio, V.; Vichi, F.; Caputo, S.L.; Trabattoni, D.; Clerici, M.; Biasin, M. Sterol metabolism modulates susceptibility to HIV-1 Infection. *AIDS Lond. Engl.* **2020**. [CrossRef]
26. Wu, T.G.; Rose, W.A.; Albrecht, T.B.; Knutson, E.P.; König, R.; Perdigão, J.R.; Nguyen, A.P.A.; Fleischmann, W.R. IFN-alpha-induced murine B16 melanoma cancer vaccine cells: Induction and accumulation of cell-associated IL-15. *J. Interf. Cytokine Res. Off. J. Int. Soc. Interf. Cytokine Res.* **2007**, *27*, 13–22. [CrossRef]
27. Nordmark, G.; Ronnblom, M.-L.E.; Primary, L. Sjogren's Syndrome and the Type I Interferon System. Available online: https://www.eurekaselect.com/101237/article (accessed on 15 August 2020).

28. Kirou, K.A.; Gkrouzman, E. Anti-interferon alpha treatment in SLE. *Clin. Immunol.* **2013**, *148*, 303–312. [CrossRef]
29. Raschi, E.; Chighizola, C.B.; Cesana, L.; Privitera, D.; Ingegnoli, F.; Mastaglio, C.; Meroni, P.L.; Borghi, M.O. Immune complexes containing scleroderma-specific autoantibodies induce a profibrotic and proinflammatory phenotype in skin fibroblasts. *Arthritis Res. Ther.* **2018**, *20*, 187. [CrossRef]
30. Lappalainen, T.; Sammeth, M.; Friedländer, M.R.; Ac't Hoen, P.; Monlong, J.; Rivas, M.A.; Gonzàlez-Porta, M.; Kurbatova, N.; Griebel, T.; Ferreira, P.G.; et al. Transcriptome and genome sequencing uncovers functional variation in humans. *Nature* **2013**, *501*, 506–511. [CrossRef]
31. The Crystal Structure of Human Endoplasmic Reticulum Aminopeptidase 2 Reveals the Atomic Basis for Distinct Roles in Antigen Processing | Biochemistry. Available online: https://pubs.acs.org/doi/10.1021/bi201230p (accessed on 17 August 2020).
32. de Castro, J.A.L.; Stratikos, E. Intracellular antigen processing by ERAP2: Molecular mechanism and roles in health and disease. *Hum. Immunol.* **2019**, *80*, 310–317. [CrossRef]
33. Evnouchidou, I.; Weimershaus, M.; Saveanu, L.; Endert, P. van ERAP1–ERAP2 Dimerization Increases Peptide-Trimming Efficiency. *J. Immunol.* **2014**. [CrossRef] [PubMed]
34. Saulle, I.; Vicentini, C.; Clerici, M.; Biasin, M. An Overview on ERAP Roles in Infectious Diseases. *Cells* **2020**, *9*, 720. [CrossRef] [PubMed]
35. Forni, D.; Cagliani, R.; Tresoldi, C.; Pozzoli, U.; Gioia, L.D.; Filippi, G.; Riva, S.; Menozzi, G.; Colleoni, M.; Biasin, M.; et al. An Evolutionary Analysis of Antigen Processing and Presentation across Different Timescales Reveals Pervasive Selection. *PLoS Genet.* **2014**, *10*, e1004189. [CrossRef] [PubMed]
36. Lori, E.M.; Cozzi-Lepri, A.; Tavelli, A.; Mercurio, V.; Ibba, S.V.; Lo Caputo, S.; Castelli, F.; Castagna, A.; Gori, A.; Marchetti, G.; et al. Evaluation of the effect of protective genetic variants on cART success in HIV-1-infected patients. *J. Biol. Regul. Homeost. Agents* **2020**, *34*. [CrossRef]
37. Genetic Risk Factors for Death with SARS-CoV-2 from the UK Biobank | medRxiv. Available online: https://www.medrxiv.org/content/10.1101/2020.07.01.20144592v1 (accessed on 15 August 2020).
38. Stamatakis, G.; Samiotaki, M.; Mpakali, A.; Panayotou, G.; Stratikos, E. Generation of SARS-CoV-2 S1 spike glycoprotein putative antigenic epitopes in vitro by intracellular aminopeptidases. *bioRxiv* **2020**. [CrossRef]
39. Hisatsune, C.; Ebisui, E.; Usui, M.; Ogawa, N.; Suzuki, A.; Mataga, N.; Takahashi-Iwanaga, H.; Mikoshiba, K. ERp44 Exerts Redox-Dependent Control of Blood Pressure at the ER. *Mol. Cell* **2015**, *58*, 1015–1027. [CrossRef]
40. Cui, X.; Rouhani, F.N.; Hawari, F.; Levine, S.J. An Aminopeptidase, ARTS-1, Is Required for Interleukin-6 Receptor Shedding. *J. Biol. Chem.* **2003**, *278*, 28677–28685. [CrossRef]
41. Michot, J.-M.; Albiges, L.; Chaput, N.; Saada, V.; Pommeret, F.; Griscelli, F.; Balleyguier, C.; Besse, B.; Marabelle, A.; Netzer, F.; et al. Tocilizumab, an anti-IL-6 receptor antibody, to treat COVID-19-related respiratory failure: A case report. *Ann. Oncol. Off. J. Eur. Soc. Med. Oncol.* **2020**, *31*, 961–964. [CrossRef]
42. Ranjit, S.; Wong, J.Y.; Tan, J.W.; Sin Tay, C.; Lee, J.M.; Yin Han Wong, K.; Pojoga, L.H.; Brooks, D.L.; Garza, A.E.; Maris, S.A.; et al. Sex-specific differences in endoplasmic reticulum aminopeptidase 1 modulation influence blood pressure and renin-angiotensin system responses. *JCI Insight* **2019**, *4*. [CrossRef]
43. Golovko, L.; Lyons, L.A.; Liu, H.; Sørensen, A.; Wehnert, S.; Pedersen, N.C. Genetic susceptibility to feline infectious peritonitis in Birman cats. *Virus Res.* **2013**, *175*, 58–63. [CrossRef]
44. Cong, F.; Liu, X.; Han, Z.; Shao, Y.; Kong, X.; Liu, S. Transcriptome analysis of chicken kidney tissues following coronavirus avian infectious bronchitis virus infection. *BMC Genom.* **2013**, *14*, 743. [CrossRef] [PubMed]

5

Development of Feline Ileum- and Colon-Derived Organoids and their Potential Use to Support Feline Coronavirus Infection

Gergely Tekes [1,*,†], Rosina Ehmann [2], Steeve Boulant [3,4] and Megan L. Stanifer [5,*]

1 Institute of Virology, Justus Liebig University Giessen, 35390 Giessen, Germany
2 Bundeswehr Institute of Microbiology, 80937 Munich, Germany; RosinaEhmann@bundeswehr.org
3 Department of Infectious Diseases, Virology, Heidelberg University Hospital, 69120 Heidelberg, Germany; s.boulant@dkfz.de
4 Research Group "Cellular Polarity and Viral Infection", German Cancer Research Center (DKFZ), 69120 Heidelberg, Germany
5 Department of Infectious Diseases, Molecular Virology, Heidelberg University Hospital, 69120 Heidelberg, Germany
* Correspondence: gergely.tekes@web.de (G.T.); m.stanifer@dkfz.de (M.L.S.)
† Current address: Elanco Animal Health, Germany.

Abstract: Feline coronaviruses (FCoVs) infect both wild and domestic cat populations world-wide. FCoVs present as two main biotypes: the mild feline enteric coronavirus (FECV) and the fatal feline infectious peritonitis virus (FIPV). FIPV develops through mutations from FECV during a persistence infection. So far, the molecular mechanism of FECV-persistence and contributing factors for FIPV development may not be studied, since field FECV isolates do not grow in available cell culture models. In this work, we aimed at establishing feline ileum and colon organoids that allow the propagation of field FECVs. We have determined the best methods to isolate, culture and passage feline ileum and colon organoids. Importantly, we have demonstrated using GFP-expressing recombinant field FECV that colon organoids are able to support infection of FECV, which were unable to infect traditional feline cell culture models. These organoids in combination with recombinant FECVs can now open the door to unravel the molecular mechanisms by which FECV can persist in the gut for a longer period of time and how transition to FIPV is achieved.

Keywords: feline coronavirus; feline enteric coronavirus; FECV; feline infectious peritonitis virus; FIPV; feline intestinal organoids

---

1. Introduction

The diverse family of the *Coronaviridae* causes infections in a wide range of mammals, birds and humans. Feline coronavirus (FCoV) is a highly prevalent member of the *Coronaviridae* family and is found in both domestic and wild cat populations worldwide [1]. FCoVs occur in two different biotypes: feline enteric coronavirus (FECV) and feline infectious peritonitis virus (FIPV). FECV infections commonly manifest as mild or asymptomatic infections of the feline enteric tract. Infections are often persistent and display intermittent shedding of virus over long periods of time which greatly contributes to the high seropositivity levels found in domestic cats. In single household cats, 20–60% of cats display signs of exposure to the virus and up to 90% of cats in multi-cat populations are seropositive [2–4].

FIPV emerges through mutations from the harmless FECV and can lead to a fatal clinical condition known as feline infectious peritonitis (FIP) [5–9]. Although the molecular pathogenesis of FIP is poorly understood [10–12], promising therapeutical approaches have recently been described [13–18]. It is important to note that both biotypes exist in two serotypes [19–21]. Serotype II FCoVs are the results of recombination between a serotype I FCoV and a closely related canine coronavirus (CCoV) [22–25] and can easily be grown in vitro. In sharp contrast, the more relevant and prevalent serotype I FCoVs cannot be propagated in cell culture. Accordingly, serotype II viruses were often used in the past to gain insight into FCoV biology instead of serotype I FCoVs. To elucidate the molecular pathogenesis of FIP, cell culture-adapted serotype I FIPV laboratory strains were obtained over time [26]. However, these viruses proved to be unsuitable to study the pathogenesis of FIP due to the loss of pathogenicity via cell culture adaptation [1,9,26]. The first reverse genetic system that enabled genetic manipulation of the entire FCoV genome was described by Tekes et al. (2008) for serotype I FIPV laboratory strain Black using a vaccinia virus vector [26–28]. However, animal experiments showed that like many other laboratory strains, serotype I FIPV Black lost its capability to induce FIP [26]. On the contrary, another commonly used serotype II FIPV laboratory strain, 79-1146, [26,29,30] is much more pathogenic and thus does not appropriately resemble most of the field strains either. Due to the lack of suitable in vitro systems for field serotype I FECVs, it is critical to establish a suitable in vitro system that enables the growth of serotype I FECV. This culture system for serotype I FECV field viruses would not only provide insight into the molecular mechanism by which FECVs persist in the gut for a longer period of time but it might also contribute to the understanding of how FIPV can evolve from FECV during a harmless persistent infection.

Over the past years, organoids have been employed as an in vitro system to support the growth of several human viruses that were unable to be cultured using standard cell culture methods [31,32]. Organoids are derived from either induced pluripotent stem cells (iPSCs) or from tissue-derived stem cells, which are grown and differentiated as three-dimensional structures that closely recapitulate the cellular composition and functions of their originating organ [33]. Tissue-derived organoids rely on the ability to isolate stem cell containing crypts. These crypts are then grown in the presence of differentiation factors (Wnt3a, R-Spondin, Noggin and EGF), allowing them to grow into three-dimensional mini-gut organoids [33]. As these complex cultures more closely resemble the multi-cell types found in their natural tissue counterparts, they often contain factors, which are required for the replication and propagation of viruses that are missing in standard cell cultures. To determine if these model systems could be used to support the relevant serotype I FECV growth, we established a cat intestinal organoid culture system and show that it is capable of supporting infection with GFP-expressing recombinant serotype I FECV generated by reverse genetics. This model will now open the doors to study the molecular mechanism of serotype I FECV-persistence in its natural enteric environment.

2. Materials and Methods

2.1. Viruses and Cell Lines

Serotype I recFECV-GFP and recFECV-S_{79}-GFP were produced in vitro using the reverse genetic system for FCoV field strains described previously [34]. Recombinant virus stocks of recFECV-S_{79}-GFP were titrated by plaque assay on routinely used felis catus whole fetus (FCWF) cells [26–28,35,36]. Virus stocks of recFECV-GFP which cannot be cultivated in standard cell culture systems were quantified by comparative Western blot analysis of the FCoV M protein together with recFECV-S_{79}-GFP [34]. The cell line FCWF was provided by the diagnostic laboratory at the Justus Liebig University Giessen and maintained in culture media (DMEM with 1× penicillin/streptomycin (Thermo, Waltham, MA, USA) and 10% FBS (Biochrom, Cambridge, UK)). According to our experience with propagation of FCoV laboratory strains, the FCWF cells were used at a confluency of approximately 90% for the infection with recFECV-S_{79}-GFP.

2.2. Animals

Handling of the animals used for enteric tissue donation was performed according to the guidelines of the Hungarian legislation on animal protection. The protocol was approved by the Pest

Megyei Kormany-hivatal, Budapest (assurance numbers PE/EA/2441-6/2016 and TMF/657-12/2016). The animals were euthanized according to the designate protocol. Female specific-pathogen-free (SPF) cats were raised and housed in pathogen free conditions for laboratory use. These were not cats taken from outside veterinary practices. The SPF cats were euthanized at an age of 25 weeks and tissue samples from the gut were collected, 6 donors were used for this study. The cats were not tested prior to isolation specifically for FCoV. Animals had access to feed ad libitum prior to euthanization and tissue collection.

2.3. Chemicals and Solutions

Conditioned media containing Wnt3a, R-Spondin and Noggin was produced from the L-WRN cell line (ATCC CRL-3276) as per manufacturer's instructions. A 293T cell line which produces only R-Spondin was a kind gift from Calvin Kuo (Standford University) and conditioned media was made as previously described [37]. All organoid media were made from a base of advanced DMEM/F12 (Thermo, Waltham, MA, USA) which contained 1% penicillin/streptomycin (Thermo), 2 mM GlutaMAX (Thermo) and 10 mM HEPES (Thermo) and is referred to as Ad DMEM/F12++. All other organoid media components are found in Table 1.

Table 1. Human and mouse media compositions tested for their ability to support feline ileum and colon organoid growth. Final concentrations and manufactures of each component are listed.

Reagent	Company	Final Concentration	Human Media	Mouse Intestine Media	Mouse Colon Media
L-WRN (Wnt3s, R-Spondin, Noggin containing conditioned media)	Made in the lab	50%	X		X
Ad DMEM/F12++	Thermo	50% (Human and mouse colon) 90% for mouse intestine	X	X	X
B27	Thermo	1X	X	X	X
Nicotinamide	Sigma (Munich, Germany)	10 mM	X		
N-acetylcysteine	Sigma	1 mM	X	X	X
A-83-01	Tocris (Bristol, UK)	500 nM	X		
SB202190	Sigma	500 nM	X		
Leu-Gastrin	Sigma	10 nM	X		
Mouse recombinant	Thermo	50 ng/mL	X	X	X
R-Spondin conditioned media	Made in the lab	10%		X	
Mouse recombinant Noggin	Peprotech (Rocky Hill, NJ, USA)	100 ng/mL		X	
Y-27632	Sigma	10 μM	X	X	X
Matrigel, Growth factor reduced (GFR), phenol free	Corning (Corning, NY, USA)	100%	X	X	X

2.4. Isolation of Feline Intestinal Cells

Ileum and colon sections (10 cm each) were harvested from 6 donors and stored in cold transport buffer (1× phosphate buffered saline (PBS), 50 ng/mL gentamicin (Thermo), 1% penicillin/streptomycin (Thermo), 1% fetal bovine serum (FBS) and 250 μg/mL Fungizone (Thermo)) until the time of isolation. The tissue was cut in half and washed 3 × 10 min with shaking in cold PBS to ensure that all fecal material had been removed. Crypts containing stem cells were isolated within 16 h of sacrificing the animals by first cutting the tissue into smaller 1 cm pieces and then either adding 2 mM EDTA to tissue sample for 1 h at 4 °C or 20 mL of Gentle Cell Dissociation Reagent (Stem Cell Technologies, Vancouver, BC, Canada) for 1 h at room temp. Tissue sections were transferred to a clean tube and 10 mL of cold PBS + 1% BSA was added to tube. Tubes were shaken to release the crypt fraction. Fractions enriched in crypts were filtered with 70 μm filters and the process was repeated to collect four to five fractions for each tissue. The fractions were observed under a light microscope and those containing the highest number of crypts were pooled and spun at 500× *g* for 5 min at 4 °C. The supernatant was removed, and crypts were washed 1× with cold DMEM/F12 (Thermo) and spun at 500× *g* for 5 min at 4 °C. The media was removed, and crypts were re-suspended in 100% Matrigel, plated in 50 μL drops in 24-well non-tissue culture treated plates (Corning) and following polymerization of the Matrigel, 500 μL organoid media was added to each well and was replaced every 48 h. Following isolation of the crypts and seeding into organoid media, the size of the organoids was monitored over time at day 3, 6, 9, 12, and 15 days by observing them under bright field microscopy with a Nikon Eclipse Ti-S microscope. Their size was measured with a 10× objective using the Nikon NIS software.

2.5. RNA Isolation, cDNA, and qPCR

RNA was harvested from cells using NucleoSpin RNA extraction kit (Macherey-Nagel, Dueren, Germany) as per manufacturer's instructions. cDNA was made using iSCRIPT reverse transcriptase (BioRad, Hercules, CA, USA) from 250 ng of total RNA as per manufacturer's instructions. Quantitative-PCR was performed using iTaq SYBR green (BioRad) as per manufacturer's instructions, GAPDH was used as normalizing gene. Pre-designed feline specific KiCqStart SYBR Green pre-designed primers were purchased from Sigma-Aldrich (Table 2).

Table 2. Feline specific primers used to control cell population in tissue and organoids.

Gene	Cell Type	Gene ID	Sequence ID
GAPDH	House keeping	493876	NM_001009307
LGR5	Stem cell	101080720	XM_003989046
SMOC2	Stem cell	101082409	XM_003986725
MUC2	Goblet cell	101096605	XM_003993797
SI	Enterocyte	100144605	NM_001123332
SYP	Enteroendocrine	101084343	XM_004000526
LYZ	Paneth cell	100127109	XM_003989032

2.6. Passaging of Feline Mini-Gut Organoids

Ileum and colon mini-gut organoids were monitored under a light microscope and were passaged when centers became dark and filled with dead cells. For passaging, media was removed and Matrigel was dissolved in cold PBS. Organoids were spun at 500× *g* for 5 min at 4 °C and PBS was removed. Subsequently three different methods were used for passaging:

i. Mechanical passaging: 1 mL of cold PBS was used to re-suspend the organoids. Using a 27-gauge needle on a 1 mL syringe, organoids were broken down by passing the solution up and down 10 times through the needle. Organoids were then spun at 500× *g* for 5 min at 4 °C. The supernatant was removed, and the crypts were re-suspended in 100% Matrigel, plated in 50 μL drops in 24-well non-tissue culture treated plates (Corning) and following polymerization of the Matrigel, organoid media 500 μL was added to each well.
ii. Trypsin-based passaging: 1 mL of cold PBS was used to re-suspend the organoids. Organoids were then spun at 500× *g* for 5 min at 4 °C. Organoids were washed a second time in cold PBS by resuspending the pellet in 1 mL of PBS and spinning at 500× *g* for 5 min at 4 °C.

The supernatant was removed and organoids were incubated in 0.05% Trypsin-EDTA (Gibco) for 5 min at 37 °C. Trypsin digestion was stopped with the addition of serum containing media and samples were spun at 500× *g* for 5 min at 4 °C. Organoids were washed a second time in cold PBS by resuspending the pellet in 1 mL of PBS and spinning at 500× *g* for 5 min at 4 °C. The supernatant was removed, and the crypts were re-suspended in 100% Matrigel, plated in 50 μL drops in 24-well non-tissue culture treated plates (Corning) and following polymerization of the Matrigel, 500 μL organoid media was added to each well.

iii. Gentle Cell Dissociation Reagent method: Media was removed and Gentle Cell Dissociation Reagent (Stem cell technologies) was added to the organoid containing pellet and incubated for 10 min at room temp. Organoids were spun at 500× *g* for 5 min at 4 °C and the supernatant was removed. Organoids were washed in DMEM/F12 and then spun at 500× *g* for 5 min at 4 °C. The supernatant was removed and the crypts were re-suspended in 100% Matrigel, plated in 50 μL drops in 24-well non-tissue culture treated plates (Corning) and following polymerization of the Matrigel, 500 μL organoid media was added to each well.

2.7. Infection of Cell Culture with Recombinant Viruses

Approximately 90% confluent monolayers of FCWF cells were washed with serum free DMEM culture media. Cells were inoculated with recFECV-GFP and recFECV-S_{79}-GFP at a multiplicity of infection (MOI) of 0.01 or serum free culture media for the mock control. The infection was incubated for one hour and subsequently the inoculum was replaced by culture media containing FBS. The formation of plaques and the GFP signal was monitored 48 hours post-infection (hpi).

2.8. Infection of Organoids

Organoids were removed from Matrigel by adding cold PBS for 5 min, liquefied Matrigel and organoids were separated by centrifugation (500× *g*, 5 min). Supernatant and Matrigel were removed and organoids were resuspended in media and gently disrupted with a 27 G needle to allow virus to access both the apical and basolateral sides of the organoids. Following disruption, organoid media containing 10^4 pfu of FCoV was added to the organoids and allowed to incubate for 6 h in suspension. Following the 6-h incubation, Matrigel was added back to the cultures and organoids were observed over a three-day period.

2.9. Statistical Methods

Statistics were calculated by Prizm using an unpaired *t*-test.

3. Results and Discussion

Currently, infectious disease research of feline intestinal pathogens has been hampered by the lack of cell-based systems allowing for the screening of viruses and antiviral compounds. To fill this gap, we developed a method to isolate and propagate ileum and colon mini-gut organoids from felines. Six SPF cats were euthanized and the ileum and colon were collected. Tissue sections were washed thoroughly to remove all fecal material and were stored at 4 °C in a transport buffer containing PBS as well as antibiotics and antifungals. Organoids were prepared within 16 h of tissue harvesting. To determine the best method to isolate crypts containing stem cells from feline intestinal tissues, samples were split in two parts and two commonly used approaches (EDTA-based dissociation and Gentle Cell Dissociation Reagent™) for harvesting human and murine intestinal crypts [38,39] were compared side by side. Feline ileum and colon samples were washed thoroughly to remove all mucus or remaining contaminants and crypts containing stem cells were then isolated following incubation in EDTA or Gentle Cell Dissociation Reagent™. Microscopic evaluation of the crypts prior to seeding showed no large difference in the quantity or quality of crypts between the two isolation methods (data not shown). Equal numbers of crypts were seeded into Matrigel and the number and size of organoids was followed over 12 days. Unlike human and murine crypts, which form organoids with 16–24 h [38–40], the feline crypts took 72 h to seal and form cyst like organoids (Figure 1A). Following initial cyst formation, the ileum and colon organoids continued to grow over the twelve-day period (Figure 1A). To determine which isolation method produced the best organoids, the number and size of organoids was followed over time. Results show that organoids isolated using the EDTA approach produced a greater number of organoids and the organoids were larger in size compared to the Gentle Cell Dissociation Reagent™ method (Figure 1B,C). The increase in the number of organoids between day 3 and 6 does not illustrate a growth in organoid number but is due to the fact that at day 3 the crypts are too small to be effectively counted. This number of organoids then remained constant over the rest of the period as they continued to grow larger (Figure 1B,C), which was consistent with murine and human organoid generation [38,39].

To determine if the organoids displayed similar cell types as the natural feline intestine, tissue and organoids samples from three donors were lysed and the relative expression of each cell type was quantified by q-RT-PCR. As the natural intestine is made of stem cells, absorptive enterocytes and secretory cells we evaluated known markers from each of these populations. As there are no markers to analyze feline organoids, we used the best-known markers for murine and human, hypothesizing that they will constitute a marker for feline tissue. As a marker of stem cells, we used LGR5 and SMOC2. LGR5 is the historical marker described by the Clevers lab, however while it constitutes an excellent marker in tissue it is known to be suboptimal for organoids [38,39,41]. On the contrary, SMOC2 represents an excellent marker for both tissue and tissue-derived organoids. As a marker for Goblet we used mucin 2 (Muc2), for enterocytes we used sucrose isomaltase (SI), for enteroendocrine cells we used synaptophysin (SYP) and for Paneth cells we used lysozyme (LYZ). Results showed that ileum from feline tissue samples displays markers for stem cells (LGR5, SMOC2), Goblet cells (Muc2), enterocytes (SI), enteroendocrine cells (SYP) and Paneth cells (LYZ) (Figure 2A). Similarly, feline colon tissues expressed similar amounts of markers for all cell types except Paneth cells which are not present in colon tissue. The ratios found in feline tissues are similar to those observed in humans [41].

Importantly, organoids derived from feline ileum expressed the same cell type-specific markers as their tissue counterpart (Figure 2B). The relative expressions of the different markers were slightly different compared to the expression in tissue (Figure 2A,B). Ileum organoids displayed a higher relative expression of the stem cells marker (SMOC2) and a lower relative expression of the enterocyte maker (SI) compared to normal feline tissue. This suggests that there are more stem cells than enterocytes. The increase in stem cell number is expected as organoids are cultured under high Wnt3a conditions which favor stem cell numbers. Similarly, colon organoids also displayed a high amount of stem cell markers and their cellular composition looked similar to those of ileum organoids (Figure 2B). These differences in expression of the different cell type specific markers are not specific to the feline ileum and colon organoids but this is also observed in both murine and human intestinal organoids [40,42]. This shows that although organoids are extremely close to the originating tissue, the fact that they are ex vivo mini-organs causes subtle differences in their expression profile and differentiation pathways. All together these data show that we have developed a novel protocol to isolate and generate intestinal organoids from feline intestinal tissue and that these organoids closely resemble their tissue counterpart.

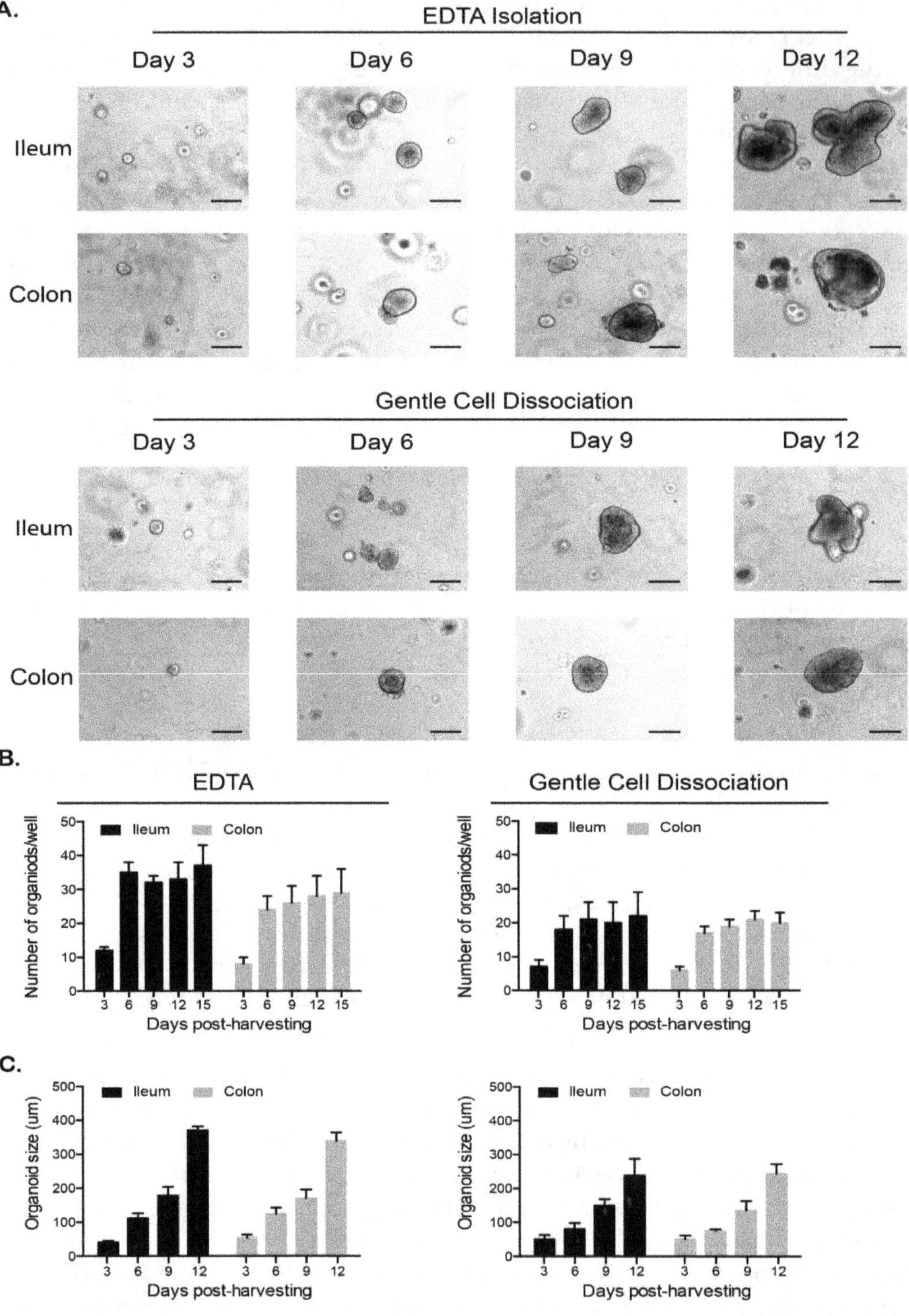

Figure 1. Ileum and colon organoids can be established from primary feline tissue. (**A–C**). The 1 cm^2 sections of feline ileum and colon tissue were incubated with EDTA or Gentle Cell Dissociation Reagent to allow for the isolation of intestinal crypts. Isolated crypts were resuspended in Matrigel and followed over a 12-day period. (**A**). Bright field images of ileum and colon organoids. Representative images are shown (n = 6 donors). Scale bar = 100 µm. (**B**). The number of organoids from EDTA and Gentle cell Dissociation Reagent isolation were counted over the indicated time course. Error bars represent standard deviation (n = 6 donors). (**C**). Following EDTA and Gentle cell Dissociation Reagent isolation, organoids were imaged in three-day intervals using a Nikon Eclipse Ti-S. Their size was measured using the Nikon NIS software. Error bars represent standard deviation (n = 6 donors).

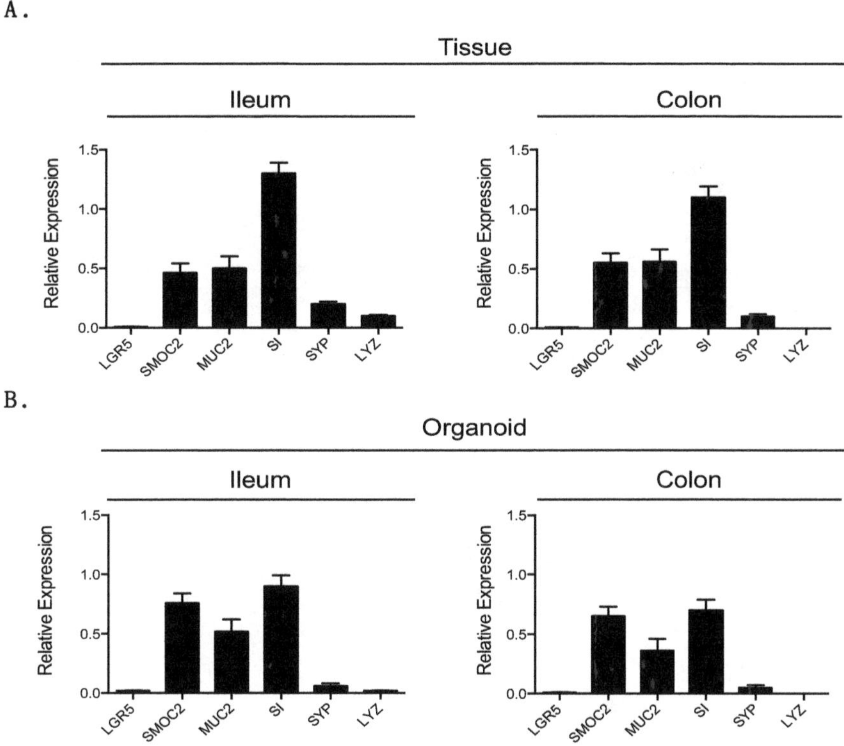

Figure 2. Feline organoids and originating tissue contain similar cell types. (**A**). The relative expression of intestinal cell type specific markers was evaluated in feline ileum and colon tissue by q-RT-PCR. (**B**). Same as A except using feline ileum- and colon-derived organoids. Results are mean +/− s.d. and are expressed as a relative expression to the house-keeping gene GAPDH. n = 3 donors and q-RT-PCR was done as technical triplicate.

3.1. Passaging and Maintenance of Feline Intestinal or Ganoids

As the ileum and colon organoids continued to grow and retain a similar identity to the natural tissue, we wanted to determine whether they could be passaged and maintained in culture. Trypsinization, mechanical disruption and passaging using Gentle Cell Dissociation Reagent™ were tested to establish the method that best supported the continued maintenance of feline ileal and colon organoids. For all methods, organoids were removed from the Matrigel and either incubated with low concentrations of trypsin, mechanical disrupted using a 27 G needle, or were incubated with Gentle Cell Dissociation Reagent™ (see materials and methods for full details). Following disruption, organoids were centrifuged to separate out organoid structures from dead cells. The organoids were then seeded into Matrigel and the formation of new organoids was followed for five days. One day post-passaging many small organoids could be seen in all conditions for both the ileum and colon organoids (Figure 3A). Organoids which were passaged by trypsinization were small and stressed, as shown by their loss of tight borders (dark rim at the periphery of the organoids under phase microscope) and the presence of many dissociated cells. On the contrary, organoids that were obtained using the mechanical disruption and Gentle Cell Dissociation Reagent™ methods were larger and their borders look more discrete (Figure 3A). Similar to the original isolation, many organoids were too small to be counted on the first day after passaging for all methods (Figure 3B). Observations five days post passaging showed that using mechanical disruption for passaging both ileum and colon organoids leads to a greater number of organoids that are larger in comparison to organoids that were passaged using trypsin (Figure 3B,C).

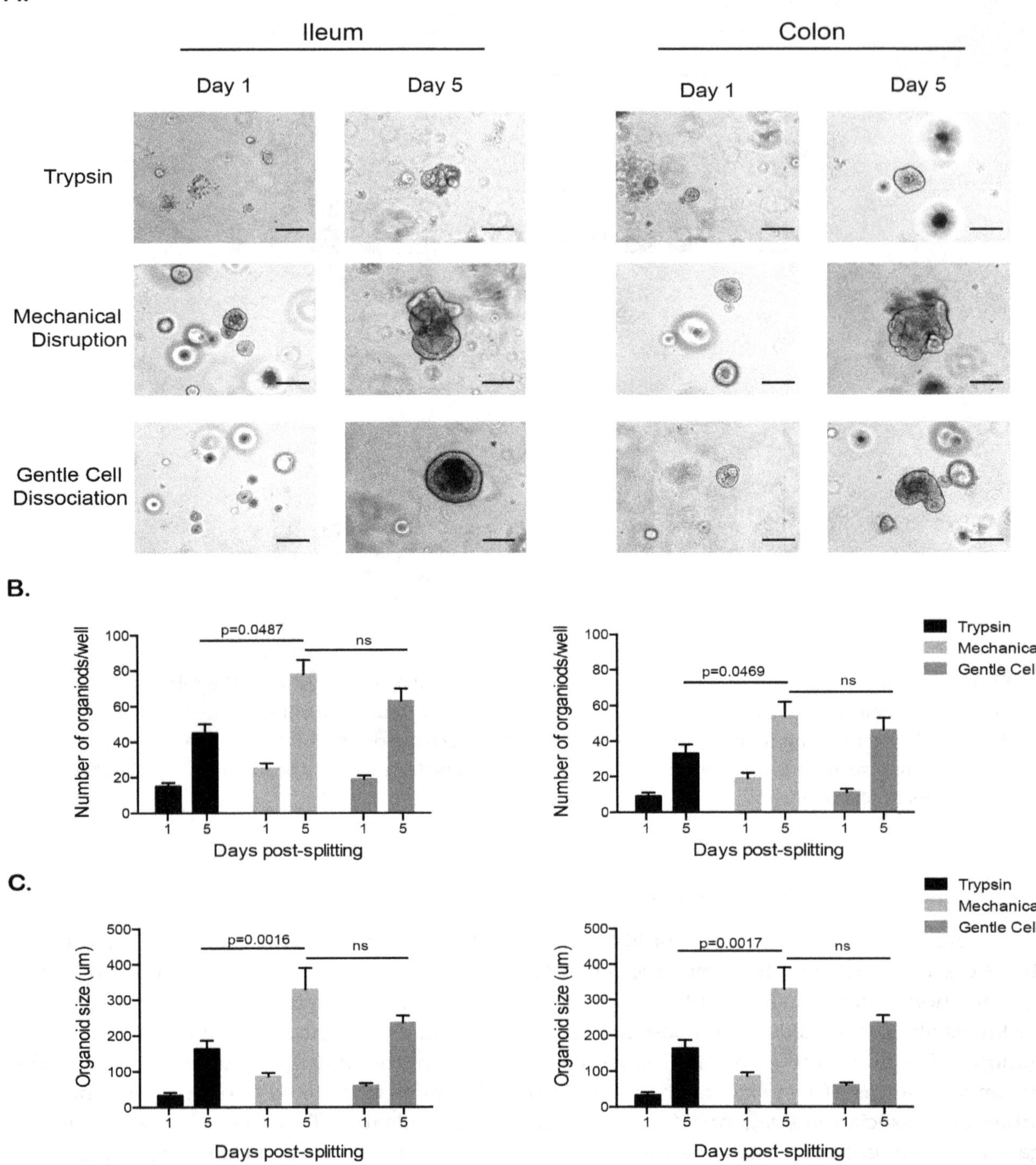

Figure 3. Mechanical disruption is the preferred method for passaging feline organoids. (**A–C**). Feline ileum- and colon-derived organoids were grown for 10 days post-harvesting prior to passaging. Three methods of passaging were used, and new organoids were followed over five days. (**A**). Bright field images of ileum and colon organoids. Representative images are shown. N = 3 donors and each donor were followed overtime as a triplicate series. Scale bar = 100 μm (**B**). The number of organoids from each passing method was counted over the indicated time course. Results are mean +/− s.d. (n = 3 donors). (**C**). The size of the organoids was determined using a Nikon Eclipse Ti-S. Results are mean +/− s.d. (n = 3 donors).

Upon isolation of feline ileum and colon crypts, organoids were maintained in a media based upon one that is commonly used to support the growth of human organoids [39,43,44]. To determine if additional media compositions would support or enhance feline intestinal organoid growth, we tested three different formulations normally used to support human intestinal and colon organoids, murine small intestine or murine colon (see material and methods) [37–40,45]. Feline ileum and colon organoids were passaged using mechanical disruption and were maintained in each of the media conditions. Ten days post passaging, the number of organoids and the size of the organoids were counted for each condition. Results show that the human media condition supported a greater number and a larger size of both ileum and colon organoids (Table 3). Additionally, to determine how many passages each media type could support, organoids were split using mechanical dissociation and followed over time. The mouse intestine media did not support long term growth of either ileum or colon organoids and both types of organoids stopped growing and died within two passages in this media (Table 3). The mouse colon media supported longer passaging than its intestine counterpart, however this was still limited to a few passages. The human media was the only one that supported longer term passaging. However, unlike human and mouse organoids [38–40], both the feline ileum and colon organoids became arrested after 14–15 passages (Table 3). This was reproduced with several animals (n = 6) and was consistent between animals suggesting that an additional media component may be required to support culturing to the extent that human and murine organoids can reach. Importantly, it seems that Wnt3A is a critical component required for the maintenance of feline organoids as the mouse intestine media, which lacks Wnt3A, supported the fewest number of passages. Overall, we could see that feline intestinal and colon organoids could be maintained using mechanical disruption and Wnt3A containing media. In recent years, intestine- and colon-derived organoids have been generated from both large farm animals (bovine and porcine) [46] and domestic animals (canine) [47,48]. Bovine, porcine and canine intestinal and colon organoids have been shown to require Wnt3a for their generation and growth [46–48]. Additionally, while many of these studies did not compare different passaging techniques, they often used mechanical disruption as a common method to passage and maintain the organoids [46,47].

Table 3. Optimization of feline ileum and colon culturing conditions. The number and size of organoids were measured 10 days post-splitting for every other passage.

	Organoid	Avg # of Organoids/Well	Avg. Size of Organoids	No of Passages
Human Media	Colon	83 +/− 11	384.2 +/− 31.5	15
Mouse Intestine Media	Colon	54 +/− 6	312.5 +/− 23.6	2
Mouse Colon Media	Colon	68 +/− 9	359.4 +/− 29.1	6
Human Media	Ileum	68 +/− 9	346.2 +/− 24.8	14
Mouse Intestine Media	Ileum	49 +/− 6	297.6 +/− 21.3	2
Mouse Colon Media	Ileum	53 +/− 7	318.9 +/− 35.4	4

3.2. *Infection of Feline Intestinal Organoids*

Currently, the molecular mechanism by which FECVs can persist in the gut could not been studied due to the lack of cell culture models supporting the replication of FECV field viruses. To evaluate if the feline intestinal and colon organoid system could support FECV infection, two different FCoV viruses (recFECV-GFP and recFECV-S79-GFP) were constructed for this study (Figure 4A). recFECV-GFP is a recombinant FCoV with the entire genome sequence of a serotype FECV field strain while recFECV-S79-GFP contains the genomic backbone of the same serotype I FECV field strains with the spike protein gene of cell-culture-adapted serotype II laboratory strain 79–1146. To determine the ability of recFECV-GFP and recFECV-S79-GFP to replicate in cell culture, these recombinant viruses were first tested on FCWF cells. Inoculation of FCWF cells with recFECV-GFP, did not lead to any formation of cytopathic effect (CPE) or fluorescence as standard cell culture systems do not support the growth of serotype I FCoV field strains (Figure 4B), while as expected, recFECV-S79-GFP growth was supported in FCWF cells demonstrated by their ability to form a characteristic plaque phenotype accompanied by green fluorescence (Figure 4B).

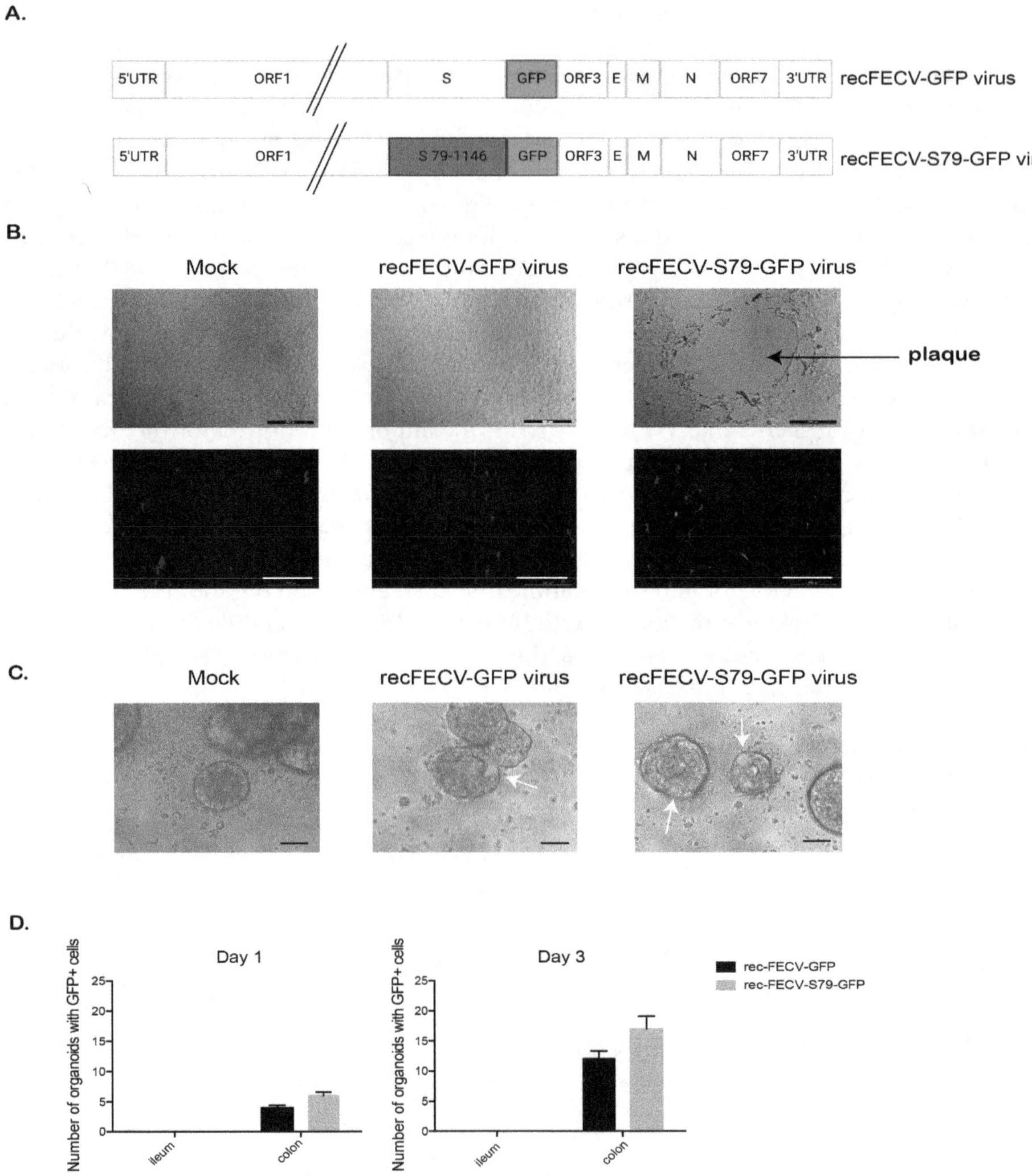

Figure 4. Feline colon organoids support infection of feline coronavirus (FCOV). (**A**). Diagram depicting the newly generated feline enteric coronavirus (FECV) recombinant viruses. (**B**). Felis catus whole fetus (FCWF) feline cells were infected with rec-FECV-GFP or recFECV-S79-GFP viruses and their ability to express GFP and produce plaques was monitored 48 h post-infection. Mock samples indicate media alone conditions. Representative images are shown. Scale bar is 200 μm. (**C**). Feline colon organoids infected with rec-FECV-GFP or recFECV-S79-GFP for 72 h. Representative images shown. White arrow heads indicate GFP positive cells. Scale bar = 100 μm (**D**). Feline ileum and colon organoids (passage 3) were infected with rec-FECV-GFP or recFECV-S79-GFP viruses. Mock samples indicate media alone conditions. Infections were monitored 1 and 3 days post-infection by wide field microscopy. Results are mean +/− s.d. n = 3 donors for each infection.

Previously, Desmarets et al. (2013) [49] established permanent feline intestinal epithelial cell cultures of ileocyte and colonocyte origin that were shown to sustain propagation of FECV field strains in vitro. However, such valuable cell lines are not broadly available to the scientific community. To test if the established feline ileum and colon organoids could be used to support feline coronavirus infection, five days post-passaging, organoids were removed from the Matrigel and incubated with feline coronavirus recFECV-GFP and recFECV-S79-GFP and followed over 72 h. Unexpectedly, no fluorescence was obtained upon infection of ileum organoids with either recFECV-GFP or recFECV-S79-GFP (Figure 4D). The lack of fluorescence of ileum organoids upon infection with both viruses is somewhat surprising, since ileum has been described as a tissue supporting FCoV-growth in vivo [50]. Whether the lack of green fluorescence of ileum organoids is due to (i) a very low viral replication rate and GFP expression which is below the detection limit or (ii) the complete lack of virus infection, cannot be ruled out in the current experimental setup. For example, luciferase-expressing recFECVs might be more suitable tools to further investigate these hypotheses. The low/no infection of ileum organoids with the recombinant viruses may reflect the limitation of ileum organoids to study the biology of serotype I FECVs in vitro. In contrast, both recombinant viruses were found to infect feline colon organoids (Figure 4C,D). One day post-infection, GFP positive cells could be detected and the number of infected cells increased over three days for both viruses (Figure 4C,D). As opposed to the FCWF cells, our feline colon organoids supported infection of recFECV-GFP. Since the number of GFP positive cells increased over time, this data strongly suggests that colon organoids support initial infection, production of de novo virus and spreading of serotype I FECV field strain-infection.

In summary, we propose that feline colon-derived organoids represent a primary intestinal cell model supporting serotype I field stain FECV infection. These cultures now open the doors for researchers to unravel the molecular mechanisms leading to FECV persistence and possibly FECV-FIPV conversion. The here described protocol provides the community with a step-by-step approach to generate feline colon-derived organoids which provides a solution for the lack of easily available culture models. Furthermore, as primary intestinal epithelium cells are often more immune-responsive than their immortalize counterparts, as such, the here described organoid model will allow the community to better study host/pathogen interactions [37] and immune response [40,51] against FCoV.

Author Contributions: M.L.S. designed the methods, prepared the organoids, and analyzed the results, R.E. constructed the GFP expressing virus. G.T. and S.B. contributed to the concept of the study and critical discussions. All authors have read and agreed to the published version of the manuscript.

References

1. Pedersen, N.C. A review of feline infectious peritonitis virus infection: 1963–2008. *J. Feline Med. Surg.* **2009**, *11*, 225–258. [CrossRef] [PubMed]
2. Addie, D.D.; Schaap, I.A.T.; Nicolson, L.; Jarrett, O. Persistence and transmission of natural type I feline coronavirus infection. *J. Gen. Virol.* **2003**, *84*, 2735–2744. [CrossRef] [PubMed]
3. Pedersen, N.C. Serologic studies of naturally occurring feline infectious peritonitis. *Am. J. Vet. Res.* **1976**, *37*, 1449–1453.
4. Sparkes, A.H.; Gruffydd-Jones, T.J.; Harbour, D.A. Feline coronavirus antibodies in UK cats. *Vet. Rec.* **1992**, *131*, 223–224. [CrossRef]
5. Kipar, A.; May, H.; Menger, S.; Weber, M.; Leukert, W.; Reinacher, M. Morphologic features and development of granulomatous vasculitis in feline infectious peritonitis. *Vet. Pathol.* **2005**, *42*, 321–330. [CrossRef] [PubMed]
6. Weiss, R.C.; Scott, F.W. Pathogenesis of feline infetious peritonitis: Pathologic changes and immunofluorescence. *Am. J. Vet. Res.* **1981**, *42*, 2036–2048.

7. Weiss, R.C.; Scott, F.W. Pathogenesis of feline infectious peritonitis: Nature and development of viremia. *Am. J. Vet. Res.* **1981**, *42*, 382–390.
8. Pedersen, N.C. Morphologic and physical characteristics of feline infectious peritonitis virus and its growth in autochthonous peritoneal cell cultures. *Am. J. Vet. Res.* **1976**, *37*, 567–572.
9. Tekes, G.; Thiel, H.J. Feline Coronaviruses: Pathogenesis of Feline Infectious Peritonitis. *Adv. Virus Res.* **2016**, *96*, 193–218. [CrossRef]
10. Pedersen, N.C.; Liu, H.; Gandolfi, B.; Lyons, L.A. The influence of age and genetics on natural resistance to experimentally induced feline infectious peritonitis. *Vet. Immunol. Immunopathol.* **2014**, *162*, 33–40. [CrossRef]
11. Pedersen, N.C. An update on feline infectious peritonitis: Diagnostics and therapeutics. *Vet. J.* **2014**, *201*, 133–141. [CrossRef] [PubMed]
12. Pedersen, N.C. An update on feline infectious peritonitis: Virology and immunopathogenesis. *Vet. J.* **2014**, *201*, 123–132. [CrossRef] [PubMed]
13. Kim, Y.; Liu, H.; Galasiti Kankanamalage, A.C.; Weerasekara, S.; Hua, D.H.; Groutas, W.C.; Chang, K.O.; Pedersen, N.C. Reversal of the Progression of Fatal Coronavirus Infection in Cats by a Broad-Spectrum Coronavirus Protease Inhibitor. *PLoS Pathog.* **2016**, *12*, e1005531. [CrossRef]
14. Dickinson, P.J.; Bannasch, M.; Thomasy, S.M.; Murthy, V.D.; Vernau, K.M.; Liepnieks, M.; Montgomery, E.; Knickelbein, K.E.; Murphy, B.; Pedersen, N.C. Antiviral treatment using the adenosine nucleoside analogue GS-441524 in cats with clinically diagnosed neurological feline infectious peritonitis. *J. Vet. Intern. Med.* **2020**, *34*, 1587–1593. [CrossRef] [PubMed]
15. Perera, K.D.; Rathnayake, A.D.; Liu, H.; Pedersen, N.C.; Groutas, W.C.; Chang, K.O.; Kim, Y. Characterization of amino acid substitutions in feline coronavirus 3C-like protease from a cat with feline infectious peritonitis treated with a protease inhibitor. *Vet. Microbiol.* **2019**, *237*, 108398. [CrossRef] [PubMed]
16. Pedersen, N.C.; Perron, M.; Bannasch, M.; Montgomery, E.; Murakami, E.; Liepnieks, M.; Liu, H. Efficacy and safety of the nucleoside analog GS-441524 for treatment of cats with naturally occurring feline infectious peritonitis. *J. Feline Med. Surg.* **2019**, *21*, 271–281. [CrossRef]
17. Murphy, B.G.; Perron, M.; Murakami, E.; Bauer, K.; Park, Y.; Eckstrand, C.; Liepnieks, M.; Pedersen, N.C. The nucleoside analog GS-441524 strongly inhibits feline infectious peritonitis (FIP) virus in tissue culture and experimental cat infection studies. *Vet. Microbiol.* **2018**, *219*, 226–233. [CrossRef]
18. Pedersen, N.C.; Kim, Y.; Liu, H.; Galasiti Kankanamalage, A.C.; Eckstrand, C.; Groutas, W.C.; Bannasch, M.; Meadows, J.M.; Chang, K.O. Efficacy of a 3C-like protease inhibitor in treating various forms of acquired feline infectious peritonitis. *J. Feline Med. Surg.* **2018**, *20*, 378–392. [CrossRef]
19. Hohdatsu, T.; Sasamoto, T.; Okada, S.; Koyama, H. Antigenic analysis of feline coronaviruses with monoclonal antibodies (MAbs): Preparation of MAbs which discriminate between FIPV strain 79-1146 and FECV strain 79-1683. *Vet. Microbiol.* **1991**, *28*, 13–24. [CrossRef]
20. Hohdatsu, T.; Okada, S.; Koyama, H. Characterization of monoclonal antibodies against feline infectious peritonitis virus type II and antigenic relationship between feline, porcine, and canine coronaviruses. *Arch. Virol.* **1991**, *117*, 85–95. [CrossRef]
21. Pedersen, N.C.; Black, J.W.; Boyle, J.F.; Evermann, J.F.; McKeirnan, A.J.; Ott, R.L. Pathogenic differences between various feline coronavirus isolates. *Adv. Exp. Med. Biol.* **1984**, *173*, 365–380. [CrossRef] [PubMed]
22. Decaro, N.; Buonavoglia, C. An update on canine coronaviruses: Viral evolution and pathobiology. *Vet. Microbiol.* **2008**, *132*, 221–234. [CrossRef] [PubMed]
23. Herrewegh, A.A.; Smeenk, I.; Horzinek, M.C.; Rottier, P.J.; de Groot, R.J. Feline coronavirus type II strains 79-1683 and 79-1146 originate from a double recombination between feline coronavirus type I and canine coronavirus. *J. Virol.* **1998**, *72*, 4508–4514. [CrossRef] [PubMed]
24. Lin, C.N.; Chang, R.Y.; Su, B.L.; Chueh, L.L. Full genome analysis of a novel type II feline coronavirus NTU156. *Virus Genes* **2013**, *46*, 316–322. [CrossRef]
25. Terada, Y.; Matsui, N.; Noguchi, K.; Kuwata, R.; Shimoda, H.; Soma, T.; Mochizuki, M.; Maeda, K. Emergence of pathogenic coronaviruses in cats by homologous recombination between feline and canine coronaviruses. *PLoS ONE* **2014**, *9*, e106534. [CrossRef]
26. Tekes, G.; Spies, D.; Bank-Wolf, B.; Thiel, V.; Thiel, H.J. A reverse genetics approach to study feline infectious peritonitis. *J. Virol.* **2012**, *86*, 6994–6998. [CrossRef]

27. Tekes, G.; Hofmann-Lehmann, R.; Stallkamp, I.; Thiel, V.; Thiel, H.J. Genome organization and reverse genetic analysis of a type I feline coronavirus. *J. Virol.* **2008**, *82*, 1851–1859. [CrossRef]
28. Tekes, G.; Hofmann-Lehmann, R.; Bank-Wolf, B.; Maier, R.; Thiel, H.J.; Thiel, V. Chimeric feline coronaviruses that encode type II spike protein on type I genetic background display accelerated viral growth and altered receptor usage. *J. Virol.* **2010**, *84*, 1326–1333. [CrossRef]
29. Thiel, V.; Thiel, H.J.; Tekes, G. Tackling feline infectious peritonitis via reverse genetics. *Bioengineered* **2014**, *5*, 396–400. [CrossRef]
30. Haijema, B.J.; Volders, H.; Rottier, P.J. Switching species tropism: An effective way to manipulate the feline coronavirus genome. *J. Virol.* **2003**, *77*, 4528–4538. [CrossRef]
31. Ettayebi, K.; Crawford, S.E.; Murakami, K.; Broughman, J.R.; Karandikar, U.; Tenge, V.R.; Neill, F.H.; Blutt, S.E.; Zeng, X.L.; Qu, L.; et al. Replication of human noroviruses in stem cell-derived human enteroids. *Science* **2016**, *353*, 1387–1393. [CrossRef] [PubMed]
32. Pyrc, K.; Sims, A.C.; Dijkman, R.; Jebbink, M.; Long, C.; Deming, D.; Donaldson, E.; Vabret, A.; Baric, R.; van der Hoek, L.; et al. Culturing the unculturable: Human coronavirus HKU1 infects, replicates, and produces progeny virions in human ciliated airway epithelial cell cultures. *J. Virol.* **2010**, *84*, 11255–11263. [CrossRef]
33. Sato, T.; Clevers, H. Growing self-organizing mini-guts from a single intestinal stem cell: Mechanism and applications. *Science* **2013**, *340*, 1190–1194. [CrossRef] [PubMed]
34. Ehmann, R.; Kristen-Burmann, C.; Bank-Wolf, B.; Konig, M.; Herden, C.; Hain, T.; Thiel, H.J.; Ziebuhr, J.; Tekes, G. Reverse Genetics for Type I Feline Coronavirus Field Isolate to Study the Molecular Pathogenesis of Feline Infectious Peritonitis. *MBio* **2018**, *9*. [CrossRef] [PubMed]
35. Mettelman, R.C.; O'Brien, A.; Whittaker, G.R.; Baker, S.C. Generating and evaluating type I interferon receptor-deficient and feline TMPRSS2-expressing cells for propagating serotype I feline infectious peritonitis virus. *Virology* **2019**, *537*, 226–236. [CrossRef] [PubMed]
36. O'Brien, A.; Mettelman, R.C.; Volk, A.; André, N.M.; Whittaker, G.R.; Baker, S.C. Characterizing replication kinetics and plaque production of type I feline infectious peritonitis virus in three feline cell lines. *Virology* **2018**, *525*, 1–9. [CrossRef] [PubMed]
37. Stanifer, M.L.; Mukenhirn, M.; Muenchau, S.; Pervolaraki, K.; Kanaya, T.; Albrecht, D.; Odendall, C.; Hielscher, T.; Haucke, V.; Kagan, J.C.; et al. Asymmetric distribution of TLR3 leads to a polarized immune response in human intestinal epithelial cells. *Nat. Microbiol.* **2020**, *5*, 181–191. [CrossRef]
38. Sato, T.; Vries, R.G.; Snippert, H.J.; van de Wetering, M.; Barker, N.; Stange, D.E.; van Es, J.H.; Abo, A.; Kujala, P.; Peters, P.J.; et al. Single Lgr5 stem cells build crypt-villus structures in vitro without a mesenchymal niche. *Nature* **2009**, *459*, 262–265. [CrossRef]
39. Sato, T.; Stange, D.E.; Ferrante, M.; Vries, R.G.; Van Es, J.H.; Van den Brink, S.; Van Houdt, W.J.; Pronk, A.; Van Gorp, J.; Siersema, P.D.; et al. Long-term expansion of epithelial organoids from human colon, adenoma, adenocarcinoma, and Barrett's epithelium. *Gastroenterology* **2011**, *141*, 1762–1772. [CrossRef]
40. Pervolaraki, K.; Stanifer, M.L.; Munchau, S.; Renn, L.A.; Albrecht, D.; Kurzhals, S.; Senis, E.; Grimm, D.; Schroder-Braunstein, J.; Rabin, R.L.; et al. Type I and Type III Interferons Display Different Dependency on Mitogen-Activated Protein Kinases to Mount an Antiviral State in the Human Gut. *Front. Immunol.* **2017**, *8*, 459. [CrossRef]
41. Triana, S.; Stanifer, M.L.; Shahraz, M.; Mukenhirn, M.; Kee, C.; Ordoñez-Rueda, D.; Paulsen, M.; Benes, V.; Boulant, S.; Alexandrov, T. Single-cell transcriptomics reveals immune response of intestinal cell types to viral infection. *bioRxiv Preprint Serv. Biol.* 2020. Available online: https://doi.org/10.1101/2020.08.19.255893 (accessed on 19 August 2020).
42. Fujii, M.; Matano, M.; Toshimitsu, K.; Takano, A.; Mikami, Y.; Nishikori, S.; Sugimoto, S.; Sato, T. Human Intestinal Organoids Maintain Self-Renewal Capacity and Cellular Diversity in Niche-Inspired Culture Condition. *Cell Stem Cell* **2018**, *23*, 787–793. [CrossRef] [PubMed]
43. Koo, B.K.; Stange, D.E.; Sato, T.; Karthaus, W.; Farin, H.F.; Huch, M.; van Es, J.H.; Clevers, H. Controlled gene expression in primary Lgr5 organoid cultures. *Nat. Methods* **2011**, *9*, 81–83. [CrossRef]
44. Jung, P.; Sato, T.; Merlos-Suarez, A.; Barriga, F.M.; Iglesias, M.; Rossell, D.; Auer, H.; Gallardo, M.; Blasco, M.A.; Sancho, E.; et al. Isolation and in vitro expansion of human colonic stem cells. *Nat. Med.* **2011**, *17*, 1225–1227. [CrossRef] [PubMed]

45. Stanifer, M.L.; Kee, C.; Cortese, M.; Zumaran, C.M.; Triana, S.; Mukenhirn, M.; Kraeusslich, H.G.; Alexandrov, T.; Bartenschlager, R.; Boulant, S. Critical Role of Type III Interferon in Controlling SARS-CoV-2 Infection in Human Intestinal Epithelial Cells. *Cell Rep.* **2020**, *32*, 107863. [CrossRef] [PubMed]
46. Derricott, H.; Luu, L.; Fong, W.Y.; Hartley, C.S.; Johnston, L.J.; Armstrong, S.D.; Randle, N.; Duckworth, C.A.; Campbell, B.J.; Wastling, J.M.; et al. Developing a 3D intestinal epithelium model for livestock species. *Cell Tissue Res.* **2019**, *375*, 409–424. [CrossRef]
47. Kramer, N.; Pratscher, B.; Meneses, A.M.C.; Tschulenk, W.; Walter, I.; Swoboda, A.; Kruitwagen, H.S.; Schneeberger, K.; Penning, L.C.; Spee, B.; et al. Generation of Differentiating and Long-Living Intestinal Organoids Reflecting the Cellular Diversity of Canine Intestine. *Cells* **2020**, *9*, 822. [CrossRef]
48. Chandra, L.; Borcherding, D.C.; Kingsbury, D.; Atherly, T.; Ambrosini, Y.M.; Bourgois-Mochel, A.; Yuan, W.; Kimber, M.; Qi, Y.; Wang, Q.; et al. Derivation of adult canine intestinal organoids for translational research in gastroenterology. *BMC Biol.* **2019**, *17*, 33. [CrossRef]
49. Desmarets, L.M.; Theuns, S.; Olyslaegers, D.A.; Dedeurwaerder, A.; Vermeulen, B.L.; Roukaerts, I.D.; Nauwynck, H.J. Establishment of feline intestinal epithelial cell cultures for the propagation and study of feline enteric coronaviruses. *Vet. Res.* **2013**, *44*, 71. [CrossRef]
50. Kipar, A.; Meli, M.L.; Baptiste, K.E.; Bowker, L.J.; Lutz, H. Sites of feline coronavirus persistence in healthy cats. *J. Gen. Virol.* **2010**, *91*, 1698–1707. [CrossRef]
51. Pervolaraki, K.; Rastgou Talemi, S.; Albrecht, D.; Bormann, F.; Bamford, C.; Mendoza, J.L.; Garcia, K.C.; McLauchlan, J.; Hofer, T.; Stanifer, M.L.; et al. Differential induction of interferon stimulated genes between type I and type III interferons is independent of interferon receptor abundance. *PLoS Pathog.* **2018**, *14*, e1007420. [CrossRef]

6

Fluorescent TAP as a Platform for Virus-Induced Degradation of the Antigenic Peptide Transporter

Magda Wąchalska [1], Małgorzata Graul [1], Patrique Praest [2], Rutger D. Luteijn [2], Aleksandra W. Babnis [1], Emmanuel J. H. J. Wiertz [2], Krystyna Bieńkowska-Szewczyk [1] and Andrea D. Lipińska [1,*]

1 Laboratory of Virus Molecular Biology, Intercollegiate Faculty of Biotechnology, University of Gdańsk, Abrahama 58, 80–307 Gdańsk, Poland; magda.wachalska@phdstud.ug.edu.pl (M.W.); malgorzata.graul@biotech.ug.edu.pl (M.G.); aleksandra.babnis@gmail.com (A.W.B.); krystyna.bienkowska-szewczyk@biotech.ug.edu.pl (K.B.-S.)

2 Department of Medical Microbiology, University Medical Center Utrecht, Heidelberglaan 100, 3584CX Utrecht, The Netherlands; P.Praest-2@umcutrecht.nl (P.P.); rdluteijn@berkeley.edu (R.D.L.); e.wiertz@umcutrecht.nl (E.J.H.J.W.)

* Correspondence: andrea.lipinska@biotech.ug.edu.pl

Abstract: Transporter associated with antigen processing (TAP), a key player in the major histocompatibility complex class I-restricted antigen presentation, makes an attractive target for viruses that aim to escape the immune system. Mechanisms of TAP inhibition vary among virus species. Bovine herpesvirus 1 (BoHV-1) is unique in its ability to target TAP for proteasomal degradation following conformational arrest by the UL49.5 gene product. The exact mechanism of TAP removal still requires elucidation. For this purpose, a TAP-GFP (green fluorescent protein) fusion protein is instrumental, yet GFP-tagging may affect UL49.5-induced degradation. Therefore, we constructed a series of TAP-GFP variants using various linkers to obtain an optimal cellular fluorescent TAP platform. Mel JuSo (MJS) cells with CRISPR/Cas9 TAP1 or TAP2 knockouts were reconstituted with TAP-GFP constructs. Our results point towards a critical role of GFP localization on fluorescent properties of the fusion proteins and, in concert with the type of a linker, on the susceptibility to virally-induced inhibition and degradation. The fluorescent TAP platform was also used to re-evaluate TAP stability in the presence of other known viral TAP inhibitors, among which only UL49.5 was able to reduce TAP levels. Finally, we provide evidence that BoHV-1 UL49.5-induced TAP removal is p97-dependent, which indicates its degradation via endoplasmic reticulum-associated degradation (ERAD).

Keywords: TAP-GFP; fluorescent TAP platform; antigen presentation; MHC I; immune evasion; BoHV-1 UL49.5

---

1. Introduction

The co-existence of a host and a virus depends on a subtle balance between the pathogen replication and the host immune response. Virus-derived peptides, originating mainly from the proteasomal degradation, are presented by the major histocompatibility complex class I (MHC I) molecules, leading to the recognition of an infected cell by cytotoxic $CD8^+$ T lymphocytes (CTLs) (reviewed in [1]). The transporter associated with antigen processing (TAP) plays a pivotal role in MHC I-restricted antigen presentation, which makes it an attractive target for viruses that aim to escape the immune system.

TAP is a heterodimer belonging to the ATP-binding cassette (ABC) family transporters. It consists of two subunits, TAP1 (ABCB2) and TAP2 (ABCB3) [2]. The core of each subunit is formed by an N-terminally-located transmembrane domain (TMD), composed of six transmembrane helices

(TMs), responsible for peptide recognition and binding [3], and a highly conserved C-terminal nucleotide-binding domain (NDB), which can bind and hydrolyze ATP [4]. Acquiring both substrates, ATP and the peptide, occurs independently [3]. This induces conformational rearrangements within TAP, resulting in a switch from an inward-open to an outward-facing conformation and release of the peptide into the lumen of endoplasmic reticulum (ER). Afterward, ATP hydrolysis triggers the release of phosphate and restores the resting state of TAP [5]. The presence of core-flanking TMD0 domains (four TMs in TAP1 and three TMs in TAP2) at the N termini of TAP1/TAP2 is not necessary for peptide transport; however, it is crucial for assembly of the peptide-loading complex (PLC) and subsequent exposure of antigenic peptides to CTLs [6].

During co-evolution with their hosts, several herpesviruses and a single known (to date) poxvirus have specialized in TAP inhibition via diverse mechanisms (reviewed in [7]). Herpes simplex virus 1 and 2 (HSV-1 and HSV-2) encode the ICP47 protein, which competes for the peptide-binding site and, through its characteristic structure, tethers the TAP-ICP47 complex in an inward-facing conformation [8–10]. In contrast, the US6 protein of human cytomegalovirus (HCMV) [11–13] and the cowpox virus (CPXV) strain Brighton Red-encoded CPXV012 protein can inhibit ATP binding to NDBs while leaving peptide binding unaffected [14–16]. Mechanisms of TAP inhibition by herpesvirus UL49.5 protein family encoded by members of the *Varicellovirus* genus are still not fully understood and seem to differ in detail between virus species. Most of the UL49.5 orthologs inhibit conformational rearrangements within TAP [17]. Bovine herpesvirus 1 (BoHV-1) UL49.5 seems to be unique in its ability to target bovine, human, and murine TAP for proteasomal degradation following the conformational arrest [7,18,19]. Varicella-zoster virus (VZV)-encoded UL49.5 can bind TAP, yet it exhibits no inhibitory properties [20]. TAP degradation activity was also described for the murine gammaherpesvirus-68 MK3 protein [21] and the rodent herpesvirus Peru pK3 ortholog [22], which both carry a cytoplasmic RING (really interesting new gene) finger domain and can act towards the murine transporter. The recently described poxvirus molluscum contagiosum virus MC80 protein can destabilize human TAP; however, in contrast to BoHV-1 UL49.5, the transporter is not the primary target of the inhibitor [23].

Recently, fluorescent tags and gene fusion technology have become indispensable in a wide range of biochemical and cell biology applications, nevertheless in some circumstances designing a functional fluorescent fusion protein remains challenging. Numerous studies have shown that a choice of a linker may have a significant impact on proper folding, yield, and functionality of the fusion protein and its interaction with other proteins. Flexible linkers are usually applied to provide a certain degree of movement, while rigid linkers are preferable to separate two bioactive domains spatially [24].

To investigate the mechanism of TAP inhibition or removal, a TAP-GFP (green fluorescent protein) fusion protein was instrumental, yet GFP-tagging was observed to abolish the susceptibility of TAP to degradation induced by the BoHV-1-encoded UL49.5 [18]. Here, we report the construction of a series of full-length TAP1 and TAP2 variants carrying either N- or C-terminal GFP with different types of linkers and evaluate the impact of the TAP-GFP fusion design on their fluorescence and functionality, as well as susceptibility to virus-induced inhibition and degradation. Such a fluorescent TAP platform may constitute a platform to explain the molecular mechanism of UL49.5 activity and potentially contribute to better characterization of the transporter itself.

2. Materials and Methods

2.1. Cells and Viruses

Madin-Darby bovine kidney (MDBK) cells (ATCC, Manassas, VA, USA, CCL-22), human melanoma Mel JuSo (MJS) cells, MJS TAP1 CRISPR/Cas9 knock-out (TAP1 KO), MJS TAP2 CRISPR/Cas9 knock-out (TAP2 KO) [25], and U937 (ATCC, CRL-1593) were cultured in RPMI 1640 (Corning, Corning, NY, USA) supplemented with 10% fetal bovine serum (FBS, Thermo Scientific (Thermo Scientific, Waltham, MA, USA)) and Antibiotic Antimycotic Solution (Thermo Scientific). Lenti-X HEK293T and GP2-293 cells (both from Takara/Clontech, Kusatsu, Japan) used for lentivirus and retrovirus production, respectively, were cultured in Iscove's modified Dulbecco's medium (IMDM, Lonza, Basel, Switzerland) supplemented as above. HEK293T (ATCC, CRL-3216) cells were cultured in Dulbecco's modified Eagle's medium (DMEM, high glucose, Lonza) supplemented as above. BoHV-1 field strain Lam (Institute for Animal Health and Science, Lelystad, The Netherlands) was propagated and titrated on MDBK cells.

2.2. DNA Constructs

All TAP constructs were cloned in lentiviral vectors downstream of an EF1α promoter.

For unmodified (wild-type, wt) TAP1 or TAP2 reconstitution, dual promoter lentiviral vectors described in [25] (pPuroR-GFP-TAP1 and pZeoR-mAmetrine-TAP2) were used. mAmetrine and marker GFP genes were removed from these vectors. Fragments of TAP1 and TAP2 sequences were ordered as synthetic genes designed for cloning in pEGFP-N3 or pEGFP-C1 (Takara/Clontech). For TAP1-N-GFP (TAP1 with the N-terminal GFP, random linker), TAP1-C-GFP (TAP1 with the C-terminal GFP, random linker), TAP2-N-GFP (TAP2 with the N-terminal GFP, random linker), and TAP2-C-GFP (TAP2 with the C-terminal GFP, random linker), fusion genes were re-cloned in the original lentiviral vectors. The amino acid sequences of random linkers resulting from the cloning procedure are depicted in Figure 1A. Fragments coding for TAP1 with helical linker sequences were ordered as synthetic genes designed for cloning in pEGFP-N3 or pEGFP-C1. TAP1-HN-GFP (TAP1 with the N-terminal GFP, helical linker) or TAP1-HC-GFP (TAP1 with the C-terminal GFP, helical linker) were re-cloned in the lentiviral vector pCDH-EF1α-MCS-(PGK-Puro) (System Biosciences, Palo Alto, CA, USA).

Genes coding for viral TAP inhibitors were cloned in retroviral vectors downstream of a retroviral promoter. The BoHV-1 UL49.5 gene was cloned from pLZRS-BoHV-1 UL49.5-IRES-GFP [18] in BamHI-EcoRI sites of pLZRS-IRES-ΔNGFR [26]. The VZV UL49.5 gene was amplified from the pLZRS-VZV UL49.5-IRES-GFP vector [20] using KOD Hot Start DNA polymerase (Merck, Darmstadt, Germany) and the following primers: forward 5′-CGGGATCCCACCATGGGATCAATTACC-3′ and reverse 5′-CCGGAATTCTTACCACGTGCTGCGTAATAC-3′. The PCR product was verified by DNA sequencing and introduced into BamHI and EcoRI sites of the pBABEpuro vector [27]. Synthetic genes encoding: myc-tagged HSV-1 ICP47 (Gene ID: 2703441), myc-tagged HCMV US6 (Gene ID: 3077555), or the FLAG-N-CPXV012 (Gene ID: 1485887) variantlacking six N-terminal amino acid residues [28] were introduced into BamHI and EcoRI restriction sites of pBABEpuro.

2.3. Retroviral and Lentiviral Transduction

For the production of recombinant lentiviruses, third-generation packaging vectors based on the pRSV-Rev and pCgpV plasmids (Cell Biolabs, San Diego, CA, USA), the obtained lentiviral expression vectors, and pCMV-VSV-G (Cell Biolabs) for pseudotyping were co-transfected into Lenti-X HEK293T cells using CalPhos mammalian transfection kit (Takara/Clontech). For recombinant retroviruses, a transfer plasmid (pBABEpuro-based or pLZRS-IRES-ΔNGFR-based) and pCMV-VSV-G were co-transfected into GP2-293 packaging cells as above. Twenty-four hours after transfection the medium was refreshed; for lentiviruses it was supplemented with 1 mM sodium butyrate (Sigma-Aldrich, Saint Louis, MO, USA). Virus-containing supernatants were collected after 48 h, concentrated with PEGit (System Biosciences), and used for transduction in the presence of 0.01 mg mL^{-1} polybrene (Sigma-Aldrich). MJS cells with TAP1 or TAP2 knock-outs were stably reconstituted with the wt or fluorescent TAP1 or TAP2 constructs using lentivirus vectors and cell-sorting for GFP- and MHC I-positive cells. The cells were subsequently transduced with a retrovirus coding for BoHV-1 UL49.5 and sorted for nerve growth receptor (NGFR)-positive cells or with a retrovirus coding for HSV-1 ICP47, HCMV US6, VZV UL49.5, or CPXV012, and selected with puromycin (2 μg mL^{-1}) (Sigma-Aldrich).

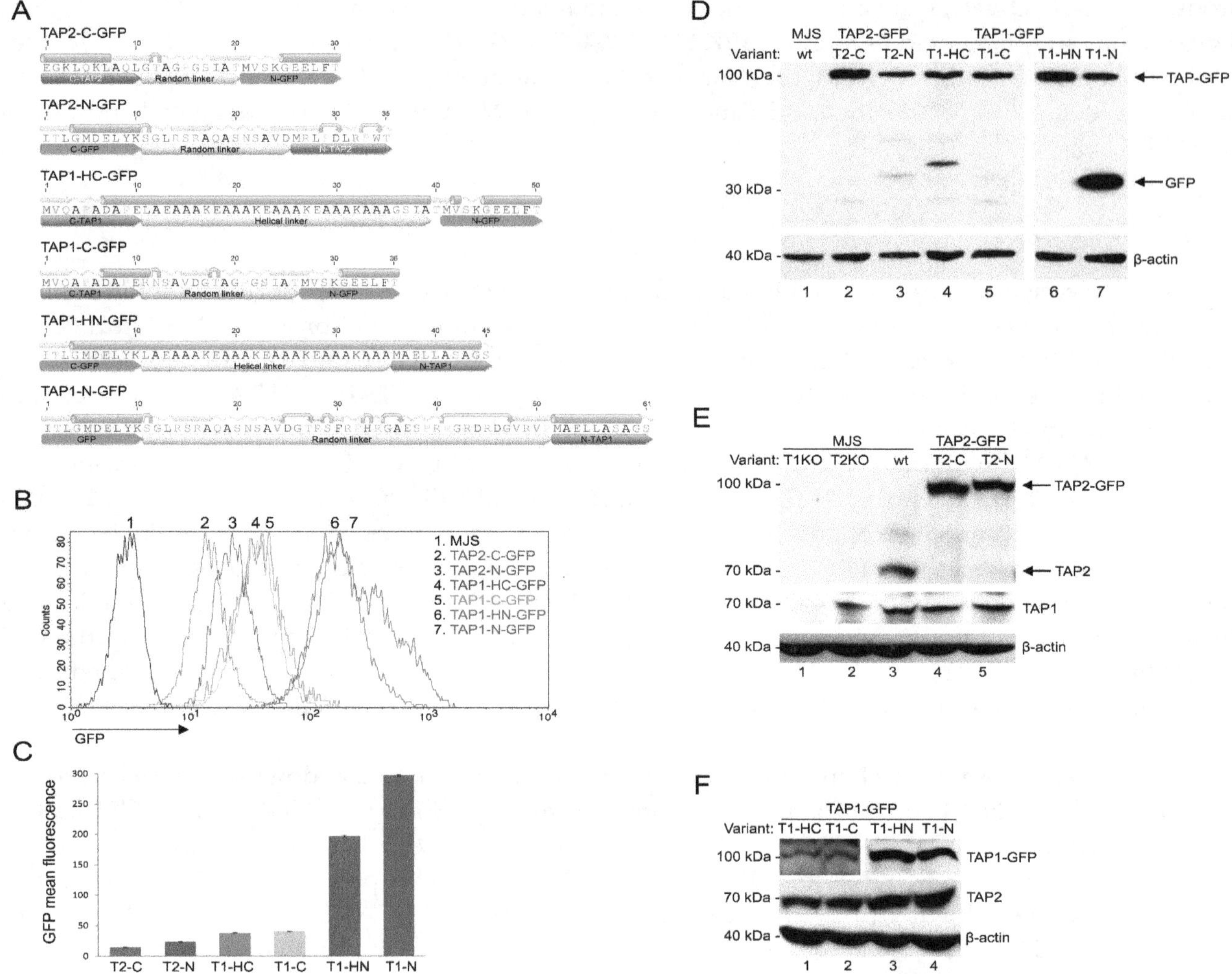

Figure 1. Construction and characterization of fluorescent transporter associated with antigen processing (TAP)-green fluorescent protein (GFP) variants. Mel JuSo (MJS) cells with CRISPR/Cas9 TAP1 or TAP2 knockouts (T1KO, T2KO) were stably reconstituted with fluorescent TAP1 or TAP2 constructs using lentivirus vectors and cell sorting. (**A**) Schematic representation of TAP-GFP constructs. Secondary structures of linkers flanked by ten amino acid residues of fused proteins were determined by the Geneious software; α-helices are depicted in pink, coiled regions in gray, β-strands in yellow, and turns in blue. (**B**) Representative histograms of TAP-GFP fluorescence intensity. (**C**) Comparative TAP-GFP analysis by flow cytometry. The mean fluorescence intensity of three independent measurements is represented as bars with standard deviations. The statistical significance was assessed by *t*-test; $p \leq 0.001$. (**D–F**) Expression of TAP-GFP variants in stable cell lines was determined by SDS-PAGE and immunoblotting using: (**D**) anti-GFP monoclonal antibody (Mab) (**E**) anti-TAP2 MAb (**F**) anti-TAP1 MAb. β-actin was used as a loading control. Abbreviations: T2-C: TAP2-C-GFP (TAP2 with the C-terminal GFP, random linker); T2-N: TAP2-N-GFP (TAP2 with the N-terminal GFP, random linker); T1-C: TAP1-C-GFP (TAP1 with the C-terminal GFP, random linker); T1-HC: TAP1-HC-GFP (TAP1 with the C-terminal GFP, helical linker); T1-N: TAP1-N-GFP (TAP1 with the N-terminal GFP, random linker); T1-HN: TAP1-HN-GFP (TAP1 with the N-terminal GFP, helical linker).

2.4. Plasmid Transfection

HEK293T cells were transfected with plasmids encoding fluorescent TAP variants using JetPRIME (Polyplus-transfection, Illkirch, France) according to the manufacturer's protocol and analyzed after 24 h by flow cytometry.

2.5. Generation of A TAP1/TAP2 Double Knock-Out U937 Cells for Reconstitution with Fluorescent TAP Variants

U937 TAP1/TAP2 KO cells were generated with a strategy described for MJS TAP1/TAP2 KO in [25] Briefly, U937 cells were transfected with pSico-CRISPR-PuroR containing the TAP2-targeting crRNA sequence 5′-GGAAGAAGAAGGCGGCAACG-3′. The cells were selected with puromycin (4 μg mL^{-1}), and cloned by limiting dilution. Individual clones were analyzed by flow cytometry to identify clones with low cell surface MHC I expression, followed by immunoblotting and DNA sequencing of the genomic target site. A clone lacking TAP2 was subsequently transfected with a pSicoR-CRISPR-PuroR vector containing the TAP1-targeting crRNA sequence 5′-GGGGTCCTCAGGGCAACGGT-3′. After selection with puromycin and cell cloning, the clones were analyzed for TAP1 expression by immunoblotting and DNA sequencing of the genomic target site. Genomic DNA sequence analysis revealed a 16-bp deletion around the TAP2 gRNA target site and multiple short deletions altering the whole TAP1 gene sequence downstream of the target site. A monoclonal cell line lacking TAP1 and TAP2 was used for reconstitution with a combination of unmodified and fluorescent TAP-encoding sequences delivered by lentivirus vectors. Reconstituted U937 cells were sorted for GFP and high MHC I expression. The cells were subsequently transduced with the BoHV-1 UL49.5-encoding retrovirus and sorted for NGFR.

2.6. Antibodies

Antibodies used for immunoblotting: mouse anti-TAP1 monoclonal antibody MAb 143.5 (kindly provided by R. Tampé, Institute of Biochemistry, The Johann Wolfgang Goethe University, Frankfurt, Germany); mouse anti-TAP2 MAb 435.3 (a kind gift from P. van Endert, INSERM U25, Institute Necker, Paris, France); rabbit anti-TAP1 (Enzo Life Sciences, Farmingdale, NY, USA); rat anti-GFP 3H9 (Chromotek, Planegg, Germany); mouse anti-myc tag 9B11 (Cell Signaling, Danvers, MA, USA); rabbit anti-β-actin (Novus Biologicals, Centennial, CO, USA); rabbit anti-β-catenin (Santa Cruz Biotechnology, Dallas, TX, USA); rabbit antibodies (H11) against a synthetic peptide derived from the N-terminal domain of BoHV-1 UL49.5 [26] and mouse anti-OctA (FLAG) G-8 (Santa Cruz Biotechnology); and mouse anti-HC10 [19] and rabbit anti-ERp57 H-220 (Santa Cruz Biotechnology). Probes used for immunofluorescence: Alexa 633-conjugated concanavalin A (ConA) (Thermo Scientific). Antibodies used for flow cytometry: mouse anti-MHC I W6/32 (Novus Biologicals); mouse anti-NGFR (Sigma-Aldrich); and Alexa 633-conjugated goat anti-mouse IgG (Thermo Scientific).

2.7. Flow Cytometry

Cell surface expression of MHC I was determined by indirect immunofluorescence using primary anti-MHC I antibodies (1:1000) and secondary antibodies (1:1000), all in phosphate buffered saline (PBS) buffer containing 1% bovine serum albumin and 0.02% sodium azide. For cell sorting, anti-NGFR antibodies (1:1000) and secondary antibodies were used. Cells were analyzed using a FACS Calibur flow cytometer (Becton Dickinson, Franklin Lakes, NJ, USA) and CellQuest software (version 5.2.1, Becton Dickinson)); for cell sorting, the sorting option of FACS Calibur was applied.

2.8. Immunoblotting and Immunoprecipitation

For immunoblotting, the cells were lysed in Cell Lytic M buffer (Sigma-Aldrich); for immunoprecipitation, the cells were lysed in a buffer containing 1% digitonin (Merck), 50 mM Tris-HCl, pH 7.5, 5 mM $MgCl_2$, and 150 mM NaCl. The buffers were supplemented with the cOmplete mini protease inhibitor cocktail (Roche, Basel, Switzerland). Cell lysates were analyzed by SDS-PAGE and immunoblotting as previously described [26] or incubated with GFP-Trap (Chromotek) according to the manufacturer's protocol to isolate protein complexes.

2.9. Peptide Transport Assay

The peptide transport assay was performed as described before [26]. Briefly, the cells were permeabilized with 2 IU mL^{-1} of Streptolysin O (Sigma-Aldrich) at 37 °C for 15 min. The cells (2×10^6 cells/sample) were subsequently incubated with 600 pmol of the fluorescein-conjugated synthetic peptide CVNKTERAY (JPT Peptide Technologies, Berlin, Germany) in the presence or absence of ATP (10 mM final concentration) at 37 °C for 10 min. Peptide translocation was terminated by adding 1 mL of ice-cold lysis buffer containing 1% Triton X-100. After 20 min of lysis, cell debris was removed by centrifugation, and supernatants were collected and incubated with 100 μL of concanavalin A (ConA)-Sepharose (Sigma-Aldrich) at 4 °C for 1 h to isolate the glycosylated peptides. After extensive washing of the beads, the peptides were eluted with elution buffer (500 mM mannopyranoside, 10 mM EDTA, 50 mM Tris-HCl, pH 8.0) by rigorous shaking for 1 h at 25 °C. Eluted peptides were separated from ConA by centrifugation at 12,000× g. The fluorescence intensity was measured using a fluorescence plate reader (EnVision, PerkinElmer, Waltham, MA, USA) with excitation and emission wavelengths of 485 nm and 530 nm, respectively.

2.10. Immunofluorescence

MJS cells were grown on microcover glass, fixed with 4% paraformaldehyde in PBS, permeabilized with 0.2% Triton X-100 in PBS, and stained with Alexa 633-ConA (1:1000), prepared in PBS containing 1% bovine serum albumin (Sigma-Aldrich). GFP booster (1:100, Chromotek) was used for MJS-TAP2-C-GFP to enhance the green fluorescence. The blue signal was electronically converted into the red during the analysis of images using Leica TCS SP8X confocal laser scanning microscope (Leica, Wetzlar, Germany).

3. Results

3.1. Construction and Characterization of Fluorescent TAP-GFP Variants

In order to develop a universal fluorescent platform for virus-induced TAP degradation, we constructed six versions of TAP-GFP fusion: four with two types of linkers, a random linker or a helical one, placed at the N- or C-terminus of TAP1, and two with a random linker at the N- or C-terminus of TAP2 (Figure 1A). A number of studies regarding fusion protein linker design have suggested that the most important features of a proper linker are its hydrophilicity and flexibility [24]. The random linkers used to join TAP and GFP have resulted from the cloning procedure into one of the pEGFP plasmid series. The analysis of their amino acid sequence revealed they were unstructured; thus, they should be more flexible. In some cases, flexible linkers may result in undesired interactions or interference between the fusion partners. In such cases, rigid linkers are preferable to separate two independently active domains spatially. An example of a rigid linker is the α helix-forming peptide AEAAAKEAAAKEAAAKA, stabilized by salt bridges between glutamate and lysine residues [29]. The distance between two separated domains can be regulated by changing the number of EAAAK motif repetitions. By using fluorescence resonance energy transfer (FRET) measurement, the helical linker with four repetitions of this motif has been demonstrated as the most efficient in separating two fluorescent proteins [30]. Therefore, this linker was selected for our studies to generate N- and C-terminal fusion of TAP1 and GFP. To introduce fluorescent TAP-GFP into human melanoma MJS cells, we first generated, by using CRISPR/Cas9-based technology, TAP1 or TAP2 knock-outs (KO). This enabled stable reconstitution of the cells with a fluorescent TAP1 or TAP2 using lentiviral vectors. The cells were subsequently sorted based on GFP to high purity (>98%). MJS cells have been shown to be permissive for BoHV-1 infection and are widely used in the research on modulation of antigen presentation by viruses [31–33].

First, we analyzed the GFP fluorescence intensity of our constructs (Figure 1B,C). Flow cytometry analysis revealed that N-terminal fusions of GFP to TAP1 exhibited the highest fluorescence, followed by TAP1 C-terminal fusion constructs. The type of linker seemed to have no crucial impact on the fluorescence intensity, although, for TAP1-N-GFP, a population of brighter green fluorescent cells could be observed when compared to TAP1-HN-GFP. TAP2 constructs exposed the lowest fluorescence, with a similar tendency of N-terminal fusion outperforming the C-terminal one. In addition, for TAP1-N-GFP, we could observe a more heterogeneous distribution of GFP fluorescence than for the other variants.

During the transient expression of fluorescent constructs in plasmid-transfected HEK293T cells, we could observe a similar range of GFP fluorescence, which indicates that differences in GFP fluorescence depend on the properties of individual fusion proteins rather than result from random incorporation of a lentivirus into a host genome (Figure S1). TAP2-C-GFP was performing noticeably better in transfected HEK293T than in the stable MJS cell line, and this time TAP1-HN-GFP slightly outperformed TAP1-N-GFP.

Next, we characterized the expression of TAP-GFP constructs by SDS-PAGE and immunoblotting. TAP1 and TAP2 have a similar apparent molecular mass of approximately 75 kDa. The fusion to GFP should yield a single protein at 100 kDa. All constructs were detected in cell lysates with anti-GFP antibodies at the predicted molecular weight of 100 kDa; only for TAP1-N-GFP did we observe an additional 30-kDa band corresponding to, most probably, cleaved free GFP or cleaved TAP-GFP fragment (Figure 1D). Fluorescent TAP could also be detected with anti-TAP1 or anti-TAP2 antibodies (Figure 1E,F). The reconstitution of TAP1KO cells with a fluorescent variant of TAP1 resulted in increased stability of endogenous TAP2 as it could be detected in higher amounts than in TAP1KO cells (compare lane 1 in Figure 1E with lanes 1–4 in Figure 1F). This is in agreement with a previous study reporting that unlike TAP1, TAP2 is unstable when expressed without the other half of the transporter [34]. The higher sensitivity of TAP2 can most probably explain why no band corresponding to TAP2 could be identified in MJS TAP1KO cells (Figure 1E, lane 1), while TAP1 could be easily detected in MJS TAP2KO cells (Figure 1E, lane 2).

3.2. TAP-GFP Localizes in the ER and Forms a Functional Transporter

TAP localizes in the cells predominantly in the ER [35]. To assess if the fluorescent TAP constructs acquired their proper localization, we stained the cells with Alexa 633-conjugated concanavalin A (ConA), an ER-cis Golgi marker, and analyzed fluorescence distribution by confocal laser scanning microscopy (Figure 2). In all cell lines, TAP-GFP (green) localized predominantly in the ER (red), as it co-localized with the ER marker (yellow). No granular localization patterns that would indicate TAP aggregates could be visualized.

Based on the currently available evidence, a lack of at least one fully functional TAP subunit results in a suppressed peptide translocation and production of empty or suboptimally loaded unstable MHC I molecules, mostly retained in the ER [36]. GFP fusion could potentially affect TAP structure and activity. Therefore, the functionality of fluorescent TAP constructs was tested by measuring cell surface expression of MHC I by flow cytometry (Figure 3). As expected, MJS cells with TAP1KO or TAP2KO exhibited strongly reduced levels of MHC I. To have a proper control for testing the effect of fluorescent tagging on TAP performance, especially during overexpression of a TAP subunit, we generated control cell lines reconstituted with unmodified TAP subunits (MJS TAP1KO + TAP1 and MJS TAP2KO + TAP2). Overexpression of individual TAP subunits might result in the formation of their homodimers, affecting the interpretation of our further experiments [19,37]. Therefore, as an additional control, we transduced TAPKO cells with vectors enabling overexpression of the existing half-transporter (MJS TAP1KO + TAP2 and MJS TAP2KO + TAP1). As expected, MHC I surface levels were reduced in those cells to a similar extent as in TAP1KO or TAP2KO cells. On the other hand, reconstitution of the missing TAP subunit with its fluorescent variant rescued surface MHC I to levels slightly higher than on parental ("wild-type") MJS cells, but similar to the ones observed on the cells reconstituted with the non-fluorescent TAP.

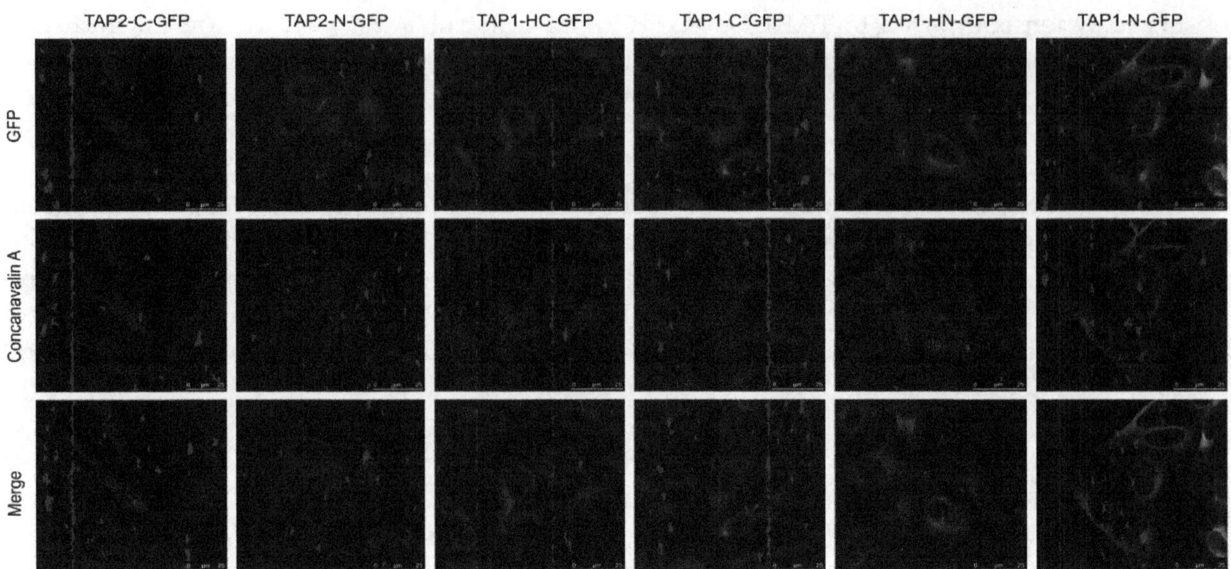

Figure 2. TAP-GFP variants demonstrate endoplasmic reticulum (ER) localization. MJS cells reconstituted with TAP-GFP variants were fixed, permeabilized, and stained with Alexa 633-conjugated concanavalin A (ER marker). Colocalization with GFP was analyzed using fluorescent confocal microscopy. For MJS TAP2-C-GFP, the GFP booster was used to enhance very weak green fluorescence.

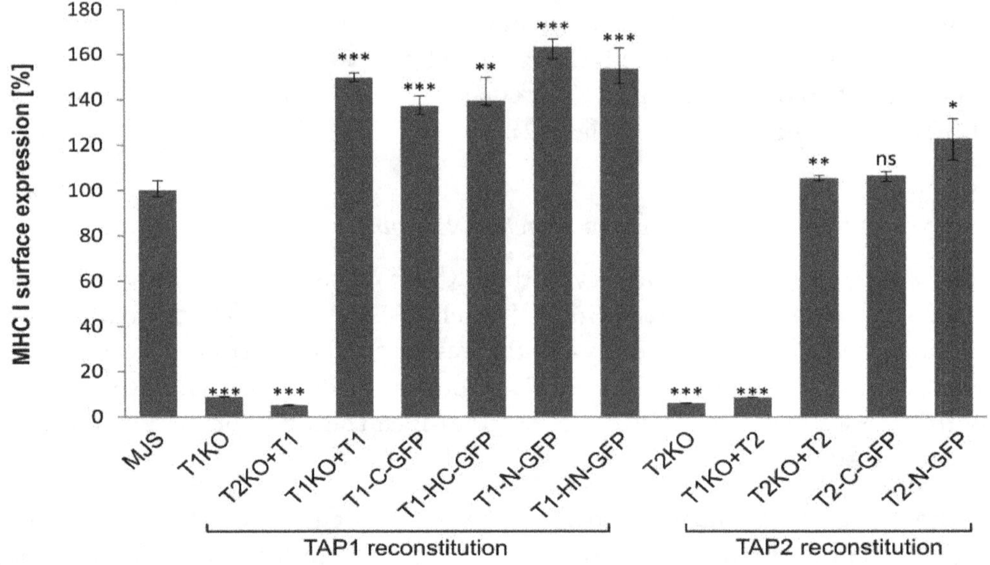

Figure 3. TAP-GFP forms a functional transporter. MJS cells with CRISPR/Cas9 TAP1 or TAP2 knockouts (T1KO or T2KO) were stably reconstituted with wild-type or fluorescent TAP1 or TAP2 variants. Surface expression of major histocompatibility complex class I (MHC I) was assessed by flow cytometry using specific antibodies (W6/32). MHC I expression on MJS cells with TAP reconstitution is presented as the percentage of MHC I mean fluorescence intensity on MJS cells (set as 100%). The analysis was performed in triplicates. The statistical significance of differences between MHC I on MJS cell with TAP reconstitution and MJS wild-type (wt) cells was estimated by *t*-test; *** $p \leq 0.001$, ** $p \leq 0.01$, * $p \leq 0.05$, ns: not significant.

3.3. *The Sensitivity of TAP-GFP Variant to UL49.5-Mediated MHC I Downregulation and TAP Degradation*

To examine the sensitivity of fluorescent TAP to BoHV-1-encoded UL49.5-mediated inhibition, UL49.5 was introduced to stable cell lines with TAP-GFP variants using a retrovirus vector. The cells were subsequently sorted for high purity based on the expression of truncated NGFR as a marker, and analyzed by flow cytometry for surface expression of MHC I (Figure 4A). According to the obtained data, all fluorescent transporters were prone to UL49.5-induced inhibition, which was illustrated as the downregulation of MHC I, to a similar extent for all the tested variants.

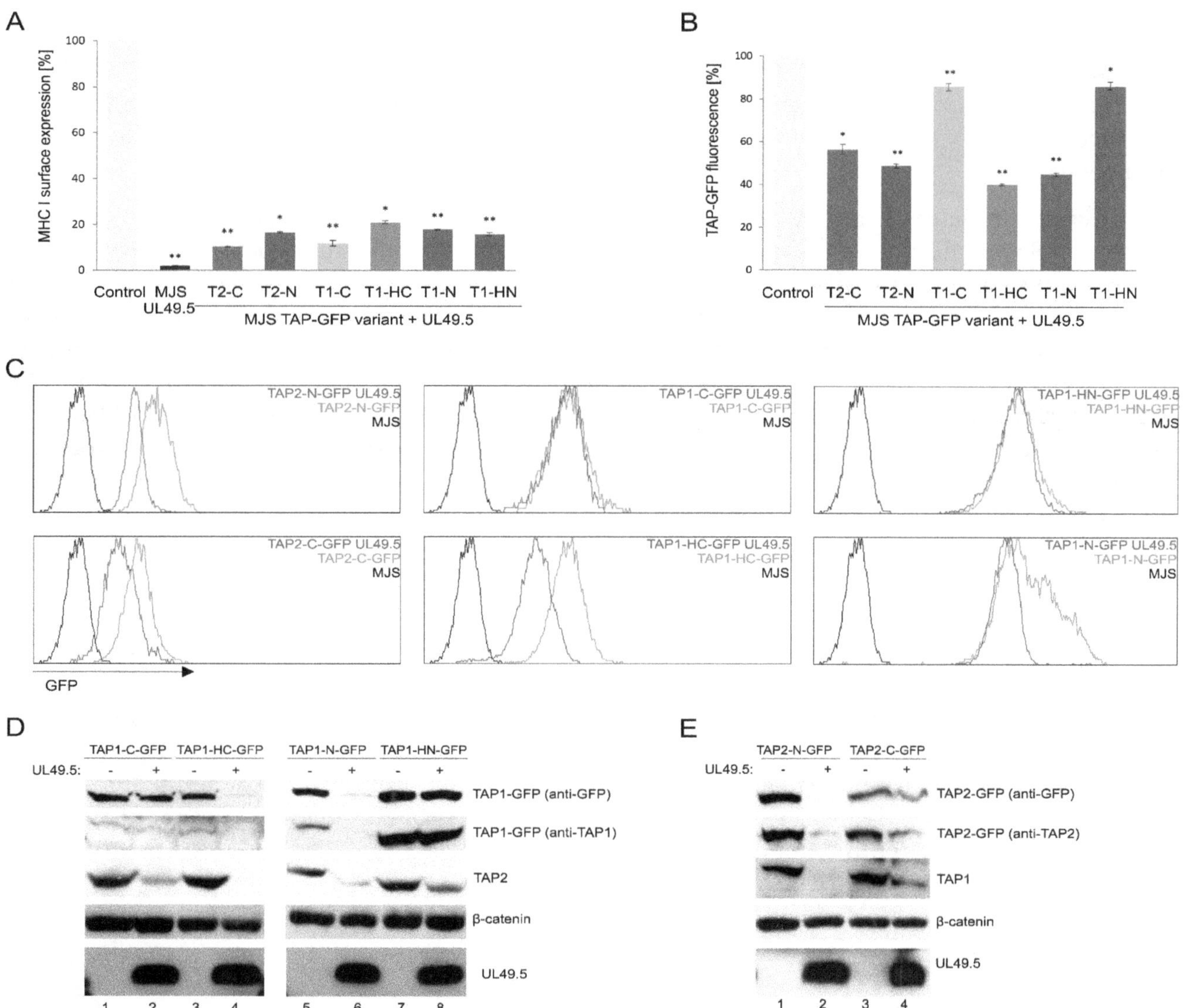

Figure 4. TAP-GFP variants differ in their sensitivity to UL49.5-mediated inhibition and degradation. MJS cells with fluorescent TAP1 or TAP2 variants were transduced with a retrovirus encoding bovine herpesvirus 1 (BoHV-1) UL49.5. (**A**) Surface expression of MHC I was assessed by flow cytometry using specific antibodies (W6/32). MHC I expression is presented as the percentage of mean fluorescence intensity; fluorescence of parental cells without UL49.5 was set as 100%. The analysis was performed in triplicates. The statistical significance of differences between MJS TAP-GFP and MJS TAP-GFP UL49.5 cell lines was estimated by *t*-test; ** $p \leq 0.001$ * $p \leq 0.005$. (**B**) GFP mean fluorescence intensity is presented as the percentage of GFP fluorescence of parental cells (set as 100%). The analysis was performed in triplicates. The statistical significance of differences between MJS TAP-GFP and MJS TAP-GFP UL49.5 cell lines was estimated by *t*-test; ** $p \leq 0.001$ * $p \leq 0.005$. (**C**) The effect of UL49.5 on GFP level in MJS TAP-GFP cells was assessed by flow cytometry. (**D**,**E**) Degradation of TAP-GFP variants in the presence of BoHV-1 UL49.5 in stable cell lines was determined by SDS-PAGE and immunoblotting using: anti-GFP, anti-TAP1, anti-TAP2, or anti-UL49.5 antibodies. β-catenin was used as a loading control.

Next, we analyzed the susceptibility of TAP-GFP variants to UL49.5-triggered degradation (Figure 4B,C). Flow cytometry analysis revealed a reduction of GFP mean fluorescence intensity by approximately 50% in the cells co-expressing UL49.5 with TAP1-N-GFP, TAP1-HC-GFP, and both TAP2 variants, in comparison to the control cells expressing the fluorescent transporter without UL49.5. There were no significant changes in GFP fluorescence observed in the cells expressing TAP1-C-GFP or TAP1-HN-GFP with UL49.5.

Susceptibility of TAP1-HC-GFP and TAP2-N-GFP constructs to UL49.5-dependent TAP inhibition and degradation was also confirmed in a reconstituted U937 cell line (Figure S2), where downregulation of surface MHC I, as well as a decrease of mean GFP fluorescence to the similar extent as in MJS cells, could be denoted.

To verify if changes in GFP fluorescence correspond with the decreased TAP protein level, immunoblotting analysis of cell lysates was performed (Figure 4D,E). A similar level of BoHV-1 UL49.5 protein could be observed in all cell lines. Decreased amounts of TAP1-N-GFP, TAP1-HC-GFP, and both TAP2-GFP fusion proteins could be detected in the presence of UL49.5, with the use of both anti-GFP and specific anti-TAP antibodies. Of note, in the case of non-degradable fluorescent TAP1 constructs (TAP1-HN-GFP and TAP1-C-GFP), the level of endogenous TAP2 was decreased (compare TAP2 in Figure 4D in lanes 1–2 and 5–6), what may suggest partial degradation of the untagged TAP subunit only. Degradation in TAP1-HC-GFP and TAP2-N-GFP cell lines was observed for both, exogenous fluorescent and endogenous TAP subunits. TAP1-N-GFP and TAP2-C-GFP characterized with only partial degradation.

3.4. *Interaction of TAP-GFP with UL49.5 and the Peptide-Loading Complex*

To assess the interaction of a non-degradable TAP with BoHV-1-encoded UL49.5 and selected components of PLC, we chose the TAP1-HN-GFP construct as demonstrating high expression. Among proteins co-immunoprecipitating with the fluorescent transporter (using GFP-Trap), we could identify UL49.5 and endogenous TAP2, as well as the known components of PLC: MHC I heavy chain and ERp57 (Figure 5). These results highlight the fact that UL49.5 is still capable of interacting with the non-degradable TAP1-HN-GFP. Unspecific binding of UL49.5 was excluded by immunoprecipitation from MJS wt cell lysate (with unmodified TAP) expressing UL49.5.

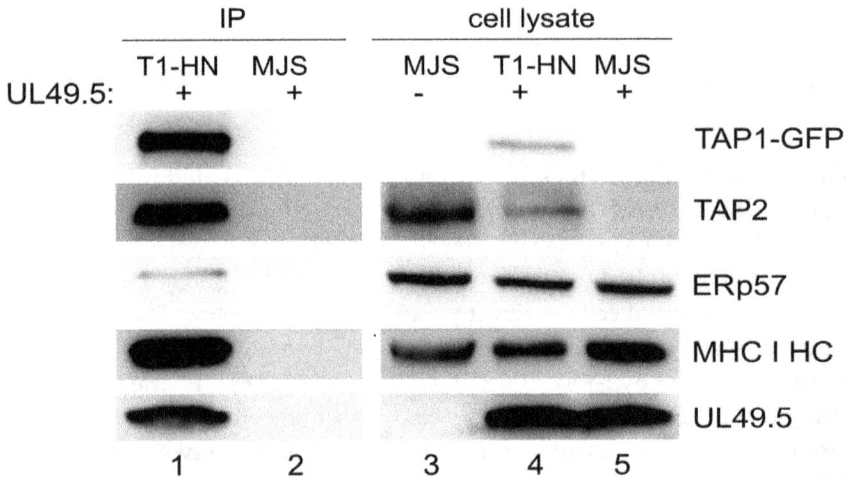

Figure 5. TAP1-GFP interacts with UL49.5 and the peptide-loading complex. TAP1-HN-GFP (T1-HN) was immunoprecipitated by GFP-Trap from lysates of MJS cells expressing TAP1-HN-GFP and UL49.5 or wt MJS with UL49.5 only. Co-precipitating proteins were analyzed by SDS-PAGE and immunoblotting using antibodies against GFP, TAP2, ERp57, MHC I HC, and UL49.5. Right panel: cell lysates were loaded on SDS-PAGE directly and analyzed by immunoblotting.

3.5. Application of the TAP2-N-GFP Variant as a Platform to Study BoHV-1 UL49.5 Activity in Virus-Infected Cells

In the more detailed studies on our fluorescent TAP platform, we focused on a single TAP-GFP variant, selecting TAP2-N-GFP. TAP1-N-GFP was excluded based on the presence of a free form of GFP. The constructs resistant to UL49.5-triggered degradation were eliminated as well, as was TAP2-C-GFP due to its very week, nearly undetectable basic green fluorescence and only partial degradation in the presence of UL49.5.

First, we confirmed that UL49.5-mediated MHC I downregulation relies on the inhibition of antigenic peptide translocation. TAP transport assay was performed in TAP2-N-GFP MJS cells and TAP2-N-GFP cells expressing UL49.5 or (as controls) in MJS wt, MJS TAP2KO cells, and MJS TAP2KO cells reconstituted with a non-fluorescent TAP2. The assay was based on cytoplasm-to-ER translocation of fluorescein-conjugated substrate peptides, in the presence of ATP or EDTA, as a passive diffusion control (Figure 6). Reconstitution with fluorescent TAP restored peptide transport when compared to the parental TAP2KO cells to a level of almost 50% higher than for MJS wt cells. This might result from high expression of exogenous TAP2 as a very similar transport activity was denoted for the non-fluorescent TAP2 reconstitution. The presence of UL49.5 inhibited peptide translocation to the level of TAP2KO cells.

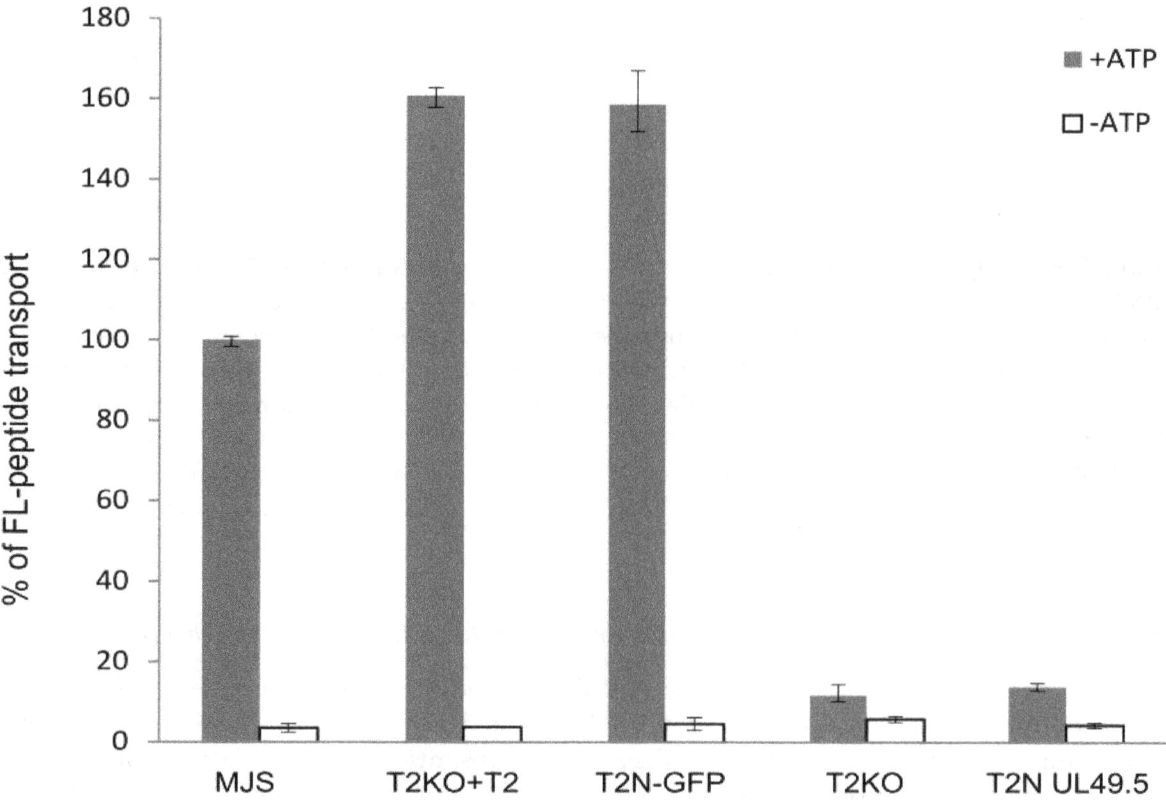

Figure 6. UL49.5-induced inhibition of peptide transport. MJS cells with CRISPR/Cas9 TAP2 knockout (T2KO) were stably reconstituted with wild-type TAP2 (T2KO+T2) or fluorescent TAP2 (T2N-GFP), subsequently transduced with a retrovirus encoding BoHV-1 UL49.5, and sorted (T2N UL49.5). Transport activity of TAP was analyzed using fluorescein-labeled peptide CVNKTERAY in the presence of ATP (gray bars) or EDTA (white bars). Peptide transport is expressed as a percentage of translocation, relative to the translocation observed in control MJS cells (set at 100%). The experiment was performed in triplicates. The statistical significance of differences between MJS controls and reconstituted or KO variants was estimated by *t*-test; for all the samples $p \leq 0.005$.

Finally, to test whether we can apply our fluorescent TAP platform to quantify the TAP level during virus infection, we infected MJS TAP2-N-GFP cells with BoHV-1 and analyzed TAP-derived GFP fluorescence by flow cytometry (Figure 7A,B). TAP2-N-GFP and endogenous TAP1 levels were assessed in infected cell lysates by immunoblotting (Figure 7C,D). The viral infection resulted in a 30% decrease of mean GFP fluorescence intensity, while approximately 70% and 50% reduction of TAP2-N-GFP and TAP1 protein levels, respectively, could be demonstrated by immunoblotting.

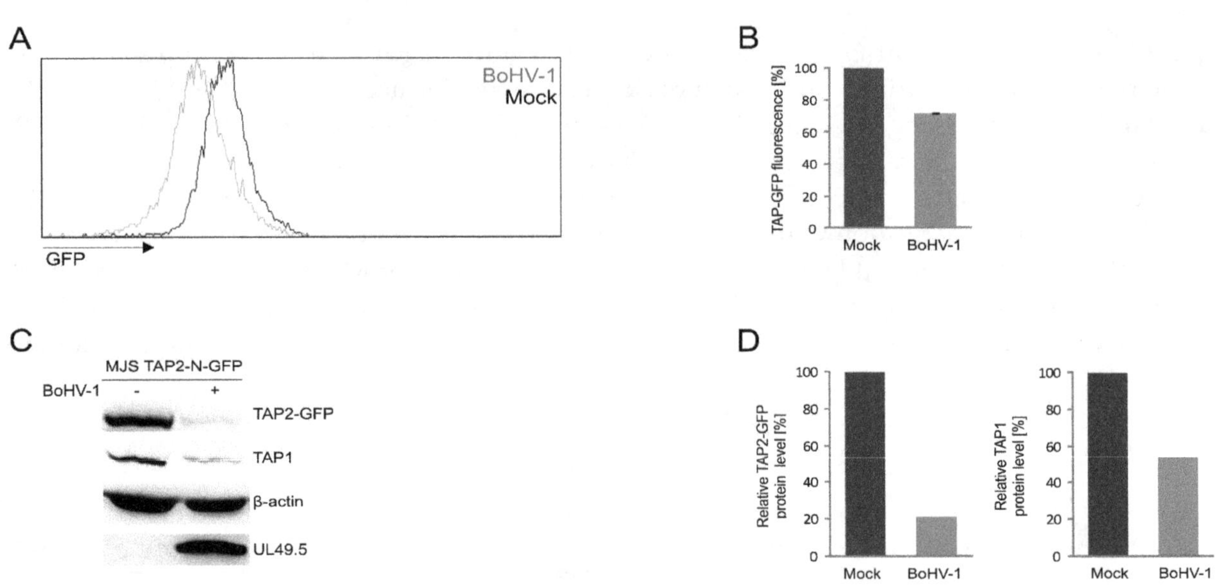

Figure 7. BoHV-1 infection results in TAP2-GFP degradation. MJS TAP2-N-GFP cells were infected with BoHV-1 at a multiplicity of infection (moi) = 10. Twenty-four hours post-infection, cells were collected and analyzed. (**A**) TAP2-GFP fluorescence was assessed by flow cytometry; histograms from a representative analysis are shown, and (**B**) depicted as the percentage of fluorescence in mock-infected MJS TAP2-N-GFP cells (set as 100%). The analysis was performed in triplicates. The statistical significance was assessed by *t*-test; $p \leq 0.0005$ (**C**) TAP2-GFP degradation was determined by SDS-PAGE and immunoblotting using anti-GFP, anti-TAP1, or anti-UL49.5 antibodies; β-actin was used as a loading control. (**D**) The relative amount of TAP2-GFP detected by immunoblotting was normalized to β-actin.

3.6. Only UL49.5 among Different Viral TAP Inhibitors Can Induce Human TAP Degradation

The effect of viral TAP inhibitors on TAP stability was reported, usually, in separate studies. TAP levels were analyzed in those reports by immunoblotting. To compare the sensitivity of fluorescent TAP to different viral inhibitors, we selected several representatives with distinct modes of action. Those were: competition for peptide (HSV-1 ICP47) or ATP (HCMV US6 and CPXV012) binding, as well as conformational arrest and degradation (BoHV-1 UL49.5) or a TAP-binding protein with no activity towards the transporter (VZV-encodedUL49.5). MJS TAP2-N-GFP cells were transduced with a retrovirus vector encoding a viral TAP inhibitor and subsequently selected to high purity. The presence of inhibitors was determined by immunoblotting (Figure 8A). Downregulation of MHC I surface expression could be observed during flow cytometry analysis, as expected, for all but the VZV-encoded protein (Figure 8B). Mean GFP fluorescence intensity measurement illustrated that BoHV-1 UL49.5 was unique in causing TAP-GFP degradation. ICP47 even seemed to slightly stabilize TAP-GFP (Figure 8C).

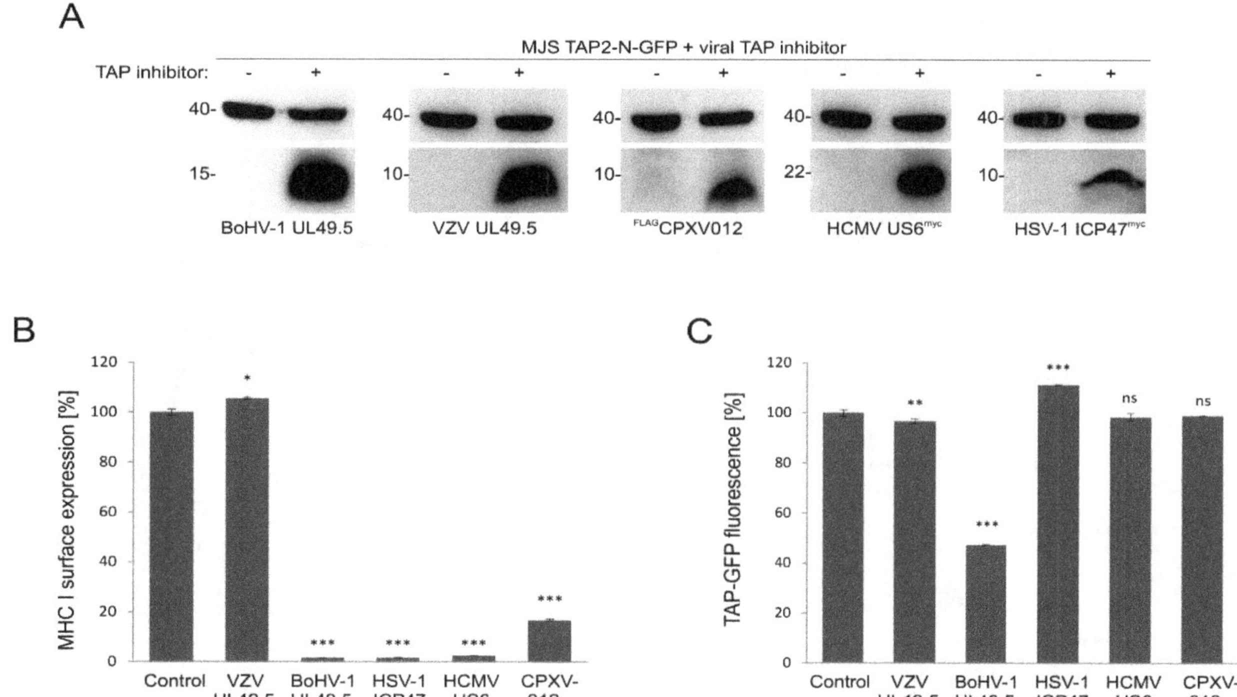

Figure 8. BoHV-1 UL49.5-mediated TAP degradation is unique among viral inhibitors of human TAP. MJS TAP2-N-GFP cells were transduced with a retrovirus encoding BoHV-1 UL49.5, varicella-zoster virus (VZV) UL49.5, human cytomegalovirus (HCMV) US6, herpes simplex virus 1 (HSV-1) ICP47, and cowpox virus CPXV012 and selected with puromycin. (**A**) The presence of viral TAP inhibitors was confirmed by SDS-PAGE and immunoblotting using anti-β, anti-BHV-1 UL49.5, anti-VZV UL49.5, anti-c-myc for HSV-1 ICP47, and HCMV US6 or anti-FLAG antibodies for CPXV012; β-actin (upper panels) was used as a loading control. Size markers are in kDa. (**B**) Surface expression of MHC I was assessed by flow cytometry using specific antibodies (W6/32). MHC I expression is presented as the percentage of MHC I on MJS TAP2-N-GFP cells (set as 100%). The analysis was performed in triplicates. The statistical significance of differences between MJS TAP2-N-GFP cells and cells with viral inhibitor was assessed by *t*-test; *** $p \leq 0.001$ * $p \leq 0.05$. (**C**) The mean fluorescence intensity of GFP was analyzed by flow cytometry and presented as the percentage of GFP fluorescence of MJS TAP2-N-GFP cells (set as 100%). The analysis was performed in triplicates. The statistical significance of differences between MJS TAP2-N-GFP cells and cells with a viral inhibitor was assessed by *t*-test; *** $p \leq 0.001$ ** $p \leq 0.01$ ns: not significant.

3.7. UL49.5-Induced TAP-GFP Degradation Is p97-Dependent

The fluorescent TAP platform can be potentially applied to search for cellular proteins involved in the activity of BoHV-1 UL49.5. According to previous studies, UL49.5-induced TAP degradation is proteasome-dependent [18]. Since TAP is an ER-resident protein, it is presumed to be removed via one of the endoplasmic reticulum-associated degradation (ERAD) pathways. To verify this hypothesis, we used NMS-873 (NMS), an allosteric inhibitor of p97/VCP (valosin-containing protein) a major AAA ATPase belonging to ERAD [38]. MJS TAP2-N-GFP cells with or without UL49.5 were treated with a 2 μM concentration of NMS for 24 h and analyzed by flow cytometry. Inhibition of p97 drastically rescued the mean fluorescence intensity of TAP2 in the presence of UL49.5 to 170% of mean fluorescence intensity in cells treated with DMSO, while in TAP2-N-GFP cells without UL49.5 we could observe only minimal increase to 115% (Figure 9A,B). Compatible results were obtained during immunoblotting analysis of cell lysates (Figure 9C, compare TAP2-GFP in lanes 1–2 and 3–4). Interestingly, inhibition of p97 seemed to stabilize also the expression of UL49.5 (Figure 9C, lanes 1–2). However, apparently, this did not result in a more pronounced TAP-GFP degradation when p97 was blocked.

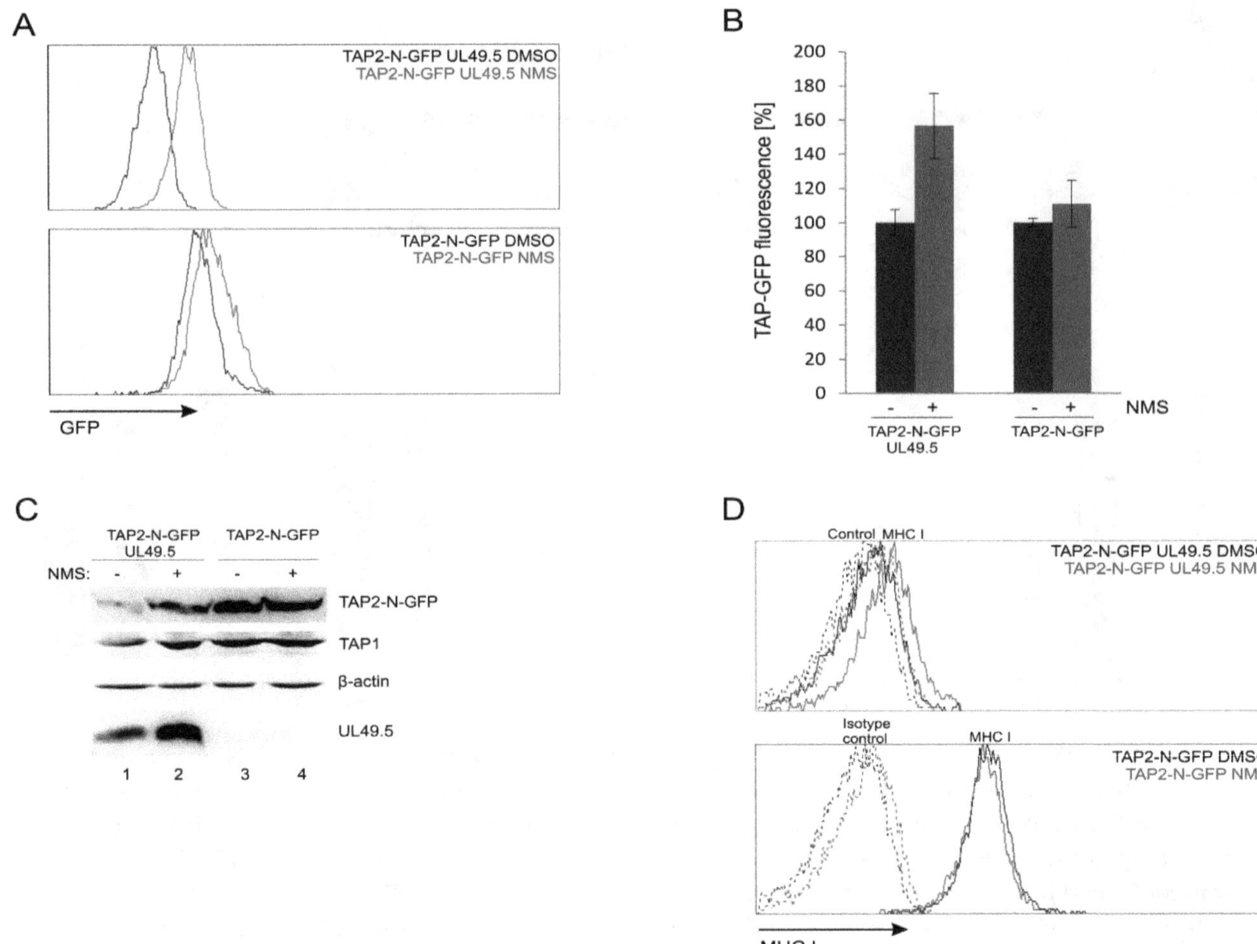

Figure 9. UL49.5 induced TAP-GFP degradation is p97-dependent. MJS TAP2-N-GFP UL49.5 cells were treated with p97 inhibitor NMS-873 (NMS) at 2 μM concentration for 24 hours. (**A**) Flow cytometry analysis of GFP fluorescence in NMS-treated and control (DMSO-treated) cells. (**B**) Relative GFP fluorescence in NMS-treated cells calculated as a percentage of GFP fluorescence in the control cells. The analysis was performed in triplicates. (**C**) The level of TAP-GFP in the presence of p97 inhibitor was determined by SDS-PAGE and immunoblotting using anti-GFP, anti-TAP1 or anti-UL49.5 antibodies; β-actin was used as a loading control. (**D**) Surface expression of MHC I assessed by flow cytometry using specific antibodies (W6/32).

Next, by using flow cytometry, we determined how inhibition of p97 influences MHC I cell surface expression in the presence (and also absence) of UL49.5. NMS-873-treated cells had only slightly improved surface MHC I levels (Figure 9D). This effect was reminiscent of MHC I levels in the presence of some UL49.5 point mutants, which lost the ability to induce TAP degradation, like, for instance, mutants in the C-terminal RGRG motif [32].

4. Discussion

Quantitative studies on protein stability and degradation may require proper tools and platforms that grant full functionality of the protein of interest and, at the same time, assess the protein levels accurately. Here, we tested the application potential of various full-length TAP to GFP fusion constructs for the studies on TAP stability in the presence of four inhibiting proteins encoded by viruses, to obtain the most suitable fluorescent TAP platform. All the tested TAP-GFP variants were functional. Nevertheless, our results point toward a critical role of GFP localization on fluorescence intensity of the tagged transporter, which in concert with the type a linker used to separate TAP and GFP may regulate its susceptibility to virally induced degradation. By using this platform, we also provide evidence that BoHV-1 UL49.5-induced TAP degradation is p97-dependent.

In a study on HSV-1 ICP47, a truncated fluorescent TAP complex (the so-called 6+6 transmembrane TAP core C-terminally fused to mVenus or mCerulean) was used to determine the effect of the viral protein on TAP thermostability [39]. However, for the studies on UL49.5, which was our primary protein of interest, the full-length TAP should constitute a better platform, as N-terminal TMD0s are required for maximum efficiency of UL49.5 binding and inhibition [19]. Full-length fluorescent TAP has been successfully used in several basic studies, some of which were to elucidate the association of H2L^{d} molecules with the TAP complex [40], follow lateral mobility of TAP in living cells [41], or illustrate its cellular distribution [42,43]. In those reports, the addition of a relatively large GFP tag to a much larger multiple membrane-spanning partner protein was tolerated to grant proper localization and functionality of the transporter. However, when exploited in a study on varicellovirus immune evasion, GFP-tagged TAP (C-terminal fusion using a random linker) failed to be degraded by BoHV-1 UL49.5, contrary to non-fluorescent wt TAP [18]. Since BoHV-1-encoded UL45.9 has been, so far, the only known viral inhibitor which can cause human TAP degradation apart from its inhibition, further investigation into this mechanism seemed very intriguing, and for this purpose, construction of fluorescent TAP was instrumental.

Designing an optimal fluorescent TAP construct was hampered by the lack of complete structural information about TAP-UL49.5 interaction, and thus, it required an experimental evaluation of different TAP-GFP variants. The latest structural study on BoHV-1 UL49.5 revealed its 3D structure, while subsequent molecular docking experiments proposed three different possible orientations of TAP-UL49.5 complex in which UL49.5 was suggested to interact simultaneously with both TAP subunits [44]. However, these models were predicted based on the structure of ICP47-arrested TAP conformation [10], and therefore the actual UL45.9-TAP binding model needs to be further confirmed.

Fluorescence analysis of constructed TAP-GFP variants in stable MJS cell lines provides evidence that the tag location, rather than the type of a linker used to separate TAP and GFP, has a pivotal impact on fluorescence intensity. N-terminal fusions generally granted stronger fluorescence (Figure 1B,C and Figure S1). It is worth mentioning here that both ends of TAP1 are present in the cytoplasm, while TAP2 incorporates its C terminus in the cytoplasm, and the N terminus localizes to the ER lumen [45,46]. Together with the fact that fluorescence of both TAP1 fusions was more intense than of TAP2, our results lead to speculations that it is the structure of both TMD0 and C-terminal NBDs that determines the fluorescent potential of the tagged constructs. For some constructs, especially for TAP1-N-GFP, we could observe additional protein products reacting with GFP-specific antibodies (Figure 1D), which might correspond to cleaved GFP and could also, most probably, explain higher and heterogeneous GFP signal of this construct observed by flow cytometry. The reason for the presence of free GFP in the case of TAP1-N-GFP is not fully understood. The length of this linker exceeds the size of other tested linkers, so it has a higher chance of affecting the stability of the protein. Another explanation might be the presence of a sequence recognized by cellular proteases, but the ExPASy PeptideCutter software analysis (https://web.expasy.org/peptide_cutter) did not reveal any significant candidates.

Fluorescent tagging did not affect the subcellular localization and function of the transporter, even upon overexpression of only one TAP subunit (Figures 2 and 3). Both TAP1 and TAP2 lack an N-terminal signal sequence for ER targeting [47], and the exact ER-targeting or ER-retention signals have not been identified to date. This encouraged us to design N-terminal GFP fusions with no additional signaling sequences preceding the tag. The localization of our constructs resembles patterns previously described for other recombinant fluorescent TAP proteins [38–41]. Our results stay in line with the studies on truncated TAP1/TAP2 [43] or functional dissection of transmembrane regions of TAP [6], which have indicated that the transmembrane segments themselves determine ER-localization. It is interesting that even genetically separated TMD0 and the core domains of TAP1 and TAP2 were previously found in the ER (TMD0 additionally localizing to the ER-Golgi intermediate compartment (ERGIC)), when co-expressed [6].

Replacing endogenous TAP1 or TAP2 with TAP-GFP or the untagged subunit restored MHC I on a cell surface equally well and to a level higher than on MJS wt cells (especially in the case of TAP1 constructs). One possible explanation could be the stronger stabilization of endogenous TAP2 by overexpression of TAP1. This effect might be especially noticeable in MJS cells since many melanoma-derived lines have lower endogenous expression of TAP, normally limiting MHC I surface levels [48]. Transduction of MJS TAP1KO or TAP2KO cells with the endogenously present subunit of the transporter did not increase MHC I level, which stays in line with the current view that although TAP1 and TAP2 can form homodimers under certain conditions, they are not functional in antigen presentation [19,35].

One of the most important results of this work provides evidence that all TAP-GFP variants were susceptible to UL49.5-induced inhibition to a similar extent, as assessed by surface MHC I downregulation (Figure 4A). However, only some of them were prone to degradation (both TAP2 fusions, TAP1-N-GFP and TAP1-HC-GFP, Figure 4B,D,E). TAP1-C-GFP remained resistant to UL49.5 what stays in agreement with the previous report [18]. As an interpretation of these data, we can suggest that the helical linker, in contrast to the random one, located at the C terminus of TAP1, effectively separates TAP from GFP to enable undisturbed TAP-UL49.5 interaction, resulting eventually in TAP degradation. Alternatively, it may also permit better access to ERAD components. In the case of the fluorescent TAP2 subunit, the location of GFP, despite the presence of random linkers, did not affect degradation, which could arise from structural differences between the TAP subunits. The TAP2 construct with GFP located in the ER lumen (TAP2-N-GFP) manifested more prominent degradation than the one with GFP in the cytoplasm. An additional observation from this experiment demonstrates that even in the case of a non-degradable fluorescent TAP variant, the second untagged endogenous TAP subunit seems to be sensitive to UL49.5-induced degradation (Figure 4D,E). This, in our opinion, supports the idea of reduced access to ERAD components in TAP1-C-GFP, whereas the access of the second untagged destabilized subunit remains, in this case, undisturbed. It is still unsolved whether UL49.5 can bind single TAP subunits, and the current mechanism points out at the heterodimer as the primary target [19]. As in MJS cells with non-degradable TAP variants, we could observe very efficient MHC I reduction, and the PLC composition in those cells seemed to be intact (Figure 5), at least with regard to the interaction of TAP with ERp57 and MHC I; our data confirm the previous report by [18], demonstrating that abolished degradation does not exclude inhibition. It even seems that TAP degradation might be only an auxiliary event, a "finish-off" effect, in the mechanism of UL49.5 action.

For further studies, we selected and validated TAP2-N-GFP as the most promising variant. TAP transport assay performed on this cell line confirmed that changes in MHC I surface levels reflect TAP transport efficiency (Figure 6). TAP transport in reconstituted cell lines, either with wt or a fluorescent version of the TAP subunit, was higher than in wt MJS. Then we demonstrated that results obtained in a stable cell line model system reflect a situation that occurs upon BoHV-1 infection, which was illustrated as loss of GFP fluorescence observed by flow cytometry and reduction of protein level shown by immunoblotting (Figure 7).

A former pulse-chase experiment with the use of proteasome inhibitor postulated co-degradation of TAP with UL49.5 [18]. In line with this working model, our data show that inhibition of p97 increases levels of both TAP and UL49.5, and demonstrate for the first time that UL49.5-induced TAP degradation requires functional p97. Most of known ER-resident substrates of this ATPase, which retrotranslocates proteins back to the cytoplasm, are ubiquitinated and targeted for proteasomal degradation [38], indicating that UL49.5 mediated TAP degradation occurs via ERAD.

Finally, the TAP2-N-GFP construct was verified as a platform for different viral TAP inhibitors, representing distinct mechanisms of transport inhibition and, most probably, binding another TAP conformation [28,39,49]. BoHV-1, HSV-1, and HCMV-encoded proteins were capable of drastic reduction of surface MHC I; CPXV012 contributed to a slightly weaker but still significant downregulation of MHC I, whereas VZV UL49.5, as expected, did not cause any changes. In terms of degradation, only BoHV-1 UL49.5 was able to decrease TAP-GFP levels, while ICP47 seemed even to stabilize TAP, which is in accordance with its reported effect on TAP thermostability. We believe that the fluorescent TAP platform provides more quantitative data in this respect when compared to previous immunoblotting analyses, which generally are more technically error-prone.

5. Conclusions

In this study, we were able to validate the application potential of fluorescent TAP as a platform for viral immune evasion studies. Our results indicate TAP-GFP variants susceptible to BoHV-1 UL49.5-induced degradation, demonstrate that this degradation is p97-dependent, and emphasize the importance of linker design in fusion protein construction. The fluorescent TAP platform can be now applied in further research on BoHV-1 UL49.5, for instance in the genome-wide search for cellular proteins responsible for UL49.5-induced degradation, where the fluorescent signal can be measured and indicate even small changes in TAP levels. TAP-GFP could be also exploited to identify the active motifs or amino acid residues of UL49.5 affecting TAP stability. The same platform with viral inhibitors can be applied, in a similar way as in the study by [28], to identify TAP conformation recognized by UL49.5.

Author Contributions: Conceptualization, M.W. and A.D.L.; methodology, M.W. and A.D.L.; investigation, M.W., M.G., P.P., A.W.B., R.D.L., and A.D.L.; resources, A.D.L. and E.J.H.J.W.; writing—original draft preparation, A.D.L. and M.W.; writing—review and editing, A.D.L., P.P., K.B.-S., and E.J.H.J.W.; visualization, M.W. and A.D.L.; supervision, A.D.L., K.B.-S., and E.J.H.J.W.; project administration, A.D.L.; funding acquisition, P.P. and A.D.L.

Acknowledgments: We would like to thank Michał Rychłowski, Laboratory of Virus Molecular Biology, University of Gdańsk, Poland, for help with fluorescent confocal microscopy and Robert-Jan Lebbink, Department of Medical Microbiology, University Medical Center Utrecht, The Netherlands, for help in preparation of this manuscript.

References

1. Jensen, P.E. Recent advances in antigen processing and presentation. *Nat. Immunol.* **2007**, *8*, 1041–1048. [CrossRef] [PubMed]
2. Dean, M.; Annilo, T. Evolution of the ATP-binding cassette (ABC) transporter superfamily in vertebrates. *Annu. Rev. Genom. Hum. Genet.* **2005**, *6*, 123–142. [CrossRef] [PubMed]
3. van Endert, P.M.; Tampé, R.; Meyer, T.H.; Tisch, R.; Bach, J.-F.; McDevitt, H.O. A sequential model for peptide binding and transport by the transporters associated with antigen processing. *Immunity* **1994**, *1*, 491–500. [CrossRef]
4. Neefjes, J.J.; Momburg, F.; Hämmerling, G.J. Selective and ATP-dependent translocation of peptides by the MHC-encoded transporter. *Science* **1993**, *261*, 769–771. [CrossRef] [PubMed]
5. Geng, J.; Sivaramakrishnan, S.; Raghavan, M. Analyses of conformational states of the transporter associated with antigen processing (TAP) protein in a native cellular membrane environment. *J. Biol. Chem.* **2013**, *288*, 37039–37047. [CrossRef]
6. Koch, J.; Guntrum, R.; Heintke, S.; Kyritsis, C.; Tampé, R. Functional dissection of the transmembrane domains of the transporter associated with antigen processing (TAP). *J. Biol. Chem.* **2004**, *279*, 10142–10147. [CrossRef]
7. Verweij, M.C.; Horst, D.; Griffin, B.D.; Luteijn, R.D.; Davison, A.J.; Ressing, M.E.; Wiertz, E.J.H.J. Viral Inhibition of the transporter associated with antigen processing (TAP): A striking example of functional convergent evolution. *PLoS Pathog.* **2015**, *11*, e1004743. [CrossRef]
8. Früh, K.; Ahn, K.; Djaballah, H.; Sempé, P.; Peter, M.; van Endert, P.M.; Tampé, R.; Peterson, P.A.; Yang, Y. A viral inhibitor of peptide transporters for antigen presentation. *Nature* **1995**, *375*, 415–418. [CrossRef]

9. Tomazin, R.; Hill, A.B.; Jugovic, P.; York, I.; van Endert, P.; Ploegh, H.L.; Andrews, D.W.; Johnson, D.C. Stable binding of the herpes simplex virus ICP47 protein to the peptide binding site of TAP. *EMBO J.* **1996**, *15*, 3256–3266. [CrossRef]
10. Oldham, M.L.; Grigorieff, N.; Chen, J. Structure of the transporter associated with antigen processing trapped by herpes simplex virus. *eLife* **2016**, *5*, e21829. [CrossRef]
11. Ahn, K.; Gruhler, A.; Galocha, B.; Jones, T.R.; Wiertz, E.J.H.J.; Ploegh, H.L.; Peterson, P.A.; Yang, Y.; Früh, K. The ER-luminal domain of the HCMV glycoprotein US6 inhibits peptide translocation by TAP. *Immunity* **1997**, *6*, 613–621. [CrossRef]
12. Kyritsis, C.; Gorbulev, S.; Hutschenreiter, S.; Pawlitschko, K.; Abele, R.; Tampé, R. Molecular mechanism and structural aspects of transporter associated with antigen processing inhibition by the cytomegalovirus protein US6. *J. Biol. Chem.* **2001**, *276*, 48031–48039. [CrossRef] [PubMed]
13. Halenius, A.; Momburg, F.; Reinhard, H.; Bauer, D.; Lobigs, M.; Hengel, H. Physical and functional interactions of the cytomegalovirus US6 glycoprotein with the transporter associated with antigen processing. *J. Biol. Chem.* **2006**, *281*, 5383–5390. [CrossRef]
14. Alzhanova, D.; Edwards, D.M.; Hammarlund, E.; Scholz, I.G.; Horst, D.; Wagner, M.J.; Upton, C.; Wiertz, E.J.; Slifka, M.K.; Früh, K. Cowpox virus inhibits the transporter associated with antigen processing to evade T cell recognition. *Cell Host Microbe* **2009**, *6*, 433–445. [CrossRef]
15. Byun, M.; Verweij, M.C.; Pickup, D.J.; Wiertz, E.J.H.J.; Hansen, T.H.; Yokoyama, W.M. Two mechanistically distinct immune evasion proteins of cowpox virus combine to avoid antiviral CD8 T Cells. *Cell Host Microbe* **2009**, *6*, 422–432. [CrossRef]
16. Luteijn, R.D.; Hoelen, H.; Kruse, E.; van Leeuwen, W.F.; Grootens, J.; Horst, D.; Koorengevel, M.; Drijfhout, J.W.; Kremmer, E.; Früh, K.; et al. Cowpox virus protein CPXV012 eludes CTLs by blocking ATP binding to TAP. *J. Immunol.* **2014**, *193*, 1578–1589. [CrossRef]
17. Verweij, M.C.; Lipinska, A.D.; Koppers-Lalic, D.; van Leeuwen, W.F.; Cohen, J.I.; Kinchington, P.R.; Messaoudi, I.; Bienkowska-Szewczyk, K.; Ressing, M.E.; Rijsewijk, F.A.M.; et al. The capacity of UL49.5 proteins to inhibit TAP is widely distributed among members of the genus varicellovirus. *J. Virol.* **2011**, *85*, 2351–2363. [CrossRef]
18. Koppers-Lalic, D.; Reits, E.A.J.; Ressing, M.E.; Lipinska, A.D.; Abele, R.; Koch, J.; Rezende, M.M.; Admiraal, P.; van Leeuwen, D.; Bienkowska-Szewczyk, K.; et al. Varicelloviruses avoid T cell recognition by UL49.5-mediated inactivation of the transporter associated with antigen processing. *Proc. Natl. Acad. Sci. USA* **2005**, *102*, 5144–5149. [CrossRef]
19. Verweij, M.C.; Koppers-Lalic, D.; Loch, S.; Klauschies, F.; de la Salle, H.; Quinten, E.; Lehner, P.J.; Mulder, A.; Knittler, M.R.; Tampé, R.; et al. The varicellovirus UL49.5 protein blocks the transporter associated with antigen processing (TAP) by inhibiting essential conformational transitions in the 6+6 transmembrane TAP core complex. *J. Immunol.* **2008**, *181*, 4894–4907. [CrossRef]
20. Koppers-Lalic, D.; Verweij, M.C.; Lipińska, A.D.; Wang, Y.; Quinten, E.; Reits, E.A.; Koch, J.; Loch, S.; Rezende, M.M.; Daus, F.; et al. Varicellovirus UL49.5 proteins differentially affect the function of the transporter associated with antigen processing, TAP. *PLoS Pathog.* **2008**, *4*, e1000080. [CrossRef]
21. Boname, J.; May, J.; Stevenson, P. The murine gamma-herpesvirus-68 MK3 protein causes TAP degradation independent of MHC class I heavy chain degradation. *Eur. J. Immunol.* **2005**, *35*, 171–179. [CrossRef] [PubMed]
22. Herr, R.A.; Wang, X.; Loh, J.; Virgin, H.W.; Hansen, T.H. Newly discovered viral E3 ligase pK3 induces endoplasmic reticulum-associated degradation of class I major histocompatibility proteins and their membrane-bound chaperones. *J. Biol. Chem.* **2012**, *287*, 14467–14479. [CrossRef] [PubMed]
23. Harvey, I.B.; Wang, X.; Fremont, D.H. Molluscum contagiosum virus MC80 sabotages MHC-I antigen presentation by targeting tapasin for ER-associated degradation. *PLoS Pathog.* **2019**, *15*, e1007711. [CrossRef] [PubMed]
24. Chen, X.; Zaro, J.L.; Shen, W.-C. Fusion protein linkers: Property, design and functionality. *Adv. Drug Deliver. Rev.* **2013**, *65*, 1357–1369. [CrossRef] [PubMed]
25. Praest, P.; Luteijn, R.D.; Brak-Boer, I.G.J.; Lanfermeijer, J.; Hoelen, H.; Ijgosse, L.; Costa, A.I.; Gorham, R.D.; Lebbink, R.J.; Wiertz, E.J.H.J. The influence of TAP1 and TAP2 gene polymorphisms on TAP function and its inhibition by viral immune evasion proteins. *Mol. Immunol.* **2018**, *101*, 55–64. [CrossRef]

26. Lipińska, A.D.; Koppers-Lalic, D.; Rychlowski, M.; Admiraal, P.; Rijsewijk, F.A.M.; Bienkowska-Szewczyk, K.; Wiertz, E.J.H.J. Bovine herpesvirus 1 UL49.5 protein inhibits the transporter associated with antigen processing despite complex formation with glycoprotein M. *J. Virol.* **2006**, *80*, 5822–5832. [CrossRef]
27. Morgenstern, J.P.; Land, H. Advanced mammalian gene transfer: High titre retroviral vectors with multiple drug selection markers and a complementary helper-free packaging cell line. *Nucleic Acid Res.* **1990**, *18*, 3587–3596. [CrossRef]
28. Lin, J.; Eggensperger, S.; Hank, S.; Wycisk, A.I.; Wieneke, R.; Mayerhofer, P.U.; Tampé, R. A negative feedback modulator of antigen processing evolved from a frameshift in the cowpox virus genome. *PLoS Pathog.* **2014**, *10*, e1004554. [CrossRef]
29. Marqusee, S.; Baldwin, R.L. Helix stabilization by Glu-Lys+ salt bridges in short peptides of de novo design. *Proc. Natl. Acad. Sci. USA* **1987**, *84*, 8898–8902. [CrossRef]
30. Arai, R.; Ueda, H.; Kitayama, A.; Kamiya, N.; Nagamune, T. Design of the linkers which effectively separate domains of a bifunctional fusion protein. *Protein Eng. Des. Sel.* **2001**, *14*, 529–532. [CrossRef]
31. Koppers-Lalic, D.; Rychlowski, M.; van Leeuwen, D.; Rijsewijk, F.A.M.; Ressing, M.E.; Neefjes, J.J.; Bienkowska-Szewczyk, K.; Wiertz, E.J.H.J. Bovine herpesvirus 1 interferes with TAP-dependent peptide transport and intracellular trafficking of MHC class I molecules in human cells. *Arch. Virol.* **2003**, *148*, 2023–2037. [CrossRef] [PubMed]
32. Verweij, M.C.; Lipińska, A.D.; Koppers-Lalic, D.; Quinten, E.; Funke, J.; van Leeuwen, H.C.; Bieńkowska-Szewczyk, K.; Koch, J.; Ressing, M.E.; Wiertz, E.J.H.J. Structural and functional analysis of the TAP-inhibiting UL49.5 proteins of varicelloviruses. *Mol. Immunol.* **2011**, *48*, 2038–2051. [CrossRef] [PubMed]
33. Praest, P.; de Buhr, H.; Wiertz, E.J.H.J. A flow cytometry-based approach to unravel viral interference with the MHC class I antigen processing and presentation pathway. In *Antigen Processing*; van Endert, P., Ed.; Springer: New York, NY, USA, 2019; Volume 1988, pp. 187–198.
34. Keusekotten, K.; Leonhardt, R.M.; Ehses, S.; Knittler, M.R. Biogenesis of functional antigenic peptide transporter TAP requires assembly of pre-existing TAP1 with newly synthesized TAP2. *J. Biol. Chem.* **2006**, *281*, 17545–17551. [CrossRef] [PubMed]
35. Kleijmeer, M.J.; Kelly, A.; Geuze, H.J.; Slot, J.W.; Townsend, A.; Trowsdale, J. Location of MHC-encoded transporters in the endoplasmic reticulum and cis-Golgi. *Nature* **1992**, *357*, 342–344. [CrossRef] [PubMed]
36. Van Kaer, L.; Ashton-Rickardt, P.G.; Ploegh, H.L.; Tonegawa, S. TAP1 mutant mice are deficient in antigen presentation, surface class I molecules, and CD4−8+ T cells. *Cell* **1992**, *71*, 1205–1214. [CrossRef]
37. Antoniou, A.N.; Ford, S.; Pilley, E.S.; Blake, N.; Powis, S.J. Interactions formed by individually expressed TAP1 and TAP2 polypeptide subunits. *Immunology* **2002**, *106*, 182–189. [CrossRef]
38. Xia, D.; Tang, W.K.; Ye, Y. Structure and function of the AAA+ ATPase p97/Cdc48p. *Gene* **2016**, *583*, 64–77. [CrossRef]
39. Herbring, V.; Bäucker, A.; Trowitzsch, S.; Tampé, R. A dual inhibition mechanism of herpesviral ICP47 arresting a conformationally thermostable TAP complex. *Sci. Rep.* **2016**, *6*, 36907. [CrossRef]
40. Marguet, D.; Spiliotis, E.T.; Pentcheva, T.; Lebowitz, M.; Schneck, J.; Edidin, M. Lateral diffusion of GFP-tagged H2Ld molecules and of GFP-TAP1 reports on the assembly and retention of these molecules in the endoplasmic reticulum. *Immunity* **1999**, *11*, 231–240. [CrossRef]
41. Reits, E.A.J.; Vos, J.C.; Grommé, M.; Neefjes, J. The major substrates for TAP in vivo are derived from newly synthesized proteins. *Nature* **2000**, *404*, 774–778. [CrossRef]
42. Kobayashi, A.; Maeda, T.; Maeda, M. Membrane localization of transporter associated with antigen processing (TAP)-like (ABCB9) visualized in vivo with a fluorescence protein-fusion technique. *Biol. Pharm. Bull.* **2004**, *27*, 1916–1922. [CrossRef] [PubMed]
43. Ghanem, E.; Fritzsche, S.; Al-Balushi, M.; Hashem, J.; Ghuneim, L.; Thomer, L.; Kalbacher, H.; van Endert, P.; Wiertz, E.; Tampe, R.; et al. The transporter associated with antigen processing (TAP) is active in a post-ER compartment. *J. Cell Sci.* **2010**, *123*, 4271–4279. [CrossRef] [PubMed]
44. Karska, N.; Graul, M.; Sikorska, E.; Zhukov, I.; Ślusarz, M.J.; Kasprzykowski, F.; Lipińska, A.D.; Rodziewicz-Motowidło, S. Structure determination of UL49.5 transmembrane protein from bovine herpesvirus 1 by NMR spectroscopy and molecular dynamics. *BBA Biomembr.* **2019**, *1861*, 926–938. [CrossRef] [PubMed]

45. Vos, J.C.; Spee, P.; Momburg, F.; Neefjes, J. Membrane topology and dimerization of the two subunits of the transporter associated with antigen processing reveal a three-domain structure. *J. Immunol.* **1999**, *163*, 6679–6685.
46. Schrodt, S.; Koch, J.; Tampé, R. Membrane topology of the transporter associated with antigen processing (TAP1) within an assembled functional peptide-loading complex. *J. Biol. Chem.* **2006**, *281*, 6455–6462. [CrossRef]
47. Lankat-Buttgereit, B.; Tampé, R. The transporter associated with antigen processing (TAP): A peptide transport and loading complex essential for cellular immune response. In *ABC Proteins: From Bacteria to Man*; Holland, B., Cole, S.P.C., Kuchler, K., Higgins, C.F., Eds.; Academic Press: Cambridge, MA, USA, 2003; pp. 533–550.
48. Kageshita, T.; Hirai, S.; Ono, T.; Hicklin, D.J.; Ferrone, S. Down-regulation of HLA class I antigen-processing molecules in malignant melanoma. *Am. J. Pathol.* **1999**, *154*, 745–754. [CrossRef]
49. Matschulla, T.; Berry, R.; Gerke, C.; Döring, M.; Busch, J.; Paijo, J.; Kalinke, U.; Momburg, F.; Hengel, H.; Halenius, A. A highly conserved sequence of the viral TAP inhibitor ICP47 is required for freezing of the peptide transport cycle. *Sci. Rep.* **2017**, *7*, 2933. [CrossRef]

7

Oncolytic Virus Encoding a Master Pro-Inflammatory Cytokine Interleukin 12 in Cancer Immunotherapy

Hong-My Nguyen †, Kirsten Guz-Montgomery † and Dipongkor Saha *

Department of Immunotherapeutics and Biotechnology, Texas Tech University Health Sciences Center School of Pharmacy, Abilene, TX 79601, USA; My.Nguyen@ttuhsc.edu (H.-M.N.); Kirsten.Montgomery@ttuhsc.edu (K.G.-M.)

* Correspondence: dipongkor.saha@ttuhsc.edu

† These authors contributed equally to this work.

Abstract: Oncolytic viruses (OVs) are genetically modified or naturally occurring viruses, which preferentially replicate in and kill cancer cells while sparing healthy cells, and induce anti-tumor immunity. OV-induced tumor immunity can be enhanced through viral expression of anti-tumor cytokines such as interleukin 12 (IL-12). IL-12 is a potent anti-cancer agent that promotes T-helper 1 (Th1) differentiation, facilitates T-cell-mediated killing of cancer cells, and inhibits tumor angiogenesis. Despite success in preclinical models, systemic IL-12 therapy is associated with significant toxicity in humans. Therefore, to utilize the therapeutic potential of IL-12 in OV-based cancer therapy, 25 different IL-12 expressing OVs (OV-IL12s) have been genetically engineered for local IL-12 production and tested preclinically in various cancer models. Among OV-IL12s, oncolytic herpes simplex virus encoding IL-12 (OHSV-IL12) is the furthest along in the clinic. IL-12 expression locally in the tumors avoids systemic toxicity while inducing an efficient anti-tumor immunity and synergizes with anti-angiogenic drugs or immunomodulators without compromising safety. Despite the rapidly rising interest, there are no current reviews on OV-IL12s that exploit their potential efficacy and safety to translate into human subjects. In this article, we will discuss safety, tumor-specificity, and anti-tumor immune/anti-angiogenic effects of OHSV-IL12 as mono- and combination-therapies. In addition to OHSV-IL12 viruses, we will also review other IL-12-expressing OVs and their application in cancer therapy.

Keywords: cancer immunotherapy; oncolytic virus; herpes simplex virus; immune checkpoint inhibitor; angiogenesis inhibitor

---

1. Introduction

Interleukin 12 (IL-12) is a powerful master regulator of both innate and adaptive anti-tumor immune responses. As a heterodimeric cytokine, it produces multifaceted anti-tumor effects [1,2], including stimulation of growth and cytotoxic activity of natural killer (NK) cells and T cells (both $CD4^+$ and $CD8^+$) [1,3–5], induction of differentiation of $CD4^+$ T cells towards Th1 phenotype [6,7], increased production of IFN-γ from NK and T cells [1,8,9], and inhibition of tumor angiogenesis [1,10]. Despite encouraging success in preclinical studies [4], the early stages of IL-12 clinical trials did not meet expectations. Severe adverse events were first reported on 15 out of 17 patients in a phase II clinical trial following intravenous IL-12 administration, and the trial was immediately terminated by the FDA following two cases of death [11,12]. Although success was observed in cutaneous T cell lymphoma variants [13,14], AIDS-related Kaposi sarcoma [15] and non–Hodgkin's lymphoma [16], severity of side effects outweighed effectiveness of IL-12 based therapies in the vast majority of oncology clinical trials [17]. In an effort to optimize efficacy and enhance the safety profile, alternative approaches are being studied to localize IL-12 expression at the tumor microenvironment.

Recent studies show that systemic toxicity of IL-12 is limited when expressed by oncolytic viruses (OVs) locally in the tumors [18–20] and in the brains of non-human primates [21]. OVs are a distinct class of anti-cancer agents with unique mechanisms of action: (i) selectively replicating in and killing cancer cells (i.e., oncolysis) without harming healthy cells or tissues [22–24], and (ii) exposing viral/tumor antigens, which promote a cascade of anti-tumor innate and adaptive immune responses (i.e., in situ vaccine effects) [25,26]. The OV-induced vaccine effects can be further enhanced through viral expression of anti-tumor cytokines such as IL-12 [18–20], as illustrated in Figure 1. Cancer immunotherapy involving OVs is an emerging and increasingly examined therapeutic approach for the treatment of cancer [27,28]. Among OVs, oncolytic herpes simplex virus (OHSV) is the furthest along in the clinic and approved by the FDA for the treatment of advanced melanoma [29]. To utilize the therapeutic potential of IL-12, there are several OHSVs encoding IL-12 which have been genetically engineered and tested in various cancers (Table 1). In addition to OHSVs, several different OVs such as adenoviruses, measles virus, maraba virus, Newcastle disease virus, Semliki forest virus, vesicular stomatitis virus and Sindbis virus are also being engineered to express IL-12 (Table 2). Our literature research found that 25 different types of OV-IL12s are either being or have recently been explored (see Tables 1 and 2). Despite this rapidly rising interest, there are no reviews on IL-12 expressing viruses that exploit their potential efficacy and safety to translate into human subjects. This review presents the most current data on this topic and provides a basic understanding of OV-IL12 as a promising treatment approach in cancer immunotherapy, which ultimately could support continued research in the future. More specifically, in this review, we will discuss safety, tumor-specificity, and anti-tumor/anti-vascular effects of OHSV-IL12 as monotherapy or combination therapy. In addition to OHSV-IL12 viruses, we will also review other IL-12 expressing OVs and their application in cancer therapy.

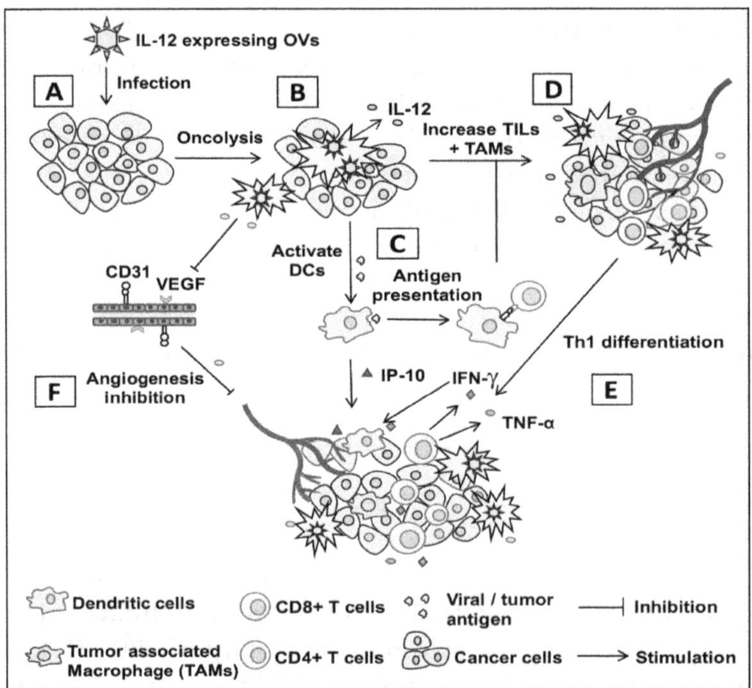

Figure 1. Graphic presentation of mechanism of action of oncolytic virus encoding IL-12. (**A**) Infection of tumor with oncolytic virus encoding IL-12 (OV-IL12). (**B**) OV-IL12 replicates in and kills cancer cells (i.e., oncolysis) and releases IL-12 in the tumor microenvironment. (**C**) Neoantigens from lysed cancer cells activate and recruit dendritic cells (DCs) into the tumor microenvironment. DCs process neoantigens, travel to nearest lymphoid organs, and present the antigen to T cells ($CD4^+$ and $CD8^+$ T cells). (**D**,**E**) T cells migrate to the site of infection (referred as tumor-infiltrating T cells or TILs), differentiate into Th1 cells, produce anti-tumor cytokines and kill cancer cells. (**F**) IL-12-induced production of IFN-γ and interferon inducible protein 10 (IP-10) produces anti-angiogenetic effect through reduction of tumoral vascular endothelial growth factor (VEGF) and $CD31^+$ tumor endothelial cells.

Table 1. List of OHSV-IL12s and their efficacy in pre-clinical cancer models.

Virus	Genomic Modification	Cancer Model	RoA	Dose (pfu)	Efficacy	Ref.
G47Δ-mIL12	ΔICP6, ΔΔICP34.5, ΔICP47, ∘LacZ, ∘mIL-12	Intracranial 005 GSC (Glioblastoma)	I.T.	5×10^5	Inhibited intracranial tumor growth and extended survival. Promoted IL-12 expression, stimulated IFN-γ production, upregulated IP-10, and inhibited VEGF. Polarized T_H1 response and inhibited T-regs function.	[19,30]
T-mfIL12	ΔICP6, ΔΔICP34.5, ΔICP47, ∘mIL-12	Intracerebral Neuro2a (neuroblastma)	I.V.	5×10^6	Prolonged survival (Mock vs. T-mfIL12, $p < 0.05$), although not statistically significant versus T-01 treatments.	[31]
NV1042	ΔICP0, ΔICP4, ΔICP34.5, ΔUL56, ΔICP47, Us11Δ, Us10Δ, UL56 (duplicated), ∘mIL-12	Subcutaneous SCC VII (Squamous Cell Carcinoma)	I.T.	1×10^7	Reduced tumor volume and improved survival (3 doses of 2×10^7 pfu). 57% of mice from NV1042 group rejected subsequent SCC re-challenge in the contralateral flank compared with 14% in NV1023 or NV1034 group.	[32]
			I.V.	5×10^7	NV1042 treatment resulted in 100% survival, in contrast to 70% of NV1023 and 0% of PBS.	[33]
M002	ΔΔICP34.5, ∘mIL-12	Intracranial X21415 (Pediatric embryonal tumor); D456 (pediatric glioblastoma); GBM-12 and UAB106 (adult glioblastoma)	I.T.	1×10^7	M002 significantly prolonged survival in mice bearing all 4 types of tumor compared to saline. No difference in survival was observed compared with G207, excluding X21415 with high levels of nectin-1	[34]
		Intracranial SCK (brain metastasized breast cancer)	I.T.	1.5×10^7	Single injection of M002 extended the survival of treated animals more effectively than a non-cytokine control virus.	[35]
		Xenograft SK-N-AS and SK-N-BE (2) (human neuroblastoma); subcutaneous Neuro-2a (murine neuroblastoma)	I.T.	1×10^7	Significant decrease in tumor growth were observed in both SK-N-AS and SK-N-BE (2) cell lines. Extended median survival compared to the parent R3659.	[36]
		HuH6 (human hepatoblastoma; G401 (human malignant rhabdoid kidney tumor); SK-NEP-1 (renal Ewing sarcoma)	I.T.	1×10^7	M002 significantly reduced tumor volume and increased survival over those treated with vehicle alone in all three different xenograft models.	[37]
R-115	Virulent with retargeted HER-2, ∘mIL-12	pLV-HER2-nectin-puro	I.P.	1×10^8 to 2×10^9	Induced greater local and systemic anti-tumor immunity and durable response than unarmed R-LM113 in both early and late schedule. All mice that survived from primary tumor challenge were protected from the distant challenge tumor and subsequent re-challenge. Increased number of CD8+ and CD141+ cells, PD-L1+ tumor cells, and Treg with a decrease in the number of CD11b+ cells. Enhanced Th1 polarization and increased expression of IFN-γ, IL-2, Granzyme B, T-bet and TNFα and tumor infiltrating lymphocytes	[38]
		Orthotopic $mHGG^{pdgf}$-hHER2 (glioblastoma)	I.T.	Low dose: 2×10^6 High dose: 1×10^8	27% of mice treated with R-115 ($n = 6$, 4 low-dose arm and 2 high-dose arm) alive 100 days after the virus treatment versus all mice died within 48 days. Increased infiltrating CD4+ and CD8+, and expression of IFN-γ	[39]

Table 1. *Cont.*

Virus	Genomic Modification	Cancer Model	RoA	Dose (pfu)	Efficacy	Ref.
vHSV-IL-12	ΔICP6, ΔΔICP34.5, ∘mIL-12	Subcutaneous Neuro2a (neuroblastoma)	I.T.	1×10^4	Significantly reduced tumor growth versus vHSV-null and other cytokine armed groups.	[40]
T2850	ΔIR 15,091bp, ∘mIL-12	Subcutaneous A20 (Murine B Lymphoma), MC38 (colon adenocarcinoma), MFC (Murine Forestomach Carcinoma)	I.T.	1×10^7	Reduced tumor volume compared to IL-12 unarmed parental group. IFN-γ level was markedly increased in the tumor bed and sera of mice infected with both T2850 and T3855 by day 4.	[41]
T3855	ΔIR 15,091bp, ∘mIL-12, ∘mPD-1	Subcutaneous B16 (melanoma)		5×10^6, 1×10^7, 3×10^7		
T3011	ΔIR15,091bp, ∘hIL-12, ∘hPD-1	Subcutaneous B16 (melanoma)	I.T.	5×10^6, 1×10^7 or 3×10^7	Reduced tumor volume as compared with control group.	[41]

Δ—deletion, ∘ insertion, RoA—Route of administration, I.T—intratumorally, IP—intraperitoneally, pfu—plaque forming unit, ref.—reference, VEGF—vascular endothelial growth factor.

Table 2. List of IL-12 expressing oncolytic viruses (other than OHSVs) and their efficacy in pre-clinical cancer models.

Virus	Strain	Cancer Model	RoA	Dose	Efficacy	Ref
Adenovirus	Ad5-yCD/ mutTKSR39 rep-mIL12	Subcutaneous TRAMP (C2 prostate adenocarcinoma)	I.T.	5×10^8 pfu	Improved local and metastatic tumor control. Increased NK and CTL cytolytic activities. Significantly increased survival, levels of IL-12 and IFN-γ in serum and tumor.	[42,43]
	Ad-TD-IL-12, Ad-TD-nsIL-12	Subcutaneous HPD1NR (pancreatic cancer)	I.T.	1×10^9 pfu	100% tumor eradication and survival of both IL-12 modified Adenovirus treated animals. Ad-TD-IL-12, but not Ad-TD-nsIL-12 resulted in a significant increase in $CD3^+CD4^-CD8^+$ populations in the spleen. Level of splenic IFN-γ, IP-10 and lymph node IFN-γ were lower in Ad-TD-nsIL-12 compared to Ad-TD-IL-12 treated hamster.	[44]
	Ad-ΔB7/IL12/GMCSF	Subcutaneous B16-F10 (melanoma)	I.T.	5×10^7 pfu	Primary tumor growth was better controlled in Ad-ΔB7/IL12/GMCSF and Ad-ΔB7/IL12 compared to Ad-ΔB7GMCSF or PBS. Increased tumor infiltrating $CD86^+$ APCs and enhanced CD4+ and CD8+ T cell-mediated Th1 antitumor immune response.Reduced VEGF expression in the tumor treated with oncolytic Ad co-expressing IL-12 and GM-CSF or IL-12 alone. IFN-γ, TNF-α and IL-6 were higher in Ad-ΔB7/IL12/GMCSF and Ad-ΔB7/IL12 compared to Ad-ΔB7GMCSF or PBS.	[45]
	RdB/IL-12/ IL-18	Subcutaneous B16-F10 (melanoma)	I.T.	1×10^8 pfu	95% and 99% tumor growth inhibition was observed in treatment with RdB/IL-12 and RdB/IL-12/IL-18, respectively. Increased Th1/Th2 cytokine ratio and increased tumor infiltration of $CD4^+$ T, $CD8^+$ T and NK cells. Promoted differentiation of T cells expressing IL-12Rβ2 or IL-18Rα.	[46]
	RdB/IL12/ DCN	Orthotopic 4T1 (Triple negative breast cancer)	I.T.	2×10^{10} VP	Both of the IL-12-expressing oncolytic Ads showed similar tumor growth inhibition up to day 9 after initial treatment. RdB/IL12/DCN increased upregulation of IFN-γ, TNF-α, infiltrating cytotoxic lymphocytes, downregulation of TGF-β expression and T-regs compared to RdB/IL12 and RdB/DCN.	[47]
	YKL-IL12/B7	Subcutaneous B16-F10 (melanoma)	I.T.	5×10^8 pfu	Tumor growth was suppressed in both YKL-IL12 and YKL-IL12/B7 treated mice vs PBS. YKL-IL12- or YKL-IL12/B7-treated mice produced a significantly greater level of IFN-γ, infiltrating APCs, CD4+, CD8+ compared with PBS.	[48]
	Ad-ΔB7/ IL-12/4-1BBL	Subcutaneous B16-F10 (melanoma)	I.T.	5×10^9 VP	100% of mice in the Ad-ΔB7/IL-12/4-1BBL group survived >30 days after initial viral injection compared with 20% of that in virus expressing either IL-12 or 4-1BBL. Mice treated with Ad-ΔB7/IL-12 or Ad-ΔB7/IL-12/4-1BBL had greater amount of tumor infiltrating CD4+ and CD8+ compared to Ad-ΔB7/4-1BBL and Ad-ΔB7.	[49]

Table 2. *Cont.*

Virus	Strain	Cancer Model	RoA	Dose	Efficacy	Ref
Measles virus	MeVac FmIL-12	Subcutaneous MC38ce (colon carcinoma)	I.T.	5×10^5–1×10^6 ciu	Tumor remissions in 90% of animals. Driven polarization of Th1-associated immune response and increased tumor infiltrating $CD8^+$ T cells. Increased IFN-γ and TNF-α, and polarization of Th1-associated immune response. Co-expression of IL-12 and IL-15 showed synergistic effect.	[20,50]
Maraba Virus	MG1-IL12-ICV	CT26 and B16F10 peritoneal carcinomatosis	I.P	Seeding dose 5×10^5, then 1×10^4 on day 3	Reduced tumor burden and improved mouse survival. Activated and matured DCs to secrete IP-10, and activated and recruited NK cells. Increased production of IFN-γ	[51]
Newcastle disease virus	rClone30–IL-12	Orthotopic H22 (hepatocarcinoma)	I.T.	1×10^7 pfu	Reduced tumor volume and improved percentage of survival. Increased IFN-γ and IP-10. Co-expression of IL-12 and IL-2 showed synergistic effect.	[52]
Semliki Forest virus	rSFV/IL12	Subcutaneous B16 (melanoma)	I.T.	10^7 IU	Single injection with SFV-IL12 resulted in significant tumor regression. 2 days after injection, IFN-γ production increased with inhibition of tumor vascularization. Splenic IP-10 and MIG expression was increased.	[53]
	SFV/IL12	Subcutaneous P815 (mastocytoma)	I.T.	10^6 IU	Significantly delayed P815 tumor growth. 40–53% of mice exhibited complete tumor regressions. Induced high levels of IFN-γ production in draining lymph nodes.	[54]
	SFV/IL12	Subcutaneous MC38 (colon adenocarcinoma)	I.T.	10^8 particles	Reduced tumor volume and improved percentage of survival. Increased tumor-specific CD8+ T lymphocytes. Enhanced the expression of CD11c, CD8α, CD40, and CD86 of tumor-infiltrating M-MDSCs in the presence of an intact endogenous IFN-I system.	[55]
	SFV-VLP-	Syngeneic RG2 (rat glioma)	I.T.	5×10^7 (low-dose) or 5×10^8 (high-dose)	Reduction in tumor volume (70%—low dose; 87%—high dose)	[56]
Vesicular stomatitis virus (VSV)	VSV-IL12	Orthotopic SCC VII (squamous cell Carcinoma)	I.T.	MOI 0.01	Significant reduction in tumor volume, and prolonged survival.	[57]
Sindbis virus	Sin/IL12	Orthotopic ES-2 cells (ovarian clear cell Carcinoma)	I.P	10^7 pfu	Reduced tumor growth and improved survival. Activated and matured DCs, activated and recruited NK cells. Increased production of IFN-γ.	[58]

Δ—deletion, RoA—Route of administration, I.T—intratumorally, IP—intraperitoneally, pfu—plaque forming unit, ref. —reference, MOI—multiplicity of infection, VP—viral particle, IU—infectious units.

2. Genetically Engineered IL-12 Expressing OHSVs and Their Therapeutic Efficacy

2.1. IL-12 Insertion Does Not Compromise Safety and Tumor Specificity of Genetically Modified OHSV-IL12 Viruses

Herpes simplex virus type 1 (HSV-1) has a large 152 kb genome, and therefore, deletion or mutation of various genes (Figure 2) does not fundamentally alter its functional properties (such as infectivity, viral replication, etc.), but rather confers tumor cell-specific replication, with no or reduced toxicity (e.g., neuropathogenicity) [59]. For example, G47Δ-IL12 (an IL-12 expressing OHSV) has three genomic modifications that endow its tumor-specificity and safety: ICP6 inactivation, γ-(ICP)34.5 and ICP47 deletion, and IL-12 insertion [30,60]. ICP6, encodes for viral ribonucleotide reductase, controls nucleotide metabolism and helps HSV to replicate in healthy or non-dividing cells that are inherently lacking sufficient nucleotide pools [59,61] (Figure 3A). ICP6 inactivation by fusion of LacZ does not hamper DNA synthesis in cancer cells (Figure 3B) [59,62] but the lack of ribonucleotide reductase results in no nucleotide metabolism and no viral DNA replication in healthy cells (Figure 3C) [59,61–63]. Therefore, mutation in the ICP6 gene makes viral infection and replication tumor-specific and thereby increases safety. Similar to ICP6 inactivation, deletion of γ-34.5 also increases safety and cancer selectivity [22,59,64]. Healthy cells have various anti-viral defense mechanisms. For example, protein kinase R (PKR) phosphorylates eukaryotic translation initiation factor eIF2α, which shuts down synthesis of foreign proteins or viral antigens (Figure 4A) [65,66]. HSV with intact γ-34.5 overturns anti-viral defense in healthy cells through γ-34.5-mediated dephosphorylation of eIF2α and helps in viral protein synthesis and viral replication in healthy cells (Figure 4B) [67,68]. Therefore, γ-34.5 deletion results in no eIF2α dephosphorylation, and thereby, no protein synthesis and viral replication (Figure 4C) [67,68]. However, in cancer cells, γ-34.5-deleted HSV can freely replicate, since cancer cells usually have defects in anti-viral pathways such as PKR-eIF2α pathway (Figure 4D). ICP47 downregulates MHC class I presentation through inhibition of transporter-associated protein (TAP) channel [69,70] and prevents detection of OHSV-infected cancer cells by virus-specific CD8$^+$ cells [71]. Thus, deletion of ICP47 enhances MHC class I expression and immune response against virus-infected tumor cells and the host's anti-tumor immunity (innate and adaptive) [59,60,72]. ICP47 deletion also complements the loss of γ-34.5 through immediate early (IE) expression of unique short sequence US11 under the control of ICP47-IE promoter [59,73]. US11 is a true late gene that binds with PKR, preventing it from phosphorylating eIF2a [59,74,75]. IE expression of US11 keeps eIF2α dephosphorylated and helps viral protein translation and synthesis [59,74,75]. Finally, in order to improve efficacy, the master anti-tumor cytokine IL-12 has been inserted into the ICP6 region of OHSV G47Δ to create G47Δ-IL12 [30]. Viral expression of IL-12 does not compromise its tumor specificity and safety, but rather significantly improves its anti-tumor properties [19,26,30,76]. T-mfIL12 (another IL-12 expressing OHSV) has the same backbone as G47Δ but with mIL-12 inserted in the ICP6 deletion region [77]. In vivo safety data showed that intravenous application of T-mfIL12 (5×10^6 pfu) in mice bearing subcutaneous, intracerebral or intravenously disseminated tumors is as safe as non-IL12 expressing OHSV [31].

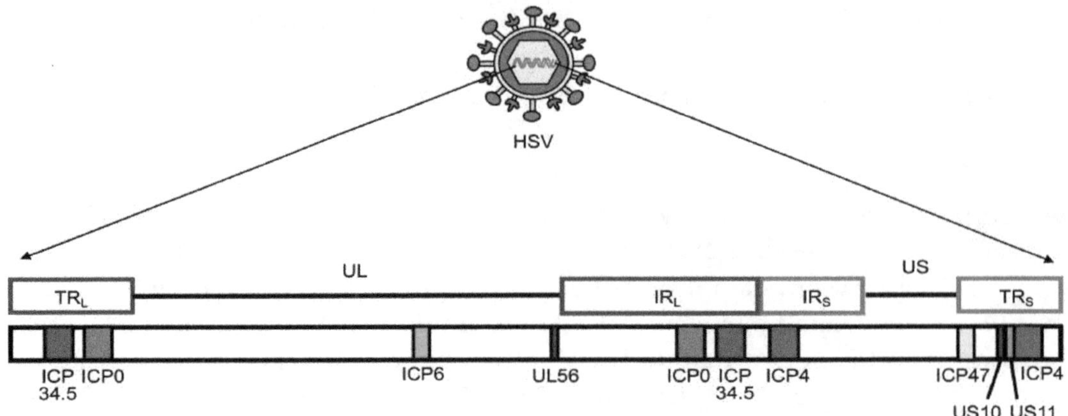

Figure 2. Schematic presentation of herpes simplex virus (HSV) genome with unique long (UL) and short (US) sequences. $TR_{L\ or\ S}$—terminal repeat long or short; $IR_{L\ or\ S}$—internal repeat long or short. Only genes that are modified and/or deleted during construction of OHSV-IL12 are presented. ICP, infected cell protein.

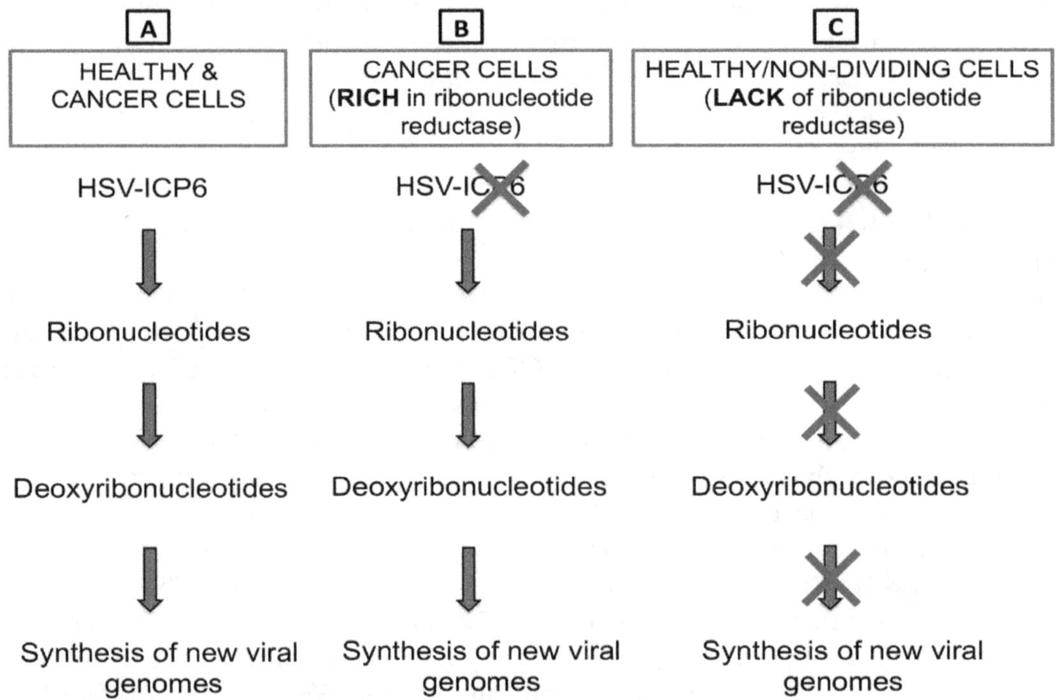

Figure 3. ICP6 inactivation drives tumor-specific replication of OHSV-IL12. (**A**) ICP6 encodes for large subunit of ribonucleotide reductase, which controls nucleotide metabolism and helps HSV to replicate in normal or healthy host cells that are inherently lacking or have insufficient nucleotide pools. (**B**) Cancer cells are rich in ribonucleotide reductase, thus HSV with an inactivated ICP6 does not hamper DNA synthesis in cancer cells. (**C**) Healthy or non-dividing cells lack ribonucleotide reductase. Thus, infection of healthy cells with an ICP6-inactivated HSV leads to no nucleotide metabolism and no viral DNA replication.

Similar to OHSV G47Δ-IL12, a HSV-1/2 recombinant vaccine strain encoding IL-12 is constructed (designated NV1042), which has also multiple deletions/mutations: (i) deletion of one copy of ICP0, ICP4, ICP34.5, and one copy of UL*56* at the $U_{L/S}$ junction, (ii) insertion of *Esherichia coli LacZ* gene under the control of the α47 promoter at the α47 locus, (iii) deletion of ICP47, and (iv) insertion of mIL-12 under the control of a hybrid a4-TK (thymidine kinase) promoter [32,59,78,79]. ICP0 is an important immediate early (IE) protein in switching viral lytic and latent phases that affects defense mechanisms of the host by blocking nuclear factor kappa B (NF-κB)-mediated transcription of immunomodulatory cytokines, inhibiting interferon regulatory factor 3 (IRF3) translocation to the nucleus, inhibiting gamma-interferon inducible protein 16 (IFI16), and degrading mature dendritic cell (DC) markers (CD83) [24,80]. After translocating to the host's nucleus, ICP0 modulates different overlapping cellular pathways to regulate intrinsic and innate antiviral defense mechanism of host cells, allowing the virus to replicate and persist [80,81]. ICP4 blocks apoptosis and positively regulates many other genes in the HSV-1 genome necessary for viral growth [82]. Function of UL56 has not been fully studied but is thought to be involved in neuro-invasiveness of HSV-1 [78]. Therefore, removal of ICP0, ICP4, ICP34.5 and UL56 attenuates virulence and ensures selective viral replication in cancer. In vivo experiment shows no toxicity after intravenous administration of NV1042 (5×10^7 pfu), as demonstrated by lack of cytopathic effects in vital organs (such as lung, brain, spleen, liver, and pancreas) during three months follow up [33]. However, its safety and tumor-selective replication is still a major concern especially for the treatment of tumors located in the central nervous system, since it has 1 intact copy of γ-34.5 (responsible for neuropathogenicity) and intact ribonucleotide reductase ICP6.

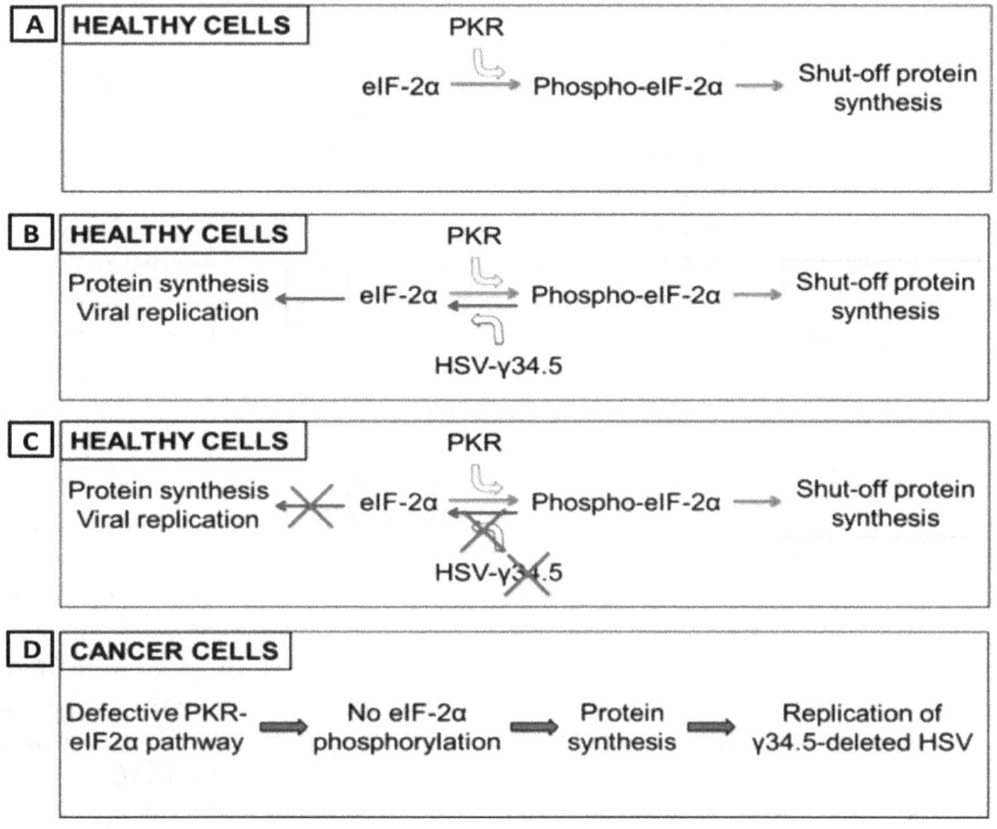

Figure 4. γ-34.5 deletion enhances safety and tumor-specificity of OHSV-IL12. (**A**) Healthy cells have inherent anti-viral defense mechanisms, such as protein kinase R (PKR). PKR phosphorylates translation initiation factor eIF2α, which shuts down synthesis of foreign proteins or viral antigens. (**B**) OHSV with an intact γ-34.5 overturns anti-viral defense in healthy cells through γ-34.5-mediated dephosphorylation of eIF2α and helps in viral protein synthesis/viral replication in healthy cells, leading to development of disease. (**C**) γ-34.5 deletion results in no eIF2α dephosphorylation in normal or healthy cells, and thereby, no protein synthesis and viral replication. (**D**) Cancer cells usually have defective PKR-eIF2α pathway, thus no inhibition of foreign protein synthesis. Therefore, γ-34.5-deleted OHSV can freely replicate in cancer cells.

The OHSV M002 and M032 have deletion of both copies of γ-34.5, with murine and human IL-12 cDNA (p35 and p40 subunits, connected by an IRES), respectively, inserted into each of the γ-34.5 deleted regions [83–86]. M002 has been reported to be safe with no significant toxicity seen after intracerebral inoculation into mice or HSV-sensitive primate Aotus nancymae, despite long-term persistence of viral DNA [87]. M032, with demonstrated safety in non-human primates [21], is now in clinical trial in patients with recurrent glioblastoma (GBM) (see clinical section) [88].

Introducing multiple mutations or deletions in the OHSV genome to confer safety and cancer selectivity may lead to over-attenuation or undermine replication efficiency in cancer cells as opposed to its wild-type or lowly mutated/deleted HSV counterparts [38]. To address this issue, a recent next-generation retargeted IL-12-expressing OHSV known as R-115 has been developed. This OHSV contains no major mutation or deletion and expresses mouse IL-12 under a CMV promoter [38,89]. IL-12-armed R-115 is a derivative of R-LM113 [90]. R-LM113 is a recombinant human epidermal growth factor receptor 2 (HER2) retargeted OHSV with no IL-12 expression, and is successfully engineered by deleting amino acid residues 6 to 38 and by moving the site of single-chain antibody insertion in front of the nectin 1 interacting surface (i.e., at residue 39) [90]. Because of retargeting, it enters and spreads from cancer cell to cell solely via HER2 receptors, and has lost the ability to enter cells through natural glycoprotein D (gD) receptors, herpes virus entry mediator (HVEM) and nectin 1 [90]. Safety profile of R-115 is evaluated in immunocompetent (wt-C57BL/6) model and HER2-transgenic/tolerant counterparts. Mice receiving R-LM113 or R-115 resist very high intraperitoneal OHSV dose of 2×10^9 PFU, which is a lethal dose for wild-type HSV that kills 83% animals [38]. In addition, 4 consecutive intratumoral injections of R-115 at 3–4 days interval shows no viral DNA in vital organs (blood, brain, heart, kidney, liver, brain and spleen) [38]. This indicates that IL12-armed R-115 is safe in mice. However, HER2 specificity makes R-115 applicable only in HER2-expressing tumors, such as mammary tumors, and merely suitable for the treatment of other lethal cancers, such as glioblastoma or other non-HER2-expressing tumors [90]. Recently, another IL-12-expressing OHSV MH1006 (ICP47 deletion and human IL-12 insertion) has been developed and found to be safe in an immunocompetent subcutaneous model of neuroblastoma [91]. However, MH1006 has an intact neurovirulence gene γ-34.5 and an intact ribonucleotide reductase ICP6, thus raising safety concerns in the brain [91]. In Table 1, we have listed all IL-12-expressing OHSVs with their genomic modifications and preclinical applications.

2.2. IL-12 Expressing OHSVs Produce Superior Anti-Tumor Immunity/Efficacy than Non-IL12 OHSVs

In an orthotopic, immunocompetent intracranial glioblastoma (GBM) stem-like cell model (005 GSC model), OHSV G47Δ-IL12 therapy provides superior results compared to OHSV G47Δ treatment alone. For example, G47Δ-IL12 causes significant extension of survival of mice compared with either G47Δ-empty (i.e., an OHSV with no IL-12 transgene expression; $p < 0.005$) or mock treatment ($p < 0.001$, >50% increase in median survival), with 10% of mice surviving long term [30]. Anti-tumor efficacy of G47Δ-IL12 is associated with a significant reduction of tumor cells (i.e., GFP^+ 005 GBM stem-like cells or GSCs) and a robust immune alteration in the tumor, following single intra-tumoral virus injection [19]. The immune alteration mainly includes, but is not limited to, increased tumor infiltration of $CD3^+$ and $CD8^+$ T cells, reduction of immunosuppressive $FoxP3^+$ regulatory T cells, and an increased ratio of $CD8^+$ T cells/regulatory T cells ($CD4^+FoxP3^+$) (a hall mark of clinical efficacy) (Figure 5) [19,30]. The role of T cells in therapeutic efficacy is further investigated in athymic nude mice (i.e., T cell-deficient mice) bearing orthotopic brain tumors [30]. In the absence of T cells, G47Δ-IL12 treatment is unable to significantly enhance survival over G47Δ-empty treatment, indicating a critical role of T cells in the IL12-mediated anti-tumor activity [30]. While replicating in vivo in the tumor, G47Δ-IL12 treatment causes increased local production of IL-12 in the tumor, which is accompanied by a marked release of downstream Th1 mediator interferon gamma (IFN-γ) in the tumors and, to a lesser extent, in the blood [30]. IL-12/IFN-γ promotes differentiation of T cells towards Th1 phenotype [92], which further produces IFN-γ and anti-tumor immune effects, as opposed to Th2 type T cells [9,93]. Similarly, T-bet^+ Th1 type cells are increased in the tumor following intra-tumoral G47Δ-IL12 treatment [19,26], though it has not been determined whether this increase is directly associated with OHSV-IL12-mediated IFN-γ production in the tumor. G47Δ-IL12 treatment promotes polarization of macrophages from pro-tumoral M2 towards anti-tumoral M1 (e.g., increased expression of $iNOS^+$ and $pSTAT1^+$ cells) without affecting total tumor-associated macrophage (TAM) population (Figure 5) [19], possibly because of IL-12 induced M1-polarizing IFN-γ expression [30]. G47Δ-IL12-mediated anti-cancer immune responses, i.e., in situ vaccine effect opens the door to combination treatment strategies involving other cancer immunotherapy drugs. In orthotopic malignant peripheral nerve sheath tumor (MPNST) models, a single intra-tumoral injection of G47Δ in sciatic nerve tumors, derived from human MPNST stem-like cells in athymic mice or mouse MPNST cells in immunocompetent mice, significantly inhibits tumor growth and prolongs survival, as compared to mock treatment [94]. Local IL-12 expression (i.e., G47Δ-IL12) further significantly improves the efficacy of G47Δ in an immunocompetent orthotopic MPNST model, indicating that IL-12 expression induces anti-MPNST immune responses and improves overall efficacy [94]. These studies support the application of G47Δ-IL12 in combination immunotherapies for MPNST tumors.

Similar to anti-tumor efficacy with G47Δ-IL12, NV1042 (i.e., another OHSV with IL-12 expression) treatment results in a striking reduction in squamous cell carcinoma (SCC) tumor volume compared with the tumors treated with NV1023 (i.e., OHSV lacking IL-12 expression) and NV1034 (i.e., OHSV lacking IL-12, but with GM-CSF expression) [32]. Fifty-seven percent of mice treated with NV1042 reject subsequent SCC re-challenge in the contralateral flank, indicating strong global anti-cancer immune response, as opposed to 14% mice treated with NV1023 or NV1034 [32]. Besides local application, NV1042 was intravenously administered for the treatment of spontaneous primary and metastatic prostate cancer in the transgenic TRAMP mice. Systemic IL12-expressing NV1042 was significantly more efficacious than non-IL12 expressing OHSV NV1023 in reducing the frequency of prostate cancer development and lung metastases [95]. NV1042 DNA was detected in primary and metastatic tumors at 2 weeks after the final systemic virus injection but not in liver or blood [95]. Similarly, anti-cancer efficacy of intravenously delivered NV1042 was also observed in disseminated pulmonary SCC. Compared to PBS and parental NV1023, the group treated with IL-12 expressing NV1042 completely showed no sign of pulmonary nodules at day 12. In a low tumor burden model, NV1042 treatment resulted in 100% survival, in contrast to 70% in NV1023-treated group and 0% in PBS-treated group [33]. Depletion of $CD4^+$ and $CD8^+$ T cells reduces anti-cancer efficacy of IL-12 expressing NV1042, which is similar to anti-cancer effects of non-IL12-expressing NV1023, indicating IL-12 expression plays an important role in enhancing oncolytic efficacy through immune modulation [33].

M002 treatment resulted in prolonged survival in both pediatric and adult intracranial patient-derived tumor xenograft models [34]. The better survival benefit is associated with OHSV receptor nectin-1 expression in tumor cells, which is usually higher in pediatric brain tumors than in adult GBMs [34]. In an immunocompetent breast cancer metastasis model, IL-12-armed M002 treatment significantly improved survival of mice over its parental unarmed OHSV R3659 (no IL-12 expression) [35], indicating IL-12 played a critical role for anti-tumor efficacy. In a syngeneic neuroblastoma model, single intracranial injection of M002 produced a minimal survival benefit over untreated mice [85], indicating the need for IL-12 in immunocompetent models. Similarly, mice bearing intracranial neuroblastoma treated intramuscularly (IM) with M002-infected irradiated neuroblastoma cells did not show any survival advantage over mice treated with non-infected irradiated tumor cells. However, a prime-boost vaccine strategy, such as IM injection of M002-infected irradiated tumor cells seven days prior to tumor implantation and seven days post-tumor implantation, produced sustained anti-tumor T-cell responses and significant survival advantage, as opposed to irradiated control tumor cells [85]. Because an important control group is missing in this experimental setup (i.e., unarmed OHSV-infected irradiated tumor cells), it is not clear whether this anti-cancer vaccine effect in neuroblastoma was due to OHSV, local IL-12 expression, or both. In syngeneic sarcoma models, M002 did not produce any survival benefit compared to its parental virus R3659 (no IL-12 expression) [96], despite M002 inducing a significant anti-tumor immune effect over R3659 treatment, such as an increased percentage of intra-tumoral $CD8^+$ T cells and activated monocytes, a decreased percentage of myeloid-derived suppressor cells (MDSCs), and increased CD8:MDSC and CD8:T regulatory cell ratios [96]. In recently performed pilot experiments in an ovarian cancer metastatic model, systemic intraperitoneal application of M002 resulted in a robust tumor-antigen specific $CD8^+$ T cell response in the peritoneal cavity and the omentum [97], which are the primary sites of ovarian cancer metastasis [98]. Because of the tumor-specific immunity, M002 treatment was more successful in controlling ovarian cancer metastasis and produced a significantly longer overall survival than mock treatment [97]. Whether the anti-tumor efficacy is minimal or better, local IL-12 expression (M002) creates a more favorable immune-active tumor microenvironment than unarmed OHSV, which makes tumors more responsive to other forms of immunotherapies, such as immune checkpoint blockades.

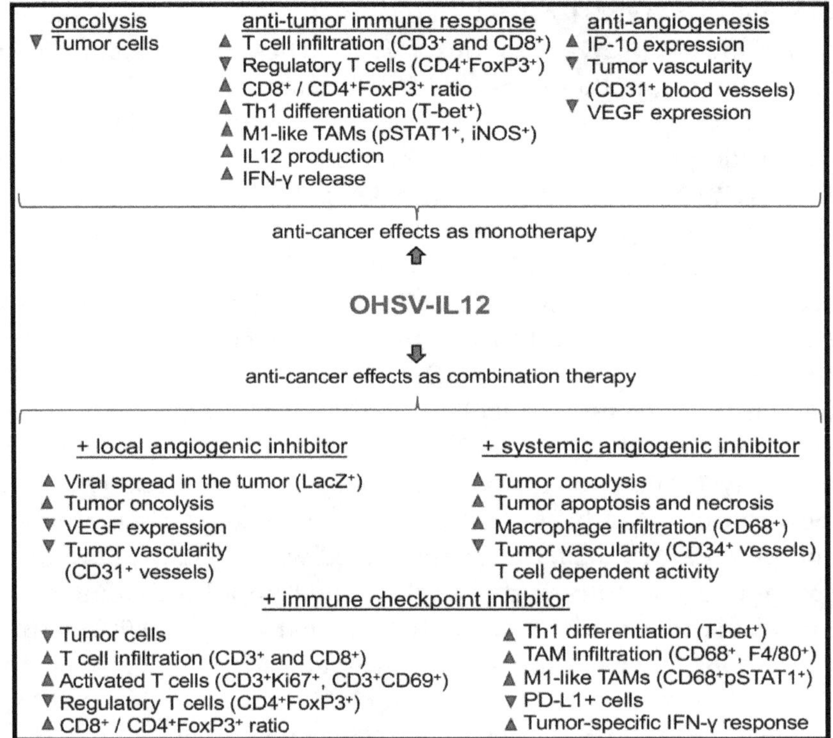

Figure 5. Anti-tumor effects of OHSV-IL12 treatment as mono- and combination-therapies. OHSV-IL12 treatment as monotherapy leads to three distinct anti-cancer effects: 1. Oncolysis, leading to reduction of cancer cells; 2. Induction of anti-tumor immunity, which is characterized by increased intratumoral infiltration of T cells, reduction of regulatory T cells, increased T effector/regulatory T cell ratio, enhanced Th1 differentiation, increased polarization of macrophages toward anti-tumoral M1-phenotype, and increased production of IL-12 and IFN-γ; and 3. Inhibition of tumor angiogenesis as demonstrated by reduced $CD31^+$ blood vessels and increased expression of vascular endothelial growth factor (VEGF) and interferon inducible protein 10 (IP-10). Because of these three aforementioned unique anti-cancer potentials, OHSV-IL12 was tested in combination with local or systemic antiangiogenic inhibitors and immune checkpoint blockade. The combination of OHSV-IL12 + local angiogenic inhibitor produces anti-tumor effects by increasing intratumoral virus spread (as determined by X-gal staining for viral LacZ expression) and oncolysis, and by reducing $CD31^+$ tumor vascularity and VEGF expression. The anti-tumor effects of the OHSV-IL12 + systemic angiogenic inhibitor are characterized by increased lysis of cancer cells and macrophage ($CD68^+$) infiltration into tumors, increased apoptosis and necrosis in the tumor microenvironment, reduced tumor vascularity, and T cell dependent anti-tumor activity. OHSV-IL12 + immune checkpoint inhibitor produces robust and multifaceted anti-cancer activities, which include: oncolysis, increased infiltration of T cells and activated T cells into tumors, reduction of immunosuppressive regulatory T cells, increased effector ($CD8^+$)/regulatory T cell ($CD4^+FoxP3^+$) ratio, Th1 differentiation, tumor-associated macrophage (TAM) infiltration and macrophage polarization towards M1-type, reduction of immune checkpoint PD-L1 positive cells, and induction of tumor-specific IFN-γ response. Upward and downward triangles indicate 'increase' and 'decrease', respectively.

IL-12-armed R115 was superior in inducing local and systemic anti-tumor immunity and durable response over unarmed R-LM113 in both early and late schedules [38]. All mice that survived the primary tumor were protected from the distant tumor challenge and subsequent re-challenge. Treatment with R115 drove Th1 polarization, increased immunomodulatory cytokines such as IFN-γ, IL-2, Granzyme B, T-bet and TNF-α, and tumor infiltrating lymphocytes [38]. Tumor microenvironment of R115 group showed an increase in number of $CD8^+$ and $CD141^+$ cells, PD-$L1^+$ tumor cells, and $FoxP3^+$ T regulatory cells with a decrease in the number of $CD11b^+$ cells [38]. In another study, a single R-115 injection in established tumors resulted in complete tumor eradication in about 30% of animals [39]. The treatment also induced a significant improvement in the overall median survival time of mice and a resistance to recurrence from the same neoplasia [39]. Interestingly, treatment with R-115 increases the number of $CD4^+$ and $CD8^+$ T cells infiltrating into the tumor microenvironment, while the vast majority of $CD4^+$ and $CD8^+$ T cells in the R-LM113 treatment group accumulated at the edge of the tumors, indicating the effects of IL-12 [39].

2.3. *Anti-Tumor Anti-Vascular Effects of OHSV-IL12*

IL-12 does not only enhances the anti-tumor immune effects of virotherapy, but it also suppresses the development of new blood vessels, a process termed angiogenesis [99], making it an anti-angiogenic cytokine. IL-12 elicits its anti-angiogenic effects through release of IFN-γ, which activates IFN-inducible protein 10 [IP-10 or CXC chemokine ligand (CXCL) 10], a chemokine that mediates chemotaxis of lymphocytes and angiostatic effects [10,17,100]. It has been demonstrated that IL-12-armed OHSV produces significantly higher level of anti-angiogenic effects (i.e., reduction of $CD31^+$ blood vessels in the tumors) through IFN-γ/IP-10 pathway in an immunocompetent brain tumor model (Figure 5) [30], compared to non-IL12 OHSV. IL12-armed OHSV treatment also causes a reduction in vascular endothelial growth factor (VEGF) expression, another likely contributor in tumor angiogenesis (Figure 5) [30].

In a model of prostate cancer in transgenic TRAMP mice, treatment with IL-12 armed NV1042 significantly reduces expression of $CD31^+$ vascularity compared to either NV1023 or mock treated tumors [95]. Anti-angiogenic property of NV1042 is confirmed by another study in a SCC model. Intratumoral delivery of NV1042 results in release of a high level of IL-12, as well as other secondary angiogenic mediators such as IFN-γ, monokine induced by gamma interferon (MIG), and IP-10. In contrast, IL-12 unarmed parental NV1023 treatment shows no increase in IL-12 expression and lower level of secondary angiogenic mediators [101]. These studies indicate that IL-12 gene transfer could significantly enhance unique anti-tumor and anti-angiogenic effects of virotherapy. These anti-angiogenic features allow OHSV-IL12 to be tested with other local or systemic angiogenesis inhibitors for an improved therapeutic outcome.

2.4. *Inhibition of Tumor Angiogenesis Enhances Anti-Tumor Potential of OHSV-IL12 Treatment*

Angiogenesis is one of the hallmarks of cancer. It plays a key role in cancer progression [102–110] and anti-angiogenic therapy has been an interesting target to control tumor growth [108,111,112]. Efforts to disrupt the vascular supply and starve the tumor from nutrients and oxygen have resulted in 11 anti-VEGF drugs approved for certain advanced cancers, either alone or in combination with chemotherapy or other targeted therapies. Unfortunately, this success has had only limited impact on overall survival of cancer patients, and rarely resulted in durable responses. Bevacizumab (Avastin), an FDA approved anti-angiogenic drug (anti-VEGF), did not show significant improvement in overall survival [112–114]. Therefore, other anti-angiogenic agents and combinatorial strategies are being tested to target complex tumor microenvironment.

Because OHSV G47Δ-IL12 does not only induce anti-tumor immunity but also produce anti-angiogenic activities [30], it is hypothesized that anti-tumor effects of G47Δ-IL12 treatment would synergize with anti-vascular drugs. Axitinib (AG-013736) is an FDA approved, orally administered potent small molecule tyrosine kinase inhibitor (TKI), which inhibits VEGF receptor

(VEGFR) 1-3, platelet-derived growth factor receptor beta (PDGFR-β) and receptor tyrosine kinase c-KIT (CD117) [115], and shows promising anti-vascular and anti-tumor activity in a variety of advanced stage cancers, including GBM [116–118]. In addition to anti-vascular effects, it also induces anti-tumor immune effects [119,120]. Therefore, anti-vascular/immune axitinib was tested in combination with anti-angiogenic/immune stimulatory G47Δ-IL12 in highly angiogenic patient and mouse GSC-derived GBM models [76]. This combination significantly extends survival in both models and involves multifaceted anti-tumor activities including: direct oncolysis of tumor cells, extensive tumor apoptosis and necrosis, increased macrophage infiltration to the tumor, greatly reduced tumor vascularity (i.e., $CD34^+$ blood vessels) and inhibition of angiogenic PDGFR/ERK pathway in patient GSC-derived GBM model, and T cell dependent activity in mouse GSC-derived GBM model (Figure 5) [76]. Since the anti-tumor efficacy of the dual combination therapy (G47Δ-IL12+axitinib) was T cell dependent, it is hypothesized that ICI (i.e., anti-PD-1 or anti-CTLA4) will improve the therapeutic outcome of G47Δ-IL12+axitinib dual combination. Interestingly, ICI did not improve anti-tumor effects of axitinib or G47Δ-IL12+axitinib combination therapy. This is in sharp contrast with the findings from other investigators, since they observed synergistic anti-tumor immune effects following axitinib + ICI combination therapy in preclinical syngeneic tumor models [121]. The underlying mechanism(s) behind the failure of ICI combination therapy in orthotopic brain tumor model is not clear and warrants further investigation. It is speculated that reduced vascular permeability after axitinib therapy may inhibit extravasation of T cells into the tumor [122]. Indeed, axitinib treatment significantly reduces T cell ($CD3^+$ and $CD8^+$) infiltration into brain tumors [76]. Because both axitinib and OHSV are already in clinical trials for brain tumors as monotherapy with limited efficacy, dual combination therapy (OHSV-IL12+axitinib) that shows anti-tumor efficacy in both immune deficient and immune competent orthotopic brain tumor models has translational relevance [76]. Because systemic anti-angiogenic therapy (i.e., axitinib) is often associated with renal toxicities [123,124], G47Δ-IL12 was tested in combination with a local OHSV expression of angiostatin (OHSV G47Δ-angio), an anti-angiogenic polypeptide, in hypervascular human GBM models [125]. The combination of two OHSVs (G47Δ-IL12+G47Δ-angio) significantly prolongs survival compare to each armed OHSV alone, which is associated with increased viral spread and reduced expression of VEGF and $CD31^+$ blood vessels in the tumor (Figure 5) [125]. This study supports further engineering of OHSV to express both IL-12 and angiostatin locally in the tumor. Use of one virus rather than two is practical in the context of future FDA approval. Similar to G47Δ-IL12 and anti-angiogenesis studies, the combination of another IL-12 expressing OHSV NV1042 with the anti-cancer chemotherapy drug vinblastine results in significant reduction of tumor burden in athymic mice bearing subcutaneous CWR22 prostate tumors, which is most likely associated with diminishing the number of $CD31^+$ endothelial cells [126].

2.5. Immune Checkpoint Inhibition Enhances OHSV-IL12 Treatment-Induced Anti-Tumor Immunity

Though anti-tumor effects of OHSV-IL12 therapy is multifaceted, virotherapy alone does not improve significant survival in cancer [19,30,76,127]. For example, G47Δ-IL12 monotherapy shows limited efficacy in preclinical immune competent models of prostate and malignant peripheral nerve sheath tumors [94]. Since OHSV-IL12 treatment changes immune phenotypes of the tumor microenvironment, it is tested in combination with other forms of immunotherapies (e.g., ICIs) to obtain a better therapeutic response [19,76]. ICIs, such as cytotoxic T lymphocyte antigen 4 (CTLA-4) and programmed death 1 (PD-1) suppress T cell-mediated anti-tumor immune responses (Figure 6), leading to tumor progression [128]. ICI antibodies (i.e., anti-CTLA-4 or anti-PD-1) are effective in unleashing tumor-induced immunosuppression and activating effector immune cells (Figure 6) [129]. Since OHSV-IL12 induces robust anti-tumor immunity [19], it is hypothesized that OHSV-IL12 will synergize with ICI antibodies and will improve the anti-cancer efficacy of G47Δ-IL12. Indeed, in 005 GSC-derived orthotopic brain tumor models, dual combination modestly extends survival compared to ICI antibody alone or G47Δ-IL12 therapy alone [19]. The modest anti-tumor efficacy of the dual combination is not due to the inability of ICI antibodies to cross the blood-brain barrier, since ICI antibodies were detected in the tumor [19]. Because CTLA-4 and PD-1 regulate anti-tumor immunity via distinct and non-redundant immune evasion mechanisms [130–132], it is hypothesized that combinatorial blockade of two immune inhibitory pathways will produce enhanced anti-tumor immune effects and will synergize with the anti-tumor efficacy of G47Δ-IL12 (i.e., triple combination therapy: G47Δ-IL12+anti-PD-1+anti-CTLA-4). Indeed, triple combination therapy leads to a significant percentage (89%) of long-term survivors (i.e., survived six months post-tumor implantation) [19]. These survivors remain protected following lethal tumor re-challenge in the contralateral hemisphere, surviving another three months until the experiment was terminated, which indicates development of long-term memory protection [19]. These unprecedented findings were reproduced in a second aggressive immune competent CT-2A GBM model [19]. The survival efficacy of the triple combination therapies is associated with a significant but complex immune alteration in the tumor microenvironment, as opposed to mock, single or dual combination therapies, which includes: i) increase tumor infiltration of T cells ($CD3^+$ and $CD8^+$); ii) increase number of proliferating T cells ($CD3^+Ki67^+$); iii) increase activated T cells ($CD8^+CD69^+$); iv) reduce regulatory T cells ($FoxP3^+$); v) increase T effector ($CD8^+$)/regulatory T cell ($CD4^+FoxP3^+$) ratio; vi) increase TAMs ($CD68^+$, $F4/80^+$); vii) skew TAMs towards M1-phenotypes ($iNOS^+$, $pSTAT1^+$, $CD68^+pSTAT1^+$); viii) increase Th1 differentiation (T-bet^+); ix) reduce PD-$L1^+$ cells; x) increase tumor-cell specific IFN-γ response; and (xi) reduce tumor cells (Figure 5) [19,26]. Depletion and inhibition of immune cell subtypes (i.e., $CD4^+$ cell depletion by anti-CD4, $CD8^+$ cell depletion by anti-CD8, peripheral macrophage depletion by liposomal clodronate, or CSF-1R inhibition by brain penetrant drug BLZ945 to target TAMs) confirms the necessity of $CD4^+$ cells, $CD8^+$ cells, and macrophages in the therapeutic efficacy, with $CD4^+$ cells playing the critical role [19]. It remains to be determined how immune cells, especially $CD4^+$ cells and M1-like macrophage polarization contribute to therapeutic efficacy. IL-12 appears to be critical for this exceptional anti-tumor efficacy, since another triple combination therapy involving the base G47Δ (without IL-12 expression) plus two systemic ICI antibodies results in only 13% long-term survivors [25], as opposed to 89% from G47Δ-IL12 + anti-PD-1 + anti-CTLA-4 combination [19].

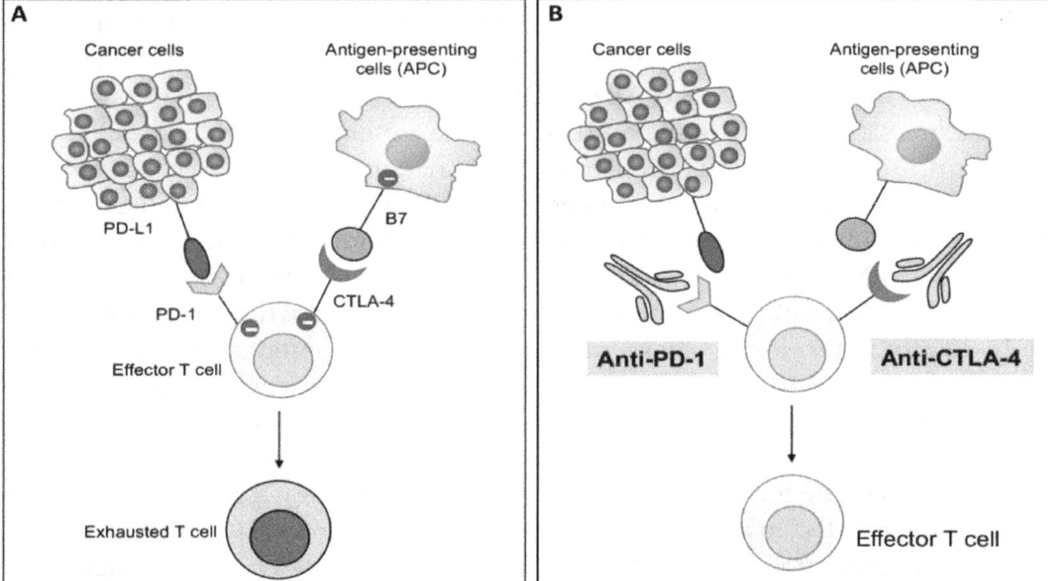

Figure 6. Graphic presentation of mechanism of action of immune checkpoint inhibitors (ICIs). (**A**) T cells express immune checkpoints programmed death 1 (PD-1) and cytotoxic T lymphocyte antigen 4 (CTLA-4), which interact with their corresponding ligands, i.e., programmed death ligand 1 (PD-L1) on cancer cells and B7 molecules on antigen presenting cells (APC), respectively. PD-1:PD-L1 and CTLA-4:B7 interactions send negative signals to immune cells, leading to exhaustion of T cells and no effector activity. (**B**) Antibodies to PD-1 (anti-PD-1) and CTLA-4 (anti-CTLA4) block those interactions and unleash anti-tumor immunity by enhancing activity of effector T cells.

3. Anti-Cancer Potential of other OVs Encoding IL-12

As IL-12 is a master anti-tumor cytokine and has distinct multifaceted anti-cancer properties [1–3,6,8,10], IL-12 expression by OVs is not limited to OHSVs. Several other OVs encoding IL-12 have been developed (Table 2), including adenoviruses [18,42,46–49,133–138], measles virus [20,50], maraba virus [51], Newcastle disease virus [52], Semliki forest virus [53–56,139–141], vesicular stomatitis virus [57,142], and Sindbis virus [58]. In these above-mentioned studies, IL-12 is expressed either alone [18,20,51,57,137–139] or co-expressed alongside with GM-CSF [134–136], pericellular matrix proteoglycan decorin [47], tumor necrosis factor-related apoptosis-inducing ligand (TRAIL) [133], IL-2 proinflammatory cytokine [52], IFN-γ inducing factor IL-18 [46], T cell co-stimulatory ligand 4-1BBL [49], $CD8^+$ co-receptor for CD28 and CTLA-4 [48], or suicide genes [42]. The anti-tumor efficacy of these engineered OVs are tested in various mouse or hamster pre-clinical cancer models and produce superior anti-tumor immunity either alone [18,20,42,46–48,51,52,57,133,135,137] or in combination with other immunotherapeutic agents such as dendritic cell (DC) vaccine [49,134,136], ICI anti-PD-1 and anti-PD-L1 [139], or cytokine-induced killer cells [138].

The tumor-targeting oncolytic adenovirus (Ad-TD) delivers either wild-type IL-12 or non-secreting IL-12 (nsIL-12) directly to pancreatic tumor cells (Table 2) [44]. In a Syrian hamster tumor model, treatment with both Ad-TD-IL12 and Ad-TD-nsIL-12 results in 100% tumor eradication and animal survival, and treatment with Ad-TD-IL12 in particular produces a significant increase in populations of $CD3^+CD4^-CD8^+$ in the spleen [44]. In addition, treatment with Ad-TD-nsIL-12 results in lower levels of lymph node IFN-γ, and splenic IFN-γ and IP-10 production [44]. Another oncolytic adenovirus expressing IL-12 (RdB/IL-12) inhibits tumor growth in murine melanoma lines by 95%, while adenovirus expressing both IL-12 and IL-18 (RdB/IL-12/IL-18) inhibits growth by 99% [46]. RdB/IL-12/IL-18 also increases the cytokine ratio of Th1/Th2, increases tumor infiltration of $CD4^+$ T, $CD8^+$ T and NK cells, and promotes differentiation of T cells expressing IL-12Rβ2 or IL-18Rα [46]. In another engineered oncolytic adenovirus (Ad-ΔB7/IL-12/4-1BBL), co-expression of both IL-12 and the cytokine 4-1BB ligand (4-1BBL) produces significantly higher survival in mice, with 100% of mice surviving more than 30 days after viral injection [49]. This is considerable when in comparison to the 20% survival rate of mice treated with virus expressing only IL-12 or 4-1BBL. In this study, mice treated with either Ad-ΔB7/IL-12 or Ad-ΔB7/IL-12/4-1BBL have a higher amount of tumor infiltrating $CD4^+$ and $CD8^+$ cells in comparison to treatment with Ad-ΔB7/4-1BBL or Ad-ΔB7 [49].

An oncolytic measles virus encoding an IL-12 fusion protein (MeVac FmIL-12) as a single agent produces potent anti-tumor immune effects in an immunocompetent colon cancer model with 90% complete remission [20]. This robust anti-cancer efficacy is dependent on T cells, particularly $CD8^+$ cells, and is associated with activation of early NK cells and effector T cells, and upregulation of effector anti-tumor cytokines IFN-γ and TNF-α [20]. The findings are similar to what we observed in GBM with OHSV G47Δ-IL12 virus [19,30]. Although it was not examined whether the long-term survivors in MeVac FmIL-12 group developed any memory protection, the potent anti-tumor immune efficacy indicates that MeVac FmIL-12 is also an attractive candidate in cancer therapy [20]. Similar to the MeVac FmIL-12 virus, IL12-expressing oncolytic maraba virus (designated MG1-IL12-ICV) treatment also leads to complete eradication of peritoneal carcinomatosis, which is associated with significant recruitment of NK cells in the tumor microenvironment [51]. The lentogenic Newcastle disease virus Clone30 strain was generated to express IL-12 (designated rClone30-IL-12) that displays improved survival and reduced tumor volume [52]. In this case, co-expression of two cytokines (IL-12 and IL-2) produced synergistic effects on treatment, with the rClone30-IL-12-IL-2 virus inducing the greatest release of IFN-γ and IP-10 [52].

The Semliki Forest virus (SFV) has a broad host range and is suicidal, causing apoptosis in infected cells, which makes it a promising viral vector [53]. In a murine tumor model using B16 cells, a single intratumoral injection of SFV expressing IL-12 (SFV-IL12) results in significant tumor regression seven days after injection. This follows a distinct inhibition of tumor vascularization and an increase in IFN-γ production at two days post-injection [53]. Furthermore, treatment with SFV-IL12 results in an increase in expression of splenic IP-10 and monokine induced by interferon-γ (MIG) [53]. Similar results are obtained on the murine colon adenocarcinoma cell line, MC38, with SFV-IL12 treatment resulting in increased tumor-specific $CD8^+$ T lymphocytes, reduced tumor volume, and improved survival [55]. Of particular notice, treatment of MC38 cells with SFV-IL12 results in tumor-infiltrating monocytic myeloid-derived suppressor cells (M-MDSCs) displaying increased expression of CD11c, CD8α, CD40, and CD86 in the presence of an intact, endogenous host type-I interferon (IFN-I) system [55].

Vesicular stomatitis virus (VSV) is another example of a viral vector [57]. Oncolytic VSV carrying IL-12 (rVSV-IL12) succeeds in effectively reducing tumor volume and prolonging survival in both human and murine SCC tumors, with 40% of mice treated with rVSV-IL12 surviving past 100 days post-injection, in comparison to mice treated with the fusogenic OV, rVSV-F, surviving only to 60 days post-injection ($p < 0.0001$) [57]. Similarly, treatment with the Sindbis viral vector carrying IL-12 (Sin/IL12) increases production of IFN-γ, and results in reduced tumor growth and improved survival in ovarian clear cell carcinoma [58]. Treatment with Sin/IL12 also modulates the regulatory functions of NK cells, increasing activation and recruitment of the cells [58]. These and afore-mentioned studies clearly suggest that IL-12 is a useful anti-cancer agent for oncolytic immunovirotherapy, and boosts anti-cancer immune properties of OVs.

4. Clinical Perspectives

Regardless of cancer types, IL-12 expressing OVs have been tested and found effective against various cancers (such as glioma, neuroblastoma, squamous cell carcinoma, metastatic breast cancer, hepatoblastoma, sarcoma, kidney cancer, lymphoma, prostate cancer, pancreatic cancer, colon cancer, ovarian cancer, melanoma, etc.) (Tables 1 and 2). This clearly suggests that IL-12 expressing OVs are an attractive therapeutic candidate for clinical translation against any forms of cancer. In general, when translating results from bench to bed side, safety remains the most concerning aspect, along with dose, route of administration, viral pharmacokinetics and resistance mechanism of host cells [143]. Recent FDA approval of OHSV T-VEC in 2015 has heightened the field of oncolytic virus-based immunotherapy. The FDA approved OHSV expresses GM-CSF instead of IL-12. The safety and efficacy of T-VEC in immune-privileged organs, such as brain, has not been extensively elucidated. Moreover, T-VEC has not demonstrated durable responses in a majority of advanced melanoma patients [29], especially those with visceral metastases [144,145], raising questions about its possible long-term efficacy in patients with difficult-to-treat metastatic cancers.

Safety and anti-tumor efficacy of OHSV-IL12 as monotherapy or combination therapy has been demonstrated in preclinical tumor models [19,30,76]. Currently, G47Δ expressing human IL-12 is under development for clinical use. Safety of M002 has been established in the brain after intracerebral administration to non-human primates [21]. M032 is now in a phase I clinical trial (NCT02062827) in patients with recurrent/progressive GBM, anaplastic astrocytoma, or GBM [88]. Similar to OHSV-IL12, an adenovirus expressing human IL-12 is also under clinical trial investigation as monotherapy in prostate cancer (NCT02555397, NCT00406939), pancreatic cancer (NCT03281382), breast cancer (NCT00849459) and melanoma (NCT01397708). In addition, Ad-RTS-hIL-12 (another adenovirus encoding for IL-12) is under clinical trial evaluation in combination with veledimex (an oral activator ligand to promote release of IL-12 by an OV) in pediatric brain tumor (NCT03330197) and adult glioblastoma or malignant glioma (NCT02026271). Triple combination of Ad-RTS-hIL-12, veledimex and nivolumab (anti-PD-1) are also in active status in glioblastoma patients (NCT03636477). In our opinion, since OV-IL12 treatment induces prominent anti-tumor immunity including increased expression of PD-L1 in the tumor microenvironment, OV-IL12 therapy may improve the response rate to anti-PD-L1 treatment, especially in cancer patients who inherently lack PD-L1 expression and/or previously unresponsive to anti-PD-L1 treatment. Thus, combination studies involving OV-IL12 and ICI warrant clinical investigation and could be an attractive treatment strategy for cancer patients.

5. Conclusions and Future Directions

Genomic manipulation and understanding of pathogenicity have made OVs an attractive candidate for cancer therapy. Among OVs, OHSV is FDA approved for cancer treatment and is the furthest along in the clinic [29,144–146]. Moreover, the availability of antiviral drugs, such as acyclovir, makes OHSV a safer anti-cancer candidate over other OVs [59,147]. However, we all have recognized that OHSV expression of IL-12 and its application as a monotherapy does not provide a desired therapeutic outcome, as demonstrated in several preclinical cancer models [30,84,86,148,149]. It requires synergistic or additive combination approach with other anti-cancer therapies for an improved therapeutic outcome [19,25,76,148,150,151]. Similar to OHSV-IL12, a combination immunotherapeutic approach is also required to enhance anti-tumor efficacy of other OVs encoding IL-12 [49,134,136,138,139].

Although ICI-based cancer immunotherapy is rapidly evolving in the field of oncology, the combination therapy involving single or multiple ICIs is often associated with significant toxicity in humans [152–156]. Moreover, ICI immunotherapy has not been successful in devastating cancer types such as GBM (CheckMate-143) [157,158] or triple-negative breast cancer (KEYNOTE-119) [159]. In contrast, OV expressing single or multiple immune stimulator does not cause toxicity when expressed locally in the tumor by the virus [21,52,160–163], even in an immune-privileged brain [21], but rather induces robust anti-tumor immunity. Thus, developing appropriately designed and stronger version of viral vectors expressing multiple immune stimulators alongside the master anti-tumor cytokine IL-12 may induce superior local anti-tumor immune responses, while reducing the need for multiple systemic ICI or other systemic drugs, and eventually thwarting the current limitation of systemic cancer immunotherapy.

Besides considering construction of new viral vectors, another issue that needs to be addressed before utilizing the full potential of OVs is limited viral spread in tumors [19] due to presence of anti-viral genes or up-regulation of anti-viral factors following OV treatment [164]. Restricted viral replication and spread in the tumor may result in reduced tumor oncolysis with limited in situ vaccine effect. Identification, followed by inhibition of anti-viral factor(s) will provide tools to develop new OV-based immunotherapeutic strategies to enhance viral replication and spread in the tumor, and to induce potent anti-tumor immunity [164]. Designing a better rationale OV-based combination treatment strategy without compromising safety will be the key for clinical success.

Author Contributions: Conceptualization, original draft preparation, review and editing, funding acquisition, D.S.; H.-M.N. and K.G.-M. contributed equally in this manuscript, prepared the figures and wrote the manuscript. All authors have read and agreed to the published version of the manuscript.

References

1. Berraondo, P.; Etxeberria, I.; Ponz-Sarvise, M.; Melero, I. Revisiting Interleukin-12 as a Cancer Immunotherapy Agent. *Clin. Cancer Res.* **2018**, *24*, 2716–2718. [CrossRef] [PubMed]
2. Lu, X. Impact of IL-12 in Cancer. *Curr. Cancer Drug Targets* **2017**, *17*, 682–697. [CrossRef] [PubMed]
3. Zeh, H.J., 3rd; Hurd, S.; Storkus, W.J.; Lotze, M.T. Interleukin-12 promotes the proliferation and cytolytic maturation of immune effectors: Implications for the immunotherapy of cancer. *J. Immunother. Emphasis Tumor Immunol.* **1993**, *14*, 155–161. [CrossRef] [PubMed]
4. Lasek, W.; Zagozdzon, R.; Jakobisiak, M. Interleukin 12: Still a promising candidate for tumor immunotherapy? *Cancer Immunol. Immunother.* **2014**, *63*, 419–435. [CrossRef]
5. Lehmann, D.; Spanholtz, J.; Sturtzel, C.; Tordoir, M.; Schlechta, B.; Groenewegen, D.; Hofer, E. IL-12 directs further maturation of ex vivo differentiated NK cells with improved therapeutic potential. *PLoS ONE* **2014**, *9*, e87131. [CrossRef]

6. Trinchieri, G.; Wysocka, M.; D'Andrea, A.; Rengaraju, M.; Aste-Amezaga, M.; Kubin, M.; Valiante, N.M.; Chehimi, J. Natural killer cell stimulatory factor (NKSF) or interleukin-12 is a key regulator of immune response and inflammation. *Prog. Growth Factor Res.* **1992**, *4*, 355–368. [CrossRef]
7. Trinchieri, G. Interleukin-12 and the regulation of innate resistance and adaptive immunity. *Nat. Rev. Immunol.* **2003**, *3*, 133–146. [CrossRef]
8. Otani, T.; Nakamura, S.; Toki, M.; Motoda, R.; Kurimoto, M.; Orita, K. Identification of IFN-gamma-producing cells in IL-12/IL-18-treated mice. *Cell Immunol.* **1999**, *198*, 111–119. [CrossRef]
9. Tugues, S.; Burkhard, S.H.; Ohs, I.; Vrohlings, M.; Nussbaum, K.; Vom Berg, J.; Kulig, P.; Becher, B. New insights into IL-12-mediated tumor suppression. *Cell Death Differ.* **2015**, *22*, 237–246. [CrossRef]
10. Angiolillo, A.L.; Sgadari, C.; Tosato, G. A role for the interferon-inducible protein 10 in inhibition of angiogenesis by interleukin-12. *Ann. N. Y. Acad. Sci.* **1996**, *795*, 158–167. [CrossRef]
11. Lamont, A.G.; Adorini, L. IL-12: A key cytokine in immune regulation. *Immunol. Today* **1996**, *17*, 214–217. [CrossRef]
12. Jenks, S. After initial setback, IL-12 regaining popularity. *J. Natl. Cancer Inst.* **1996**, *88*, 576–577. [CrossRef]
13. Duvic, M.; Sherman, M.L.; Wood, G.S.; Kuzel, T.M.; Olsen, E.; Foss, F.; Laliberte, R.J.; Ryan, J.L.; Zonno, K.; Rook, A.H. A phase II open-label study of recombinant human interleukin-12 in patients with stage IA, IB, or IIA mycosis fungoides. *J. Am. Acad. Dermatol.* **2006**, *55*, 807–813. [CrossRef] [PubMed]
14. Rook, A.H.; Wood, G.S.; Yoo, E.K.; Elenitsas, R.; Kao, D.M.; Sherman, M.L.; Witmer, W.K.; Rockwell, K.A.; Shane, R.B.; Lessin, S.R.; et al. Interleukin-12 therapy of cutaneous T-cell lymphoma induces lesion regression and cytotoxic T-cell responses. *Blood* **1999**, *94*, 902–908. [CrossRef] [PubMed]
15. Little, R.F.; Pluda, J.M.; Wyvill, K.M.; Rodriguez-Chavez, I.R.; Tosato, G.; Catanzaro, A.T.; Steinberg, S.M.; Yarchoan, R. Activity of subcutaneous interleukin-12 in AIDS-related Kaposi sarcoma. *Blood* **2006**, *107*, 4650–4657. [CrossRef]
16. Younes, A.; Pro, B.; Robertson, M.J.; Flinn, I.W.; Romaguera, J.E.; Hagemeister, F.; Dang, N.H.; Fiumara, P.; Loyer, E.M.; Cabanillas, F.F.; et al. Phase II clinical trial of interleukin-12 in patients with relapsed and refractory non-Hodgkin's lymphoma and Hodgkin's disease. *Clin. Cancer Res.* **2004**, *10*, 5432–5438. [CrossRef] [PubMed]
17. Del Vecchio, M.; Bajetta, E.; Canova, S.; Lotze, M.T.; Wesa, A.; Parmiani, G.; Anichini, A. Interleukin-12: Biological properties and clinical application. *Clin. Cancer Res.* **2007**, *13*, 4677–4685. [CrossRef]
18. Poutou, J.; Bunuales, M.; Gonzalez-Aparicio, M.; Garcia-Aragoncillo, E.; Quetglas, J.I.; Casado, R.; Bravo-Perez, C.; Alzuguren, P.; Hernandez-Alcoceba, R. Safety and antitumor effect of oncolytic and helper-dependent adenoviruses expressing interleukin-12 variants in a hamster pancreatic cancer model. *Gene Ther.* **2015**, *22*, 696–706. [CrossRef]
19. Saha, D.; Martuza, R.L.; Rabkin, S.D. Macrophage Polarization Contributes to Glioblastoma Eradication by Combination Immunovirotherapy and Immune Checkpoint Blockade. *Cancer Cell* **2017**, *32*, 253–267.e255. [CrossRef]
20. Veinalde, R.; Grossardt, C.; Hartmann, L.; Bourgeois-Daigneault, M.C.; Bell, J.C.; Jager, D.; von Kalle, C.; Ungerechts, G.; Engeland, C.E. Oncolytic measles virus encoding interleukin-12 mediates potent antitumor effects through T cell activation. *Oncoimmunology* **2017**, *6*, e1285992. [CrossRef]
21. Roth, J.C.; Cassady, K.A.; Cody, J.J.; Parker, J.N.; Price, K.H.; Coleman, J.M.; Peggins, J.O.; Noker, P.E.; Powers, N.W.; Grimes, S.D.; et al. Evaluation of the safety and biodistribution of M032, an attenuated herpes simplex virus type 1 expressing hIL-12, after intracerebral administration to aotus nonhuman primates. *Hum. Gene Ther. Clin Dev.* **2014**, *25*, 16–27. [CrossRef]
22. Saha, D.; Ahmed, S.S.; Rabkin, S.D. Exploring the Antitumor Effect of Virus in Malignant Glioma. *Drugs Future* **2015**, *40*, 739–749. [CrossRef]
23. Saha, D.; Martuza, R.L.; Curry, W.T. Viral oncolysis of glioblastoma. In *Neurotropic viral infections*, 2nd ed.; Reiss, C.S., Ed.; Springer: New York, NY, USA, 2016; Volume 2, pp. 481–517.
24. Saha, D.; Wakimoto, H.; Rabkin, S.D. Oncolytic herpes simplex virus interactions with the host immune system. *Curr. Opin. Virol.* **2016**, *21*, 26–34. [CrossRef]
25. Saha, D.; Martuza, R.L.; Rabkin, S.D. Oncolytic herpes simplex virus immunovirotherapy in combination with immune checkpoint blockade to treat glioblastoma. *Immunotherapy* **2018**, *10*, 779–786. [CrossRef] [PubMed]

26. Saha, D.; Martuza, R.L.; Rabkin, S.D. Curing glioblastoma: Oncolytic HSV-IL12 and checkpoint blockade. *Oncoscience* **2017**, *4*, 67–69. [CrossRef] [PubMed]
27. Kaufman, H.L.; Bommareddy, P.K. Two roads for oncolytic immunotherapy development. *J. Immunother. Cancer* **2019**, *7*, 26. [CrossRef]
28. Raja, J.; Ludwig, J.M.; Gettinger, S.N.; Schalper, K.A.; Kim, H.S. Oncolytic virus immunotherapy: Future prospects for oncology. *J. Immunother. Cancer* **2018**, *6*, 140. [CrossRef]
29. Andtbacka, R.H.; Kaufman, H.L.; Collichio, F.; Amatruda, T.; Senzer, N.; Chesney, J.; Delman, K.A.; Spitler, L.E.; Puzanov, I.; Agarwala, S.S.; et al. Talimogene Laherparepvec Improves Durable Response Rate in Patients With Advanced Melanoma. *J. Clin. Oncol.* **2015**, *33*, 2780–2788. [CrossRef]
30. Cheema, T.A.; Wakimoto, H.; Fecci, P.E.; Ning, J.; Kuroda, T.; Jeyaretna, D.S.; Martuza, R.L.; Rabkin, S.D. Multifaceted oncolytic virus therapy for glioblastoma in an immunocompetent cancer stem cell model. *Proc. Natl. Acad. Sci. USA* **2013**, *110*, 12006–12011. [CrossRef]
31. Guan, Y.; Ino, Y.; Fukuhara, H.; Todo, T. Antitumor Efficacy of Intravenous Administration of Oncolytic Herpes Simplex Virus Expressing Interleukin 12. *Mol. Ther.* **2006**, *13*, S108. [CrossRef]
32. Wong, R.J.; Patel, S.G.; Kim, S.; DeMatteo, R.P.; Malhotra, S.; Bennett, J.J.; St-Louis, M.; Shah, J.P.; Johnson, P.A.; Fong, Y. Cytokine gene transfer enhances herpes oncolytic therapy in murine squamous cell carcinoma. *Hum. Gene. Ther.* **2001**, *12*, 253–265. [CrossRef]
33. Wong, R.J.; Chan, M.K.; Yu, Z.; Kim, T.H.; Bhargava, A.; Stiles, B.M.; Horsburgh, B.C.; Shah, J.P.; Ghossein, R.A.; Singh, B.; et al. Effective intravenous therapy of murine pulmonary metastases with an oncolytic herpes virus expressing interleukin 12. *Clin. Cancer Res.* **2004**, *10*, 251–259. [CrossRef] [PubMed]
34. Friedman, G.K.; Bernstock, J.D.; Chen, D.; Nan, L.; Moore, B.P.; Kelly, V.M.; Youngblood, S.L.; Langford, C.P.; Han, X.; Ring, E.K.; et al. Enhanced Sensitivity of Patient-Derived Pediatric High-Grade Brain Tumor Xenografts to Oncolytic HSV-1 Virotherapy Correlates with Nectin-1 Expression. *Sci. Rep.* **2018**, *8*, 13930. [CrossRef] [PubMed]
35. Cody, J.J.; Scaturro, P.; Cantor, A.B.; Yancey Gillespie, G.; Parker, J.N.; Markert, J.M. Preclinical evaluation of oncolytic deltagamma(1)34.5 herpes simplex virus expressing interleukin-12 for therapy of breast cancer brain metastases. *Int. J. Breast Cancer* **2012**, *2012*, 628697. [CrossRef] [PubMed]
36. Gillory, L.A.; Megison, M.L.; Stewart, J.E.; Mroczek-Musulman, E.; Nabers, H.C.; Waters, A.M.; Kelly, V.; Coleman, J.M.; Markert, J.M.; Gillespie, G.Y.; et al. Preclinical evaluation of engineered oncolytic herpes simplex virus for the treatment of neuroblastoma. *PLoS ONE* **2013**, *8*, e77753. [CrossRef] [PubMed]
37. Megison, M.L.; Gillory, L.A.; Stewart, J.E.; Nabers, H.C.; Mroczek-Musulman, E.; Waters, A.M.; Coleman, J.M.; Kelly, V.; Markert, J.M.; Gillespie, G.Y.; et al. Preclinical evaluation of engineered oncolytic herpes simplex virus for the treatment of pediatric solid tumors. *PLoS ONE* **2014**, *9*, e86843. [CrossRef] [PubMed]
38. Leoni, V.; Vannini, A.; Gatta, V.; Rambaldi, J.; Sanapo, M.; Barboni, C.; Zaghini, A.; Nanni, P.; Lollini, P.L.; Casiraghi, C.; et al. A fully-virulent retargeted oncolytic HSV armed with IL-12 elicits local immunity and vaccine therapy towards distant tumors. *PLoS Pathog* **2018**, *14*, e1007209. [CrossRef]
39. Alessandrini, F.; Menotti, L.; Avitabile, E.; Appolloni, I.; Ceresa, D.; Marubbi, D.; Campadelli-Fiume, G.; Malatesta, P. Eradication of glioblastoma by immuno-virotherapy with a retargeted oncolytic HSV in a preclinical model. *Oncogene* **2019**, *38*, 4467–4479. [CrossRef]
40. Ino, Y.; Saeki, Y.; Fukuhara, H.; Todo, T. Triple combination of oncolytic herpes simplex virus-1 vectors armed with interleukin-12, interleukin-18, or soluble B7-1 results in enhanced antitumor efficacy. *Clin. Cancer Res.* **2006**, *12*, 643–652. [CrossRef]
41. Yan, R.; Zhou, X.; Chen, X.; Liu, X.; Ma, J.; Wang, L.; Liu, Z.; Zhan, B.; Chen, H.; Wang, J.; et al. Enhancement of Oncolytic Activity of oHSV Expressing IL-12 and Anti PD-1 Antibody by Concurrent Administration of Exosomes Carrying CTLA-4 miRNA. *Immunotherapy* **2019**, *5*, 1–10. [CrossRef]
42. Freytag, S.O.; Barton, K.N.; Zhang, Y. Efficacy of oncolytic adenovirus expressing suicide genes and interleukin-12 in preclinical model of prostate cancer. *Gene Ther.* **2013**, *20*, 1131–1139. [CrossRef] [PubMed]
43. Freytag, S.O.; Zhang, Y.; Siddiqui, F. Preclinical toxicology of oncolytic adenovirus-mediated cytotoxic and interleukin-12 gene therapy for prostate cancer. *Mol. Ther. Oncolytics.* **2015**, *2*. [CrossRef] [PubMed]
44. Wang, P.; Li, X.; Wang, J.; Gao, D.; Li, Y.; Li, H.; Chu, Y.; Zhang, Z.; Liu, H.; Jiang, G.; et al. Re-designing Interleukin-12 to enhance its safety and potential as an anti-tumor immunotherapeutic agent. *Nat. Commun.* **2017**, *8*, 1395. [CrossRef] [PubMed]

45. Choi, K.J.; Zhang, S.N.; Choi, I.K.; Kim, J.S.; Yun, C.O. Strengthening of antitumor immune memory and prevention of thymic atrophy mediated by adenovirus expressing IL-12 and GM-CSF. *Gene Ther.* **2012**, *19*, 711–723. [CrossRef]
46. Choi, I.K.; Lee, J.S.; Zhang, S.N.; Park, J.; Sonn, C.H.; Lee, K.M.; Yun, C.O. Oncolytic adenovirus co-expressing IL-12 and IL-18 improves tumor-specific immunity via differentiation of T cells expressing IL-12Rbeta2 or IL-18Ralpha. *Gene Ther.* **2011**, *18*, 898–909. [CrossRef]
47. Oh, E.; Choi, I.K.; Hong, J.; Yun, C.O. Oncolytic adenovirus coexpressing interleukin-12 and decorin overcomes Treg-mediated immunosuppression inducing potent antitumor effects in a weakly immunogenic tumor model. *Oncotarget* **2017**, *8*, 4730–4746. [CrossRef]
48. Lee, Y.S.; Kim, J.H.; Choi, K.J.; Choi, I.K.; Kim, H.; Cho, S.; Cho, B.C.; Yun, C.O. Enhanced antitumor effect of oncolytic adenovirus expressing interleukin-12 and B7-1 in an immunocompetent murine model. *Clin. Cancer Res.* **2006**, *12*, 5859–5868. [CrossRef]
49. Huang, J.H.; Zhang, S.N.; Choi, K.J.; Choi, I.K.; Kim, J.H.; Lee, M.G.; Kim, H.; Yun, C.O. Therapeutic and tumor-specific immunity induced by combination of dendritic cells and oncolytic adenovirus expressing IL-12 and 4-1BBL. *Mol. Ther.* **2010**, *18*, 264–274. [CrossRef]
50. Backhaus, P.S.; Veinalde, R.; Hartmann, L.; Dunder, J.E.; Jeworowski, L.M.; Albert, J.; Hoyler, B.; Poth, T.; Jager, D.; Ungerechts, G.; et al. Immunological Effects and Viral Gene Expression Determine the Efficacy of Oncolytic Measles Vaccines Encoding IL-12 or IL-15 Agonists. *Viruses* **2019**, *11*, 914. [CrossRef]
51. Alkayyal, A.A.; Tai, L.H.; Kennedy, M.A.; de Souza, C.T.; Zhang, J.; Lefebvre, C.; Sahi, S.; Ananth, A.A.; Mahmoud, A.B.; Makrigiannis, A.P.; et al. NK-Cell Recruitment Is Necessary for Eradication of Peritoneal Carcinomatosis with an IL12-Expressing Maraba Virus Cellular Vaccine. *Cancer Immunol. Res.* **2017**, *5*, 211–221. [CrossRef]
52. Ren, G.; Tian, G.; Liu, Y.; He, J.; Gao, X.; Yu, Y.; Liu, X.; Zhang, X.; Sun, T.; Liu, S.; et al. Recombinant Newcastle Disease Virus Encoding IL-12 and/or IL-2 as Potential Candidate for Hepatoma Carcinoma Therapy. *Technol. Cancer Res. Treat.* **2016**, *15*, NP83-94. [CrossRef] [PubMed]
53. Asselin-Paturel, C.; Lassau, N.; Guinebretiere, J.M.; Zhang, J.; Gay, F.; Bex, F.; Hallez, S.; Leclere, J.; Peronneau, P.; Mami-Chouaib, F.; et al. Transfer of the murine interleukin-12 gene in vivo by a Semliki Forest virus vector induces B16 tumor regression through inhibition of tumor blood vessel formation monitored by Doppler ultrasonography. *Gene Ther.* **1999**, *6*, 606–615. [CrossRef] [PubMed]
54. Colmenero, P.; Chen, M.; Castanos-Velez, E.; Liljestrom, P.; Jondal, M. Immunotherapy with recombinant SFV-replicons expressing the P815A tumor antigen or IL-12 induces tumor regression. *Int. J. Cancer* **2002**, *98*, 554–560. [CrossRef]
55. Melero, I.; Quetglas, J.I.; Reboredo, M.; Dubrot, J.; Rodriguez-Madoz, J.R.; Mancheno, U.; Casales, E.; Riezu-Boj, J.I.; Ruiz-Guillen, M.; Ochoa, M.C.; et al. Strict requirement for vector-induced type I interferon in efficacious antitumor responses to virally encoded IL12. *Cancer Res.* **2015**, *75*, 497–507. [CrossRef]
56. Roche, F.P.; Sheahan, B.J.; O'Mara, S.M.; Atkins, G.J. Semliki Forest virus-mediated gene therapy of the RG2 rat glioma. *Neuropathol. Appl. Neurobiol.* **2010**, *36*, 648–660. [CrossRef]
57. Shin, E.J.; Wanna, G.B.; Choi, B.; Aguila, D., 3rd; Ebert, O.; Genden, E.M.; Woo, S.L. Interleukin-12 expression enhances vesicular stomatitis virus oncolytic therapy in murine squamous cell carcinoma. *Laryngoscope* **2007**, *117*, 210–214. [CrossRef]
58. Granot, T.; Venticinque, L.; Tseng, J.C.; Meruelo, D. Activation of cytotoxic and regulatory functions of NK cells by Sindbis viral vectors. *PLoS ONE* **2011**, *6*, e20598. [CrossRef]
59. Peters, C.; Rabkin, S.D. Designing Herpes Viruses as Oncolytics. *Mol Ther Oncolytics* **2015**, *2*. [CrossRef]
60. Todo, T.; Martuza, R.L.; Rabkin, S.D.; Johnson, P.A. Oncolytic herpes simplex virus vector with enhanced MHC class I presentation and tumor cell killing. *Proc. Natl. Acad. Sci. USA* **2001**, *98*, 6396–6401. [CrossRef]
61. Goldstein, D.J.; Weller, S.K. Factor(s) present in herpes simplex virus type 1-infected cells can compensate for the loss of the large subunit of the viral ribonucleotide reductase: characterization of an ICP6 deletion mutant. *Virology* **1988**, *166*, 41–51. [CrossRef]
62. Aghi, M.; Visted, T.; Depinho, R.A.; Chiocca, E.A. Oncolytic herpes virus with defective ICP6 specifically replicates in quiescent cells with homozygous genetic mutations in p16. *Oncogene* **2008**, *27*, 4249–4254. [CrossRef] [PubMed]

63. Cameron, J.M.; McDougall, I.; Marsden, H.S.; Preston, V.G.; Ryan, D.M.; Subak-Sharpe, J.H. Ribonucleotide reductase encoded by herpes simplex virus is a determinant of the pathogenicity of the virus in mice and a valid antiviral target. *J. Gen. Virol.* **1988**, *69*, 2607–2612. [CrossRef] [PubMed]
64. Markert, J.M.; Razdan, S.N.; Kuo, H.C.; Cantor, A.; Knoll, A.; Karrasch, M.; Nabors, L.B.; Markiewicz, M.; Agee, B.S.; Coleman, J.M.; et al. A phase 1 trial of oncolytic HSV-1, G207, given in combination with radiation for recurrent GBM demonstrates safety and radiographic responses. *Mol. Ther.* **2014**, *22*, 1048–1055. [CrossRef] [PubMed]
65. Pasieka, T.J.; Baas, T.; Carter, V.S.; Proll, S.C.; Katze, M.G.; Leib, D.A. Functional genomic analysis of herpes simplex virus type 1 counteraction of the host innate response. *J. Virol.* **2006**, *80*, 7600–7612. [CrossRef] [PubMed]
66. Wylie, K.M.; Schrimpf, J.E.; Morrison, L.A. Increased eIF2alpha phosphorylation attenuates replication of herpes simplex virus 2 vhs mutants in mouse embryonic fibroblasts and correlates with reduced accumulation of the PKR antagonist ICP34.5. *J. Virol.* **2009**, *83*, 9151–9162. [CrossRef]
67. He, B.; Gross, M.; Roizman, B. The gamma(1)34.5 protein of herpes simplex virus 1 complexes with protein phosphatase 1alpha to dephosphorylate the alpha subunit of the eukaryotic translation initiation factor 2 and preclude the shutoff of protein synthesis by double-stranded RNA-activated protein kinase. *Proc. Natl. Acad. Sci. USA* **1997**, *94*, 843–848.
68. Li, Y.; Zhang, C.; Chen, X.; Yu, J.; Wang, Y.; Yang, Y.; Du, M.; Jin, H.; Ma, Y.; He, B.; et al. ICP34.5 protein of herpes simplex virus facilitates the initiation of protein translation by bridging eukaryotic initiation factor 2alpha (eIF2alpha) and protein phosphatase 1. *J. Biol. Chem.* **2011**, *286*, 24785–24792. [CrossRef]
69. Orr, M.T.; Edelmann, K.H.; Vieira, J.; Corey, L.; Raulet, D.H.; Wilson, C.B. Inhibition of MHC class I is a virulence factor in herpes simplex virus infection of mice. *PLoS Pathog* **2005**, *1*, e7. [CrossRef]
70. Eggensperger, S.; Tampe, R. The transporter associated with antigen processing: A key player in adaptive immunity. *Biol. Chem.* **2015**, *396*, 1059–1072. [CrossRef]
71. Goldsmith, K.; Chen, W.; Johnson, D.C.; Hendricks, R.L. Infected cell protein (ICP)47 enhances herpes simplex virus neurovirulence by blocking the CD8+ T cell response. *J. Exp. Med.* **1998**, *187*, 341–348. [CrossRef]
72. Liu, B.L.; Robinson, M.; Han, Z.Q.; Branston, R.H.; English, C.; Reay, P.; McGrath, Y.; Thomas, S.K.; Thornton, M.; Bullock, P.; et al. ICP34.5 deleted herpes simplex virus with enhanced oncolytic, immune stimulating, and anti-tumour properties. *Gene Ther.* **2003**, *10*, 292–303. [CrossRef] [PubMed]
73. Wollmann, G.; Ozduman, K.; van den Pol, A.N. Oncolytic virus therapy for glioblastoma multiforme: concepts and candidates. *Cancer J.* **2012**, *18*, 69–81. [CrossRef] [PubMed]
74. Mulvey, M.; Poppers, J.; Sternberg, D.; Mohr, I. Regulation of eIF2alpha phosphorylation by different functions that act during discrete phases in the herpes simplex virus type 1 life cycle. *J. Virol.* **2003**, *77*, 10917–10928. [CrossRef] [PubMed]
75. Cassady, K.A.; Gross, M.; Roizman, B. The herpes simplex virus US11 protein effectively compensates for the gamma1(34.5) gene if present before activation of protein kinase R by precluding its phosphorylation and that of the alpha subunit of eukaryotic translation initiation factor 2. *J. Virol.* **1998**, *72*, 8620–8626. [CrossRef] [PubMed]
76. Saha, D.; Wakimoto, H.; Peters, C.W.; Antoszczyk, S.J.; Rabkin, S.D.; Martuza, R.L. Combinatorial effects of VEGFR kinase inhibitor axitinib and oncolytic virotherapy in mouse and human glioblastoma stem-like cell models. *Clin. Cancer Res.* **2018**. [CrossRef] [PubMed]
77. Todo, T. "Armed" oncolytic herpes simplex viruses for brain tumor therapy. *Cell Adh. Migr.* **2008**, *2*, 208–213. [CrossRef]
78. Kelly, K.J.; Wong, J.; Fong, Y. Herpes simplex virus NV1020 as a novel and promising therapy for hepatic malignancy. *Expert Opin. Investig. Drugs* **2008**, *17*, 1105–1113. [CrossRef]
79. Bennett, J.J.; Malhotra, S.; Wong, R.J.; Delman, K.; Zager, J.; St-Louis, M.; Johnson, P.; Fong, Y. Interleukin 12 secretion enhances antitumor efficacy of oncolytic herpes simplex viral therapy for colorectal cancer. *Ann. Surg.* **2001**, *233*, 819–826. [CrossRef]
80. Smith, M.C.; Boutell, C.; Davido, D.J. HSV-1 ICP0: Paving the way for viral replication. *Future Virol.* **2011**, *6*, 421–429. [CrossRef]
81. Lanfranca, M.P.; Mostafa, H.H.; Davido, D.J. HSV-1 ICP0: An E3 Ubiquitin Ligase That Counteracts Host Intrinsic and Innate Immunity. *Cells* **2014**, *3*, 438–454. [CrossRef]

82. Jacobs, A.; Breakefield, X.O.; Fraefel, C. HSV-1-based vectors for gene therapy of neurological diseases and brain tumors: Part I. HSV-1 structure, replication and pathogenesis. *Neoplasia* **1999**, *1*, 387–401. [CrossRef] [PubMed]
83. Parker, J.N.; Pfister, L.A.; Quenelle, D.; Gillespie, G.Y.; Markert, J.M.; Kern, E.R.; Whitley, R.J. Genetically engineered herpes simplex viruses that express IL-12 or GM-CSF as vaccine candidates. *Vaccine* **2006**, *24*, 1644–1652. [CrossRef] [PubMed]
84. Parker, J.N.; Gillespie, G.Y.; Love, C.E.; Randall, S.; Whitley, R.J.; Markert, J.M. Engineered herpes simplex virus expressing IL-12 in the treatment of experimental murine brain tumors. *Proc. Natl. Acad. Sci. USA* **2000**, *97*, 2208–2213. [CrossRef] [PubMed]
85. Bauer, D.F.; Pereboeva, L.; Gillespie, G.Y.; Cloud, G.A.; Elzafarany, O.; Langford, C.; Markert, J.M.; Lamb, L.S., Jr. Effect of HSV-IL12 Loaded Tumor Cell-Based Vaccination in a Mouse Model of High-Grade Neuroblastoma. *J. Immunol. Res.* **2016**, *2016*, 2568125. [CrossRef] [PubMed]
86. Hellums, E.K.; Markert, J.M.; Parker, J.N.; He, B.; Perbal, B.; Roizman, B.; Whitley, R.J.; Langford, C.P.; Bharara, S.; Gillespie, G.Y. Increased efficacy of an interleukin-12-secreting herpes simplex virus in a syngeneic intracranial murine glioma model. *Neuro. Oncol.* **2005**, *7*, 213–224. [CrossRef] [PubMed]
87. Markert, J.M.; Cody, J.J.; Parker, J.N.; Coleman, J.M.; Price, K.H.; Kern, E.R.; Quenelle, D.C.; Lakeman, A.D.; Schoeb, T.R.; Palmer, C.A.; et al. Preclinical evaluation of a genetically engineered herpes simplex virus expressing interleukin-12. *J. Virol.* **2012**, *86*, 5304–5313. [CrossRef]
88. Patel, D.M.; Foreman, P.M.; Nabors, L.B.; Riley, K.O.; Gillespie, G.Y.; Markert, J.M. Design of a Phase I Clinical Trial to Evaluate M032, a Genetically Engineered HSV-1 Expressing IL-12, in Patients with Recurrent/Progressive Glioblastoma Multiforme, Anaplastic Astrocytoma, or Gliosarcoma. *Hum. Gene Ther. Clin. Dev.* **2016**, *27*, 69–78. [CrossRef]
89. Menotti, L.; Avitabile, E.; Gatta, V.; Malatesta, P.; Petrovic, B.; Campadelli-Fiume, G. HSV as A Platform for the Generation of Retargeted, Armed, and Reporter-Expressing Oncolytic Viruses. *Viruses* **2018**, *10*, 352. [CrossRef]
90. Menotti, L.; Cerretani, A.; Hengel, H.; Campadelli-Fiume, G. Construction of a fully retargeted herpes simplex virus 1 recombinant capable of entering cells solely via human epidermal growth factor receptor 2. *J. Virol.* **2008**, *82*, 10153–10161. [CrossRef]
91. Liu, X.J.; Wang, X.Y.; Guo, J.X.; Zhu, H.J.; Zhang, C.R.; Ma, Z.H. Oncolytic property of HSV-1 recombinant viruses carrying the human IL-12. *Zhonghua Yi Xue Za Zhi* **2017**, *97*, 2135–2140. [CrossRef]
92. Athie-Morales, V.; Smits, H.H.; Cantrell, D.A.; Hilkens, C.M. Sustained IL-12 signaling is required for Th1 development. *J. Immunol.* **2004**, *172*, 61–69. [CrossRef] [PubMed]
93. Grivennikov, S.I.; Greten, F.R.; Karin, M. Immunity, inflammation, and cancer. *Cell* **2010**, *140*, 883–899. [CrossRef] [PubMed]
94. Antoszczyk, S.; Spyra, M.; Mautner, V.F.; Kurtz, A.; Stemmer-Rachamimov, A.O.; Martuza, R.L.; Rabkin, S.D. Treatment of orthotopic malignant peripheral nerve sheath tumors with oncolytic herpes simplex virus. *Neuro. Oncol.* **2014**, *16*, 1057–1066. [CrossRef] [PubMed]
95. Varghese, S.; Rabkin, S.D.; Nielsen, G.P.; MacGarvey, U.; Liu, R.; Martuza, R.L. Systemic therapy of spontaneous prostate cancer in transgenic mice with oncolytic herpes simplex viruses. *Cancer Res.* **2007**, *67*, 9371–9379. [CrossRef]
96. Ring, E.K.; Li, R.; Moore, B.P.; Nan, L.; Kelly, V.M.; Han, X.; Beierle, E.A.; Markert, J.M.; Leavenworth, J.W.; Gillespie, G.Y.; et al. Newly Characterized Murine Undifferentiated Sarcoma Models Sensitive to Virotherapy with Oncolytic HSV-1 M002. *Mol. Ther. Oncolytics* **2017**, *7*, 27–36. [CrossRef]
97. Thomas, E.D.; Meza-Perez, S.; Bevis, K.S.; Randall, T.D.; Gillespie, G.Y.; Langford, C.; Alvarez, R.D. IL-12 Expressing oncolytic herpes simplex virus promotes anti-tumor activity and immunologic control of metastatic ovarian cancer in mice. *J. Ovarian Res.* **2016**, *9*, 70. [CrossRef]
98. Krist, L.F.; Kerremans, M.; Broekhuis-Fluitsma, D.M.; Eestermans, I.L.; Meyer, S.; Beelen, R.H. Milky spots in the greater omentum are predominant sites of local tumour cell proliferation and accumulation in the peritoneal cavity. *Cancer Immunol. Immunother.* **1998**, *47*, 205–212. [CrossRef]
99. Sorensen, E.W.; Gerber, S.A.; Frelinger, J.G.; Lord, E.M. IL-12 suppresses vascular endothelial growth factor receptor 3 expression on tumor vessels by two distinct IFN-gamma-dependent mechanisms. *J. Immunol.* **2010**, *184*, 1858–1866. [CrossRef]

100. Cicchelero, L.; Denies, S.; Haers, H.; Vanderperren, K.; Stock, E.; Van Brantegem, L.; de Rooster, H.; Sanders, N.N. Intratumoural interleukin 12 gene therapy stimulates the immune system and decreases angiogenesis in dogs with spontaneous cancer. *Vet. Comp. Oncol.* **2017**, *15*, 1187–1205. [CrossRef]
101. Wong, R.J.; Chan, M.K.; Yu, Z.; Ghossein, R.A.; Ngai, I.; Adusumilli, P.S.; Stiles, B.M.; Shah, J.P.; Singh, B.; Fong, Y. Angiogenesis inhibition by an oncolytic herpes virus expressing interleukin 12. *Clin. Cancer Res.* **2004**, *10*, 4509–4516. [CrossRef]
102. Das, S.; Marsden, P.A. Angiogenesis in glioblastoma. *N Engl. J. Med.* **2013**, *369*, 1561–1563. [CrossRef]
103. Gatson, N.N.; Chiocca, E.A.; Kaur, B. Anti-angiogenic gene therapy in the treatment of malignant gliomas. *Neurosci. Lett.* **2012**, *527*, 62–70. [CrossRef]
104. Hardee, M.E.; Zagzag, D. Mechanisms of glioma-associated neovascularization. *Am. J. Pathol.* **2012**, *181*, 1126–1141. [CrossRef] [PubMed]
105. Jain, R.K.; di Tomaso, E.; Duda, D.G.; Loeffler, J.S.; Sorensen, A.G.; Batchelor, T.T. Angiogenesis in brain tumours. *Nat. Rev. Neurosci.* **2007**, *8*, 610–622. [CrossRef] [PubMed]
106. Zirlik, K.; Duyster, J. Anti-Angiogenics: Current Situation and Future Perspectives. *Oncol. Res. Treat.* **2018**, *41*, 166–171. [CrossRef] [PubMed]
107. Ramjiawan, R.R.; Griffioen, A.W.; Duda, D.G. Anti-angiogenesis for cancer revisited: Is there a role for combinations with immunotherapy? *Angiogenesis* **2017**, *20*, 185–204. [CrossRef] [PubMed]
108. Rajabi, M.; Mousa, S.A. The Role of Angiogenesis in Cancer Treatment. *Biomedicines* **2017**, *5*, 34. [CrossRef]
109. Melegh, Z.; Oltean, S. Targeting Angiogenesis in Prostate Cancer. *Int. J. Mol. Sci.* **2019**, *20*, 2676. [CrossRef]
110. van Moorselaar, R.J.; Voest, E.E. Angiogenesis in prostate cancer: Its role in disease progression and possible therapeutic approaches. *Mol. Cell Endocrinol.* **2002**, *197*, 239–250. [CrossRef]
111. Batchelor, T.T.; Reardon, D.A.; de Groot, J.F.; Wick, W.; Weller, M. Antiangiogenic therapy for glioblastoma: Current status and future prospects. *Clin. Cancer Res.* **2014**, *20*, 5612–5619. [CrossRef]
112. Wick, W.; Chinot, O.L.; Bendszus, M.; Mason, W.; Henriksson, R.; Saran, F.; Nishikawa, R.; Revil, C.; Kerloeguen, Y.; Cloughesy, T. Evaluation of pseudoprogression rates and tumor progression patterns in a phase III trial of bevacizumab plus radiotherapy/temozolomide for newly diagnosed glioblastoma. *Neuro. Oncol.* **2016**, *18*, 1434–1441. [CrossRef] [PubMed]
113. Lombardi, G.; Pambuku, A.; Bellu, L.; Farina, M.; Della Puppa, A.; Denaro, L.; Zagonel, V. Effectiveness of antiangiogenic drugs in glioblastoma patients: A systematic review and meta-analysis of randomized clinical trials. *Crit. Rev. Oncol. Hematol.* **2017**, *111*, 94–102. [CrossRef] [PubMed]
114. Kelly, W.K.; Halabi, S.; Carducci, M.; George, D.; Mahoney, J.F.; Stadler, W.M.; Morris, M.; Kantoff, P.; Monk, J.P.; Kaplan, E.; et al. Randomized, double-blind, placebo-controlled phase III trial comparing docetaxel and prednisone with or without bevacizumab in men with metastatic castration-resistant prostate cancer: CALGB 90401. *J. Clin. Oncol.* **2012**, *30*, 1534–1540. [CrossRef]
115. Hu-Lowe, D.D.; Zou, H.Y.; Grazzini, M.L.; Hallin, M.E.; Wickman, G.R.; Amundson, K.; Chen, J.H.; Rewolinski, D.A.; Yamazaki, S.; Wu, E.Y.; et al. Nonclinical antiangiogenesis and antitumor activities of axitinib (AG-013736), an oral, potent, and selective inhibitor of vascular endothelial growth factor receptor tyrosine kinases 1, 2, 3. *Clin. Cancer Res.* **2008**, *14*, 7272–7283. [CrossRef] [PubMed]
116. Ho, A.L.; Dunn, L.; Sherman, E.J.; Fury, M.G.; Baxi, S.S.; Chandramohan, R.; Dogan, S.; Morris, L.G.; Cullen, G.D.; Haque, S.; et al. A phase II study of axitinib (AG-013736) in patients with incurable adenoid cystic carcinoma. *Ann. Oncol.* **2016**, *27*, 1902–1908. [CrossRef] [PubMed]
117. McNamara, M.G.; Le, L.W.; Horgan, A.M.; Aspinall, A.; Burak, K.W.; Dhani, N.; Chen, E.; Sinaei, M.; Lo, G.; Kim, T.K.; et al. A phase II trial of second-line axitinib following prior antiangiogenic therapy in advanced hepatocellular carcinoma. *Cancer* **2015**, *121*, 1620–1627. [CrossRef] [PubMed]
118. Schiller, J.H.; Larson, T.; Ou, S.H.; Limentani, S.; Sandler, A.; Vokes, E.; Kim, S.; Liau, K.; Bycott, P.; Olszanski, A.J.; et al. Efficacy and safety of axitinib in patients with advanced non-small-cell lung cancer: Results from a phase II study. *J. Clin. Oncol.* **2009**, *27*, 3836–3841. [CrossRef]
119. Du Four, S.; Maenhout, S.K.; Benteyn, D.; De Keersmaecker, B.; Duerinck, J.; Thielemans, K.; Neyns, B.; Aerts, J.L. Disease progression in recurrent glioblastoma patients treated with the VEGFR inhibitor axitinib is associated with increased regulatory T cell numbers and T cell exhaustion. *Cancer Immunol. Immunother.* **2016**, *65*, 727–740. [CrossRef]

120. Du Four, S.; Maenhout, S.K.; De Pierre, K.; Renmans, D.; Niclou, S.P.; Thielemans, K.; Neyns, B.; Aerts, J.L. Axitinib increases the infiltration of immune cells and reduces the suppressive capacity of monocytic MDSCs in an intracranial mouse melanoma model. *Oncoimmunology* **2015**, *4*, e998107. [CrossRef]
121. Laubli, H.; Muller, P.; D'Amico, L.; Buchi, M.; Kashyap, A.S.; Zippelius, A. The multi-receptor inhibitor axitinib reverses tumor-induced immunosuppression and potentiates treatment with immune-modulatory antibodies in preclinical murine models. *Cancer Immunol. Immunother.* **2018**, *67*, 815–824. [CrossRef]
122. Wilmes, L.J.; Pallavicini, M.G.; Fleming, L.M.; Gibbs, J.; Wang, D.; Li, K.L.; Partridge, S.C.; Henry, R.G.; Shalinsky, D.R.; Hu-Lowe, D.; et al. AG-013736, a novel inhibitor of VEGF receptor tyrosine kinases, inhibits breast cancer growth and decreases vascular permeability as detected by dynamic contrast-enhanced magnetic resonance imaging. *Magn Reson Imaging* **2007**, *25*, 319–327. [CrossRef] [PubMed]
123. Chen, Y.; Rini, B.I.; Motzer, R.J.; Dutcher, J.P.; Rixe, O.; Wilding, G.; Stadler, W.M.; Tarazi, J.; Garrett, M.; Pithavala, Y.K. Effect of Renal Impairment on the Pharmacokinetics and Safety of Axitinib. *Target Oncol.* **2016**, *11*, 229–234. [CrossRef]
124. Gross-Goupil, M.; Francois, L.; Quivy, A.; Ravaud, A. Axitinib: A review of its safety and efficacy in the treatment of adults with advanced renal cell carcinoma. *Clin. Med. Insights Oncol.* **2013**, *7*, 269–277. [CrossRef] [PubMed]
125. Zhang, W.; Fulci, G.; Wakimoto, H.; Cheema, T.A.; Buhrman, J.S.; Jeyaretna, D.S.; Stemmer Rachamimov, A.O.; Rabkin, S.D.; Martuza, R.L. Combination of oncolytic herpes simplex viruses armed with angiostatin and IL-12 enhances antitumor efficacy in human glioblastoma models. *Neoplasia* **2013**, *15*, 591–599. [CrossRef] [PubMed]
126. Passer, B.J.; Cheema, T.; Wu, S.; Wu, C.L.; Rabkin, S.D.; Martuza, R.L. Combination of vinblastine and oncolytic herpes simplex virus vector expressing IL-12 therapy increases antitumor and antiangiogenic effects in prostate cancer models. *Cancer Gene Ther.* **2013**, *20*, 17–24. [CrossRef]
127. Chen, C.Y.; Hutzen, B.; Wedekind, M.F.; Cripe, T.P. Oncolytic virus and PD-1/PD-L1 blockade combination therapy. *Oncolytic Virother* **2018**, *7*, 65–77. [CrossRef]
128. Huang, J.; Liu, F.; Liu, Z.; Tang, H.; Wu, H.; Gong, Q.; Chen, J. Immune Checkpoint in Glioblastoma: Promising and Challenging. *Front. Pharmacol.* **2017**, *8*, 242. [CrossRef]
129. Sharma, P.; Allison, J.P. Immune checkpoint targeting in cancer therapy: Toward combination strategies with curative potential. *Cell* **2015**, *161*, 205–214. [CrossRef]
130. Twyman-Saint Victor, C.; Rech, A.J.; Maity, A.; Rengan, R.; Pauken, K.E.; Stelekati, E.; Benci, J.L.; Xu, B.; Dada, H.; Odorizzi, P.M.; et al. Radiation and dual checkpoint blockade activate non-redundant immune mechanisms in cancer. *Nature* **2015**, *520*, 373–377. [CrossRef]
131. Curran, M.A.; Montalvo, W.; Yagita, H.; Allison, J.P. PD-1 and CTLA-4 combination blockade expands infiltrating T cells and reduces regulatory T and myeloid cells within B16 melanoma tumors. *Proc. Natl. Acad. Sci. USA* **2010**, *107*, 4275–4280. [CrossRef]
132. Topalian, S.L.; Drake, C.G.; Pardoll, D.M. Immune checkpoint blockade: A common denominator approach to cancer therapy. *Cancer Cell* **2015**, *27*, 450–461. [CrossRef]
133. El-Shemi, A.G.; Ashshi, A.M.; Na, Y.; Li, Y.; Basalamah, M.; Al-Allaf, F.A.; Oh, E.; Jung, B.K.; Yun, C.O. Combined therapy with oncolytic adenoviruses encoding TRAIL and IL-12 genes markedly suppressed human hepatocellular carcinoma both in vitro and in an orthotopic transplanted mouse model. *J. Exp. Clin. Cancer Res.* **2016**, *35*, 74. [CrossRef]
134. Zhang, S.N.; Choi, I.K.; Huang, J.H.; Yoo, J.Y.; Choi, K.J.; Yun, C.O. Optimizing DC vaccination by combination with oncolytic adenovirus coexpressing IL-12 and GM-CSF. *Mol. Ther.* **2011**, *19*, 1558–1568. [CrossRef] [PubMed]
135. Kim, W.; Seong, J.; Oh, H.J.; Koom, W.S.; Choi, K.J.; Yun, C.O. A novel combination treatment of armed oncolytic adenovirus expressing IL-12 and GM-CSF with radiotherapy in murine hepatocarcinoma. *J. Radiat. Res.* **2011**, *52*, 646–654. [CrossRef] [PubMed]
136. Oh, E.; Oh, J.E.; Hong, J.; Chung, Y.; Lee, Y.; Park, K.D.; Kim, S.; Yun, C.O. Optimized biodegradable polymeric reservoir-mediated local and sustained co-delivery of dendritic cells and oncolytic adenovirus co-expressing IL-12 and GM-CSF for cancer immunotherapy. *J. Control. Release* **2017**, *259*, 115–127. [CrossRef] [PubMed]

137. Bortolanza, S.; Bunuales, M.; Otano, I.; Gonzalez-Aseguinolaza, G.; Ortiz-de-Solorzano, C.; Perez, D.; Prieto, J.; Hernandez-Alcoceba, R. Treatment of pancreatic cancer with an oncolytic adenovirus expressing interleukin-12 in Syrian hamsters. *Mol. Ther.* **2009**, *17*, 614–622. [CrossRef]
138. Yang, Z.; Zhang, Q.; Xu, K.; Shan, J.; Shen, J.; Liu, L.; Xu, Y.; Xia, F.; Bie, P.; Zhang, X.; et al. Combined therapy with cytokine-induced killer cells and oncolytic adenovirus expressing IL-12 induce enhanced antitumor activity in liver tumor model. *PLoS ONE* **2012**, *7*, e44802. [CrossRef]
139. Quetglas, J.I.; Labiano, S.; Aznar, M.A.; Bolanos, E.; Azpilikueta, A.; Rodriguez, I.; Casales, E.; Sanchez-Paulete, A.R.; Segura, V.; Smerdou, C.; et al. Virotherapy with a Semliki Forest Virus-Based Vector Encoding IL12 Synergizes with PD-1/PD-L1 Blockade. *Cancer Immunol. Res.* **2015**, *3*, 449–454. [CrossRef]
140. Schirmacher, V.; Forg, P.; Dalemans, W.; Chlichlia, K.; Zeng, Y.; Fournier, P.; von Hoegen, P. Intra-pinna anti-tumor vaccination with self-replicating infectious RNA or with DNA encoding a model tumor antigen and a cytokine. *Gene Ther.* **2000**, *7*, 1137–1147. [CrossRef]
141. Quetglas, J.I.; Dubrot, J.; Bezunartea, J.; Sanmamed, M.F.; Hervas-Stubbs, S.; Smerdou, C.; Melero, I. Immunotherapeutic synergy between anti-CD137 mAb and intratumoral administration of a cytopathic Semliki Forest virus encoding IL-12. *Mol. Ther.* **2012**, *20*, 1664–1675. [CrossRef]
142. Klas, S.D.; Robison, C.S.; Whitt, M.A.; Miller, M.A. Adjuvanticity of an IL-12 fusion protein expressed by recombinant deltaG-vesicular stomatitis virus. *Cell Immunol.* **2002**, *218*, 59–73. [CrossRef]
143. Lawler, S.E.; Speranza, M.C.; Cho, C.F.; Chiocca, E.A. Oncolytic Viruses in Cancer Treatment: A Review. *JAMA Oncol.* **2017**, *3*, 841–849. [CrossRef] [PubMed]
144. Senzer, N.N.; Kaufman, H.L.; Amatruda, T.; Nemunaitis, M.; Reid, T.; Daniels, G.; Gonzalez, R.; Glaspy, J.; Whitman, E.; Harrington, K.; et al. Phase II clinical trial of a granulocyte-macrophage colony-stimulating factor-encoding, second-generation oncolytic herpesvirus in patients with unresectable metastatic melanoma. *J. Clin. Oncol* **2009**, *27*, 5763–5771. [CrossRef]
145. Johnson, D.B.; Puzanov, I.; Kelley, M.C. Talimogene laherparepvec (T-VEC) for the treatment of advanced melanoma. *Immunotherapy* **2015**, *7*, 611–619. [CrossRef]
146. Bommareddy, P.K.; Patel, A.; Hossain, S.; Kaufman, H.L. Talimogene Laherparepvec (T-VEC) and Other Oncolytic Viruses for the Treatment of Melanoma. *Am. J. Clin. Dermatol* **2017**, *18*, 1–15. [CrossRef] [PubMed]
147. Vere Hodge, R.A.; Field, H.J. Antiviral agents for herpes simplex virus. *Adv. Pharmacol.* **2013**, *67*, 1–38. [CrossRef] [PubMed]
148. Parker, J.N.; Meleth, S.; Hughes, K.B.; Gillespie, G.Y.; Whitley, R.J.; Markert, J.M. Enhanced inhibition of syngeneic murine tumors by combinatorial therapy with genetically engineered HSV-1 expressing CCL2 and IL-12. *Cancer Gene Ther.* **2005**, *12*, 359–368. [CrossRef]
149. Ghouse, S.M.; Bommareddy, P.K.; Nguyen, H.M.; Guz-Montgomery, K.; Saha, D. Oncolytic herpes simplex virus encoding IL12 controls triple-negative breast cancer growth and metastasis in CD8-dependent manner. In Proceedings of the International Oncolytic Virus Conference, Rochester, MN, USA, 9–12 October 2019.
150. Esaki, S.; Nigim, F.; Moon, E.; Luk, S.; Kiyokawa, J.; Curry, W., Jr.; Cahill, D.P.; Chi, A.S.; Iafrate, A.J.; Martuza, R.L.; et al. Blockade of transforming growth factor-beta signaling enhances oncolytic herpes simplex virus efficacy in patient-derived recurrent glioblastoma models. *Int. J. Cancer* **2017**, *141*, 2348–2358. [CrossRef]
151. Ning, J.; Wakimoto, H.; Peters, C.; Martuza, R.L.; Rabkin, S.D. Rad51 Degradation: Role in Oncolytic Virus-Poly(ADP-Ribose) Polymerase Inhibitor Combination Therapy in Glioblastoma. *J. Natl. Cancer Inst.* **2017**, *109*, 1–13. [CrossRef]
152. Johnson, D.B.; Chandra, S.; Sosman, J.A. Immune Checkpoint Inhibitor Toxicity in 2018. *JAMA* **2018**, *320*, 1702–1703. [CrossRef]
153. Lyon, A.R.; Yousaf, N.; Battisti, N.M.L.; Moslehi, J.; Larkin, J. Immune checkpoint inhibitors and cardiovascular toxicity. *Lancet Oncol.* **2018**, *19*, e447–e458. [CrossRef]
154. Pai, C.S.; Simons, D.M.; Lu, X.; Evans, M.; Wei, J.; Wang, Y.H.; Chen, M.; Huang, J.; Park, C.; Chang, A.; et al. Tumor-conditional anti-CTLA4 uncouples antitumor efficacy from immunotherapy-related toxicity. *J. Clin. Investig.* **2019**, *129*, 349–363. [CrossRef] [PubMed]
155. Palmieri, D.J.; Carlino, M.S. Immune Checkpoint Inhibitor Toxicity. *Curr. Oncol Rep.* **2018**, *20*, 72. [CrossRef] [PubMed]

156. Rota, E.; Varese, P.; Agosti, S.; Celli, L.; Ghiglione, E.; Pappalardo, I.; Zaccone, G.; Paglia, A.; Morelli, N. Concomitant myasthenia gravis, myositis, myocarditis and polyneuropathy, induced by immune-checkpoint inhibitors: A life-threatening continuum of neuromuscular and cardiac toxicity. *eNeurologicalSci* **2019**, *14*, 4–5. [CrossRef]
157. Filley, A.C.; Henriquez, M.; Dey, M. Recurrent glioma clinical trial, CheckMate-143: The game is not over yet. *Oncotarget* **2017**, *8*, 91779–91794. [CrossRef]
158. Omuro, A.; Vlahovic, G.; Lim, M.; Sahebjam, S.; Baehring, J.; Cloughesy, T.; Voloschin, A.; Ramkissoon, S.H.; Ligon, K.L.; Latek, R.; et al. Nivolumab with or without ipilimumab in patients with recurrent glioblastoma: Results from exploratory phase I cohorts of CheckMate 143. *Neuro. Oncol.* **2018**, *20*, 674–686. [CrossRef]
159. Emens, L.A. Breast Cancer Immunotherapy: Facts and Hopes. *Clin. Cancer Res.* **2018**, *24*, 511–520. [CrossRef]
160. Conry, R.M.; Westbrook, B.; McKee, S.; Norwood, T.G. Talimogene laherparepvec: First in class oncolytic virotherapy. *Hum. Vaccin. Immunother.* **2018**, *14*, 839–846. [CrossRef]
161. Siurala, M.; Havunen, R.; Saha, D.; Lumen, D.; Airaksinen, A.J.; Tahtinen, S.; Cervera-Carrascon, V.; Bramante, S.; Parviainen, S.; Vaha-Koskela, M.; et al. Adenoviral Delivery of Tumor Necrosis Factor-alpha and Interleukin-2 Enables Successful Adoptive Cell Therapy of Immunosuppressive Melanoma. *Mol. Ther.* **2016**, *24*, 1435–1443. [CrossRef]
162. Tahtinen, S.; Blattner, C.; Vaha-Koskela, M.; Saha, D.; Siurala, M.; Parviainen, S.; Utikal, J.; Kanerva, A.; Umansky, V.; Hemminki, A. T-Cell Therapy Enabling Adenoviruses Coding for IL2 and TNFalpha Induce Systemic Immunomodulation in Mice With Spontaneous Melanoma. *J. Immunother.* **2016**, *39*, 343–354. [CrossRef]
163. Martinet, O.; Divino, C.M.; Zang, Y.; Gan, Y.; Mandeli, J.; Thung, S.; Pan, P.Y.; Chen, S.H. T cell activation with systemic agonistic antibody versus local 4-1BB ligand gene delivery combined with interleukin-12 eradicate liver metastases of breast cancer. *Gene Ther.* **2002**, *9*, 786–792. [CrossRef]
164. Kurokawa, C.; Iankov, I.D.; Anderson, S.K.; Aderca, I.; Leontovich, A.A.; Maurer, M.J.; Oberg, A.L.; Schroeder, M.A.; Giannini, C.; Greiner, S.M.; et al. Constitutive Interferon Pathway Activation in Tumors as an Efficacy Determinant Following Oncolytic Virotherapy. *J. Natl. Cancer Inst.* **2018**, *110*, 1123–1132. [CrossRef]

8

The Combination of IFN β and TNF Induces an Antiviral and Immunoregulatory Program via Non-Canonical Pathways Involving STAT2 and IRF9

Mélissa K. Mariani [1], Pouria Dasmeh [2,3], Audray Fortin [1], Elise Caron [1], Mario Kalamujic [1], Alexander N. Harrison [1,4], Diana I. Hotea [1,2], Dacquin M. Kasumba [1,2], Sandra L. Cervantes-Ortiz [1,5], Espérance Mukawera [1], Adrian W. R. Serohijos [2,3] and Nathalie Grandvaux [1,2,*]

1 CRCHUM—Centre Hospitalier de l'Université de Montréal, Montréal, QC H2X 0A9, Canada
2 Department of Biochemistry and Molecular Medicine, Faculty of Medicine, Université de Montréal, Montréal, QC H3T 1J4, Canada
3 Centre Robert Cedergren en Bioinformatique et Génomique, Université de Montréal, Montréal, QC H3T 1J4, Canada
4 Department of Microbiology and Immunology, McGill University, Montréal, QC H3A 2B4, Canada
5 Department of Microbiology, Infectiology and Immunology, Faculty of Medicine, Université de Montréal, Montréal, QC H3T 1J4, Canada
* Correspondence: nathalie.grandvaux@umontreal.ca

Abstract: Interferon (IFN) β and Tumor Necrosis Factor (TNF) are key players in immunity against viruses. Compelling evidence has shown that the antiviral and inflammatory transcriptional response induced by IFNβ is reprogrammed by crosstalk with TNF. IFNβ mainly induces interferon-stimulated genes by the Janus kinase (JAK)/signal transducer and activator of transcription (STAT) pathway involving the canonical ISGF3 transcriptional complex, composed of STAT1, STAT2, and IRF9. The signaling pathways engaged downstream of the combination of IFNβ and TNF remain elusive, but previous observations suggested the existence of a response independent of STAT1. Here, using genome-wide transcriptional analysis by RNASeq, we observed a broad antiviral and immunoregulatory response initiated in the absence of STAT1 upon IFNβ and TNF costimulation. Additional stratification of this transcriptional response revealed that STAT2 and IRF9 mediate the expression of a wide spectrum of genes. While a subset of genes was regulated by the concerted action of STAT2 and IRF9, other gene sets were independently regulated by STAT2 or IRF9. Collectively, our data supports a model in which STAT2 and IRF9 act through non-canonical parallel pathways to regulate distinct pool of antiviral and immunoregulatory genes in conditions with elevated levels of both IFNβ and TNF.

Keywords: interferon; tumor necrosis factor; STAT; interferon regulatory factor; antiviral; autoimmunity; inflammation

---

1. Introduction

Interferon (IFN) β plays a critical role in the first line of defense against viruses, through its ability to induce a broad antiviral transcriptional response in virtually all cell types [1]. IFNβ also possesses key immunoregulatory functions that determine the outcome of the adaptive immune response against pathogens [1,2]. Over the years, in vitro and in vivo studies aimed at characterizing the mechanisms and the functional outcomes of IFNβ signaling were mostly performed in relation to single cytokine stimulation. This unlikely reflects physiological settings, as a plethora of cytokines are secreted in a specific situation. As a consequence, a cell rather simultaneously responds to a cocktail of cytokines to foster the appropriate transcriptional program. Response to IFNβ is no exception and is very context-dependent, particularly regarding the potential crosstalk with other cytokines. Elevated levels of IFNβ and Tumor Necrosis Factor (TNF) are found during the host response to viruses. Aberrant increased levels of both cytokines is also associated with a number of autoinflammatory and autoimmune diseases such as Systemic Lupus Erythematosus (SLE), psoriasis, and Sjögren's syndrome [3]. While the cross-regulation of IFNβ and TNF is well documented [4–6], the functional crosstalk between these two cytokines remains poorly known.

IFNβ typically acts through binding to the IFNAR receptor (IFNAR1 and IFNAR2) leading to the Janus kinase (JAK)/signal transducer and activator of transcription (STAT) pathway involving JAK1- and Tyk2-mediated phosphorylation of STAT1 and STAT2, and to a lesser extent other STAT members in a cell-specific manner [7,8]. Phosphorylated STAT1 and STAT2, together with IFN Regulatory Factor (IRF) 9, form the IFN-stimulated gene factor 3 (ISGF3) complex that binds to the consensus IFN-stimulated response element (ISRE) sequences in the promoter of hundreds of IFN stimulated genes (ISGs) [9]. Formation of the ISGF3 complex is considered a hallmark of the engagement of the type I IFN response and, consequently, the requirement of STAT1 in a specific setting has become a marker of the engagement of type I IFN signaling [7,10]. However, in recent years, this paradigm has started to be challenged with accumulating evidence demonstrating the existence of non-canonical JAK-STAT signaling that mediates type I IFN responses [8,11].

Synergism between IFNβ and TNF was shown to enhance the antiviral response to Vesicular Stomatitis Virus (VSV), Myxoma virus, and paramyxovirus infections [12–14]. Gene expression analyses showed that IFNβ and TNF synergistically regulate hundreds of genes induced by individual cytokines alone, but also drive a specific delayed transcriptional program composed of genes that are either not responsive to IFNβ or TNF separately or are only responsive to either one of the cytokine [4,13]. The signaling mechanisms engaged downstream of the costimulation with IFNβ and TNF remained elusive, but it is implicitly assumed that the fate of the gene expression response requires that both IFNβ- and TNF-induced signaling pathways exhibit significant crosstalk. Analysis of the enrichment of specific transcription factors binding sites in the promoters of a panel of genes synergistically induced by IFNβ and TNF failed to give a clue about the specificity of the transcriptional regulation of these genes [13]. We previously showed that the *DUOX2* gene belongs to the category of delayed genes that are remarkably induced to high levels in response to the combination of IFNβ and TNF in lung epithelial cells [8]. We found that *DUOX2* expression required STAT2 and IRF9 but not STAT1, suggesting that STAT2 and IRF9 activities might segregate in an alternative STAT1-independent pathway that could be involved in gene regulation downstream of IFNβ and TNF [14].

In the present study, we aimed to fully characterize the transcriptional profile of the delayed response to IFNβ and TNF that occurs independently of STAT1 and evaluate the role of STAT2 and IRF9 in the regulation of this response. We found that the costimulation by IFNβ and TNF induces a broad set of antiviral and immunoregulatory genes in the absence of STAT1. We also report the differential regulation of distinct subsets of IFNβ− and TNF-induced genes by STAT2 and IRF9. While IFNβ and TNF act in part through the concerted action of STAT2 and IRF9, specific sets of genes were only regulated by either STAT2 or IRF9. Altogether, our findings uncovered non-canonical STAT2 and/or IRF9-dependent pathways that coexist to regulate distinct pools of antiviral and immunoregulatory genes in a context of IFNβ and TNF crosstalk.

2. Materials and Methods

2.1. Cell Culture and Stimulation

A549 cells (American Type Culture Collection, ATCC) were grown in F-12 nutrient mixture (Ham) medium supplemented with 10% heat-inactivated fetal bovine serum (HI-FBS) and 1% L-glutamine. The 2ftGH fibrosarcoma cell line and the derived STAT1-deficient U3A cell line, a generous gift from Dr. G. Stark, Cleveland, USA [15], were grown in DMEM medium supplemented with 10% HI-FBS or

HI-Fetal Clone III (HI-FCl) and 1% L-glutamine. U3A cells stably expressing STAT1 were generated by transfection of the STAT1 alpha flag pRc/CMV plasmid (Addgene plasmid #8691; a generous gift from Dr. J. Darnell, Rockfeller University, USA [16,17]) and selection with 800 μg/ml Geneticin (G418). Monoclonal populations of U3A stably expressing STAT1 cells were isolated. A pool of two clones, referred to as U3A-STAT1, was used in the experiments to mitigate the clonal effects. U3A-STAT1 cells were maintained in culture in DMEM supplemented with 10% HI-FCl, 1% Glu, and 200 μg/mL G418. All cell lines were cultured without antibiotics except for the selection of stable cells. All media and supplements were from Gibco (Life Technologies, Grand Island, NY, USA), with the exception of HI-FCl, which was from HyClone (Logan, UT, USA). Mycoplasma contamination was excluded by regular analysis using the MycoAlert Mycoplasma Detection Kit (Lonza, Basel, Switzerland). Cells were stimulated with IFNβ (1000 U/mL, PBL Assay Science, Piscataway, NJ, USA), TNF (10 ng/mL, R&D Systems, Minneapolis, MN, USA) or IFNβ (1000 U/mL) +TNF (10 ng/mL) for the indicated times.

2.2. siRNA Transfection

The sequences of non-targeting control (Ctrl) and STAT2- and IRF9-directed RNAi oligonucleotides (Dharmacon, Lafayette, CO, USA) have previously been described in [14]. U3A cells at 30% confluency were transfected using the Oligofectamine transfection reagent (Life Technologies-Thermofisher, Carlsbad, CA, USA). RNAi transfection was pursued for 48 h before stimulation.

2.3. Immunoblot Analysis

Cells were lysed on ice using Nonidet P-40 lysis buffer as fully detailed in [18]. Whole-cell extracts (WCE) were quantified using the Bradford protein assay (Bio-Rad, Hercules, CA, USA), resolved by SDS-PAGE and transferred to nitrocellulose membrane before analysis by immunoblot. Membranes were incubated with the following primary antibodies: anti-actin Cat #MAB1501 from Millipore (Burlington, MA, USA), anti-IRF9 Cat #610285 from BD Transduction Laboratories (San Jose, CA, USA), and anti-STAT1-P-Tyr701 Cat #9171, anti-STAT2-P-Tyr690 Cat #4441, anti-STAT1 Cat #9172, anti-STAT2 Cat #4594, all from Cell Signaling (Danvers, MA, USA), before further incubation with horseradish peroxidase (HRP)-conjugated secondary antibodies (KPL, Gaithersburg, MD, USA or Jackson Immunoresearch Laboratories, West Grove, PA, USA). Antibodies were diluted in PBS containing 0.5% Tween and either 5% nonfat dry milk or BSA. Immunoreactive bands were visualized by enhanced chemiluminescence (Western Lightning Chemiluminescence Reagent Plus, Perkin-Elmer Life Sciences Waltham, MA, USA) using a LAS4000mini CCD camera apparatus (GE Healthcare, Mississauga, ON, Canada).

2.4. RNA Isolation and qRT-PCR Analyses

Total RNA was prepared using the RNAqueous-96 Isolation Kit (Invitrogen-Thermo Fisher, Carlsbad, CA, USA) following the manufacturer's instructions. Total RNA (1 μg) was subjected to reverse transcription using the QuantiTect Reverse Transcription Kit (Qiagen, Toronto, ON, Canada). Quantitative PCR were performed using either Fast start SYBR Green Kit (Roche, Indianapolis, IN, USA) for *MX1, IDO, APOBEC3G, CXCL10, NOD2, PKR, IRF1, IFIT1* and *IL8* or TaqMan Gene Expression Assays (Life Technologies-Thermo Fisher) for *DUOX2, IFI27, SERPINB2, IL33, CCL20, ISG20*. Sequences of oligonucleotides and probes used in PCR reactions are described in Supplemental Table S4. Data collection was performed on a Rotor-Gene 3000 Real Time Thermal Cycler (Corbett Research, Mortlake, Australia). Gene inductions were normalized over S9 levels, measured using Fast start SYBR Green Kit or TaqMan probe as necessary. Fold induction of genes was determined using the ΔΔCt method [19]. All qRT-PCR data are presented as the mean ± standard error of the mean (SEM).

2.5. RNA-Sequencing (RNASeq)

Total RNA prepared as described above was quantified using a NanoDrop Spectrophotometer ND-1000 (NanoDrop Technologies, Inc., Wilmington, DE, USA) and its integrity was assessed using a 2100 Bioanalyzer (Agilent Technologies, Santa Clara, CA, USA). Libraries were generated from 250 ng of total RNA using the NEBNext poly(A) magnetic isolation module and the KAPA stranded RNA-Seq library preparation kit (Kapa Biosystems, Wilmington, MA, USA), as per the manufacturer's recommendations. TruSeq adapters and PCR primers were purchased from IDT. Libraries were quantified using the Quant-iT™ PicoGreen® dsDNA Assay Kit (Molecular Probes, Eugene, OR, USA) and the Kapa Illumina GA with Revised Primers-SYBR Fast Universal kit (Kapa Biosystems). Average size fragment was determined using a LabChip GX (Perkin-Elmer Life Sciences, Waltham, MA, USA) instrument. Massively parallel sequencing was carried out on an Illumina HiSeq 2500 sequencer (Illumina Inc., San Diego, CA, USA). Read counts were obtained using HTSeq. Reads were trimmed from the 3′ end to have a Phred score of at least 30. Illumina sequencing adapters were removed from the reads and all reads were required to have a length of at least 32bp. Trimming and clipping was performed using Trimmomatic [20]. The filtered reads were aligned to the Homo-sapiens assembly GRCh37 reference genome. Each read set was aligned using STAR [21] and merged using Picard (http://broadinstitute.github.io/picard/). For all samples, the sequencing resulted in more than 29 million clean reads (ranging from 29 to 44 million reads) after removing low quality reads and adaptors. The reads were mapped to the total of 63,679 gene biotypes including 22,810 protein-coding genes. The non-specific filter for 1 count-per million reads (CPM) in at least three samples was applied to the reads and 14,254 genes passed this criterion.

2.6. Bioinformatics Analysis

Differential transcripts analysis. A reference-based transcript assembly was performed, which allows the detection of known and novel transcripts isoforms, using Cufflinks [22], merged using Cuffmerge (cufflinks/AllSamples/merged.gtf) and used as a reference to estimate transcript abundance and perform differential analysis using Cuffdiff and Cuffnorm tool to generate a normalized data set that includes all the samples. The fragments per kilobase million (FPKM) values calculated by Cufflinks were used as input. The transcript quantification engine of Cufflinks, Cuffdiff, was used to calculate transcript expression levels in more than one condition and test them for significant differences. To identify a transcript as being differentially expressed, Cuffdiff tests the observed log-fold-change in its expression against the null hypothesis of no change (i.e., the true log-fold-change is zero). Because of measurement errors, technical variability, and cross-replicate biological variability might result in an observed log-fold-change that is non-zero, Cuffdiff assesses significance using a model of variability in the log-fold-change under the null hypothesis. This model is described in detail in [23]. The differential gene expression analysis was performed using DESeq [24] and edgeR [25] within the R Bioconductor packages. Genes were considered differentially expressed between two group if they met the following requirement: fold change (FC) > ±1.5, $p < 0.05$, false discovery rate (FDR) < 0.05.

Enrichment of gene ontology (GO). GO enrichment analysis amongst differentially expressed genes (DEGs) was performed using Goseq [26] against the background of full human genome (hg19). GO-terms with adjusted p value < 0.05 were considered significantly enriched.

Clustering of DEGs. We categorized the DEGs according to their response upon silencing of siSTAT2 and siIRF9; categories are listed as A to I (Figure 2E). Then to determine relationship between these categories, we calculated the distance of centers of different categories. For each gene, we transformed siSTAT2 and siIRF9 FC to deviation from the mean FC of the category the respective gene belongs to using the equation: $FC_{new} = FC_{old} - \varepsilon\ (FC_{category})$. The parameter ε was estimated to give the perfect match between predefined categories (A to I) and clustering based on Euclidean distance. Results were plotted as a heatmap.

Modular transcription analysis. The *tmod* package in R [27] was used for modular transcription analysis. In brief, each transcriptional module is a set of genes that shows coherent expression across many biological samples [28,29]. Modular transcription analysis then calculates significant enrichment of a set of foreground genes, here DEGs, in pre-defined transcriptional module compared to a reference set. For transcriptional modules, we used a combined list of 606 distinct functional modules encompassing 12,712 genes, defined by Chaussabel et al. [30] and Li et al. [31], as the reference set in *tmod* package (Supplemental Table S5). The hypergeometric test devised in *tmodHGtest* was used to calculate enrichments and p-values employing Benjamini-Hochberg correction [32] for multiple sampling. All the statistical analyses and graphical presentations were performed in R [33].

2.7. Virus Titration by Plaque Assay

Quantification of VSV infectious virions was achieved through methylcellulose plaque forming unit assays. U3A and U3A-STAT1 cells were either left untreated or stimulated with IFNβ or IFNβ + TNF for 30 h. Cells were then infected with Vesicular Stomatitis Virus (VSV)-GFP (kindly provided by Dr. J. Bell, University of Ottawa, Canada) at a multiplicity of infection (MOI) of 5 for 1 h in serum-free medium (SFM). Cells were then washed twice with SFM and further cultured in DMEM medium containing 2% HI-FCl. The supernatants were harvested at 12 h post-infection and serial dilutions were used to infect confluent Vero cells (ATCC) for 1 h in SFM. The medium was then replaced with 1% methylcellulose in DMEM containing 10% HI-FCl. Two days post-infection, GFP-positive plaques were detected using a Typhoon Trio apparatus and quantified using the ImagequantTL software (GE Healthcare, Mississauga, ON, Canada).

2.8. Luciferase Gene Reporter Assay

U3A or U3A-STAT1 cells at 90% confluency were cotransfected with 100 ng of one of the following CXCL10 promoter containing firefly luciferase reporter plasmids (generously donated by Dr. David Proud, Calgary, [34]), CXCL10prom-972pb-pGL4 (full length −875/+97 promoter), CXCL10prom-376pb-pGL4 (truncated −279/+97 promoter), CXCL10prom972pb-ΔISRE(3)-pGL4 (full length promoter with ISRE(3) site mutated), together with 50 ng of pRL-null renilla-luciferase expressing plasmid (internal control). Transfection was performed using Lipofectamine 2000 (Life Technologies-Thermo Fisher) using a 1:2 DNA to lipofectamine ratio. At 8 h post-transfection, cells were stimulated for 16 h with either IFNβ or IFNβ + TNF. Firefly and renilla luciferase activities were quantified using the Dual-luciferase reporter assay system (Promega Corporation, Madison, WI, USA). Luciferase activities were calculated as the luciferase/renilla ratio and were expressed as fold over the non-stimulated condition.

2.9. Statistical Analyses

Statistical analyses of qRT-PCR and luciferase assay results were performed using the Prism v7 or v8 software (GraphPad Software, San Diego, CA, USA) using the tests indicated in the figure legends. Statistical significance was evaluated using the following p values: $p < 0.05$ (*), $p < 0.01$ (**), $p < 0.001$ (***) or $p < 0.0001$ (****). Differences with a p-value < 0.05 were considered significant. Statistical analysis of the RNASeq data is described in the Bioinformatics analysis section above.

2.10. Data and Software Availability

The entire set of RNAseq data has been submitted to the Gene Expression Omnibus (GEO) database (http://www.ncbi.nlm.nih.gov/geo) under accession number GEO: GSE111195.

3. Results

3.1. Distinct Induction Profiles of Antiviral and Immunoregulatory Genes in Response to IFNβ, TNF and IFNβ + TNF

First, to validate previous observations, we sought to determine the induction profile by the combination of IFNβ and TNF of a selected panel of immunoregulatory and antiviral genes that had previously been shown to be regulated by IFNβ or TNF alone. A549 cells, in which we previously documented the synergistic action of IFNβ and TNF on the *DUOX2* gene, were stimulated either with IFNβ, TNF, or IFNβ + TNF for various times between 3–24 h and the relative mRNA expression levels were quantified by qRT-PCR. Analysis of the expression of the selected genes revealed distinct profiles of response to IFNβ, TNF, or IFNβ + TNF (Figure 1). *IDO, DUOX2, CXCL10, APOBEC3G, ISG20,* and *IL33* exhibited synergistic induction in response to IFNβ + TNF compared to IFNβ or TNF alone. Expression in response to IFNβ + TNF increased over time, with maximum expression levels observed between 16 and 24 h. While *NOD2* and *IRF1* induction following stimulation with IFNβ + TNF was also significantly higher than upon IFNβ or TNF single cytokine stimulation, they exhibited a steady-state induction profile starting as early as 3 h. *MX1* and *PKR*, two typical IFNβ-inducible ISGs, were found induced by IFNβ + TNF similarly to IFNβ alone. *CCL20* responded to IFNβ + TNF with a kinetic and amplitude similar to TNF, but was not responsive to IFNβ alone. *IL8* expression was not induced by IFNβ, but was increased by TNF starting at 3 h and remained steady until 24 h. In contrast to other genes, *IL8* induction in response to IFNβ + TNF was significantly decreased compared to TNF alone. Overall, these results confirm previous reports that a subset of antiviral and immunoregulatory genes is greatly increased in response to IFNβ + TNF compared to either IFNβ or TNF alone.

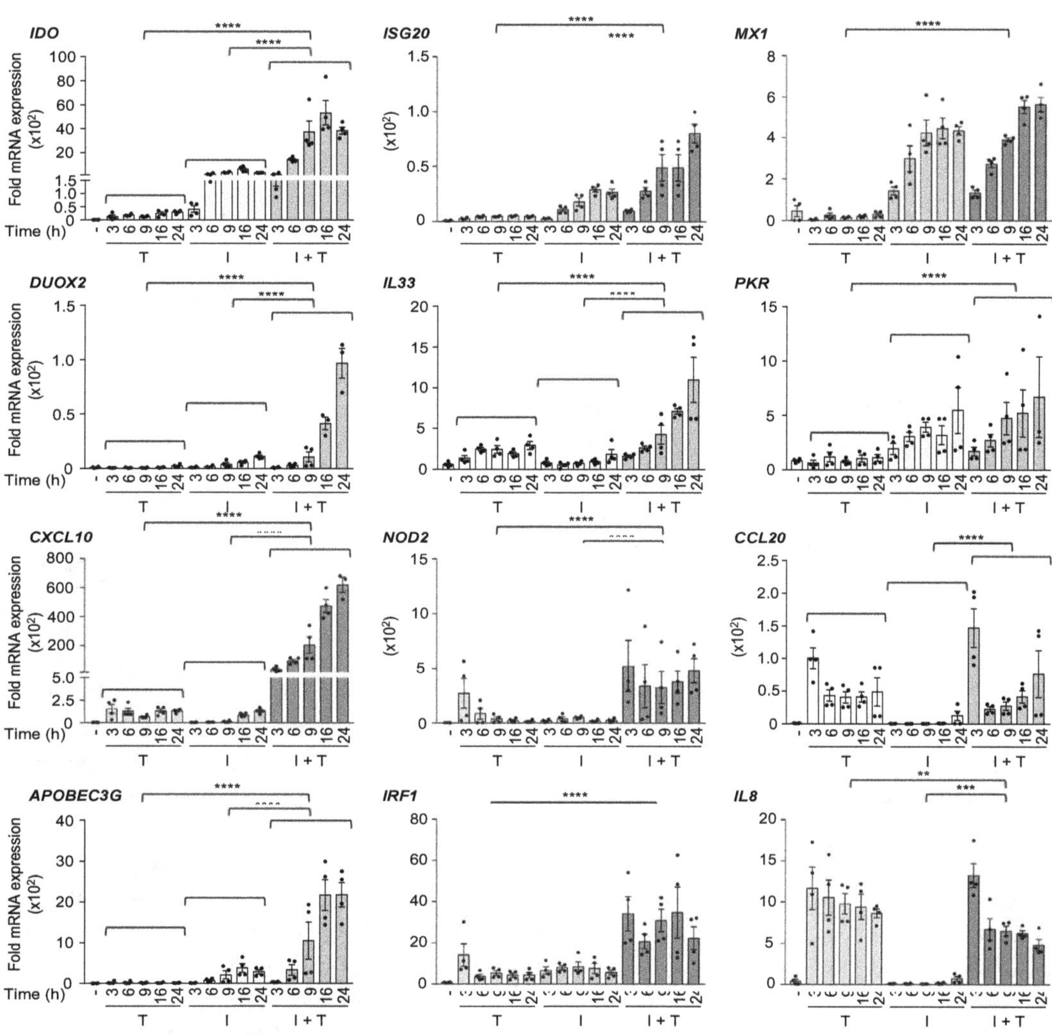

Figure 1. Expression of a panel of antiviral and immunoregulatory genes in response to IFNβ, TNF, and IFNβ + TNF. A549 cells were stimulated with either TNF (T), IFNβ (I), or costimulated with IFNβ + TNF (I + T) for the indicated times. Quantification of mRNA was performed by qRT-PCR and expressed as fold expression after normalization to the S9 mRNA levels using the ΔΔCt method. Mean +/− SEM, $n \geq 3$. Statistical comparison of TNF vs. IFNβ + TNF and IFNβ vs. IFNβ + TNF was conducted using two-way ANOVA with Tukey's post-test. $p < 0.01$ (**), $p < 0.001$ (***) or $p < 0.0001$ (****).

3.2. *Workflow for Genome-Wide Characterization of the Delayed Transcriptional Program Induced by IFNβ+TNF in the Absence of STAT1*

In a previous study, we provided evidence supporting the existence of a STAT1-independent, but STAT2- and IRF9-dependent, pathway engaged downstream of IFNβ + TNF [14]. Here, using the STAT1-deficient human U3A cell line [15], we aimed to fully characterize the STAT1-independent transcriptional program induced by IFNβ + TNF. The human U3A cell line was derived by mutagenesis from the 2ftGH cells [15], in which a synergistic response to IFNβ + TNF can be observed on a subset of genes (Figure 2A). Two hallmarks of STAT2 and IRF9 activation, i.e., STAT2 Tyr690 phosphorylation and induction of IRF9, were observed in the U3A cells following stimulation with IFNβ + TNF, although to reduced levels compared to the parental 2ftGH cells expressing endogenous STAT1 (Figure 2B). This observation implies that the activation of STAT2 and IRF9 depends to a large extent on the STAT1-dependent canonical pathway, but that a significant response occurs in the absence of STAT1. Therefore, the human U3A cell model is suitable for specifically studying STAT1-independent, but STAT2- and IRF9-dependent, gene expression in response to IFNβ + TNF.

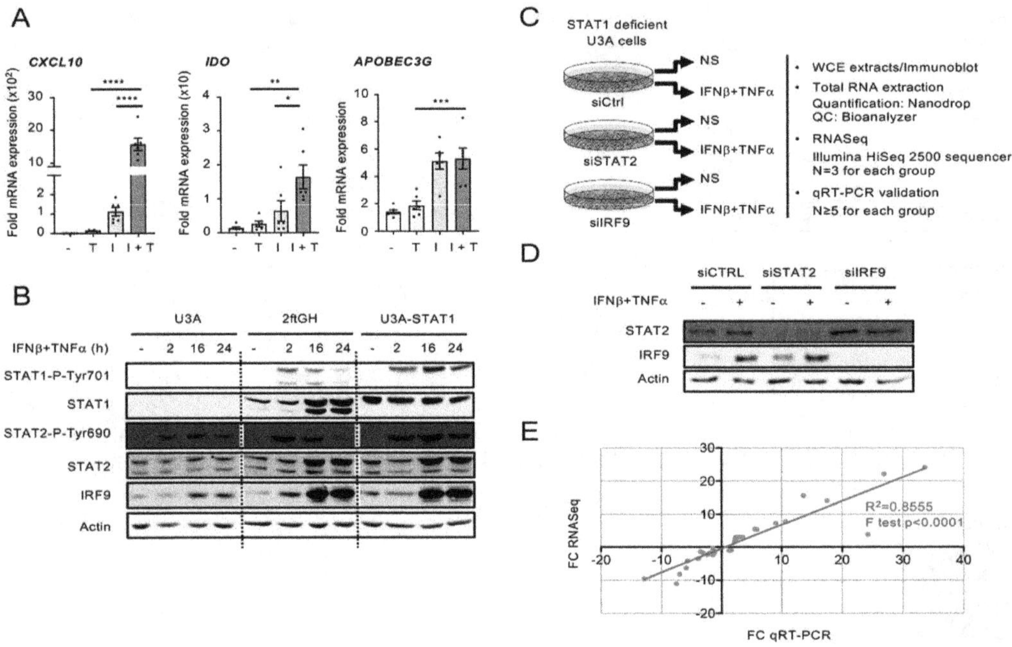

Figure 2. Experimental design used to study the STAT1-independent delayed transcriptional program induced by the combination of IFNβ and TNF. (**A**) 2ftGH cells were stimulated with either TNF (T), IFNβ (I), or costimulated with IFNβ + TNF (I + T) for 24 h. Quantification of mRNA was performed by qRT-PCR and expressed as fold expression after normalization to the S9 mRNA levels using the ΔΔCt method. Mean +/− SEM, $n \geq 5$. Statistical comparison was conducted using one-way ANOVA with Tukey's post-test. $p < 0.05$ (*), $p < 0.01$ (**), $p < 0.001$ (***), or $p < 0.0001$ (****). (**B**) U3A (STAT1-deficient), 2ftGH (parental STAT1-positive) cells and U3A-STAT1 cells (U3A cells stably reconstituted with STAT1) were left untreated or stimulated with IFNβ + TNF for the indicated times. WCE (whole cell extracts) were analyzed by SDS-PAGE followed by immunoblot using anti STAT1-P-Tyr701, total STAT1, STAT2-P-Tyr690, total STAT2, IRF9, or actin antibodies. (**C–E**) U3A cells were transfected with siCTRL, siSTAT2, or siIRF9 before being left untreated (NS) or stimulated with IFNβ + TNF for 24 h. (**C**) The schematic describes the workflow of sample preparation and analysis. (**D**) WCE were analyzed by SDS-PAGE followed by immunoblot using anti STAT2, IRF9, and actin antibodies. (**E**) Graph showing the correlation between fold-changes (FC) measured by RNASeq and qRT-PCR for 13 randomly selected genes. Data from siCTRL NS vs. siCTRL IFNβ +TNF, siSTAT2 IFNβ +TNF vs. siCTRL IFNβ + TNF, siIRF9 IFNβ + TNF vs. siCTRL IFNβ + TNF conditions were used.

To profile the genome wide transcriptional program induced by the combination of IFNβ and TNF in the absence of STAT1 and define the role of STAT2 and IRF9, the U3A cells were transfected with Control (Ctrl)−, STAT2- or IRF9-RNAi and further left untreated or stimulated with IFNβ (1000 U/mL) + TNF (10 ng/mL) for 24 h (Figure 2C). Efficient silencing was confirmed by immunoblot (Figure 2D). Total RNA was isolated and analyzed by RNA sequencing (n = 3 for each group detailed in Figure 2B) on an Illumina HiSeq2500 platform. To validate the expression profile of genes in RNASeq results, the fold changes (FC) of 13 genes randomly selected were analyzed by qRT-PCR in each experimental groups, i.e., siCTRL non-stimulated (NS) vs. siCTRL IFNβ + TNF, siCTRL IFNβ + TNF vs. siSTAT2 IFNβ + TNF and siCTRL IFNβ + TNF vs. siIRF9 IFNβ + TNF. A positive linear relationship between RNASeq and qRT-PCR FC was observed (Figure 2E).

3.3. A Broad Antiviral and Immunoregulatory Transcriptional Signature is Induced by IFNβ + TNF in the Absence of STAT1

To identify differentially expressed genes (DEGs) upon IFNβ + TNF stimulation in the absence of STAT1, comparison between non-stimulated (NS) and IFNβ + TNF-stimulated control cells was performed (Figure 3A). In total, 612 transcripts, including protein-coding transcripts, pseudogenes and long non-coding RNA (lncRNA), were significantly different (FC > 1.5, $p < 0.05$, FDR < 0.05) in IFNβ + TNF vs. NS. Among these, 482 DEGs were upregulated and 130 were downregulated (Figure 3B; See Supplemental Table S1 for a complete list of DEGs).

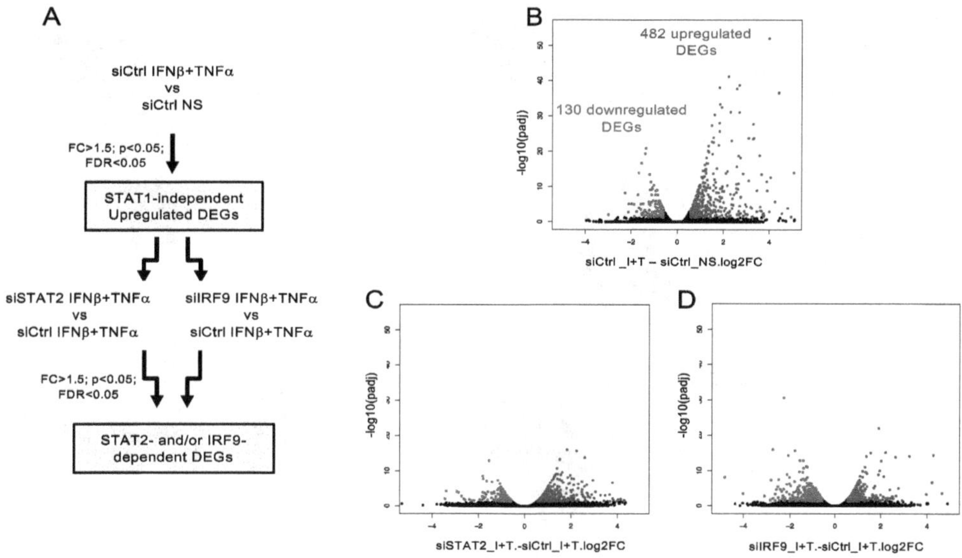

Figure 3. Analysis of STAT1-independent IFNβ + TNF-induced DEGs. (**A**) Diagram describing the bioinformatics analysis strategy used to determine STAT1-independent differentially expressed genes (DEGs) and their regulation by STAT2 and IRF9. (**B**) Volcano plots of the fold-change (FC) vs. adjusted p-value of IFNβ + TNF (I + T) vs. non-stimulated (NS) siCtrl conditions. (**C**) Volcano plots of the fold-change vs. adjusted p-value of siSTAT2 IFNβ + TNF vs. siCTRL IFNβ + TNF (I + T) conditions. (**D**) Volcano plots of the fold-change vs. adjusted p-value of siIRF9 IFNβ + TNF vs. siCTRL IFNβ + TNF conditions.

To identify the Biological Processes (BP) and Molecular Functions (MF) induced by IFNβ + TNF independently of STAT1, we further analyzed the upregulated DEGs. The top 40 upregulated DEGs are shown in Figure 4A. We subjected the complete list of upregulated DEGs (Supplemental Table S1) through Gene Ontology (GO) enrichment analysis. The top enriched GO BP ($p < 10^{-10}$) and MF, are depicted in Figure 4B (See Supplemental Table S2 for a complete list of enriched GO). The majority of the top enriched BPs were related to cytokine production and function (response to cytokine, cytokine-mediated signaling pathway, cytokine production, and regulation of cytokine production), immunoregulation (immune response, immune system process, innate immune response, and regulation of immune system process) and host defense response (defense response, response to other organism, 2′-5′-oligoadenylate synthetase activity and dsRNA binding). Fourteen MF categories were found enriched in IFNβ + TNF. The top enriched MFs were related to cytokine and chemokine functions (cytokine activity, cytokine receptor binding, chemokine activity, and Interleukin 1-receptor binding). Other enriched MF included peptidase related functions (endopeptidase inhibitor activity, peptidase regulator activity, and serine-type endopeptidase activity).

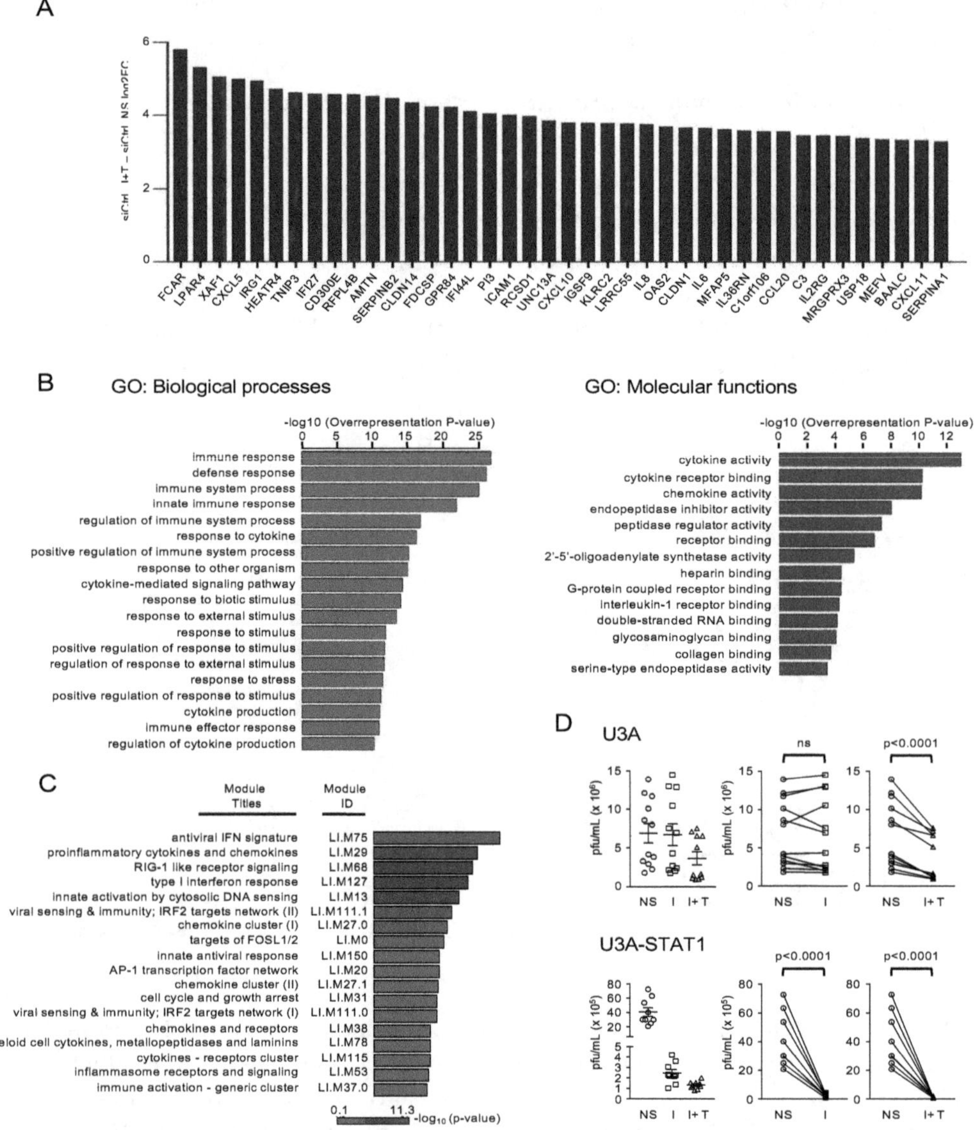

Figure 4. Functional characterization of STAT1-independent IFNβ + TNF-induced DEGs. (**A**) Top forty IFNβ + TNF- upregulated DEGs. (**B**) Gene ontology (GO) enrichment analysis of the differentially upregulated genes in IFNβ + TNF vs. non-stimulated siCtrl conditions based on the Biological Processes and Molecular Function categories. Top enriched terms are shown and the full list is available in Supplemental Table S2. (**C**) Modular transcription analysis of upregulated DEGs. Eighteen enriched modules are shown. The full list of enriched modules is available in Supplemental Table S3. (**D**) U3A and STAT1-rescued U3A-STAT1 cells were stimulated with IFNβ (I) or IFNβ + TNF (I + T) for 30 h before infection with VSV at a MOI of 5 for 12 h. The release of infectious viral particles was quantified by plaque forming unit (pfu) assay. The left graphs show dot-plots of all stimulations. Statistical comparisons were performed on the "before and after" plots (displayed on the right of dot-plot graphs) using ratio paired *t*-tests.

To gain deeper insight into the relevance of the STAT1-independent IFNβ + TNF-induced transcripts towards immunological and host defense responses, we conducted a modular transcription analysis of upregulated DEGs against 606 immune-related functional modules. These modules were previously defined from co-clustered gene sets built via an unbiased data-driven approach as detailed in the material and methods section [30,31]. STAT1-independent IFNβ + TNF-induced DEGs showed significant enrichment in 37 modules (See Supplemental Table S3 for a complete list of enriched modules). Six of these modules were associated with virus sensing/Interferon antiviral response, including LI.M75 (antiviral IFN signature), LI.M68 (RIG-I-like receptor signaling), LI.M127 (type I interferon response), LI.M111.0 and LI.M111.1 (IRF2 target network), and LI.M150 (innate antiviral response) (Figure 4C). Additionally, six modules were associated with immunoregulatory functions, including LI.M29 (proinflammatory cytokines and chemokines), LI.M27.0 and LI.M27.1 (chemokine cluster I and II), LI.M38 (chemokines and receptors), LI.M115 (cytokines receptors cluster), and LI.M37.0 (immune activation - generic cluster) (Figure 4C). Module analysis also showed enriched AP-1 transcription factor-related network modules, LI.M20 and LI.M0, and cell cycle and growth arrest LI.M31 module.

GO enrichment and modular transcription analyses were suggestive of an antiviral response being induced by IFNβ + TNF. To demonstrate the physiological relevance of STAT1-independent gene expression, we evaluated the capacity of IFNβ + TNF to restrict virus replication in the absence of STAT1. Previous study has shown that infection by Vesicular Stomatitis Virus (VSV) is sensitive to IFNβ, but not to TNF, in 2ftGH-derived cells [35]. Thus, U3A cells were stimulated with IFNβ alone or IFNβ + TNF and further infected with (VSV). While VSV replicated similarly in untreated U3A cells and in cells treated with IFNβ, treatment with IFNβ + TNF significantly limited VSV replication (Figure 4D). As a control of the efficiency of IFNβ alone treatment, we observed a significant antiviral response when STAT1 expression was stably restored to endogenous levels in U3A cells (U3A-STAT1) (Figures 2B and 4D). Collectively, these data demonstrate that the combination of IFNβ + TNF induces an efficient antiviral and immunoregulatory transcriptional response in the absence of STAT1.

3.4. Clustering of STAT1-Independent IFNβ + TNF Induced DEGs According to their Regulation by STAT2 and IRF9

Having shown that IFNβ + TNF induce a broad antiviral and immunoregulatory transcriptional response independently of STAT1, we next sought to gain insight into the role of STAT2 and IRF9. To do so, we compared transcripts levels in siSTAT2_IFNβ + TNF vs. siCTRL_IFNβ + TNF and siIRF9_IFNβ + TNF vs. siCTRL_IFNβ + TNF conditions (Figure 3A and Supplemental Table S1). Volcano plots revealed that a fraction of IFNβ + TNF-induced DEGs were significantly (FC > 1.5, $p < 0.05$, FDR < 0.05) downregulated or upregulated upon silencing of STAT2 (Figure 3C) or IRF9 (Figure 3D). Nine distinct theoretical categories of DEGs can be defined based on their potential individual behavior across siSTAT2 and siIRF9 groups (Categories A–I, Figure 5A): a gene can either be downregulated upon STAT2 and IRF9 silencing, indicative of a positive regulation by STAT2 and IRF9 (Category A); conversely, a gene negatively regulated by STAT2 and IRF9 would exhibit upregulation upon STAT2 and IRF9 silencing (Category B); genes that do not exhibit significant differential expression in siSTAT2 and siIRF9 groups would be classified as STAT2 and IRF9 independent (Category C); IRF9-independent genes could exhibit positive (Category D) or negative (Category E) regulation by STAT2; conversely, STAT2-independent genes might be positively (Category F) or negatively (Category G) regulated by IRF9; lastly, STAT2 and IRF9 could have opposite effects on a specific gene regulation (Category H and I). Based on a priori clustering of RNASeq data we found that in the absence of STAT1 IFNβ + TNF-induced DEGs clustered into only seven of the nine possible categories (Figure 5B). The top 15 upregulated DEGs of each category are shown in Figure 5C. The complete list of genes in each category is available in Supplemental Table S1. Amongst the 482 DEGs, 163 genes exhibited either inhibition or upregulation following silencing of STAT2 and/or IRF9 (Categories B–G). A large majority of upregulated DEGs, i.e., 319 out of the 482 DEGs, were not significantly affected by either STAT2 or

IRF9 silencing, and were therefore classified as STAT2/IRF9-independent (Figure 5B). No genes were found in Category H and only one gene was found in Category I.

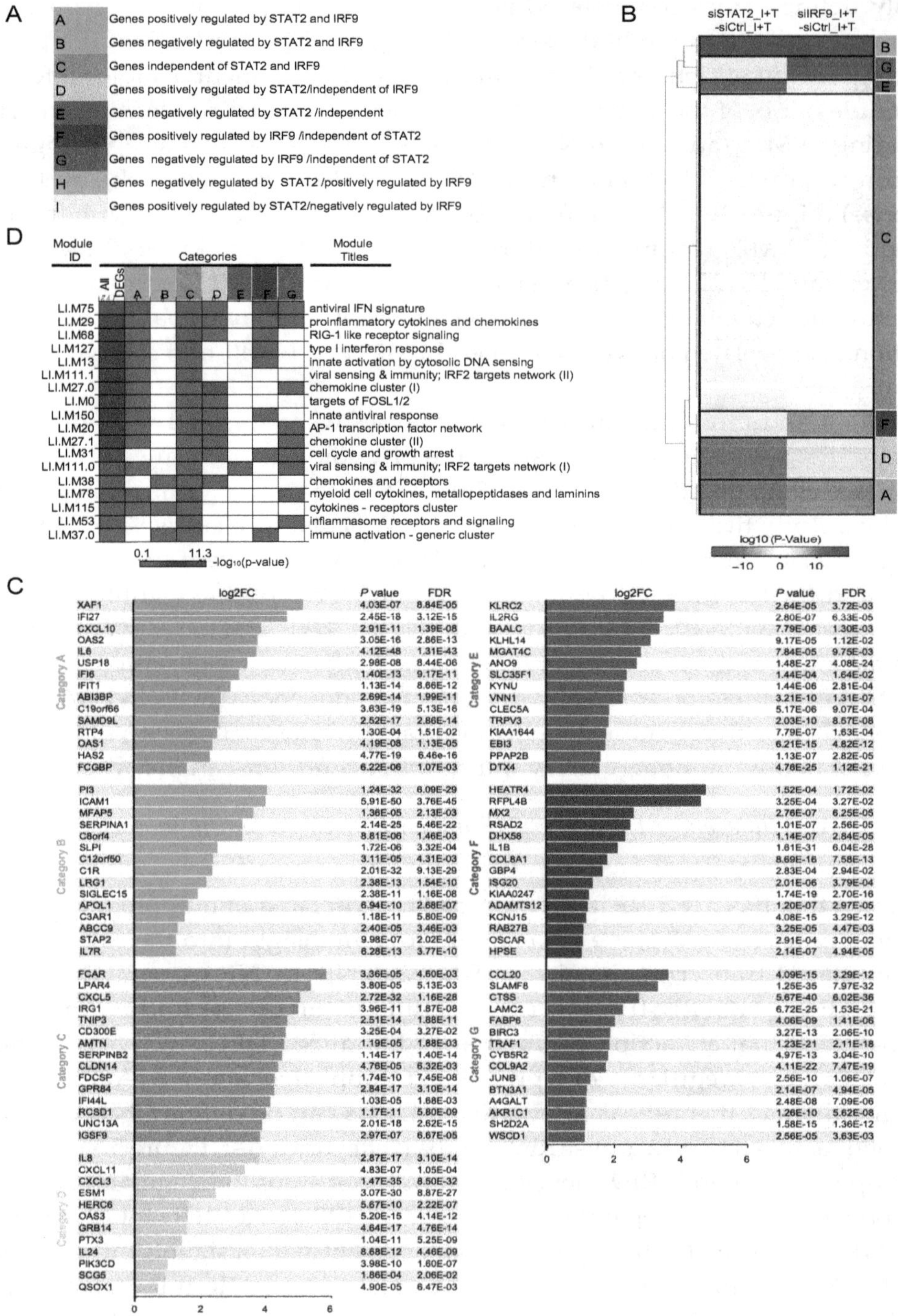

Figure 5. Clustering of IFNβ + TNF-induced DEGs according to their regulation by STAT2 and IRF9. (**A**) Theoretical categories in which IFNβ + TNF-induced DEGs can be segregated based on their potential individual regulation by STAT2 and IRF9. (**B**) Hierarchical clustering of the categories of DEG responses according to their regulation by STAT2 and IRF9. Euclidean distance metric is used for the construction of distance matrix and the categories are used as a priori input into clustering algorithm as detailed in Materials and Methods. (**C**) Top 15 induced DEGs (Log2FC, siCTRL NS vs. siCTRL IFNβ + TNF) in each category identified in (**B**) are displayed. Note that category D contains only twelve genes, so all genes are shown. The full list of genes is available in Supplemental Table S1. (**D**) Diagram showing enriched transcription modules in each gene category.

To functionally interpret these clusters, we applied the modular transcription analysis to each of the categories to assess for the specific enrichment of the functional modules found associated with IFNβ + TNF-upregulated DEGs (Figure 5D). First, most modules, except LI.M31 (cell cycle and growth arrest), LI.M38 (chemokines and receptors), LI.M37.0 (immune activation - generic cluster), and LI.M53 (inflammasome receptors and signaling), were enriched in the category of DEGs positively regulated by STAT2 and IRF9 (Category A). Conversely, the cluster negatively regulated by STAT2 and IRF9 (Category B) exclusively contains enriched LI.M38 (chemokines and receptors), LI.M37.0 (immune activation - generic cluster) and LI.M115 (cytokines receptors cluster). The cluster of IRF9-independent genes that are negatively regulated by STAT2 (Category E) only exhibited enrichment in the virus sensing/IRF2 target network LI.M111.0 module, while the IRF9-independent/STAT2-positively regulated cluster (Category D) encompasses antiviral and immunoregulatory functions. The STAT2-independent but IRF9 positively regulated transcripts (Category F) were mainly enriched in modules associated with the IFN antiviral response, including LI.M75 (antiviral IFN signature), LI.M68 (RIG-I-like receptor signaling), LI.M127 (type I interferon response), and LI.M150 (innate antiviral response). Finally, the STAT2-independent but IRF9 negatively regulated cluster (Category G) was mostly enriched in modules associated with immunoregulatory functions, including LI.M29 (proinflammatory cytokines and chemokines), LI.M27.0 and LI.M27.1 (chemokine cluster I and II), LI.M78 (myeloid cell cytokines), but also with cell cycle and growth arrest (LI.M31) and inflammasome receptors and signaling (LI.M53). Of note, all modules were found enriched in the cluster of genes induced independently of STAT2 and IRF9 (Category C), pointing to a broad function of this pathway(s) in the regulation of the antiviral and immunoregulatory program elicited by IFNβ + TNF. Altogether, these observations reveal that STAT2 and IRF9 are involved in the regulation of a subset of the genes induced in response to the co-stimulation by IFNβ and TNF in the absence of STAT1. Importantly, our results also reveal that STAT2 and IRF9 act in part in a concerted manner, but also independently in distinct non-canonical pathways, to regulate specific subsets of the IFNβ + TNF-induced DEGs.

3.5. Differential Regulation of CXCL10 in Response to IFNβ and IFNβ+TNF

The canonical ISGF3 complex mediates ISGs transcriptional regulation through binding to ISRE consensus sequences [7]. Identification of DEGs upregulated by IFNβ+TNF in a STAT1-independent, but STAT2 and IRF9-dependent, manner raised the question of the ISRE site usage compared to the ISGF3 pathway. To address this question, we studied the regulation of the *CXCL10* gene promoter that was found induced by STAT2 and IRF9 in the absence of STAT1 in response to IFNβ and TNF (Supplemental Table S1 and Figure 5C), but is also inducible by IFNβ alone in the presence of STAT1 (Figures 1 and 2A). The *CXCL10* promoter contains three ISRE sites. We used full-length wild-type (972bp containing the three ISRE sites), truncated (376bp containing only the ISRE (3) site) or mutated (972bp containing a mutated ISRE (3) site) *CXCL10* promoter luciferase (*CXCL10*prom-Luc) reporter constructs (Figure 6A) to determine the ISRE consensus site(s) requirement. U3A and STAT1-rescued U3A-STAT1 cells were transfected with the *CXCL10*prom-Luc constructs and further stimulated with IFNβ or IFNβ+TNF to monitor the canonical ISGF3-dependent and the STAT1-independent responses. In STAT1-deficient U3A cells, IFNβ was unable to activate the *CXCL10*prom reflecting the dependence on the ISFG3 pathway. In contrast, induction of *CXCL10*prom was significantly induced when cells were stimulated with IFNβ +TNF. This induction in the absence of STAT1 was not altered by the deletion of ISRE (1) and (2) sites, but was significantly impaired when the ISRE (3) site was mutated (Figure 6B). STAT1 expression rescue led to a higher induction of *CXCL10*prom by IFNβ + TNF. Additionally, IFNβ-mediated induction of *CXCL10*prom was restored, albeit to a much lower extent than IFNβ +TNF. The activation of the *CXCL10*prom in the presence of STAT1 involved both the distal ISRE (1) and/or ISRE (2) sites and the proximal ISRE (3) site (Figure 6B). Altogether, this shows that the STAT1-independent, but STAT2/IRF9-dependent, pathway mediates gene expression through a restricted ISRE site usage compared to the ISGF3-dependent regulation.

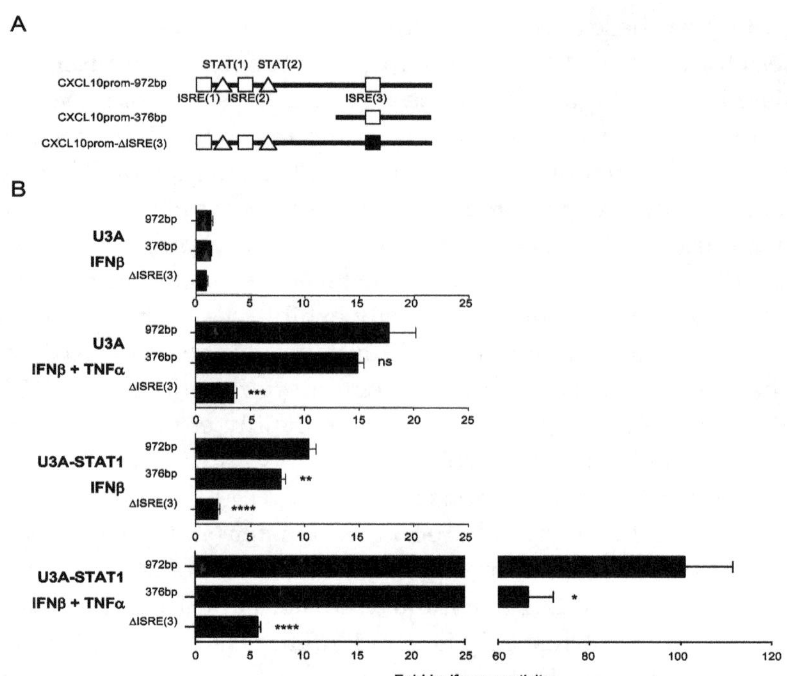

Figure 6. Analysis of the *CXCL10* promoter regulation in response to IFNβ + TNF vs. IFNβ. (**A**) Schematic representation of the *CXCL10* promoter (CXCL10prom) luciferase constructs used in this study indicating the main transcription factors consensus binding sites. (**B**) U3A and U3A-STAT1 cells were transfected with the indicated CXCL10prom-luciferase reporter constructs and either left untreated or stimulated with IFNβ or IFNβ + TNF. Relative luciferase activities were measured at 16 h post-stimulation and expressed as fold over the corresponding unstimulated condition. Mean +/− SEM, n = 6. Statistical analyses were performed using an unpaired t-test comparing each promoter to the CXCL10prom-972bp construct. $p < 0.01$ (**), $p < 0.001$ (***), and $p < 0.0001$ (****).

4. Discussion

Previous studies have described that IFNβ and TNF synergize to elicit a specific delayed transcriptional program that differs from the one induced by either cytokine alone [13,36]. The mechanisms underlying the transcriptional induction of genes specifically regulated by IFNβ and TNF remain poorly defined. The present study was specifically designed to document the functional relevance of a previously observed delayed gene expression induced by IFNβ in the presence of TNF in the absence of STAT1 [14] and to document the role of STAT2 and IRF9 in this response.

The observation that STAT2 and IRF9 activation in response to IFNβ + TNF is reduced in STAT1-deficient U3A cells compared to the wild-type 2ftGH parental cells and that IFNβ + TNF-mediated induction of the STAT2- and IRF9-dependent *CXCL10* promoter exhibits partial dependence on STAT1 support a model in which a canonical ISGF3 pathway is engaged downstream of the costimulation. Importantly, it also implied the existence of a STAT2- and/or IRF9-dependent transcriptional response occurring in the absence of STAT1. The human STAT1-deficient U3A cell model offered a unique opportunity to specifically pinpoint this STAT1-independent response. In this model, genome wide RNA sequencing highlighted that the transcriptional program induced by IFNβ + TNF in the absence of STAT1 encompasses a wide range of immunoregulatory and antiviral functions. The functional relevance of this response was confirmed by the observation that the treatment with IFNβ + TNF induced an antiviral state capable of restricting VSV replication in the absence of STAT1. This points to a significant role of the STAT1-independent pathway in the establishment of the antiviral state induced by the synergistic action of IFNβ and TNF that enhances the restriction of VSV (Figure 4D and [12]), Myxoma virus [13], and paramyxoviruses [14]. Although previous reports have shown that type I IFNs, mostly IFNα, alone can trigger STAT1-independent responses [8], we neither observed establishment of an IFNβ-induced antiviral state against VSV, nor activation of the *CXCL10* promoter in the absence of STAT1 in our model (Figures 4D and 6).

We previously reported that IFNβ + TNF induces the *DUOX2* gene via a STAT2- and IRF9-dependent pathway in the absence of STAT1 [14]. To what extent this pathway contributes to the STAT1-independent transcriptional response elicited by IFNβ + TNF remained to be addressed. Here, we demonstrate that IFNβ + TNFα-induced DEGs segregate into seven categories that reflect distinct contributions of STAT2 and/or IRF9, thereby highlighting an unexpected heterogeneity of the STAT1-independent pathways engaged downstream of IFNβ + TNF. Importantly, only one anecdotic gene was found in categories implying inverse regulation by STAT2 and IRF9 (categories H and I) pointing to convergent functions of STAT2 and IRF9 when both are engaged in gene regulation. We can rule out that these distinct regulation mechanisms reflect specific induction profiles by IFNβ + TNF as *CXCL10, IL33, CCL20,* and *ISG20* all exhibit synergistic induction by IFNβ + TNF, but are differentially regulated by STAT2 and/or IRF9; while *CXCL10* is dependent on STAT2 and IRF9, *IL33* is independent on STAT2 and IRF9, and *CCL20* and *ISG20* are STAT2-independent but IRF9-dependent (Figure 5; Supplemental Table S1). Consistent with our previous observation [14], we found several STAT1-independent genes positively regulated by STAT2 and IRF9 (Category A). DEGs in this category encompass most of the functions induced in response to IFNβ + TNF, with the notable exception of cell cycle and growth arrest and inflammasome and receptor signaling functions. Genes negatively regulated by STAT2 and IRF9 were also identified (Category B). Formation of an alternative STAT2/IRF9-containing complex mediating gene expression in the absence of STAT1 [37–41] has been reported, but with limited DNA-binding affinity for the typical ISRE sequence [37]. The existence of a STAT2/IRF9 complex is also supported by our recent observation of a high affinity of IRF9 for STAT2 with an equilibrium dissociation constant (Kd) of 10 nM [42]. A recent report of experiments, performed in murine bone marrow-derived macrophages proposes a model in which murine STAT2/IRF9 complex drives basal expression of ISGs, while IFNβ-inducible expression of ISGs depends on a switch to the ISGF3 complex [43]. This differs from our results as silencing of either STAT2 or IRF9 did not alter basal gene expression (Supplemental Figure S1). Further analysis of the *CXCL10* promoter demonstrates a restricted usage of ISRE sites by the STAT2/IRF9 pathway compared to the ISGF3 pathway. Further large-scale studies will be needed to identify the parameters allowing binding of ISGF3, but not STAT2/IRF9, to specific ISRE sequences upon IFNβ + TNF.

The observation that some IFNβ + TNF-induced genes were solely dependent on STAT2 (either positively or negatively) but not on IRF9, (Categories D and E) is a rare genome wide demonstration of gene regulation by STAT2 independently of STAT1 and IRF9. Previous reports have identified ISGF3-independent, STAT2-dependent genes but the association with IRF9 was not formally excluded [44–47]. STAT2 was shown to associate with STAT3 and STAT6, but it is not clear whether IRF9 is also part of these alternative complexes [44,46]. Transcriptional module analyses demonstrated that the functional distribution of genes negatively regulated by STAT2 is very limited compared to other categories; only a virus-sensing module was enriched in this category. In contrary, IRF9-independent genes positively regulated by STAT2 mediate broader antiviral and immunoregulatory functions.

ISGF3-independent functions of IRF9 have been proposed based on the study of IRF9 deficiencies [11,48]. However, IRF9 target genes in these contexts have been barely documented. Intriguingly, Li et al. [49] studied IFNα-induced genes and their dependency on the ISGF3 subunits. While they confirmed previous studies showing that IFNα can trigger a delayed and sustained ISG response via an ISGF3-independent pathway, it is very striking that they did not find STAT1- and STAT2-independent but IRF9-dependent genes. All identified IRF9-dependent genes were either STAT2- or STAT1-dependent. This result greatly differs with our study. Here, we found several IFNβ + TNF-induced DEGs independent of STAT1 and STAT2, but positively or negatively regulated by IRF9 (Categories F and G). Typically, IRF9 is considered a positive regulator of gene transcription. However, our findings are consistent with recent reports documenting the role of IRF9 in the negative regulation of the TRIF/NF-κB transcriptional response [50] or the expression of SIRT1 in acute myeloid leukemia cells [51]. The molecular mechanisms underlying gene regulation by IRF9 without association with either STAT1 or STAT2 remain to be elucidated. To the best of our knowledge, no alternative IRF9-containing complex has yet been described.

Our analysis showed that a large number of genes were induced by IFNβ + TNF independently of STAT2 and IRF9 (Category C). All transcriptional modules were enriched in this category pointing to a major role of this pathway in the establishment of a host defense and immunoregulatory response. The STAT2 and IRF9 independent genes does not solely reflects induction by TNF alone. For instance, *APOBEC3G* that is amongst the STAT2- and IRF9- independent genes is not induced by TNF alone (Figure 2A). While NF-κB, a downstream effector of the TNF receptor, is an obvious candidate for the regulation of these DEGs, this might fall short in explaining the synergistic action of IFNβ + TNF as we did not observe enhanced NF-κB activation compared to TNF alone [14]. Alternatively, the potential role of AP-1 is supported by the finding that the AP-1 transcription network module is enriched amongst IFNβ + TNF-induced DEGs. However, this module is not restricted to genes regulated independently of STAT2 and IRF9. It is also worth noting that two modules of IRF2-target genes were enriched, although again not specifically in the STAT2- and IRF9-independent category. A similar crosstalk was reported between IFNα and TNF in macrophages resulting in increased colocalized recruitment of IRF1 and p65 to the promoter of a subset of genes [52]. However, while IRF1 was found synergistically induced by IFNβ + TNF at early stages (Figure 1), we did not observe significant induction of IRF1 in the absence of STAT1 by RNASeq (Supplemental Table S1) or qRT-PCR (data not shown), suggesting that IRF1 is unlikely to be involved in our system. Further studies will be required to uncover these STAT2- and IRF9-independent pathways.

This study provides novel insight into the molecular pathways leading to delayed antiviral and immunoregulatory gene expression in conditions where elevated levels of both IFNβ and TNF are present. Altogether our results demonstrate that in addition to the engagement of an ISGF3-dependent canonical response, a broad transcriptional program is elicited independently of STAT1, and support a model in which STAT2 and IRF9 contribute to the regulation of this response through non-canonical parallel pathways involving their concerted or independent action (Figure 7).

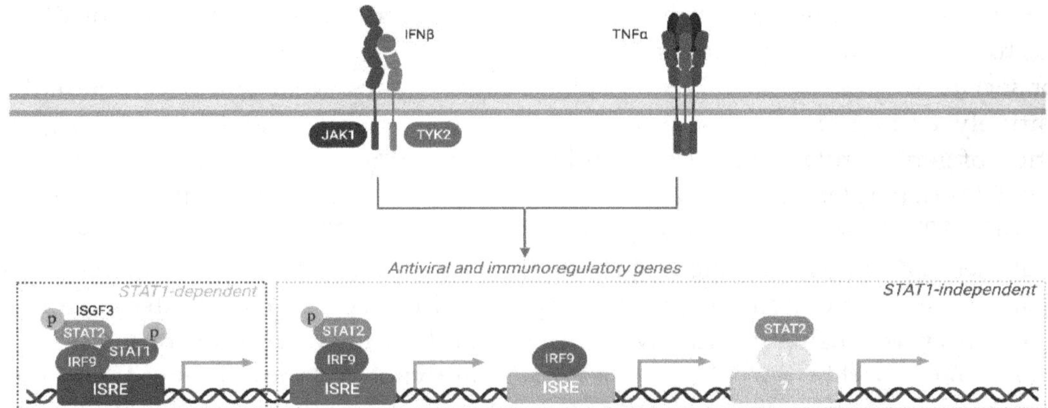

Figure 7. Role of distinct STAT2 and/or IRF9-dependent pathways in the regulation of distinct subset of antiviral and immunoregulatory genes in response to IFNβ and TNF. Our data supports a model in which multiple pathways participate to the synergistic action of IFNβ + TNF. While the STAT1-dependent pathway, likely ISFG3, is engaged downstream of IFNβ and TNF, STAT1-independent pathways are also involved in the control of the delayed gene expression. STAT2 and IRF9 act not only in a concerted fashion, likely as a complex, but also independently. IRF9 is known to act as the DNA-binding subunit of the ISGF3 complex and therefore likely mediates binding of STAT2/IRF9 complexes and of alternative complexes devoid of STAT2. The mechanisms of STAT2-dependent regulation of gene expression remains to be characterized.

Consistent with accumulating evidence [8], these distinct STAT2 and IRF9 actions most likely result from the formation of specific complexes that coexist with ISGF3 upon IFNβ and TNF stimulation. Studies are underway to biochemically solve the complexity of the dynamic and specific mechanisms of activation of the alternative STAT2 and/or IRF9-containing complexes in a wild-type cell context to further characterize the transcriptional response induced by IFNβ and TNF.

Author Contributions: Conceptualization, M.K.M. and N.G.; formal analysis, M.K.M., N.G., P.D., and A.W.R.S.; investigation, M.K.M., A.F., E.C., M.K., A.N.H., D.M.K., D.I.H., S.L.C.-O., and E.M.; writing—original draft preparation, N.G.; writing—review and editing, P.D., D.M.K. and D.I.H.

Acknowledgments: We are very thankful to D. Proud (University of Calgary, Canada), G. Stark (Cleveland clinic, USA), and J. Bell (University of Ottawa, Canada) for sharing reagents used in this study. RNASeq analyses were performed at McGill University and Génome Québec Innovation Centre. Figure 6 was generated using biorender.com.

References

1. McNab, F.; Mayer-Barber, K.; Sher, A.; Wack, A.; O'Garra, A. Type I interferons in infectious disease. *Nat. Rev. Immunol.* **2015**, *15*, 87–103. [CrossRef] [PubMed]
2. Tomasello, E.; Pollet, E.; Vu Manh, T.P.; Uze, G.; Dalod, M. Harnessing Mechanistic Knowledge on Beneficial Versus Deleterious IFN-I Effects to Design Innovative Immunotherapies Targeting Cytokine Activity to Specific Cell Types. *Front. Immunol.* **2014**, *5*, 526. [CrossRef] [PubMed]
3. Lee-Kirsch, M.A. The Type I Interferonopathies. *Annu. Rev. Med.* **2017**, *68*, 297–315. [CrossRef] [PubMed]
4. Yarilina, A.; Park-Min, K.H.; Antoniv, T.; Hu, X.; Ivashkiv, L.B. TNF activates an IRF1-dependent autocrine loop leading to sustained expression of chemokines and STAT1-dependent type I interferon-response genes. *Nat. Immunol.* **2008**, *9*, 378–387. [CrossRef] [PubMed]
5. Palucka, A.K.; Blanck, J.P.; Bennett, L.; Pascual, V.; Banchereau, J. Cross-regulation of TNF and IFN-alpha in autoimmune diseases. *Proc. Natl. Acad. Sci. USA* **2005**, *102*, 3372–3377. [CrossRef] [PubMed]
6. Tliba, O.; Tliba, S.; Da Huang, C.; Hoffman, R.K.; DeLong, P.; Panettieri, R.A., Jr.; Amrani, Y. Tumor necrosis factor alpha modulates airway smooth muscle function via the autocrine action of interferon beta. *J. Biol. Chem.* **2003**, *278*, 50615–50623. [CrossRef] [PubMed]
7. Au-Yeung, N.; Mandhana, R.; Horvath, C.M. Transcriptional regulation by STAT1 and STAT2 in the interferon JAK-STAT pathway. *JAKSTAT* **2013**, *2*, e23931. [CrossRef]
8. Fink, K.; Grandvaux, N. STAT2 and IRF9: Beyond ISGF3. *JAKSTAT* **2013**, *2*, e27521. [CrossRef]
9. Schneider, W.M.; Chevillotte, M.D.; Rice, C.M. Interferon-stimulated genes: A complex web of host defenses. *Annu. Rev. Immunol.* **2014**, *32*, 513–545. [CrossRef]
10. Levy, D.E.; Marie, I.J.; Durbin, J.E. Induction and function of type I and III interferon in response to viral infection. *Curr. Opin. Virol.* **2011**, *1*, 476–486. [CrossRef]
11. Majoros, A.; Platanitis, E.; Kernbauer-Holzl, E.; Rosebrock, F.; Muller, M.; Decker, T. Canonical and Non-Canonical Aspects of JAK-STAT Signaling: Lessons from Interferons for Cytokine Responses. *Front. Immunol.* **2017**, *8*, 29. [CrossRef] [PubMed]
12. Mestan, J.; Brockhaus, M.; Kirchner, H.; Jacobsen, H. Antiviral activity of tumour necrosis factor. Synergism with interferons and induction of oligo-2′, 5′-adenylate synthetase. *J. Gen. Virol.* **1988**, *69*, 3113–3120. [CrossRef] [PubMed]
13. Bartee, E.; Mohamed, M.R.; Lopez, M.C.; Baker, H.V.; McFadden, G. The addition of tumor necrosis factor plus beta interferon induces a novel synergistic antiviral state against poxviruses in primary human fibroblasts. *J. Virol.* **2009**, *83*, 498–511. [CrossRef] [PubMed]
14. Fink, K.; Martin, L.; Mukawera, E.; Chartier, S.; De Deken, X.; Brochiero, E.; Miot, F.; Grandvaux, N. IFNbeta/TNFalpha synergism induces a non-canonical STAT2/IRF9-dependent pathway triggering a novel DUOX2 NADPH oxidase-mediated airway antiviral response. *Cell Res.* **2013**, *23*, 673–690. [CrossRef] [PubMed]

15. McKendry, R.; John, J.; Flavell, D.; Muller, M.; Kerr, I.M.; Stark, G.R. High-frequency mutagenesis of human cells and characterization of a mutant unresponsive to both alpha and gamma interferons. *Proc. Natl. Acad. Sci. USA* **1991**, *88*, 11455–11459. [CrossRef] [PubMed]
16. Horvath, C.M.; Wen, Z.; Darnell, J.E., Jr. A STAT protein domain that determines DNA sequence recognition suggests a novel DNA-binding domain. *Genes Dev.* **1995**, *9*, 984–994. [CrossRef]
17. Schindler, C.; Fu, X.Y.; Improta, T.; Aebersold, R.; Darnell, J.E., Jr. Proteins of transcription factor ISGF-3: One gene encodes the 91-and 84-kDa ISGF-3 proteins that are activated by interferon alpha. *Proc. Natl. Acad. Sci. USA* **1992**, *89*, 7836–7839. [CrossRef]
18. Robitaille, A.C.; Mariani, M.K.; Fortin, A.; Grandvaux, N. A High Resolution Method to Monitor Phosphorylation-dependent Activation of IRF3. *J. Vis. Exp.* **2016**, *107*, e53723. [CrossRef]
19. Dussault, A.A.; Pouliot, M. Rapid and simple comparison of messenger RNA levels using real-time PCR. *Biol. Proced. Online* **2006**, *8*, 1–10. [CrossRef]
20. Bolger, A.M.; Lohse, M.; Usadel, B. Trimmomatic: A flexible trimmer for Illumina sequence data. *Bioinformatics* **2014**, *30*, 2114–2120. [CrossRef]
21. Dobin, A.; Davis, C.A.; Schlesinger, F.; Drenkow, J.; Zaleski, C.; Jha, S.; Batut, P.; Chaisson, M.; Gingeras, T.R. STAR: Ultrafast universal RNA-seq aligner. *Bioinformatics* **2013**, *29*, 15–21. [CrossRef]
22. Roberts, A.; Pimentel, H.; Trapnell, C.; Pachter, L. Identification of novel transcripts in annotated genomes using RNA-Seq. *Bioinformatics* **2011**, *27*, 2325–2329. [CrossRef]
23. Trapnell, C.; Hendrickson, D.G.; Sauvageau, M.; Goff, L.; Rinn, J.L.; Pachter, L. Differential analysis of gene regulation at transcript resolution with RNA-seq. *Nat. Biotechnol.* **2013**, *31*, 46–53. [CrossRef]
24. Anders, S.; Huber, W. Differential expression analysis for sequence count data. *Genome Biol.* **2010**, *11*, R106. [CrossRef]
25. Robinson, M.D.; McCarthy, D.J.; Smyth, G.K. edgeR: A Bioconductor package for differential expression analysis of digital gene expression data. *Bioinformatics* **2010**, *26*, 139–140. [CrossRef]
26. Young, M.D.; Wakefield, M.J.; Smyth, G.K.; Oshlack, A. Gene ontology analysis for RNA-seq: Accounting for selection bias. *Genome Biol.* **2010**, *11*, R14. [CrossRef]
27. Weiner, J.; Domaszewska, T. tmod: An R package for general and multivariate enrichment analysis. *PeerJ Prepr.* **2016**, *4*, e2420v1. [CrossRef]
28. Bar-Joseph, Z.; Gerber, G.K.; Lee, T.I.; Rinaldi, N.J.; Yoo, J.Y.; Robert, F.; Gordon, D.B.; Fraenkel, E.; Jaakkola, T.S.; Young, R.A.; et al. Computational discovery of gene modules and regulatory networks. *Nat. Biotechnol.* **2003**, *21*, 1337–1342. [CrossRef]
29. Chaussabel, D.; Baldwin, N. Democratizing systems immunology with modular transcriptional repertoire analyses. *Nat. Rev. Immunol.* **2014**, *14*, 271–280. [CrossRef]
30. Chaussabel, D.; Quinn, C.; Shen, J.; Patel, P.; Glaser, C.; Baldwin, N.; Stichweh, D.; Blankenship, D.; Li, L.; Munagala, I.; et al. A modular analysis framework for blood genomics studies: Application to systemic lupus erythematosus. *Immunity* **2008**, *29*, 150–164. [CrossRef]
31. Li, S.; Rouphael, N.; Duraisingham, S.; Romero-Steiner, S.; Presnell, S.; Davis, C.; Schmidt, D.S.; Johnson, S.E.; Milton, A.; Rajam, G.; et al. Molecular signatures of antibody responses derived from a systems biology study of five human vaccines. *Nat. Immunol.* **2014**, *15*, 195–204. [CrossRef]
32. Benjamini, Y.; Hochberg, Y. Controlling the False Discovery Rate: A Practical and Powerful Approach to Multiple Testing. *J. R. Stat. Soc. Ser. B (Methodol.)* **1995**, *57*, 289–300. [CrossRef]
33. Ross, I.; Gentleman, R. R: A language for data analysis and graphics. *J. Comput. Graph. Stat.* **1996**, *5*, 299–314.
34. Zaheer, R.S.; Koetzler, R.; Holden, N.S.; Wiehler, S.; Proud, D. Selective transcriptional down-regulation of human rhinovirus-induced production of CXCL10 from airway epithelial cells via the MEK1 pathway. *J. Immunol.* **2009**, *182*, 4854–4864. [CrossRef]
35. Davis, A.M.; Hagan, K.A.; Matthews, L.A.; Bajwa, G.; Gill, M.A.; Gale, M., Jr.; Farrar, J.D. Blockade of virus infection by human CD4+ T cells via a cytokine relay network. *J. Immunol.* **2008**, *180*, 6923–6932. [CrossRef]
36. Bartee, E.; McFadden, G. Human cancer cells have specifically lost the ability to induce the synergistic state caused by tumor necrosis factor plus interferon-beta. *Cytokine* **2009**, *47*, 199–205. [CrossRef]
37. Bluyssen, H.A.; Levy, D.E. Stat2 is a transcriptional activator that requires sequence-specific contacts provided by stat1 and p48 for stable interaction with DNA. *J. Biol. Chem.* **1997**, *272*, 4600–4605. [CrossRef]

38. Martinez-Moczygemba, M.; Gutch, M.J.; French, D.L.; Reich, N.C. Distinct STAT structure promotes interaction of STAT2 with the p48 subunit of the interferon-alpha-stimulated transcription factor ISGF3. *J. Biol. Chem.* **1997**, *272*, 20070–20076. [CrossRef]
39. Gupta, S.; Jiang, M.; Pernis, A.B. IFN-alpha activates Stat6 and leads to the formation of Stat2:Stat6 complexes in B cells. *J. Immunol.* **1999**, *163*, 3834–3841.
40. Lou, Y.J.; Pan, X.R.; Jia, P.M.; Li, D.; Xiao, S.; Zhang, Z.L.; Chen, S.J.; Chen, Z.; Tong, J.H. IRF-9/STAT2 [corrected] functional interaction drives retinoic acid-induced gene G expression independently of STAT1. *Cancer Res.* **2009**, *69*, 3673–3680. [CrossRef]
41. Abdul-Sater, A.A.; Majoros, A.; Plumlee, C.R.; Perry, S.; Gu, A.D.; Lee, C.; Shresta, S.; Decker, T.; Schindler, C. Different STAT Transcription Complexes Drive Early and Delayed Responses to Type I IFNs. *J. Immunol.* **2015**, *195*, 210–216. [CrossRef] [PubMed]
42. Rengachari, S.; Groiss, S.; Devos, J.; Caron, E.; Grandvaux, N.; Panne, D. Structural basis of STAT2 recognition by IRF9 reveals molecular insights into ISGF3 function. *Proc. Natl. Acad. Sci. USA* **2018**, *115*, E601–E609. [CrossRef] [PubMed]
43. Platanitis, E.; Demiroz, D.; Schneller, A.; Fischer, K.; Capelle, C.; Hartl, M.; Gossenreiter, T.; Muller, M.; Novatchkova, M.; Decker, T. A molecular switch from STAT2-IRF9 to ISGF3 underlies interferon-induced gene transcription. *Nat. Commun.* **2019**, *10*, 2921. [CrossRef] [PubMed]
44. Ghislain, J.J.; Fish, E.N. Application of genomic DNA affinity chromatography identifies multiple interferon-alpha-regulated Stat2 complexes. *J. Biol. Chem.* **1996**, *271*, 12408–12413. [CrossRef] [PubMed]
45. Brierley, M.M.; Marchington, K.L.; Jurisica, I.; Fish, E.N. Identification of GAS-dependent interferon-sensitive target genes whose transcription is STAT2-dependent but ISGF3-independent. *FEBS J.* **2006**, *273*, 1569–1581. [CrossRef]
46. Wan, L.; Lin, C.W.; Lin, Y.J.; Sheu, J.J.; Chen, B.H.; Liao, C.C.; Tsai, Y.; Lin, W.Y.; Lai, C.H.; Tsai, F.J. Type I IFN induced IL1-Ra expression in hepatocytes is mediated by activating STAT6 through the formation of STAT2: STAT6 heterodimer. *J. Cell Mol. Med.* **2008**, *12*, 876–888. [CrossRef]
47. Perry, S.T.; Buck, M.D.; Lada, S.M.; Schindler, C.; Shresta, S. STAT2 mediates innate immunity to Dengue virus in the absence of STAT1 via the type I interferon receptor. *PLoS Pathog.* **2011**, *7*, e1001297. [CrossRef]
48. Suprunenko, T.; Hofer, M.J. The emerging role of interferon regulatory factor 9 in the antiviral host response and beyond. *Cytokine Growth Factor Rev.* **2016**, *29*, 35–43. [CrossRef]
49. Li, W.; Hofer, M.J.; Songkhunawej, P.; Jung, S.R.; Hancock, D.; Denyer, G.; Campbell, I.L. Type I interferon-regulated gene expression and signaling in murine mixed glial cells lacking signal transducers and activators of transcription 1 or 2 or interferon regulatory factor 9. *J. Biol. Chem.* **2017**, *292*, 5845–5859. [CrossRef]
50. Zhao, X.; Chu, Q.; Cui, J.; Huo, R.; Xu, T. IRF9 as a negative regulator involved in TRIF-mediated NF-kappaB pathway in a teleost fish, Miichthys miiuy. *Mol. Immunol.* **2017**, *85*, 123–129. [CrossRef]
51. Tian, W.L.; Guo, R.; Wang, F.; Jiang, Z.X.; Tang, P.; Huang, Y.M.; Sun, L. IRF9 inhibits human acute myeloid leukemia through the SIRT1-p53 signaling pathway. *FEBS Lett.* **2017**, *591*, 2951. [CrossRef]
52. Park, S.H.; Kang, K.; Giannopoulou, E.; Qiao, Y.; Kang, K.; Kim, G.; Park-Min, K.H.; Ivashkiv, L.B. Type I interferons and the cytokine TNF cooperatively reprogram the macrophage epigenome to promote inflammatory activation. *Nat. Immunol.* **2017**, *18*, 1104–1116. [CrossRef]

9

Teratogenic Rubella Virus Alters the Endodermal Differentiation Capacity of Human Induced Pluripotent Stem Cells

Nicole C. Bilz [1,†], Edith Willscher [2,†], Hans Binder [2], Janik Böhnke [1,‡], Megan L. Stanifer [3], Denise Hübner [1], Steeve Boulant [3,4], Uwe G. Liebert [1] and Claudia Claus [1,*]

1 Institute of Virology, University of Leipzig, 04103 Leipzig, Germany
2 Interdisciplinary Center for Bioinformatics, University of Leipzig, 04107 Leipzig, Germany
3 Schaller Research Group at CellNetworks, Department of Infectious Diseases, Virology, Heidelberg University Hospital, 69120 Heidelberg, Germany
4 Research Group "Cellular Polarity and Viral Infection" (F140), German Cancer Research Center (DKFZ), 69120 Heidelberg, Germany
* Correspondence: claudia.claus@medizin.uni-leipzig.de
† These authors contributed equally.
‡ Current address: Institute for Biomedical Engineering, Department of Cell Biology, RWTH Aachen University, Medical School, 52074 Aachen, Germany.

Abstract: The study of congenital virus infections in humans requires suitable ex vivo platforms for the species-specific events during embryonal development. A prominent example for these infections is rubella virus (RV) which most commonly leads to defects in ear, heart, and eye development. We applied teratogenic RV to human induced pluripotent stem cells (iPSCs) followed by differentiation into cells of the three embryonic lineages (ecto-, meso-, and endoderm) as a cell culture model for blastocyst- and gastrulation-like stages. In the presence of RV, lineage-specific differentiation markers were expressed, indicating that lineage identity was maintained. However, portrait analysis of the transcriptomic expression signatures of all samples revealed that mock- and RV-infected endodermal cells were less related to each other than their ecto- and mesodermal counterparts. Markers for definitive endoderm were increased during RV infection. Profound alterations of the epigenetic landscape including the expression level of components of the chromatin remodeling complexes and an induction of type III interferons were found, especially after endodermal differentiation of RV-infected iPSCs. Moreover, the eye field transcription factors RAX and SIX3 and components of the gene set vasculogenesis were identified as dysregulated transcripts. Although iPSC morphology was maintained, the formation of embryoid bodies as three-dimensional cell aggregates and as such cellular adhesion capacity was impaired during RV infection. The correlation of the molecular alterations induced by RV during differentiation of iPSCs with the clinical signs of congenital rubella syndrome suggests mechanisms of viral impairment of human development.

Keywords: ectoderm; mesoderm; human development; embryogenesis; interferon response; interferon-induced genes; self-organizing map (SOM) data portrayal; epigenetic signature; embryoid body; TGF-β and Wnt/β-catenin pathway

---

1. Introduction

The enveloped, single stranded (positive-sense) RNA virus rubella virus (RV) of the genus *Rubivirus* within the family *Togaviridae* is one of the few viruses that can cause an intrauterine infection. How these viruses are transmitted vertically from the infected mother to the fetus and how they impact human development is only partially resolved. In the case of the very efficient teratogen

RV, the human-specific symptoms are categorized as congenital rubella syndrome (CRS) with the classical triad of clinical symptoms being sensorineural deafness, congenital heart disease (including cardiovascular and vascular anomalies), and cataracts [1,2]. Heart defects in CRS may comprise ventricular/atrial septal defects, patent ductus arteriosus, and patent foramen ovale. In congenital rubella, ocular (ophthalmic) pathologies include cataract, microphthalmia, glaucoma, and pigmentary retinopathy [1,2]. Furthermore, in tissue samples from three fatal CRS cases RV was detected in cardiac and adventitia (aorta and pulmonary artery) fibroblasts in association with vascular lesions [3]. The risk for the development of congenital defects is especially prevalent during maternal rubella until gestational week 11 and 12 [4–6]. Thus, intrauterine RV infection is only of concern during the first trimester. While congenital malformations are common, premature delivery and stillbirths are not markedly increased after intrauterine RV infection [1].

There are a number of ethical constraints associated with the study of human embryogenesis and congenital malformations, especially as early implantation stages of human embryos are inaccessible [7]. With embryonic stem cells (ESCs) and induced pluripotent stem cells (iPSCs), as the two types of human pluripotent stem cells (PSCs), these novel ex vivo cell culture platforms allow for the analysis of human embryonic germ layer segregation and as well as for developmental toxicity testing [8]. As a cell culture model, they represent a blastocyst-like stage, which can be extended to gastrulation-like stages through their differentiation into derivatives of the embryonic germ layers (ectoderm, mesoderm and endoderm). Additionally, their suitability as a developmental model has been demonstrated for cardiac commitment during development [9] as the heart is the first organ to develop and cardiac cell fate decisions occur very early. Furthermore, cultivation of ESCs in combination with suitable 3D matrices or together with trophoblast cells enables the formation of blastoids, gastruloids, and even embryoids (or embryo-like entities) as culture dish models for human embryogenesis [7,10].

PSCs and PSC-based differentiation models, especially the mouse (m) ESC test, are already validated for testing of teratogenic and embryotoxic substances such as thalidomide (brand name Contergan®), [11,12]. However, their potential for the study of infections during pregnancy is just at the beginning of evaluation [13,14]. In line with the limited number of viruses that can cause perinatal infection, iPSCs possess intrinsic mechanisms that restrict virus infections. In addition, compared to differentiated somatic cells, iPSCs have a higher expression level of a distinct set of interferon (IFN)-induced genes [14]. This appears to counterbalance the absence of a type I IFN response in iPSCs as an essential component of antiviral innate immunity [15].

Teratogenic RV can be maintained in iPSCs over several passages followed by directed differentiation into embryonic germ layer cells [13], highlighting iPSCs as a promising model for the very early mechanisms involved in rubella embryopathy. As a follow-up to this study we aimed at the identification of RV-induced molecular alterations in these cells before and after initiation of directed differentiation through transcriptomics. The most profound effects associated with RV infection were detected in endodermal cells derived from RV-infected iPSCs. Markers for definitive endoderm were upregulated, which occurred in association with profound epigenetic changes, an upregulation of factors involved in vasculogenesis, and reduced activity of the TGF-β signaling pathway. Additionally, ectodermal cells revealed an altered expression profile of essential transcription factors for eye field development during RV infection. Thus, the study of RV infection on iPSCs and derived lineages provides insights into viral alterations of early developmental pathways and as such into congenital diseases in general.

2. Materials and Methods

2.1. Cell Lines and Cultivation

Vero (green monkey kidney epithelial cell line, ATCC CCL-81) and A549 (human lung carcinoma epithelial cells, ATCC, LGC Standards GmbH, Wesel, Germany) were cultured in Dulbecco's modified Eagle's medium (DMEM; Thermo Fisher Scientific, Darmstadt, Germany) with high glucose, GlutaMAX,

10% fetal calf serum (FCS) and 100 U/mL penicillin–streptomycin. If not otherwise indicated, the vector-free human episomal A18945 iPS cell line (alias TMOi001-A), (Thermo Fisher Scientific) was maintained in mTeSR™1 medium (StemCell Technologies, Cologne, Germany) with 10 μg/mL gentamycin on Matrigel™ (BD Biosciences, dispensed in DMEM/F-12)-coated culture plates with daily medium change. They were passaged enzymatically at a ratio of 1:6 to 1:10 every 3 to 5 days with collagenase type IV (Thermo Fisher Scientific) in DMEM-F12 with the addition of 10 μM Y-27632 ROCK inhibitor.

2.2. Directed and Undirected Differentiation of iPSCs

Directed differentiation was performed as an endpoint differentiation assay through the STEMdiff™ trilineage differentiation kit (StemCell Technologies). The differentiation protocol was performed according to the manufacturer's instructions and required cultivation of A18945 iPSCs in mTeSR™1 medium. Single cells, as obtained after treatment with Accutase (Merck/Sigma-Aldrich Chemie GmbH, Taufkirchen, Germany), were plated on Matrigel. Every 24 h medium change of the respective STEMdiff™ trilineage differentiation medium for ectoderm, mesoderm, and endoderm was performed. Samples were collected after 5 days (mesoderm and endoderm) and 7 days (ectoderm) of cultivation. Undirected differentiation was initiated 24 h after collagenase-passaging of iPSC cultures at a ratio of 1:4 through application of undirected differentiation medium (DMEM-F12, 1x MEM-NEAA, 0.2 mM L-glutamine, 20% FBS, 0.11 mM β-mercaptoethanol, and 100 U/mL penicillin) followed by further cultivation for 5 days.

2.3. Embryoid Body Formation

EB formation as based on a previous publication [16] and (http://www.biolamina.com/media.ashx/instructions-bl010.pdf) was carried out in suspension culture and single cell suspensions were obtained after Accutase (Sigma-Aldrich) treatment. A total of 1×10^6 cells was seeded in 200 μL of EB culture medium (DMEM-F12, 20% KnockOut™ Serum Replacement (Thermo Fisher Scientific), 1× MEM-NEAA, 0.2 mM L-glutamine, 0.11 mM β-mercaptoethanol and 1 mg/mL Gentamicin) medium into one well of a nontreated conical 96-well plate and centrifuged at 600× *g* for 5 min. After cultivation for 2 days the EBs were transferred according to the protocol to a low attachment flat-bottom six-well plate and medium was changed every third day.

2.4. Virus Infection and Interferon Assays

The supernatant of infected Vero cells was collected and cleared from cellular debris by centrifugation at 350× *g* for 10 min at 4 °C and filtration through a 0.45 μm syringe filter. Thereafter ultracentrifugation with a 20% sucrose cushion (*w*/*v* in PBS) was performed for 2 h at 25,000 rpm and 4 °C. The obtained pellets were resuspended in mTeSR1. Viral titers were determined by standard plaque assay. As described previously [13], iPSC cultures with a 40–50% confluency were acutely infected with 7.5×10^5 plaque forming units (PFU) of RV per well of a 24-well plate. This corresponds approximately to an MOI of 20. The applied MOI can only be estimated as iPSCs were passaged enzymatically in clumps. The inoculum was replaced with fresh mTesR1 medium after 2 h of incubation [13]. After 4 to 5 days of cultivation, RV-infected iPSCs were passaged.

For exogenous (or paracrine) IFN treatment human recombinant IFN lambda 1 (IL-29, #300-02L) and 2 (IL28A, #300-02K), were purchased from Peprotech (Hamburg, Germany), and 3 (IL-28B, #CS26) from Novoprotein (Novoprotein, PELOBIOTECH GmbH, Planegg/Martinsried, Germany). The Accuri C6 flow cytometer (BD Bioscience, Heidelberg, Germany) was used for IFN measurement by the LEGENDplex human type 1/2/3 IFN panel (BioLegend, San Diego, CA, USA). The double-stranded (ds) RNA analogue polyinosinic-polycytidylic acid (poly I:C; Santa Cruz Biotechnology, Heidelberg, Germany) was added either directly to the cell culture or transfected at a concentration of 1 μg using Lipofectamine 2000 (Thermo Fisher Scientific) as transfection reagent.

2.5. *Calcein Live Cell Staining*

For live cell staining, EBs were incubated with mTeSR1 plus calcein FM (Sigma-Aldrich) at 1 μM. After an incubation period for 30 min, EBs were washed twice with PBS and analyzed on an inverted fluorescence microscope.

2.6. *RNA Isolation*

Total RNA was extracted from mock- and RV-infected cells by Trizol reagent (Thermo Fisher Scientific). The purification was performed with the Direct-zol RNA kit (Zymo Research, Freiburg, Germany) according to manufacturer's instructions. The integrity of the RNA samples was confirmed through analysis on a fragment analyzer (Advanced Analytical). Only samples with a RIN (RNA integrity number as a means of quality assessment) equal to 7 or greater were subjected to further analysis.

2.7. *Microarray Gene Expression Analysis and SOM Portrayal*

Isolated RNA was processed and hybridized to Illumina HT-12 v4 Expression BeadChips (Illumina, San Diego, CA, USA) and measured on the Illumina HiScan. Raw intensity data of 47,323 gene probes was extracted by Illumina GenomeStudio and subsequently background corrected, transformed into $\log_{10}$-scale, quantile normalized, and centralized to obtain gene expression estimates. Two independent samples per condition and cell type were processed.

Expression data were then further processed using self-organizing map (SOM) machine learning. The method distributes the gene-centered expression values among 2500 microclusters called meta-genes, which were arranged in a two-dimensional 50×50 lattice and colored in maroon-to-blue for high-to-low meta-gene expression values. These mosaic images visualize the transcriptome patterns of each individual sample and therefore can be understood as their molecular portraits exhibiting clusters of coexpressed genes in the samples studied [17]. Mean portraits over replicates were calculated by averaging the meta-gene landscapes of replicated samples while difference portraits between different cell types were obtained by subtracting the respective metagene values to highlight differentially expressed genes. Clusters of coexpressed genes were identified by selecting so-called 'spot-areas' in the SOM portraits using overexpression criteria as described previously [17]. For functional interpretation of the expression-modules, we applied gene set enrichment analysis using the gene set Z-score (GSZ), [17]. Enrichment of functional gene sets in the spot cluster was calculated by applying Fisher's exact test. We considered gene sets related to biological processes (BP) of the gene ontology (GO) classification, standard literature sets [17,18], and literature sets curated by our group. Downstream analysis methods were described previously [17,19] and are implemented in the R-package 'oposSOM' used for analysis [20].

Pathway activity was analyzed based on pathway topologies and gene expression data using the pathway signal flow method as implemented in oposSOM [21].

2.8. *Quantitative Real-Time PCR Analysis of Viral and Cellular RNA*

For determination of the mRNA expression level of selected cellular genes, 1.2 μg of total RNA were reverse transcribed with Oligo$(dT)_{18}$ primer and AMV reverse transcriptase (Promega, Mannheim, Germany) at 42 °C for 1 h. This was followed by an incubation step at 70 °C for 10 min. The carousel-based LightCycler 2.0 (Roche, Mannheim, Germany) was used for quantitative real-time PCR (qRT-PCR) experiments. These experiments included a 1:5 dilution of the respective cDNA samples together with 1 μg BSA and the *GoTaq® qPCR master mix* (Promega). Supplement Table S1 lists oligonucleotides and probes targeting viral p90 gene that were used for quantification of viral RNA as described [22]. Two different approaches for relative expression analysis were pursued. For direct comparison of one sample type after mock- and RV-infection, comparative delta delta Ct (ΔΔCt) was used. For comparison of gene expression levels among different cell types within a large data set, a modified version of the comparative delta delta Ct (ΔΔCt) method was used. The normalized relative quantity (NRQ) values were derived from qbase+ software (Biogazelle, Zulte, Belgium) which are based on the mean expression values of all samples and replicates within a given data set [23].

2.9. *Immunofluorescence*

For assessment of viral proteins, immunofluorescence was carried out as described [13]. Briefly, cells were fixed with 2% (*w*/*v*) paraformaldehyde in PBS and permeabilized with 0.1 Triton X-100 followed by incubation with mAb anti-E1 from Viral Antigens (Viral Antigens Incorporation, Memphis, TN, USA) at a 1:200 dilution as primary antibody.

2.10. *Statistical Analysis*

All statistical calculations were done with Graph Pad Prism software (GraphPad Software, Inc., La Jolla, CA, USA). Asterisks (* $p < 0.05$, ** $p < 0.01$, *** $p < 0.001$, **** $p < 0.0001$) highlight the level of significance in diagrams which include data as means ± standard deviation (SD). For comparison of normalized mRNA expression levels in RV-infected samples with the corresponding mock controls, a paired Student's *t* test (consistent ratios of paired values) was applied. Statistical analysis for different samples was based on one-way ANOVA followed by Bonferroni's multiple comparison test.

3. Results

3.1. *In the Presence of RV, iPSCs Maintain Pluripotent Properties and Lineage Identity after Initiation of Differentiation*

Specification to one of the three germ lineages is the first critical step in directing differentiation to downstream cellular phenotypes. Therefore, directed as well as undirected differentiation was induced in RV-infected iPSCs which were subcultured for two to five passages. Passaging of infected iPSCs results in a homogenous level of infection within iPSC cultures without affecting the protein expression level of the pluripotency marker OCT4 [13]. During passaging of RV-infected iPSCs, replication occurred at a rather constant rate as assessed by viral titer and E1 protein expression rate [13]. Furthermore, passaging allows for adaptation of RV to iPSCs and excludes any possible effects of the differentiation process itself on the otherwise acute infection with RV (Figure 1A). Undirected differentiation is spontaneous and was thus induced to assess whether RV, without a specific differentiation stimulus, directs a gene expression profile different from the mock-infected population. Directed differentiation of RV-infected iPSCs into ecto-, endo-, or mesodermal cells (thereafter referred to as RV-infected) was initiated with the STEMdiff trilineage differentiation kit as an endpoint differentiation approach to determine which of the early cell fate decision pathways could be affected by RV.

RV establishes a noncytopathic infection of iPSCs with a homogenous distribution of infected cells within the respective colony (Figure 1B), [13]. Differentiation into all three embryonic germ layers supported RV replication at a comparable rate (Figure 1C [i,ii]), [13]. As a next step, we generated an expression heatmap of selected marker genes (based on microarray whole transcriptome data) for assessment of pluripotency and lineage identity (Figure 1D). In agreement with the maintenance of OCT4 (octamer-binding transcription factor 4, also known as POU5F1) expression in RV-infected iPSCs [13], high expression of pluripotency markers CDH1 and OCT4 was noted. Their expression was maintained to some degree in endodermal cells, which is in agreement with the conditions of the STEMdiff trilineage differentiation kit. The same applies to the expression of the pluripotency marker SOX2 (SRY (Sex Determining Region Y)-Box 2) in ectodermal cells. The expression profile of lineage-specific markers confirmed ectodermal (PAX6 (Paired Box 6), DLK1 (Delta-Like 1 Homolog), and FABP7 (Fatty Acid Binding Protein 7)), mesodermal (HAND1 (Heart and Neural Crest-Derived Transcript 1), CDX2 (Caudal Type Homeobox 2), APLNR (Apelin Receptor)) and endodermal (LEFTY1 (Left-Right Determination Factor 1), EOMES (Eomesodermin), NODAL (Nodal Growth Differentiation Factor)) identity after initiation of directed differentiation in mock- and RV-infected iPSCs (Figure 1D). Additionally, some overlap between lineages, especially between mesoderm and endoderm, was noted for RV-infected samples (Figure 1D). Transcriptomic data was confirmed by RT-qPCR of pluripotency marker OCT4 and of selected markers for ectoderm (PAX6), mesoderm (HAND1), and endoderm (NODAL) lineages followed by relative quantification by qbase+ method (Figure 1E). Figure 1D and E indicate that during undirected differentiation, especially mesodermal markers were expressed, which occurred at a comparable level between mock- and RV-infected cells. Among the lineage-specific markers, the expression level of HAND1 was significantly downregulated in mesodermal cells after RV infection. Additionally, RV infection did not alter stemness-related expression signatures as indicated by the transcriptomic activity of the GO gene set telomere maintenance (Figure 1F). Telomere maintenance is active in stem cells, but gets deactivated in differentiated somatic cells [24].

In summary, comparable to the mock-control, RV maintained the pluripotent properties of iPSCs and enabled initiation of differentiation into embryonic germ layer cells as indicated by expression of essential germ layer markers.

3.2. High-Resolution Transcriptomic Maps Reveal Modules of Coregulated Genes Promoted by RV Infection during Endodermal Differentiation

As we found out that RV infection did not affect unspecific differentiation of iPSCs and enabled their lineage-specific differentiation, we wanted to focus on the effect of RV on lineage identity. The self-organizing map (SOM) transcriptome data portrayal provides a high-resolution visualization of the transcriptome landscape of each cell system studied in terms of a quadratic mosaic image and decomposes into clusters of coregulated genes. They are represented as colored spot-like areas where red and blue colors code activated and deactivated gene clusters, respectively. These transcriptomic portraits were then used to evaluate the mutual relatedness between the cell systems by means of a phylogenetic similarity tree (Figure 2A). The tree structure results from the fact that common and different spot patterns in the portraits reflect mutual similarities and differences of the activated cellular programs which enable judging the effect of RV-infection on the different lineages (see the portraits in Figure 2A). For an overview, we generated a spot-summary map in Figure 2B which shows the activated spots observed in any of the samples together with their functional context as extracted by means of gene set enrichment analysis of the genes in each of the spot-clusters of coexpressed genes (see also Supplement Table S1). In total, we identified five relevant spots labeled with capital letters A–E. Each of the spots is characterized by a specific expression profile (Supplement Figures S1 and S2) which, in turn, shows close similarities with the expression profiles of distinct gene sets (shown as 'barcode' plots in Figure 2C).

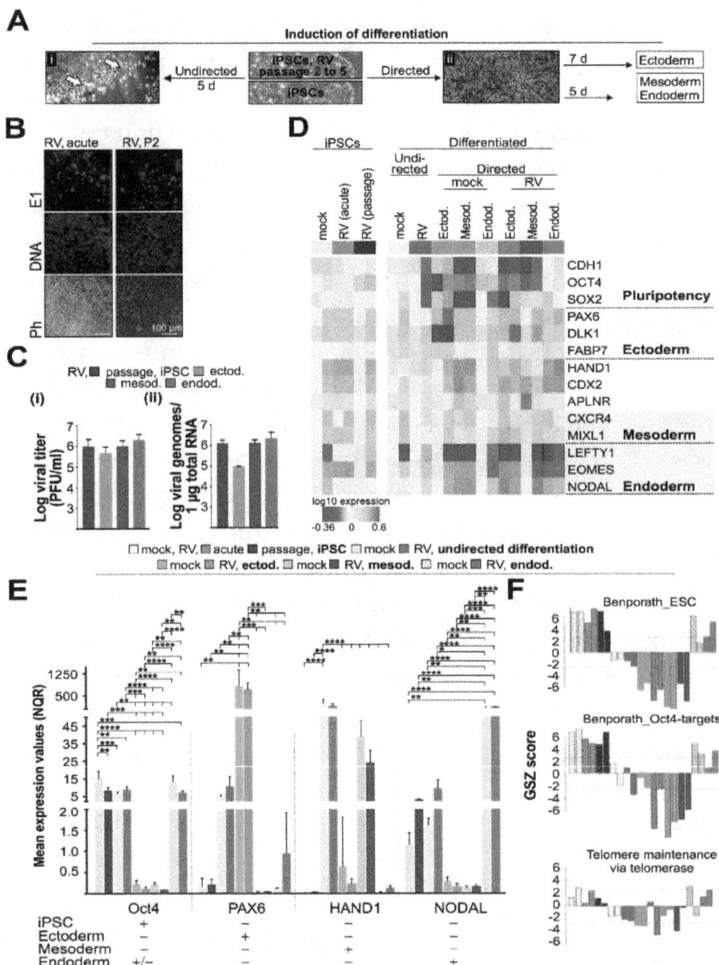

Figure 1. The identity of iPSCs and derived lineages was maintained in the presence of rubella virus (RV). (**A**) Overview of the methods applied to initiate differentiation in mock- and RV-infected iPSCs. (**A**) [**i**] Undirected differentiation in the respective induction medium started at the rim region of iPSC colonies (indicated by white arrows) and extended to their center over time of incubation. (**A**) [**ii**] Additionally, directed differentiation into the primary germ layers ectoderm, mesoderm, and endoderm was induced with the STEMdiff differentiation kit. (**B**) Immunofluorescence analysis with anti-E1 antibody (shown in red) was performed to monitor distribution of RV-positive cells within iPSC colonies. Nuclei are shown in blue. Ph, phase contrast (**C**) To assess RV replication in iPSCs and derived lineages, (**i**) virus progeny, and (**ii**) the amount of genomic viral RNAs was determined by standard plaque assay (n = 11 for passaged iPSCs, otherwise n = 3) and TaqMan-based reverse transcription-quantitative PCR (passaged iPSCs and ectodermal cells n = 3, mesodermal cells n = 7, endodermal cells n = 4), respectively. (**D**) Expression heatmap of selected marker genes of pluripotency and lineage identity (based on microarray whole transcriptome data) in mock- and RV-infected iPSCs and iPSC-derived lineages. Shading indicates overlap in the expression of some of the marker genes between iPSCs and iPSC-derived lineage cells, respectively. (**E**) The expression of indicated target genes in mock- and Wb-12-infected iPSCs and derived cells after initiation of undirected and directed differentiation was determined by real-time quantitative PCR (RT-qPCR) and analyzed by qbase+ software. For normalization, chromosome 1 open reading frame 43 (C1orf43) and hypoxanthine guanine phosphoribosyl transferase 1 (HPRT1) were used. Relative gene expression was calculated as normalized relative quantity (NRQ) and given as means ± SD (n = 3 to 5). As a reference, lineage-specific expression levels were assigned based on literature data and transcriptomics data by Stemcell Technologies for the Trilineage differentiation kit. (**F**) Analysis of stemness-related expression signatures was based on the GO gene set telomere maintenance. An embryonal stem cell signature and a gene set collecting OCT4 targets were both taken from [25].

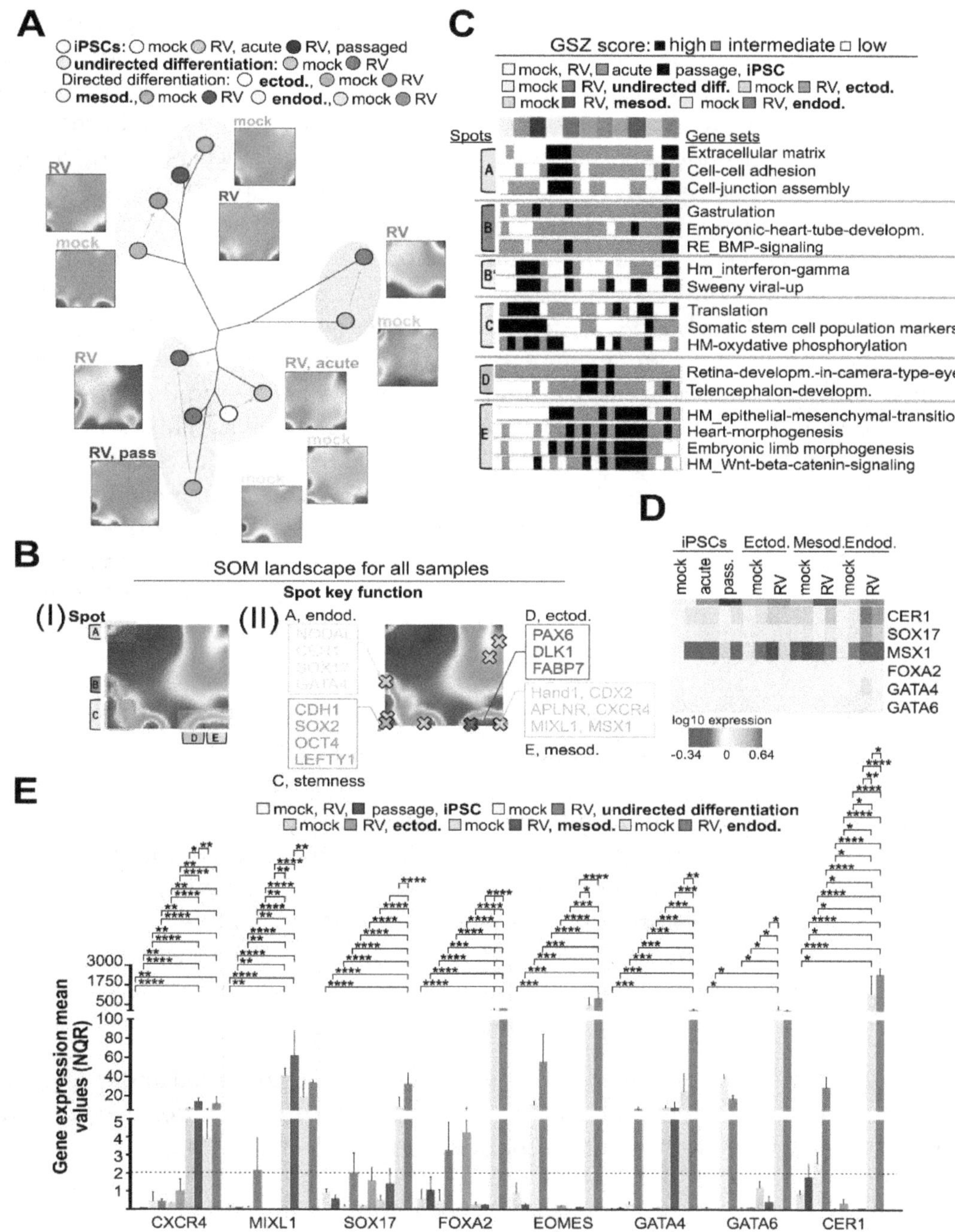

Figure 2. RV supports expression of markers for definite endoderm. (**A**) Similarity tree of the gene expression portraits of the cell systems studied. Both, mock- and RV-infected mesodermal and ectodermal cells share relatively high mutual similarity of their transcriptomes. In contrast, mock- and RV-infected endodermal cells form a separate branch that is closer to the mock- and RV-infected iPSCs. (**B**) Spot summary map (**I**) provides an overview of activated cellular programs and their functional context, which is depicted in more detail in (**II**). (**C**) Barplot representation of the expression profiles of gene sets related to different functions. Their genes were enriched in the spots that were identified in the SOM portraits. They are thus indicated accordingly (see also the overview map in part B of the Figure and Table S2). (**D**) Selected marker genes for definite endoderm are illustrated in the heatmap of the transcriptome of mock- and RV-infected iPSCs and iPSC-derived lineages. (**E**) The expression of marker genes for definite endoderm was determined by RT-qPCR in mock- and Wb-12-infected iPSCs and iPSC-derived lineages by qbase+ software. For normalization, C1orf43 and HPRT1 were used. Relative gene expression was calculated as normalized relative quantity (NRQ) and given as means ± SD (n = 3 to 5). See also Figures S1 and S2.

Mapping of the position of genes into the SOM enables us to deduce their expression profile in the cell systems studied according to the gene's location in or near the spots and the respective spot profiles (Figure 2B and Supplement Figure S1). Importantly, key genes of ectoderm and mesoderm development are found in spots D and E, respectively, and are confirmed to be upregulated in the ectoderm and mesoderm cells. On the other hand, genes related to gastrulation and heart-tube development were enriched in spot B and found to be activated in RV-infected endoderm cells. This occurred together with genes that are activated upon interferon response (see next subsection). Stemness key genes locate in spot C together with part of the developmental genes of the endoderm (Supplement Figure S1). Difference portraits clearly indicate that spot B associates with RV infection (Figure 2B), which contains the genes NODAL, CER1, SOX17, and GATA4, reflecting their activation upon RV infection in endodermal cells. In addition, we are able to show that mesoderm and ectoderm cells share similar expression patterns characterized by upregulated spots D and E and downregulated stemness genes (spot C), which, in contrast, are upregulated in endoderm cells. This is in agreement with the expression level of pluripotency genes in the endodermal lineage which was higher than in the remaining two embryonal germ layers (Figure 1D,E). As a consequence, mesoderm and ectoderm cells occupy neighboring positions at one end of the similarity tree, while endoderm cells are found at the opposite end. These findings highlight two important aspects: (I) there is a close similarity of transcriptional patterns among ecto-, mesoderm, and iPSC samples without and with RV infection; and (II) the endodermal lineage is an exception, where RV infection induces a notable shift away from the corresponding mock sample.

To further elucidate this specific effect of RV, we analyzed the transcriptomic data for markers for definitive endoderm (CXCR4 (C-X-C Motif Chemokine Receptor 4), MIXL1, SOX17, FOXA2 (Forkhead Box A2), EOMES, GATA6, CER1 (Cerberus, DAN family BMP Antagonist), and LEFTY, [26]. The expression heatmap shown in Figure 2D indicates an increased expression level of CER1 and SOX17 after RV infection. Figure 2E shows qPCR analysis followed by relative quantification by qbase+ method and highlights that the definite endoderm markers CXCR4, MIXL1, SOX17, GATA4/6, and CER1 were significantly higher expressed in RV-infected endodermal cells as compared to their mock-infected counterparts. While RV infection did not induce these markers above a cut-off of a two-fold increase during mesodermal and ectodermal differentiation, EOMES, GATA4, and CER1 were specifically induced by RV during undirected differentiation.

In conclusion, similarity analysis supports the hypothesis that ectodermal and mesodermal lineage identity was maintained after infection with RV, while endodermal cells derived from RV-infected iPSCs were enriched in markers for definitive endoderm.

3.3. RV Infection Activates IFN Type III Response Pathways on iPSCs and Derived Lineages

Difference portraits (Figure 3A) indicated that RV-infection specifically upregulated genes in spot B, which were associated with "IFN response" characteristics. This is supported by the profiles of gene expression signatures of viral infections such as by influenza virus and pneumonia that is accompanied by interferon activation. These genes were consistently upregulated in RV-infected samples with the largest observed effect in endodermal cells (Supplement Figure S3), [27–29]. Spot B highlighted in the SOM landscapes contained genes involved in IFN and viral response mechanisms (Figure 3A, a list of genes is given in Figure 3B). This is further emphasized by the expression heatmap shown in Figure 3C. Whereas genes involved in IFN-sensing, including the type III IFN receptor IFNLR1 (IFN lambda receptor 1), were not altered in their expression level, the IFN-signaling components STAT1 (signal transducer and activator of transcription 1) and IRF9 (interferon regulatory factor 9) appeared to be slightly upregulated at their mRNA expression level after infection with RV, especially in endodermal cells. The highest level of upregulation was found for IFN-stimulated genes (ISGs), notably for MX1 (MX dynamin like GTPase 1), IFITM2 (interferon induced transmembrane protein 2), and ISG15 (interferon-stimulated gene 15). Mapping of these genes into SOM space further underlines these findings: The IFN-signaling genes and the ISGs accumulate in and around spots B and D, respectively, while IFN-sensing genes are located outside these spot regions (Figure 3B). The increase in the expression level of selected marker genes of the IFN pathway in the presence of RV was confirmed by RT-qPCR (Figure 3D). Compared to RV-infected iPSCs and ecto- and mesodermal cells, the expression level of IRF9 and STAT1 and selected ISGs (IFITM1/2, IFIT1, and ISG15) was significantly higher in RV-infected endodermal cells. The highest increase in mRNA expression after RV infection was noted for the ISGs IFIT1 and ISG15. Therefore, we determined whether this gene expression pattern was indeed associated with IFN generation during RV infection through quantification of type I (α and β), type II (γ), and type III (λ1 and λ2/3) IFNs by the LEGENDplex assay from the supernatants of RV-infected cells (Figure 3E). In iPSCs as well as iPSC-derived lineage cells, RV infection induced secretion of type III IFNs, namely IFN λ2/3 (Figure 3E). As a positive control, the synthetic dsRNA analog poly I:C was used, which was either transfected into iPSCs or added directly to the supernatant. Either application of poly I:C did not lead to secretion of any type I, II, or III interferons (Figure 3E). The activation of type III IFNs by RV was also confirmed at the mRNA level by RT-qPCR (Figure 3F). Compared to the mock-infected control, RV induced a significant increase in the mRNA expression of IFN λ2/3 in endodermal cells (Figure 3F). Thereafter, we addressed the discrepancy between IFN λ2/3 protein (Figure 3E) and mRNA level (Figure 3F). Gene set analysis revealed that mRNAs associated with the KH type-splicing regulatory protein (KHSRP) were specifically enriched in endodermal cells (Figure 3G). KHSRP is involved in post-transcriptional regulation of mRNA expression, including IFN λ3 [30]. This could explain the discrepancy between IFN λ2/3 protein and mRNA expression level.

To address the influence of the type III IFNs secreted during RV to the cell culture supernatant on the gene expression landscape of iPSCs, type III IFNs were added exogenously for two weeks of cultivation during daily medium change of iPSCs. The zoom-in similarity tree shown in Figure 3H highlights the relatedness between the expression portraits of mock- and RV-infected iPSCs as well as iPSCs after application of type III IFNs. The gene expression profile of passaged RV-infected cells shifted away from iPSCs after exogenous IFN type III application, but closer to the mock-infected cells, suggesting an adaptation of RV to iPSCs.

In conclusion, in iPSCs and iPSC-derived embryonic lineages, RV infection induced a type III IFN response together with activation of ISGs, notably MX1 and ISG15. This activation appeared to be specifically profound in endodermal cells.

3.4. *RV Infection Is Associated with Chromatin Remodeling*

Alterations of gene expression patterns during development are governed by epigenetic mechanisms in cooperation with regulation via transcription factor networks [32,33]. Particularly, we found that gene signatures of epigenetic impact, such as targets of the polycomb repressive complex 2 (PRC2), of H3K27me3, and of bivalently (H3K4me3 and H3K27me3) marked gene promoters, have an almost antagonistic expression profile as compared to the stemness signatures (Figure 4A, in comparison to Figure 1F).

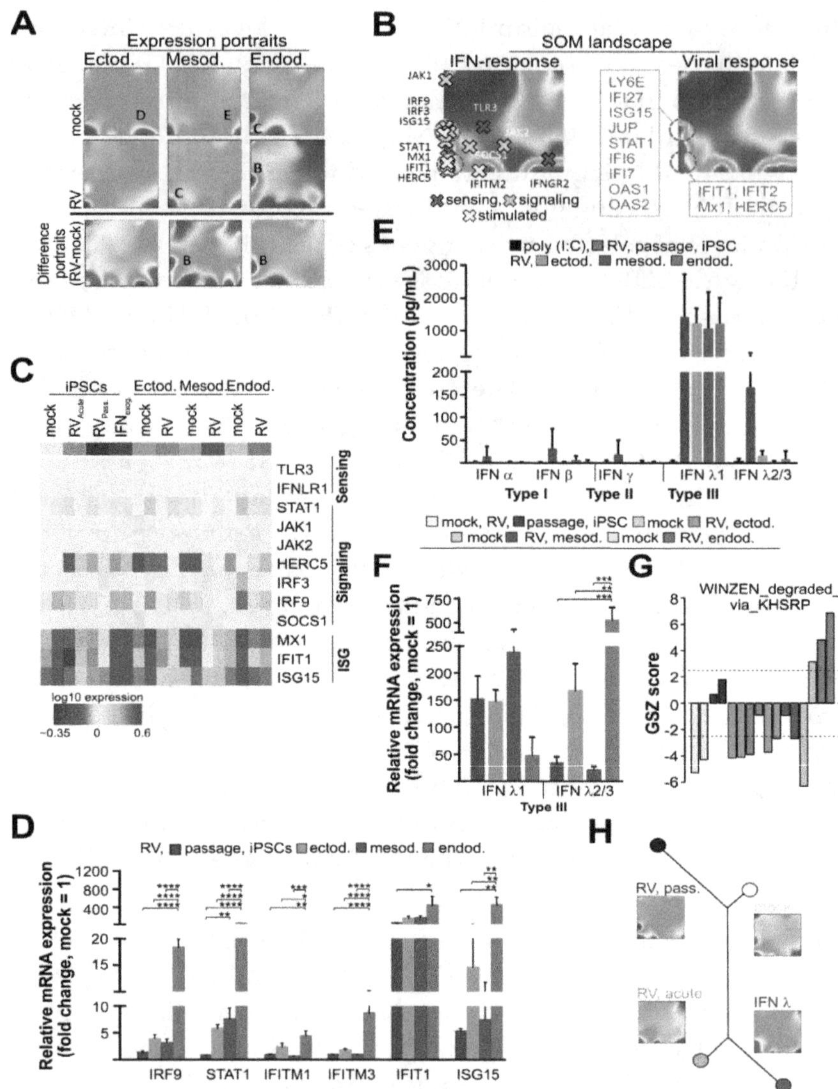

Figure 3. RV infection activates an IFN response in iPSCs and derived lineages. (**A**) Expression portraits of the embryonal germ layers before and after RV infection and the respective difference portraits reveal characteristic spot patterns, where spots C, D, and E are specific for endoderm, ectoderm, and mesoderm cells, respectively. Spot B appears after RV infection mainly in endodermal cells. (**B**) IFN response genes with signaling and stimulated functions in the IFN response pathways accumulate in spots B and C and were upregulated predominantly in iPS and endodermal cells after RV infection as also indicated by the IFN and viral response gene signature profiles. (**C**) Expression heatmap of selected marker genes involved in IFN-sensing and -signaling in mock- and RV-infected iPSCs and iPSC-derived lineages. Interferon-stimulated genes (ISGs) that were identified by the SOM analysis shown in (**A**) are included. (**D**) The mRNA expression level of the IFN-signaling components IRF9 and STAT1 and indicated ISGs was verified by RT-qPCR analysis. Data are given as means ± SD (n = 3 and n = 5 for RV-infected mesoderm, IRF9 and STAT1). (**E**) The IFN profile for RV-infected iPSCs and derived lineages was determined by the LEGENDplex IFN panel for undiluted supernatants collected after five (iPSCs and mesodermal cells) and seven days (endodermal and ectodermal) of cultivation. (**F**) The mRNA expression level of type III IFNs was verified by qPCR analysis and given as means ± SD (n = 3). (**G**) Mean expression of a gene set (gene set Z-score, GSZ) that is controlled by KSRP, which appears to keep inflammatory gene expression within defined limits [31]. (**H**) Similarity tree of the gene expression portraits of mock- and RV-infected iPSCs in comparison to iPSCs after cultivation in the presence of exogenous type III IFNs. (**D**,**F**) For normalization of qRT-PCR data in the $2^{-\Delta\Delta Ct}$ method, the HPRT1 gene was used. See also Figure S3.

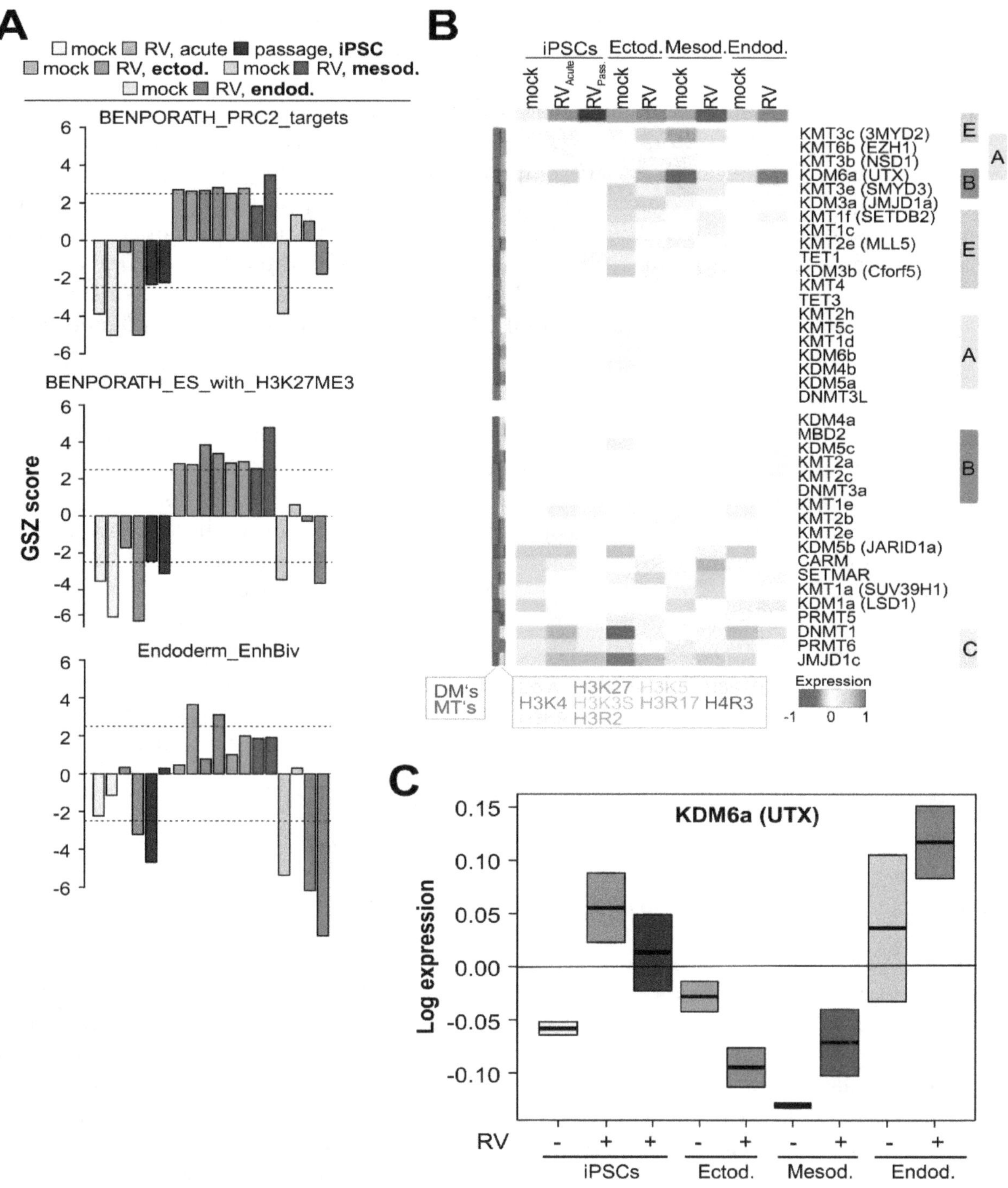

Figure 4. RV infection is accompanied by an altered expression of components of the SWI/SNF and NURF chromatin remodeling complexes. (**A**) Mean GSZ expression signatures of stemness-related transcriptional programs which act via epigenetic programming (PRC2 targets, repressive and bivalent chromatin marks). (**B**) Expression heatmap of chromatin modifying enzymes in the cell systems studied. The gene expression data of methyltransferases (MTs) and demethylases (DMs) of DNA cytosines, histone lysine and arginine side chains were assigned to the expression spot-cluster A–E according to their expression profiles. This suggests their involvement in the regulation of chromatin structure as writers and erasers of methylation marks at histone lysine and arginine side chains and affecting DNA methylation. Notably, a very strong variability was observed for KDM6a (alias UTX), a constituent of the SWI/SNF ATP-dependent chromatin remodeling machinery. (**C**) Expression of KDM6a directly relates to stemness programs (Figure 1F) and inversely relates to programs repressing stemness functions (part A of the Figure). See also Figure S4.

Next, we focused on chromatin remodeling which is essential for lineage segregation [9,34]. The analysis of transcription factor networks that act in regions of euchromatin with transcriptionally active genes and regulatory elements revealed that genes in repressed and bivalent states of endoderm progenitors become more rigorously deactivated in RV-infected endoderm cells than genes from these states of mesoderm progenitors in RV-infected mesoderm cells (Supplement Figure S4A). This suggests that RV infection specifically dedifferentiates endoderm cells by suppressing developmental suppressors. We then studied enzymes affecting methylation of DNA and of arginine and lysine side chains of histones such as H3K4, H3K9, H3K27, and H3K36, with potential impact on chromatin structure. The profiles of the DNA methylation maintenance methyltransferase DNMT1, of PRMT6, a methyltransferase of histone arginine side chains, and of JMJD1c, a H3K9 demethylase, and KDM5b demethylating H3K4me3 correlate with the 'stemness' spot cluster C upregulated in iPS and endodermal cells.

Notably, the gene encoding the H3K4 demethylase KDM6a (alias UTX) was markedly upregulated in RV-infected iPSCs and, especially, endoderm-derived cells (Figure 4B). KDM6a is a constituent of the SWI/SNF ATP-dependent chromatin remodeling machinery. Figure 4C highlights that in comparison to iPSCs and ecto- and mesodermal cells, the upregulation of KDM6a was highest in endodermal cells. The alterations of its expression suggest its role in chromatin remodeling after RV infection described above. This motivated us to estimate the expression patterns of other genes encoding components of the SWI/SNF and of the NURF chromatin remodeling complexes [9] by mapping them into the SOM (Supplement Figure S4B). We found that, indeed, Smarcc2, Smarcd3, and Btpf were all upregulated in endoderm-derived cells after RV infection, which further supports the assumption that SWI/SNF and possibly also NURF contribute to chromatin remodeling during RV infection. In conclusion, expression changes of different sets of genes involved in epigenetic regulation and of constituents of the ATP-dependent chromatin remodeling complexes such as KDM6a-UTX were detected in association with RV-infection, especially during endodermal differentiation.

3.5. RV Infection Impairs Aggregation of iPSCs into Embryoid Bodies

The progression of embryogenesis does not only involve the activation of developmental pathways, but also requires cell–cell interactions based on adhesive forces [35]. The relevance of these observations for RV-infected iPSCs was emphasized by gene ontology analysis regarding focal adhesion and regulation of cell adhesion (Figure 5A). Transcriptomic analysis revealed that THY-1 (also known as cluster of differentiation (CD) 90) was among the targets affected by RV (Figure 5B). The relevance of THY-1 (CD90) for cellular adhesion capacity was highlighted for CD90 negative carcinoma, which compared to their CD90 positive counterparts lack the ability to form spheres [36]. Accordingly, we have addressed whether RV alters the spontaneous aggregation capacity of iPSCs into 3D aggregates called embryoid bodies (EBs). EB formation relies on cell–cell adhesive interactions [35]. Compared to the mock control, EBs generated from RV-infected iPSCs were reduced in diameter and of irregular shape (Figure 5C). Furthermore, during cultivation they lost stability and small-sized debris was generated (Figure 5C). In contrast to RV-infected iPSCs, the mock-infected controls generated viable EBs as indicated by staining with calcein performed after two weeks of cultivation (Figure 5D). In conclusion, RV infection impaired the adhesion capacity of iPSCs as shown by their reduced ability to assemble into EBs. This suggests an impaired cell–cell interaction capacity during lineage segregation.

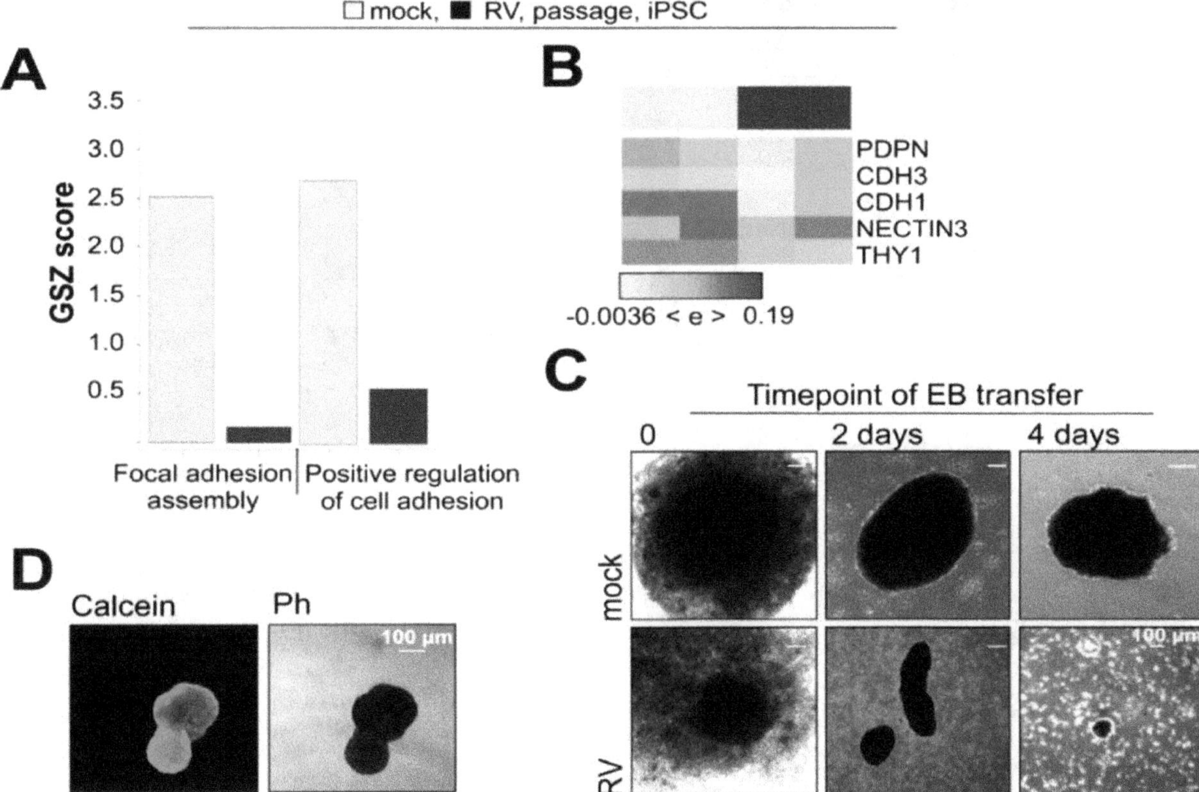

Figure 5. RV impairs the cellular adhesion capacity of iPSCs. (**A**) Gene signatures related to focal adhesion in mock- and RV-infected iPSCs. (**B**) Expression heatmap of selected marker genes involved in cellular adhesion. (**C**) Assessment of 3D stability through embryoid body (EB) formation. Shown are images before and after cultivation in suspension. (**D**) To verify viability, EBs were stained with calcein after cultivation for two weeks.

3.6. RV Infection Specifically Affects Developmental Pathways during Endodermal Differentiation

For assessment of the effect of RV on global cellular signaling networks, we focused on two important signaling pathways, namely transforming growth factor β (TGF-β) and Wnt/β-catenin (Wnt), (Figure 6A,B, respectively). The TGF-β signaling pathway is involved in cell growth and differentiation during embryogenesis [37]. The Wnt signaling pathway regulates the interaction between cellular pathways involved in primary germ layer formation and is required for mesodermal differentiation from pluripotent stem cells [38]. A more detailed view of the TGF-β and Wnt signaling pathways is provided in Supplementary Figures S5 and S6, respectively. As expected, the highest TGF-β signaling pathway activity was observed in endodermal cells, while the Wnt signaling pathway was most active in mesodermal cells (Supplement Figures S5 and S6). Thus, these two lineages were depicted in Figure 6A,B, respectively, to highlight the effect of RV on their activity in comparison to the respective controls. Within the TGF-β signaling pathway, mock-infected endodermal cells show high cell cycle activity induced by CDKN2B and its downstream interaction partners, which became deactivated during RV infection (Figure 6A and Supplement Figure S5). Additionally, RV infection in endodermal cells was specifically accompanied by a strong activation of NODAL, an essential component of the TGF-β signaling pathway (Figure 6A and Supplement Figure S5). During RV infection, the transcriptional activity of Wnt signaling pathway was reduced in mesodermal cells (Figure 6B), whereas for ectodermal and endodermal cells, almost no alteration in its activity was detected (Supplement Figure S6).

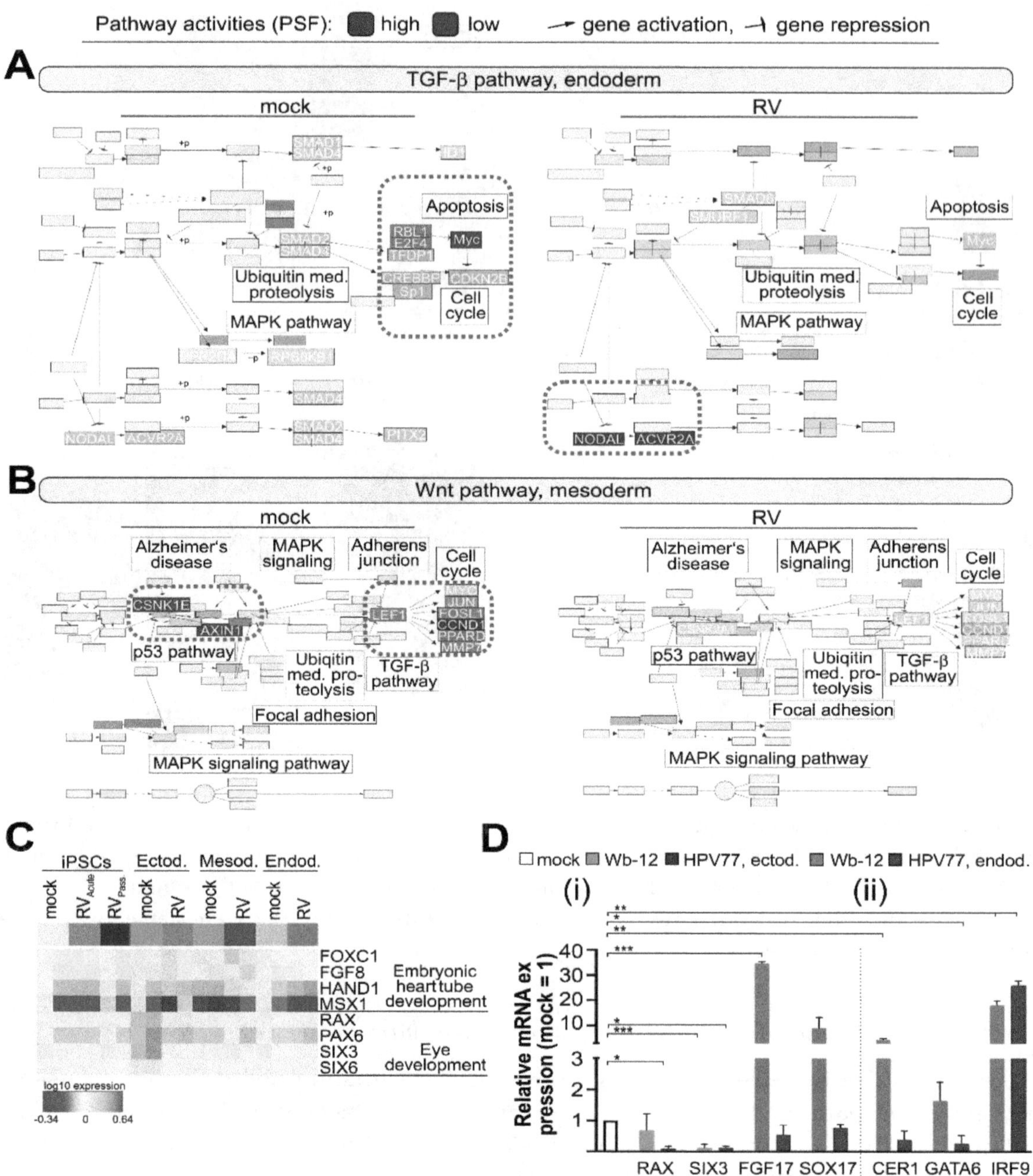

Figure 6. RV exerts lineage-specific effects on developmental signaling pathways and alters expression of transcription and growth factors. Pathway signal flow (PSF) activity plot of (**A**) the TGF-β signaling pathway in endodermal cells (highlighted are genes with higher (NODAL, ACVR2A, Myc) and lower (RBL1 and E2F4) activity after RV infection) and (**B**) of the Wnt signaling pathway in mesodermal cells (highlighted are genes with lower (the CSNK1E/AXIN1E and the LEF1/CCND1 axis) activity after RV infection). The calculation of the activity of the nodes was based on the PSF algorithm using the respective gene expression values and the wirings between the nodes [21]. (**C**) Selected genes within pathways that were specifically affected by RV infection are illustrated in the heatmap of the transcriptome of mock- and RV-infected iPSCs and iPSC-derived lineages. (**D**) For qRT-PCR expression analysis, the $2^{-\Delta\Delta Ct}$ method based on normalization to HPRT1 gene was used. Values are given as means ± SD (n = 4 for Wb-12-infected ectoderm, n = 2 for HPV77-infected ectoderm, otherwise n = 3). See also Figures S5–S7.

As congenital rubella leads to defects in heart and eye development, we analyzed the impact of RV infection on the underlying molecular pathways. For members of the gene annotation embryonic heart tube development and the gene set heart morphogenesis only a slight effect of RV infection was noted (Supplement Figure S7), suggesting the involvement of other factors. In mesodermal cells, expression of HAND1, which is involved in embryonic heart tube development, was reduced after RV infection as compared to the mock controls (Figures 6C and 1E). The ectoderm gives rise to components of the eye. At the molecular level, RV infection of ectodermal cells impaired the gene set eye development (Figure 6C), (Supplement Figure S7). Specifically, SIX3 and SIX6, as key transcription factors for mammalian eye development [39,40], were reduced in their expression level (Figure 6C). Among others, SIX3, together with RAX, initiates transcription of genes required for lens placode formation [39].

To further assess developmental pathways with relevance for the teratogenic outcome of RV infection, we determined the mRNA expression of RAX and SIX3 (as important factors for eye development) besides FGF17 (Fibroblast Growth Factor 17) and SOX17 (as contributing factors for endodermal differentiation and cardiovascular development) by RT-qPCR in ectodermal and endodermal cells, respectively (Figure 6D). Here, the vaccine strain HPV77 was used in addition to Wb-12 strain. Attenuated vaccine strains such as HPV77 are not teratogenic as revealed after immunization of unknowingly pregnant women [4]. Thus, any alteration at the molecular level that is present during wild-type Wb-12, but not HPV77 infection, emphasizes its possible contribution to congenital rubella. In comparison to the mock control, a similar reduction in the expression of RAX and SIX3 was detected in ectodermal cells after infection with both RV strains. However, a different picture emerged for FGF17 and SOX17. Figure 6D [i] highlights an increase in the expression of FGF17 and SOX17, which was significant for FGF17 compared to the mock control, but only for endodermal cells derived from Wb-12-infected iPSCs, not for endodermal cells derived from HPV77-infected iPSCs. Moreover, the increase in the expression of the definitive endoderm markers CER and GATA6 after infection with Wb-12 as shown in Figure 2E was not detected after infection with HPV77 (Figure 6D [ii]). However, both RV strains induced a significant increase of the IFN-signaling component IRF9 (Figure 6D [ii]). The endoderm plays an essential role in the crosstalk between the lineages and contributes to the epithelial lining of many organs, including the vascular network. Accordingly, the gene set vasculogenesis, but not angiogenesis, was affected by RV infection (Supplement Figure S7). This emphasizes our notion on the correlation between the impact of RV infection on endodermal cells and congenital rubella.

In summary, specific signatures including the TGF-β signaling pathway were affected by RV infection, but in a lineage-specific manner. In ectodermal cells, RV infection significantly reduced expression of SIX3 as key transcription factors for eye field development. Only for the clinical isolate Wb-12, but not for the vaccine strain HPV77, was an impact on the growth factor FGF17 and the endodermal transcription factor SOX17 noted.

Figure 7 summarizes the findings of this study in correlation to the main CRS symptoms. The noncytolytic course of infection of RV during directed differentiation is in agreement with its persistence in multiple organs and tissues during congenital rubella. We have not identified any indication at the molecular level that could contribute to the defects in ear development during congenital rubella. However, sensorineural deafness is often a late-onset symptom and could be associated with pathological alterations in the brain of the infected infants [3].

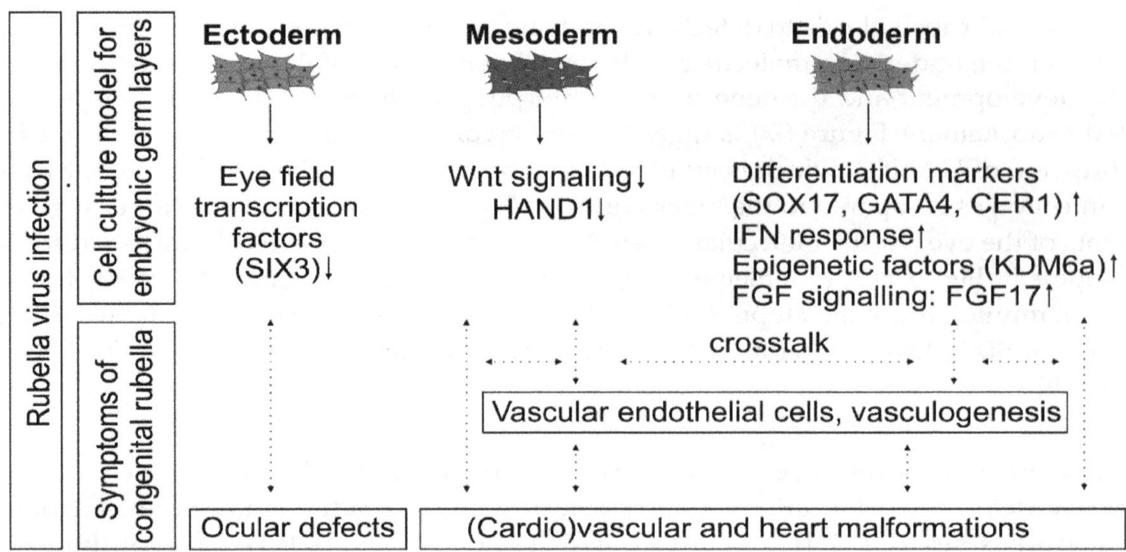

Figure 7. Graphical summary of the identified molecular alterations induced by RV during directed differentiation of iPSCs into the three embryonic germ layer cells. The data was set in a possible relation to the prevailing symptoms of congenital rubella embryopathy. Especially endodermal cells were characterized by profound alterations in their gene expression landscape, including the expression of markers for definitive endoderm and epigenetic factors. This could impair the crosstalk between endodermal and mesodermal cells during differentiation.

4. Discussion

Knowledge on developmental signaling networks is an essential prerequisite to understand congenital abnormalities, either caused by pathogenic, hereditary or environmental risk factors. Models for developmental toxicity testing range from iPSCs to iPSC-derived EBs and three-dimensional organoids. They have different properties regarding high-throughput screening capacity and relevance for in vivo developmental processes [41]. Their proper assessment requires compounds or pathogens with well-known symptoms arising from embryotoxic or teratogenic alterations during embryonal development. Here, we used RV to correlate clinical observations for congenital rubella syndrome with its impact on the differentiation capacity of iPSCs. Although iPSC-based cell culture models reflect only transient stages during human embryogenesis, they allow us to recapitulate essential developmental pathways that are otherwise inaccessible [42].

Among human pathogens, RV is rather exceptional in its ability to replicate noncytopathically in iPSCs, which in general represent a rather restrictive environment to most viral infections [43]. The protection of human development from a pathogenic insult involves several mechanisms, including transcriptional silencing of viruses in pluripotent stem cells [44] and an intrinsic high expression level of IFN-induced genes [14]. This includes interferon-induced transmembrane protein 1 (IFITM1) and its capacity to restrict the potentially harmful reactivation of human endogenous retroviruses [43]. Otherwise, the antiviral innate immune response in iPSCs is rather refractive [15]. The constitutive overexpression of an active IRF7 as a master regulator of the type I IFN system revealed the harmful effects an activated type I IFN response would have on the expression of pluripotency and lineage specific genes, especially of endodermal cells [45]. In contrast to the engineered type I IFN response in iPSCs through overexpression of IRF7, no morphological changes were noted after infection of iPSCs with RV [13]. However, in agreement with the study on the effect of type I IFNS on differentiation capacity of iPSCs [45], the impact of RV on directed differentiation was most profound during endodermal differentiation. The differences in the signaling cascades of type I and III IFNs [46] might explain the milder effects noted after RV-associated type III IFN activation as compared to the severe effects of an engineered type I IFN response [45]. Our data complements a recent study on the impact of Influenza A virus (IAV) on the pluripotency and proteome of hiPSCs [47]. Whereas, in contrast to RV,

IAV reduces the pluripotency of iPSCs, both virus infections induce ISG15 and IFN λ1 [47], highlighting this observation as an innate immune mechanism that is already developed in iPSCs. Further studies need to address whether the impact of RV infection on endodermal differentiation is correlated with the activation of the type III IFN signaling pathway and how this affects the course of infection of RV in iPSCs. In ectodermal cells, RV infection was associated with the downregulation of SIX3, an essential transcription factor for early eye development [48]. Together with SIX6, SIX3 suppresses Wnt signaling, which could contribute to the slight activation of this essential developmental signaling pathway in ectodermal cells derived from RV-infected iPSCs [40]. Their functional importance during retinal development and eye field specification was recently shown by the use of iPSC-derived retinal organoids [39]. Our study complements a previous study on the gene expression profile of fetal (HUVEC originating from umbilical cord veins) and adult (HSaVEC derived from the saphenous vein) endothelial cells which revealed a specific enrichment of 18 downregulated genes within the GO terms "sensory organ development", "eye development", and "ear development" [49].

Among the embryonic germ layers, especially differentiation to definite endoderm appeared to be affected by RV infection. In addition to its role in formation of organs of the digestive tract, the interaction of endodermal cells with precardiac mesoderm drives specification and differentiation of cardiac myocytes and cells of endocardial endothelium [50]. This is supported by studies on the contribution of signals from endodermal cells and the interactive crosstalk between the endoderm and mesoderm to differentiation of ESCs to a cardiomyogenic lineage [9]. RV infection does not only target the endoderm, but also signals that facilitate this interactive crosstalk. This includes Cerberus as a bone morphogenetic protein (BMP) antagonist [51]. The secretion of Cerberus from endodermal cells initiates differentiation of the neighboring tissue, namely the overlying cardiac mesoderm [51,52]. Furthermore, the analysis of endoderm-depleted frog and avian embryos revealed that the endoderm contributes to vasculogenesis and vascular tube formation [53]. Thus, as summarized in Figure 7, the molecular events identified in RV-infected endodermal cells could contribute to cardiovascular defects during congenital rubella [2].

Besides the mere expression level of essential components of developmental pathways, post-translational histone modifications are involved in the regulation of gene expression during development. The balance between H3K4me as an active and H3K27 as an inactive state histone modification directs the switch between active and inactive pathways during differentiation [54]. The activity of the KDM6A (UTX) demethylase was especially upregulated in endodermal cells during RV infection. KDM6A demethylase activity was reported to counteract DNA damage response and cell death induction in differentiating ESCs [55], which could also apply to RV-infected endodermal cells.

RV infection was associated with an upregulation of definitive endoderm-enriched transcription factors, including GATA4, EOMES, and SOX17 [56]. In a context- and dose-dependent manner, the transcription factor EOMES directs cardiac development as well as endoderm specification [57]. Whereas SOX17-null mice revealed a downregulation of several genes involved in heart development [58], the ectopic overexpression of SOX17 during hematopoiesis impaired survival of early hematopoietic precursors due to induction of apoptosis [59]. This indicates that normal embryonal development, especially cardiac specification, requires fine-tuned expression of several factors [60], which appears to be affected by RV infection.

The characterization of teratogens such as RV on iPSC-based models is an essential requirement to emphasize their suitability for the assessment of embryotoxicants and to identify relevant parameters to increase their predictive power. Congenital heart malformations are not only caused by pathogens such as RV, they are the most common among human developmental defects identified for human births. iPSC-based models enable valuable insights into human development and processes that might disturb its normal progression, which will broaden our diagnostic and treatment options for congenital defects. Further studies are needed to correlate the identified transcriptional changes with functional consequences for pathways directing embryonal development.

Supplementary Materials:
Table S1: Related to description of quantitative real-time PCR analysis, Table S2: Related to Figure 2C. Spot cluster characteristics, Figure S1: Mean expression profiles of 'spot'-clusters of genes which were denoted with capital letters A–E., Figure S2: Gene set enrichment for pathway analysis of mock- and RV-infected iPSCs and iPSC-derived lineages., Figure S3: Characterization of the IFN-response gene signature in RV-infected iPSCs and iPSC-derived lineages., Figure S4: Characterization of gene expression signatures related to epigenetic regulation., Figure S5: Pathway signal flow (PSF) activity plot of the TGF-beta signaling pathway in ecto-, meso-, and endodermal cells derived from mock- and RV-infected iPSCs., Figure S6: Pathway signal flow (PSF) activity plot of the Wnt signaling pathway in ecto-, meso-, and endodermal cells derived from mock- and RV-infected iPSCs., Figure S7: Gene set expression signatures of developmental programs in mock- and RV-infected iPSCs and iPSC-derived lineages.

Author Contributions: C.C. supervision of the study; funding acquisition, C.C., H.B. wrote the first draft of the manuscript, all authors read and revised the final manuscript, N.C.B., J.B., D.H. performed experiments, E.W. and H.B. performed mathematical modelling and processing and analysis of transcriptomic data. M.L.S., S.B. and U.G.L. provided resources.

Acknowledgments: For provision of Wb-12 strain the authors want to thank B. Weißbrich (University of Wuerzburg, Germany). We want to thank Knut Krohn and Petra Süptitz from the core unit DNA technologies, IZKF Leipzig, Medical Faculty of the University of Leipzig, Leipzig, Germany for RNA quality assessment and technical support of the microarray experiments. The authors also want to thank Sandra Bergs for technical support.

References

1. Dudgeon, J.A. Congenital rubella. *J. Pediatr.* **1975**, *87*, 1078–1086. [CrossRef]
2. Duszak, R.S. Congenital rubella syndrome—Major review. *Optometry* **2009**, *80*, 36–43. [CrossRef] [PubMed]
3. Lazar, M.; Perelygina, L.; Martines, R.; Greer, P.; Paddock, C.D.; Peltecu, G.; Lupulescu, E.; Icenogle, J.; Zaki, S.R. Immunolocalization and distribution of rubella antigen in fatal congenital rubella syndrome. *EBioMedicine* **2016**, *3*, 86–92. [CrossRef] [PubMed]
4. Freij, B.J.; South, M.A.; Sever, J.L. Maternal rubella and the congenital rubella syndrome. *Clin. Perinatol.* **1988**, *15*, 247–257. [CrossRef]
5. Bouthry, E.; Picone, O.; Hamdi, G.; Grangeot-Keros, L.; Ayoubi, J.M.; Vauloup-Fellous, C. Rubella and pregnancy: Diagnosis, management and outcomes. *Prenat. Diagn.* **2014**, *34*, 1246–1253. [CrossRef] [PubMed]
6. Enders, G.; Nickerl-Pacher, U.; Miller, E.; Cradock-Watson, J.E. Outcome of confirmed periconceptional maternal rubella. *Lancet* **1988**, *1*, 1445–1447. [CrossRef]
7. Rossant, J.; Tam, P.P.L. Exploring early human embryo development. *Science* **2018**, *360*, 1075–1076. [CrossRef] [PubMed]
8. Kugler, J.; Huhse, B.; Tralau, T.; Luch, A. Embryonic stem cells and the next generation of developmental toxicity testing. *Expert Opin. Drug Metab. Toxicol.* **2017**, *13*, 833–841. [CrossRef]
9. Van Vliet, P.; Wu, S.M.; Zaffran, S.; Puceat, M. Early cardiac development: A view from stem cells to embryos. *Cardiovasc. Res.* **2012**, *96*, 352–362. [CrossRef]
10. Simunovic, M.; Brivanlou, A.H. Embryoids, organoids and gastruloids: New approaches to understanding embryogenesis. *Development* **2017**, *144*, 976–985. [CrossRef]
11. Meganathan, K.; Jagtap, S.; Wagh, V.; Winkler, J.; Gaspar, J.A.; Hildebrand, D.; Trusch, M.; Lehmann, K.; Hescheler, J.; Schluter, H.; et al. Identification of thalidomide-specific transcriptomics and proteomics signatures during differentiation of human embryonic stem cells. *PLoS ONE* **2012**, *7*, e44228. [CrossRef]
12. Luz, A.L.; Tokar, E.J. Pluripotent stem cells in developmental toxicity testing: A review of methodological advances. *Toxicol. Sci.* **2018**, *165*, 31–39. [CrossRef]
13. Hubner, D.; Jahn, K.; Pinkert, S.; Bohnke, J.; Jung, M.; Fechner, H.; Rujescu, D.; Liebert, U.G.; Claus, C. Infection of ipsc lines with miscarriage-associated coxsackievirus and measles virus and teratogenic rubella virus as a model for viral impairment of early human embryogenesis. *ACS Infect. Dis.* **2017**, *3*, 886–897. [CrossRef]
14. Wu, X.; Dao Thi, V.L.; Huang, Y.; Billerbeck, E.; Saha, D.; Hoffmann, H.H.; Wang, Y.; Silva, L.A.V.; Sarbanes, S.; Sun, T.; et al. Intrinsic immunity shapes viral resistance of stem cells. *Cell* **2018**, *172*, 423–438. [CrossRef]

15. Hong, X.X.; Carmichael, G.G. Innate immunity in pluripotent human cells: Attenuated response to interferon-beta. *J. Biol. Chem.* **2013**, *288*, 16196–16205. [CrossRef]
16. Dziedzicka, D.; Markouli, C.; Barbe, L.; Spits, C.; Sermon, K.; Geens, M. A high proliferation rate is critical for reproducible and standardized embryoid body formation from laminin-521-based human pluripotent stem cell cultures. *Stem Cell Rev.* **2016**, *12*, 721–730. [CrossRef]
17. Wirth, H.; Loffler, M.; von Bergen, M.; Binder, H. Expression cartography of human tissues using self organizing maps. *BMC Bioinform.* **2011**, *12*, 306. [CrossRef]
18. Subramanian, A.; Tamayo, P.; Mootha, V.K.; Mukherjee, S.; Ebert, B.L.; Gillette, M.A.; Paulovich, A.; Pomeroy, S.L.; Golub, T.R.; Lander, E.S.; et al. Gene set enrichment analysis: A knowledge-based approach for interpreting genome-wide expression profiles. *Proc. Natl. Acad. Sci. USA* **2005**, *102*, 15545–15550. [CrossRef]
19. Wirth, H.; von Bergen, M.; Murugaiyan, J.; Rosler, U.; Stokowy, T.; Binder, H. Maldi-typing of infectious algae of the genus prototheca using som portraits. *J. Microbiol. Methods* **2012**, *88*, 83–97. [CrossRef]
20. Loffler-Wirth, H.; Kalcher, M.; Binder, H. Opossom: R-package for high-dimensional portraying of genome-wide expression landscapes on bioconductor. *Bioinformatics* **2015**, *31*, 3225–3227. [CrossRef]
21. Nersisyan, L.; Löffler-Wirth, H.; Arakelyan, A.; Binder, H. Gene set- and pathway- centered knowledge discovery assigns transcriptional activation patterns in brain, blood, and colon cancer: A bioinformatics perspective. *Int. J. Knowl. Discov. Bioinform.* **2014**, *4*, 46–69. [CrossRef]
22. Claus, C.; Bergs, S.; Emmrich, N.C.; Hubschen, J.M.; Mankertz, A.; Liebert, U.G. A sensitive one-step taqman amplification approach for detection of rubella virus clade i and ii genotypes in clinical samples. *Arch. Virol.* **2017**, *162*, 477–486. [CrossRef]
23. Hellemans, J.; Mortier, G.; De Paepe, A.; Speleman, F.; Vandesompele, J. Qbase relative quantification framework and software for management and automated analysis of real-time quantitative pcr data. *Genome Biol.* **2007**, *8*, R19. [CrossRef]
24. Wang, H.; Zhang, K.; Liu, Y.; Fu, Y.; Gao, S.; Gong, P.; Wang, H.; Zhou, Z.; Zeng, M.; Wu, Z.; et al. Telomere heterogeneity linked to metabolism and pluripotency state revealed by simultaneous analysis of telomere length and rna-seq in the same human embryonic stem cell. *BMC Biol.* **2017**, *15*, 114. [CrossRef]
25. Ben-Porath, I.; Thomson, M.W.; Carey, V.J.; Ge, R.; Bell, G.W.; Regev, A.; Weinberg, R.A. An embryonic stem cell-like gene expression signature in poorly differentiated aggressive human tumors. *Nat. Genet.* **2008**, *40*, 499–507. [CrossRef]
26. Chu, L.F.; Leng, N.; Zhang, J.; Hou, Z.; Mamott, D.; Vereide, D.T.; Choi, J.; Kendziorski, C.; Stewart, R.; Thomson, J.A. Single-cell rna-seq reveals novel regulators of human embryonic stem cell differentiation to definitive endoderm. *Genome Biol.* **2016**, *17*, 173. [CrossRef]
27. Hopp, L.; Loeffler-Wirth, H.; Nersisyan, L.; Arakelyan, A.; Binder, H. Footprints of sepsis framed within community acquired pneumonia in the blood transcriptome. *Front Immunol* **2018**, *9*, 1620. [CrossRef]
28. Andres-Terre, M.; McGuire, H.M.; Pouliot, Y.; Bongen, E.; Sweeney, T.E.; Tato, C.M.; Khatri, P. Integrated, multi-cohort analysis identifies conserved transcriptional signatures across multiple respiratory viruses. *Immunity* **2015**, *43*, 1199–1211. [CrossRef]
29. Sweeney, T.E.; Wong, H.R.; Khatri, P. Robust classification of bacterial and viral infections via integrated host gene expression diagnostics. *Sci. Transl. Med.* **2016**, *8*, 346ra91. [CrossRef]
30. Schmidtke, L.; Schrick, K.; Saurin, S.; Kafer, R.; Gather, F.; Weinmann-Menke, J.; Kleinert, H.; Pautz, A. The kh-type splicing regulatory protein (ksrp) regulates type iii interferon expression post-transcriptionally. *Biochem. J.* **2019**, *476*, 333–352. [CrossRef]
31. Winzen, R.; Thakur, B.K.; Dittrich-Breiholz, O.; Shah, M.; Redich, N.; Dhamija, S.; Kracht, M.; Holtmann, H. Functional analysis of ksrp interaction with the au-rich element of interleukin-8 and identification of inflammatory mrna targets. *Mol. Cell. Biol.* **2007**, *27*, 8388–8400. [CrossRef]
32. Dambacher, S.; Hahn, M.; Schotta, G. Epigenetic regulation of development by histone lysine methylation. *Heredity* **2010**, *105*, 24–37. [CrossRef]
33. Thalheim, T.; Hopp, L.; Binder, H.; Aust, G.; Galle, J. On the cooperation between epigenetics and transcription factor networks in the specification of tissue stem cells. *Epigenomes* **2018**, *2*, 20. [CrossRef]
34. Grandy, R.A.; Whitfield, T.W.; Wu, H.; Fitzgerald, M.P.; VanOudenhove, J.J.; Zaidi, S.K.; Montecino, M.A.; Lian, J.B.; van Wijnen, A.J.; Stein, J.L.; et al. Genome-wide studies reveal that h3k4me3 modification in bivalent genes is dynamically regulated during the pluripotent cell cycle and stabilized upon differentiation. *Mol. Cell. Biol.* **2016**, *36*, 615–627. [CrossRef]

35. Bratt-Leal, A.M.; Carpenedo, R.L.; McDevitt, T.C. Engineering the embryoid body microenvironment to direct embryonic stem cell differentiation. *Biotechnol. Prog.* **2009**, *25*, 43–51. [CrossRef]
36. Zhang, K.T.; Che, S.Y.; Su, Z.; Zheng, S.Y.; Zhang, H.Y.; Yang, S.L.; Li, W.D.; Liu, J.P. Cd90 promotes cell migration, viability and sphere-forming ability of hepatocellular carcinoma cells. *Int. J. Mol. Med.* **2018**, *41*, 946–954. [CrossRef]
37. Liu, C.; Peng, G.; Jing, N. Tgf-beta signaling pathway in early mouse development and embryonic stem cells. *Acta Biochim. Biophys. Sin.* **2018**, *50*, 68–73. [CrossRef]
38. Lindsley, R.C.; Gill, J.G.; Kyba, M.; Murphy, T.L.; Murphy, K.M. Canonical wnt signaling is required for development of embryonic stem cell-derived mesoderm. *Development* **2006**, *133*, 3787–3796. [CrossRef]
39. Weed, L.S.; Mills, J.A. Strategies for retinal cell generation from human pluripotent stem cells. *Stem Cell Investig.* **2017**, *4*, 65. [CrossRef]
40. Diacou, R.; Zhao, Y.; Zheng, D.; Cvekl, A.; Liu, W. Six3 and six6 are jointly required for the maintenance of multipotent retinal progenitors through both positive and negative regulation. *Cell Rep.* **2018**, *25*, 2510–2523. [CrossRef]
41. Worley, K.E.; Rico-Varela, J.; Ho, D.; Wan, L.Q. Teratogen screening with human pluripotent stem cells. *Integr. Biol.* **2018**, *10*, 491–501. [CrossRef]
42. Rathjen, J. The states of pluripotency: Pluripotent lineage development in the embryo and in the dish. *ISRN Stem Cells* **2014**, *2014*, 19. [CrossRef]
43. Fu, Y.; Zhou, Z.; Wang, H.; Gong, P.; Guo, R.; Wang, J.; Lu, X.; Qi, F.; Liu, L. Ifitm1 suppresses expression of human endogenous retroviruses in human embryonic stem cells. *FEBS Open Bio* **2017**, *7*, 1102–1110. [CrossRef]
44. Wolf, D.; Goff, S.P. Embryonic stem cells use zfp809 to silence retroviral dnas. *Nature* **2009**, *458*, 1201–1204. [CrossRef]
45. Eggenberger, J.; Blanco-Melo, D.; Panis, M.; Brennand, K.J.; tenOever, B.R. Type i interferon response impairs differentiation potential of pluripotent stem cells. *Proc. Natl. Acad. Sci. USA* **2019**, *116*, 1384–1393. [CrossRef]
46. Pervolaraki, K.; Talemi, S.R.; Albrecht, D.; Bormann, F.; Bamford, C.; Mendoza, J.L.; Garcia, K.C.; McLauchlan, J.; Hofer, T.; Stanifer, M.L.; et al. Differential induction of interferon stimulated genes between type i and type iii interferons is independent of interferon receptor abundance. *PLoS Pathog.* **2018**, *14*, e1007420. [CrossRef]
47. Zahedi-Amiri, A.; Sequiera, G.L.; Dhingra, S.; Coombs, K.M. Influenza a virus-triggered autophagy decreases the pluripotency of human-induced pluripotent stem cells. *Cell Death Dis.* **2019**, *10*, 337. [CrossRef]
48. Heavner, W.; Pevny, L. Eye development and retinogenesis. *Cold Spring Harb. Perspect. Biol.* **2012**, *4*, a008391. [CrossRef]
49. Geyer, H.; Bauer, M.; Neumann, J.; Ludde, A.; Rennert, P.; Friedrich, N.; Claus, C.; Perelygina, L.; Mankertz, A. Gene expression profiling of rubella virus infected primary endothelial cells of fetal and adult origin. *Virol. J.* **2016**, *13*, 21. [CrossRef]
50. Lough, J.; Sugi, Y. Endoderm and heart development. *Dev. Dyn.* **2000**, *217*, 327–342. [CrossRef]
51. Mulloy, B.; Rider, C.C. The bone morphogenetic proteins and their antagonists. *Vitam. Horm.* **2015**, *99*, 63–90.
52. Foley, A.C.; Korol, O.; Timmer, A.M.; Mercola, M. Multiple functions of cerberus cooperate to induce heart downstream of nodal. *Dev. Biol.* **2007**, *303*, 57–65. [CrossRef]
53. Vokes, S.A.; Krieg, P.A. Endoderm is required for vascular endothelial tube formation, but not for angioblast specification. *Development* **2002**, *129*, 775–785.
54. Mikkelsen, T.S.; Ku, M.; Jaffe, D.B.; Issac, B.; Lieberman, E.; Giannoukos, G.; Alvarez, P.; Brockman, W.; Kim, T.K.; Koche, R.P.; et al. Genome-wide maps of chromatin state in pluripotent and lineage-committed cells. *Nature* **2007**, *448*, 553–560. [CrossRef]
55. Hofstetter, C.; Kampka, J.M.; Huppertz, S.; Weber, H.; Schlosser, A.; Muller, A.M.; Becker, M. Inhibition of kdm6 activity during murine esc differentiation induces DNA damage. *J. Cell Sci.* **2016**, *129*, 788–803. [CrossRef]
56. Fisher, J.B.; Pulakanti, K.; Rao, S.; Duncan, S.A. Gata6 is essential for endoderm formation from human pluripotent stem cells. *Biol Open* **2017**, *6*, 1084–1095. [CrossRef]
57. Pfeiffer, M.J.; Quaranta, R.; Piccini, I.; Fell, J.; Rao, J.; Ropke, A.; Seebohm, G.; Greber, B. Cardiogenic programming of human pluripotent stem cells by dose-controlled activation of eomes. *Nat. Commun.* **2018**, *9*, 440. [CrossRef]

58. Pfister, S.; Jones, V.J.; Power, M.; Truisi, G.L.; Khoo, P.L.; Steiner, K.A.; Kanai-Azuma, M.; Kanai, Y.; Tam, P.P.; Loebel, D.A. Sox17-dependent gene expression and early heart and gut development in sox17-deficient mouse embryos. *Int. J. Dev. Biol.* **2011**, *55*, 45–58. [CrossRef]
59. Serrano, A.G.; Gandillet, A.; Pearson, S.; Lacaud, G.; Kouskoff, V. Contrasting effects of sox17- and sox18-sustained expression at the onset of blood specification. *Blood* **2010**, *115*, 3895–3898. [CrossRef]
60. George, R.M.; Firulli, A.B. Hand factors in cardiac development. *Anat. Rec.* **2019**, *302*, 101–107. [CrossRef]

10

Ebola Virus Nucleocapsid-Like Structures Utilize Arp2/3 Signaling for Intracellular Long-Distance Transport

Katharina Grikscheit [1,2], Olga Dolnik [1], Yuki Takamatsu [1,3], Ana Raquel Pereira [4] and Stephan Becker [1,2,*]

[1] Institute of Virology, Philipps University Marburg, Hans-Meerwein-Str. 2, 35043 Marburg, Germany; katharina.grikscheit@posteo.de (K.G.); dolnik@staff.uni-marburg.de (O.D.); yukiti@niid.go.jp (Y.T.)
[2] German Center for Infection Research (DZIF), Partner Site: Giessen-Marburg-Langen, Hans-Meerwein-Str. 2, 35043 Marburg, Germany
[3] Department of Virology I, National Institute of Infectious Diseases, Tokyo 208-0011, Japan
[4] Oxford Nano Imager, Linacre House, Banbury Rd, Oxford OX2 8TA, UK; pereira.arr@gmail.com
[*] Correspondence: becker@staff.uni-marburg.de

Abstract: The intracellular transport of nucleocapsids of the highly pathogenic Marburg, as well as Ebola virus (MARV, EBOV), represents a critical step during the viral life cycle. Intriguingly, a population of these nucleocapsids is distributed over long distances in a directed and polar fashion. Recently, it has been demonstrated that the intracellular transport of filoviral nucleocapsids depends on actin polymerization. While it was shown that EBOV requires Arp2/3-dependent actin dynamics, the details of how the virus exploits host actin signaling during intracellular transport are largely unknown. Here, we apply a minimalistic transfection system to follow the nucleocapsid-like structures (NCLS) in living cells, which can be used to robustly quantify NCLS transport in live cell imaging experiments. Furthermore, in cells co-expressing LifeAct, a marker for actin dynamics, NCLS transport is accompanied by pulsative actin tails appearing on the rear end of NCLS. These actin tails can also be preserved in fixed cells, and can be visualized via high resolution imaging using STORM in transfected, as well as EBOV infected, cells. The application of inhibitory drugs and siRNA depletion against actin regulators indicated that EBOV NCLS utilize the canonical Arp2/3-Wave1-Rac1 pathway for long-distance transport in cells. These findings highlight the relevance of the regulation of actin polymerization during directed EBOV nucleocapsid transport in human cells.

Keywords: Ebola virus; actin cytoskeleton; nucleocapsid transport; Arp2/3 complex

---

1. Introduction

The dynamic actin cytoskeleton is commonly utilized by pathogens for entry, exit and their intracellular assembly [1]. While for some intracellular bacteria and DNA viruses, the mechanism of hijacking the actin cytoskeleton is described in detail, the remodeling of actin through highly pathogenic RNA viruses such as filoviruses remains poorly understood [1,2]. Recently, it has been demonstrated that the intracytoplasmic transport of the highly pathogenic Marburg virus (MARV) and the Ebola virus (EBOV) depends on actin polymerization, but the detailed mechanisms and cellular interaction partners remain largely unknown [3–5].

MARV and EBOV belong to the family of *Filoviridae*, which are filamentous, enveloped viruses containing a single strand, negative-sense RNA genome, and which cause severe fevers in humans with very high fatality rates [6]. Strikingly, filoviral RNA encodes for only seven structural proteins, which are multifunctional, and diversely hijack, disable or reorganize cellular pathways [7]. For example,

EBOV VP24 and VP35 are able to block the cell's interferon system, resulting in the viruses' escape from the innate immune response [8].

A hallmark of the filoviral life cycle is the formation of perinuclear inclusion bodies, where transcription and replication occur and de novo viral nucleocapsids are formed [9]. These nucleocapsids are mainly composed of the nucleoprotein (NP) that encapsidates the viral RNA in highly organized helical structures (approximately 1000 nm long and 50 nm in diameter) [10]. Furthermore, nucleocapsids contain the NP-binding proteins VP24 and VP35, as well as polymerase L and the transcription initiation factor VP30 [10–12].

Following their assembly in the inclusion bodies, individual nucleocapsids have to be transported over long distances through the cytoplasm to reach the cell periphery [3,5,9]. It has been shown that they accumulate in filopodia structures. Finally, nucleocapsids acquire their envelope at the plasma membrane, containing the viral trans-membrane glycoprotein (GP) and the viral matrix protein VP40, triggering the budding of the filamentous infectious virus particles. [8]. The directed transport is required for the rapid distancing away from the inclusion bodies, and to transfer functional nucleocapsids to the plasma membrane. It has been shown that individual single mutations within the nucleocapsid proteins critically affect long-distance transport, thereby leading to a significant delay and reduction of the release of filamentous nucleocapsids [13–15].

While other viruses utilize the microtubule network for intracellular distribution, [16] the transport of nucleocapsids of MARV as well as EBOV is entirely blocked by the pharmacological inhibition of actin polymerization [3,4]. However, the role of actin nucleators and the regulation of their upstream effectors in this process remains elusive. One major obstacle in studying filoviruses is the requirement of high containment laboratories, restricting cell biological experiments that could accelerate our understanding of the critical steps in the virus life cycle. To overcome this problem, we recently established a novel transfection-based system that only requires three viral proteins to produce nucleocapsid-like structures (NCLS), which highly resemble EBOV nucleocapsids in their structure as well as their intracellular dynamics [5]. Through the co-expression of GFP-tagged VP30, which faithfully labels NCLS, it is possible to monitor intracellular transport mechanisms with real time imaging under normal laboratory conditions (biosafety level 1) conditions (Figure 1A,B and Figure 2A).

The host cell contains a highly dynamic actin cytoskeleton that is required for many essential cellular processes, such as cytokinesis, contractility and motility [17,18]. Filamentous actin (called F-actin) is assembled from monomeric globular G-actin, and this polymerization, as well as depolymerization, is highly regulated through a plethora of different cellular factors [19–21]. With this high power of regulation, the cell is able to assemble and coordinate diverse structures, such as the strong cortical actin network stabilizing the cell cortex, short and highly dynamic filaments that are involved in vesicular trafficking, and filament networks that form membrane protrusions such as the filopodia and lamellipodia required for cell motility or cell–cell contact formation [22–25]. The Arp2/3 (actin-related protein) complex efficiently nucleates the actin filaments typically attached to mother filaments to form the highly branched networks that (amongst other things) enable lamellipodia to rapidly adapt during cell migration [26]. This active protein complex has to be tightly regulated, and requires activation for polymerization [27]. So-called nucleation promoting factors (NFPs), such as WAVE1 and WASP proteins, directly induce Arp2/3 complex activity [27,28], and are themselves activated downstream of RhoGTPase signaling in a highly spatial–temporal manner [29].

The Arp2/3 complex is highly conserved, and different pathogens, including *Listeria monocytogenes* and vaccinia virus, utilize its activity for viral intracellular transport steps [30]. For example, the membrane-integrated *Listeria* protein ActA mimics the NPF WASP, thereby recruiting and activating the Arp2/3 complex [31–33]. The Arp2/3 complex in turn induces local actin polymerization, resulting in so-called actin comet tails that efficiently propel the bacterium through the cytoplasm, pushing it into neighboring cells. Furthermore, actin comet tails have been previously observed at EBOV nucleocapsids [3]; however, the mechanism by which viral nucleocapsids use actin dynamics for their transport has not been described in detail.

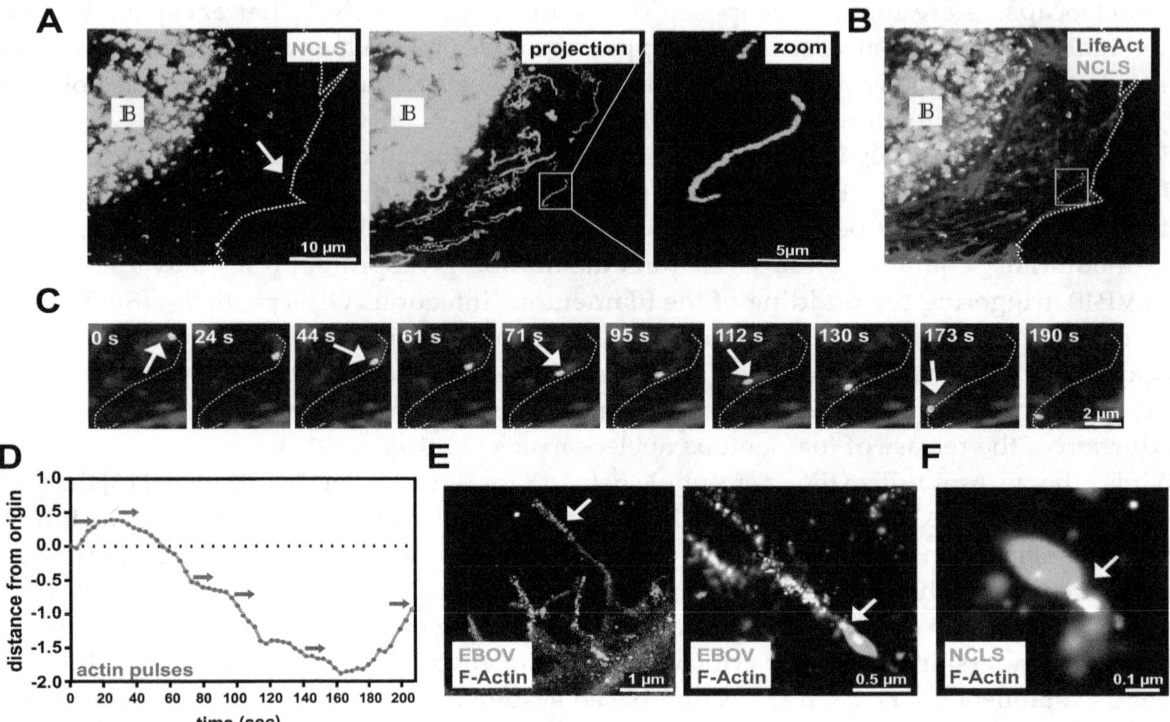

Figure 1. Ebola virus (EBOV) nucleocapsid-like structures induce polar actin tails during transport. (**A**) Exemplary still image from a movie of a Huh7 cell expressing NP, VP24, VP35 and VP30-GFP (left panel). During imaging, NCLS are tracked for 2 min producing long-distance trajectories as depicted in the maximum intensity projection of the movie (right panel). The white arrow highlights an individual track. IB labels inclusion bodies. (**B**) Co-imaging of actin using LifeAct-CLIP (stained with Alexa-657 dye) reveals a dense and dynamic actin network. White inset is magnified in (**C**) showing the time course of an individual NCLS (white arrow, Figure 1**B**), revealing a pulsative actin tail located on one site of the subviral particle during movement (see also Movie S1). (**D**) Graph showing the movement of this NCLS over time. The red arrows indicate actin pulses observed. (**E**,**F**) STORM images showing NCLS or EBOV nucleocapsids immunolabelled with NP (green) and stained with Phalloidin (magenta). (**D**) The left panel shows a filamentous virus particle likely prior to budding (white arrow). Individual intracellular nucleocapsids reveal a preserved actin tail (white arrow). (**E**) Huh7 cells were transfected with NP, VP24 and VP35, and fixed after 24 h. The samples were stained with anti-NP and Phalloidin, and prepared for STORM microscopy. The zoomed image shows an individual NCLS with a preserved actin tail (white arrow).

In this study, we employ our recently established live cell imaging approach to delineate the cellular pathways by which EBOV exploits host actin signaling, and extend the previously applied manual quantification approach to a semi-automatic high throughput method. Using small inhibitory compounds and siRNA-mediated knockdown, we demonstrate that Arp2/3 complex activity downstream of Rac1 is critically involved in the directed long-distance transport of EBOV nucleocapsid structures. Furthermore, through co-visualization of NCLS transport with the actin marker LifeAct, we detected pulsative actin tails accompanying the movement of NCLS through the cytoplasm, which requires Arp2/3 activity.

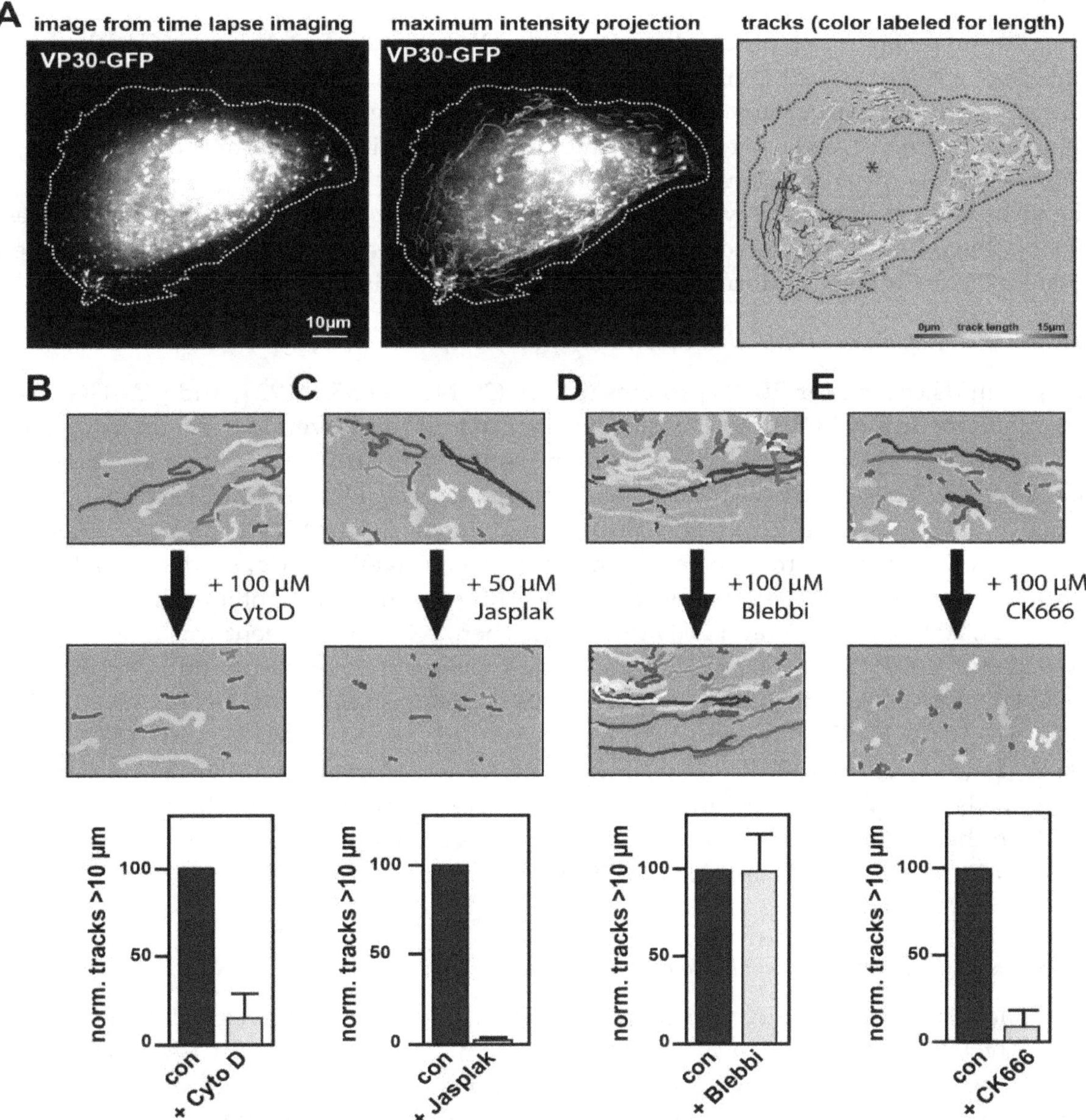

Figure 2. Analysis of NCLS transport using actin-modulating drugs. (**A**) Exemplary still image from a movie of a Huh7 cell expressing NP, VP24, VP35 and VP30-GFP (left panel). For live cell imaging, NCLS are tracked through the cell over 2 min with images captured every 900 ms. While the middle panel shows the maximum-intensity projection of this movie, the right panel depicts the same movie quantified using the spot algorithm (Imaris). The tracks are color-coded for length. To exclude artefacts, we semi-manually deleted areas with strong accumulation of VP30-GFP, such as the inclusion bodies (asterisk). (**B–E**) Huh7 cells transfected with NCLS (VP30-GFP) were recorded for 2 min. This was followed by a short incubation with cytoskeletal-modulating drugs after which the cells were re-recorded. Per experiment three cells were recorded and all tracks > 20 s were quantified. The pictures show a magnified area after quantification with Imaris. Cells were incubated with (**B**) 100 µM Cytochalasin D, (**C**) 50 µM Jasplakinolide, (**D**) 100 µM para-nitro-Blebbistatin or (**E**) 100 µM CK666 following the first imaging. The graphs depict the normalized number of tracks >10 µm, error bars show the SD of $n = 3$ experiments.

2. Materials and Methods

2.1. Cells and Viruses

Huh7 (human hepatoma) cells were cultured in DMEM (Life Technologies, Carlsbad, CA, USA) supplemented with 10% (*v*/*v*) fetal calf serum (FCS) (PAN Biotech), 5 mM L-glutamine, 50 U/mL penicillin and 50 µg/mL streptomycin (Life Technologies) and grown at 37 °C with 5% CO_2. For live cell imaging experiments, cells were kept in phenol-free Leibovitz's medium (Life Technologies) with PS/Q, non-essential amino acid solution and 20% (*v*/*v*) FCS.

The virus used in this study was based on EBOV Zaire (Strain Mayinga; GenBank accession no AF27200 (National Center for Biotechnology Information, Bethesda, MD USA)). The experiments with infectious EBOV virus were performed in the BSL-4 facility at the University of Marburg.

2.2. Transfections, Plasmids, siRNA and Inhibitors

The plasmids coding for EBOV proteins (pCAGGS-NP, -VP35, -VP24) and pCAGGS-VP30-GFP were described previously [34]. Transient transfections of Huh7 cells were carried out using TransIT-LT1 (Mirus, Madison, WI, USA) according to the manufacturers' instructions with 3 µL reagent per 1 µg plasmid DNA [5]. Transfection of siRNAs was performed using DharmaFECT (Horizon, Waterbeach, UK) using 25 nM siRNA and 5 µL transfection reagent in Opti-MEM (Thermo Fisher, Waltham, MA, USA). Huh7 cells were transfected with siRNA in a 6-well µ-slide, seeded to a 4-well µ-slide (IBIDI) and then transfected with the plasmids encoding for the viral proteins (200 ng/µL VP30-GFP, 30 ng/µL VP24, 200 ng/µL VP35 and 200 ng/µL NP). Then, NCLS movement was monitored 24 h later through detection of VP30-GFP. The FlexiTube siRNA used were purchased from Qiagen and diluted to 10 µM. The siRNAs used in the study were HS-ACTR3_5 (AAAGTGGGTGATCAAGCTCAA), HS_RAC1_6 (ATGCATTTCCTGGAGAATATA), HS_WASF1_3 (CAAGAACGTGTGGACCGTTTA) and HS-CDC42_16 (5′-CATCAGATTTTGAAAATATTTAA 3′). For treatment with inhibitors, individual transfected cells (NP, VP24, VP35 and VP30-GFP) were monitored, then the inhibitor was added to the cells in the appropriate dilution and after a short incubation the very same cell was imaged. Cytochalasin D, Jasplakinolide and NSC 23766 were obtained from Sigma Aldrich, and CK666 was purchased from EMD Millipore.

2.3. Antibodies and Reagents for Microscopy

The following primary antibodies were used in this study: polyclonal anti-chicken EBOV NP [35], polyclonal rabbit anti-Wave1 (Sigma Aldrich, St. Louis, MO, USA), mouse anti-Arp3 (Sigma, A5979), mouse anti-Rac1 (Cytoskeleton) and mouse anti-Tubulin (Sigma). The corresponding secondary antibodies were anti-mouse-HRP and anti-rabbit-HRP (DAKO, Jena, Germany), and anti-chicken-Alexa555 (Invitrogen, Carlsbad, CA, USA). Western Blot analyses were performed as described previously [5]. SNAP/CLIP technology was used to visualize actin by using LifeAct-CLIP. LifeAct-CLIP was cloned using standard cloning procedures and then transfected with the NCLS system. For co-visualization of actin and NCLS transport, transfected cells cells were incubated with the dye CLIP-647 (1:500 diluted in media, NEB) for 30 min prior to the experiment, then washed and replaced with Leibovitz medium for live cell imaging.

2.4. Confocal Microscopy, Live Cell Imaging and STORM Microscopy

Huh7 cells were fixed using 4% PFA/DMEM, permeabilized with 0.3%Triton-X100/PBS and blocked with blocking buffer containing 3% glycerol, 2% BSA, 0.2% Tween and 0.05% NaN_3. Primary antibodies as well as secondary antibodies were diluted in blocking buffer and both were incubated for 1 h at RT. The actin cytoskeleton was visualized with Phalloidin conjugated with Alexa647 (Cytoskeleton). For STORM analysis of transfected cells, cells were fixed with 4% PFA/PBS, permeabilized with 0.3% Triton-X100 and then stained with Phalloidin-Alexa647 (1:50) in PBS for 48 h at 4 °C, before being immunolabelled for NP using anti-chicken NP (1:1000) in blocking buffer followed by an incubation with anti-chicken-Alexa555 (Invitrogen, 1:250). Every instance of Phalloidin or antibody staining was followed by a post-fixation step using 4% PFA/PBS. Cells infected with EBOV were fixed with 4% PFA/PBS and permeabilized, then stained with Phalloidin for 1 h in the BSL4 lab, followed by 48 h incubation with 4% PFA/DMEM. Cells were then blocked and immunolabeled as described above and re-stained with Phalloidin for 48 h. For STORM analysis, the coverslip was mounted in switching buffer (1 M MEA-HCl in Glucose buffer containing catalase, TCEP and glucose oxidase).

Dual color live-cell imaging was performed on a SP8 confocal laser scanning microscope (Leica) equipped with a 64 × 1.4 oil objective, and movies for single color live cell imaging were recorded using a Nikon ECLIPSE TE2000-E microscope with a 64x-oil objective. The movies for quantification were acquired every 900 ms for 120 s. All live cell imaging was performed at 37 °C. STORM images were acquired using the Oxford Nanoimaging system (ONI). The light was collected by the 100× objective and imaged onto the EM-CCD camera at 30 ms per frame. Images were processed using the NimOS localization software (Oxford Nanoimaging).

2.5. *Particle Tracking, Quantification and Statistics*

Image processing was performed with Imaris (Bitplane, Oxford Instruments, Abingdon, UK), FIJI (NIH) and Photoshop CS6 (Adobe, San Jose, CA, USA). For figure design, we used Adobe Illustrator. NCLS tracking was performed using Imaris (Bitplane). To this end, VP30-GFP signal was monitored every 900 ms over 2 min and then moving NCLS were analyzed using the "spots feature", a presetting that has been optimized for cell and particle tracking within the software. Within the "spots" feature, we determined an algorithm that automatically classifies objects of over 0.4-μm diameter appearing continuously for at least 20 s as NCLS. The positions of the structures were then calculated in each image of the 2-min movie. From these data, the software is able to calculate trajectories using an autoregressive motion algorithm. After the computational calculation and visualization of all tracks, we semi-manually filtered out particle aggregations, as well as GFP signals derived from inclusion bodies, by deleting spots within certain areas within the cell (see Figure 2A), resulting in approximately 400–800 tracks per cell. We filtered the data (e.g., for tracks >10 μm or straightness > 0.7) which were then used for quantification. This algorithm was then applied to all files of a set of experiments. Each experiment was performed in at least three independent repetitions. Statistical analyses were performed with Prism x (GraphPad, San Diego, CA, USA).

3. Results

3.1. *Nucleocapsid Movement Is Accompanied by Pulsative Actin Tails*

Here, we employ a novel live cell imaging system to monitor the long-distance movement of EBOV NCLS, in which only three viral proteins are expressed, namely, NP, VP24 and VP35 (Figure 1A, left panel) [5]. Using VP30-GFP, we monitor the intracellular transport of NCLS for two minutes, and the corresponding maximum-intensity projection reveals that a robust number of NCLS (6.9% ± 2.7%) are transported in a highly directed fashion over long distances (>10μm–20μm), now referred to as long-distance trajectories (Figure 1A, right panel). However, due to defective assembly or saturation of the cellular transport machinery, we also detect many VP30-GFP-positive NCLS that do not move, or travel only short distances (Figure S1A).

Actin tail formation at EBOV nucleocapsids was detected in an earlier study [3]. Therefore, we aimed here to determine whether or not the established transfection system can recapitulate the detection of actin tails at NCLS as well. In cells that co-express the actin marker LifeAct to visualize the actin cytoskeleton (Figure 1B), we observed that NCLS induces actin tails during their movement through the cell (movie S1), thereby further highlighting the relevance of this transfection system to the study of the transport of EBOV nucleocapsids. Here, we further assessed the movement of an individual NCLS-track over time, starting from its origin and following it through the cell (inset of Figure 1B is magnified in Figure 1C, white arrows). In Figure 1D, the red arrows indicate actin pulses during movement, which appear to correlate with the altered movement of NCLS, such as changes in direction (Figure 1C, Movie S1). While with the live cell imaging setting we used approximately 80% of the tracks that moved in a directed fashion revealed detectable actin tails at their rear end (Figure 3E, control siRNA), we cannot exclude the possibility that smaller NCLS induce actin polymerization below the detection limit, due to the high intracellular background level of LifeAct.

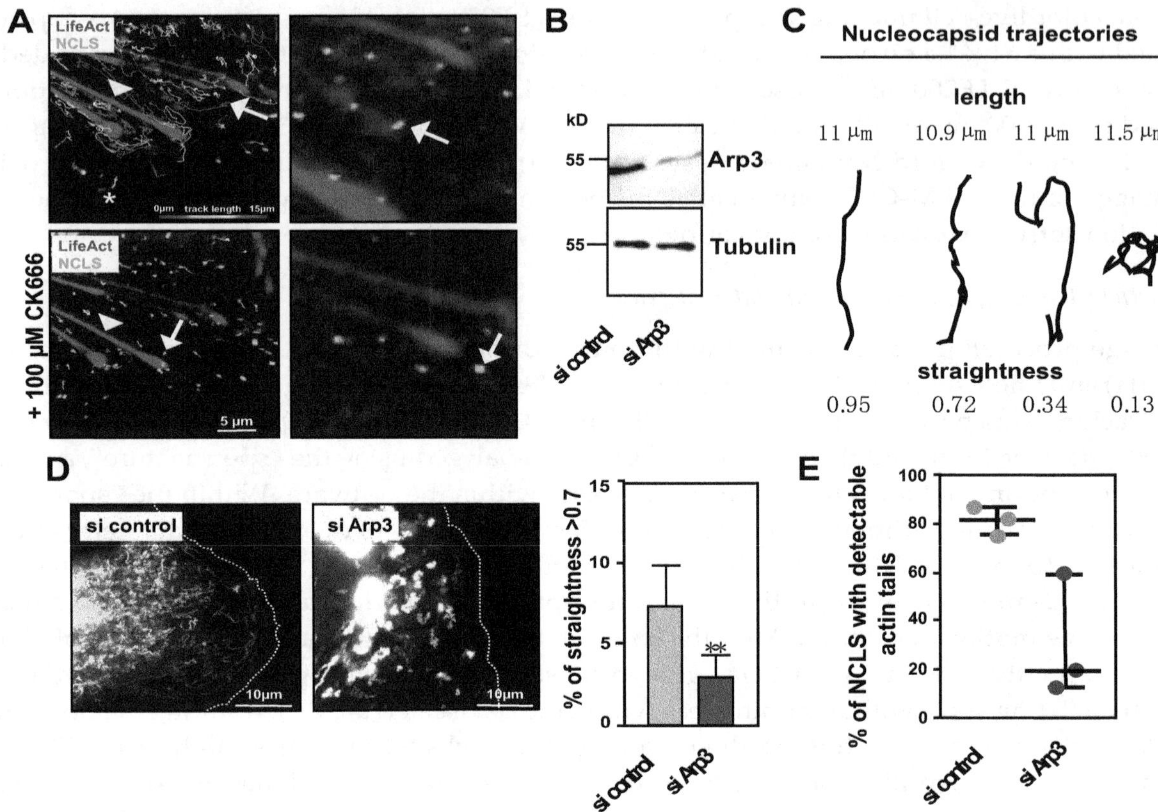

Figure 3. NCLS induction of actin tails depends on Arp2/3 activity. (**A**) Huh7 cells transfected with NP, VP24 and VP35 and LifeAct-CLIP were monitored over time. The left panels show an overlay image of the tracks as calculated by Imaris (color coded for length) and LifeAct-CLIP (still image from movie). Note, CK666 incubation does not interfere with filopodia (asterisk) or stress fibers (arrow head) (see Movie S2). The zoomed image (right panels) show still images of the movie, revealing that CK666 treatment abolishes actin tail formation (white arrows). (**B**) Western Blot showing the depletion of Arp3 after siRNA treatment. (**C**) Cartoon showing different movements of trajectories and their corresponding straightness value. (**D**) Left panel shows maximum intensity projections of time lapse imaging of Huh7 cells transfected either with control or Arp3 siRNA. The right panel shows the graphical analysis of live cell imaging comparing all tracks > 10 μm in cells transfected either with control or Arp3 siRNA. Note, the relative amount of tracks with a straightness > 0.7, representing directed long-distance transport, is reduced in cells transfected with Arp3 siRNA (taken from $n = 3$ experiments, * $p = 0.0027$, Mann–Whitney Test). (**E**) Graph shows the percentage of trajectories with pulsative actin tails in live cell imaging ($n = 3$, error bars indicate mean with standard deviation).

While live cell imaging was applied to capture rapid and transient phenomena, the image resolution remained low. To precisely determine the localization of actin tails at filoviral nucleocapsids, we decided to use a high resolution, single molecule detection technique (STORM). To this end, we either transfected cells with NP, VP35 and VP24, or infected cells with EBOV and fixed them after

24 h. This was followed by the staining of the actin cytoskeleton and immunolabelling for the viral NP protein. Intracellular NCLS or EBOV nucleocapsids appear as elongated NP-positive structures (Figure 1E,F). However, compared to NCLS in transfected cells, filamentous nucleocapsids in cells infected with EBOV were commonly detected along the plasma membrane and within filopodia, likely representing virus particles prior to or during budding, thereby supporting previous observations from live-cell imaging and electron microscopy (Figure 1D, left panel) [3]. Co-staining with Phalloidin revealed actin tail structures in the vicinity of the nucleocapsids in infected cells and transfected cells (Figure 1D,E), which were less frequent in infected cells. This finding is likely a consequence of the very high spatial-temporal dynamics of actin tail formation, which do not allow a quantifiable detection in fixed cells. Further, in the minimalistic transfection system, NCLS remain in the cytosol and are not released, thus the detection of NCLS with actin tails may occur more frequently. Additionally, it cannot be excluded that in infected cells, the long fixation procedure required for the handling of BSL4 samples might interfere with the fixation of these transient actin structures.

3.2. *Characterization of NCLS Transport Dynamics*

To measure and characterize the long-distance movement of NCLS in a semi-automated manner, we analyzed our time-lapse movies with a tracking tool within the software Imaris. To this end, we set up an algorithm to follow NCLS, and filtered for trajectories > 10 μm that show mean velocities comparable to those measured in previous reports (here, 167 ± 58 nm/s compared to 187 ± 67 nm/s as in Takamatsu et al., 2018 [5]) (Figure 2A).

These studies demonstrated that the long-distance movement of EBOV nucleocapsids, as well as NCLS, is sensitive to the inhibition of actin polymerization [3,5]. Here, we recapitulated these experiments by monitoring NCLS movement followed by an incubation with an actin-modulating drug, and then re-recorded the very same cell. Using this system, we ensured that the imaged cells were capable of showing long-distance trajectories, and would consequently enable us to analyze alterations in NCLS movement. As observed previously, the application of the inhibitor of actin polymerization, Cytochalasin D, results in a rapid and strong decrease in any long-distance trajectories of NCLS, quantified as a highly reduced number of tracks over 10 μm (Figure 2B) [3,5]. Expanding on previous results, we further incubated the cells with the actin polymerization and stabilization agent Jasplakinolide, resulting in a similar phenotype, thereby indicating that indeed not only actin polymerization, but also depolymerization, is required for the long-distance transport of NCLS (Figure 2C). Treatment with Blebbistatin (here we used para-nitro-Blebbistatin to avoid phototoxicity in live cell experiments) did not interfere with long-distance movement, suggesting that myosin II activity is not required for the long-distance transport of NCLS (Figure 2D). Taken together, these findings reinforce the importance of actin regulation and dynamics in the long-distance transport of NCLS.

3.3. *Arp2/3 Complex Activity Is Required for Actin Tail Formation and Directed Long-Distance Transport*

Previous experiments showed that inhibitors of Arp2/3 affected the long-distance movement of EBOV nucleocapsids [3]. To confirm and further elaborate on these results, we used the minimalistic transfection system to analyze how treatment with an Arp2/3 inhibitor (CK666) affects NCLS movement [36]. Movie S2 shows that blocking Arp2/3 activity rapidly abolishes any directed long-distance movement of NCLS, and that minimal movement likely derives from Brownian motion (Figure 2E, Movie S2). When dissecting different pools of tracks, we observed that treatment with CK666 strongly decreases tracks >10 μm, subsequently resulting in increases in tracks < 5 μm, which also show altered mean speeds (Figure S1A,B). In cells treated with CK666, we observed a slight reduction in the overall NCLS (−19 ± 7%), which might have resulted from the decreased NCLS transport from the inclusion bodies into the cytoplasm (Figure S1C). Furthermore, Movie S2 reveals that the abolishment of NCLS transport, using CK666, coincides with the loss of actin tail formation (Figure 3A, tracks are color-coded for length and actin is visualized with stills from the confocal movie). Time-lapse imaging further demonstrates that Arp2/3-independent structures, such as filopodia or stress fibers, are not compromised during the observed time frame, thereby highlighting a specific role for Arp2/3-dependent actin polymerization in the directed long-distance transport of NCLS (Movie S2).

Next, we continued with the depletion of one protein within the Arp2/3 complex, namely Arp3, using siRNA transfection to further evaluate its role in actin tail-dependent NCLS transport (Figure 3B). The siRNA depletion of Arp3 likely results in the destabilization or reduction of functional Arp2/3 complexes, and does not entirely block all long-distance movement of NCLS, as observed under treatment with CK666. Thus, we assessed whether the remaining tracks showed altered directionality in cells depleted for Arp3, through the calculation of trajectory straightness (ratio of distance to length) (Figure 3C). We determined that tracks with a straightness > 0.7 (Figure 3C), and indeed the depletion of Arp3, interferes with the total number of straight tracks, further highlighting a role for Arp2/3 activity in the long-distance transport of NCLS (Figure 3D,E). Next, we compared the formation of actin tails in control and Arp3-depleted cells. While approximately 80% of the NCLS that travel over long distances showed pulsative actin tails, cells depleted for Arp3 showed a decreased level of detectable actin polymerization at their rear end (Figure 3E,F). Taken together, these results indicate that Arp2/3 activity is required for the efficient directed long-distance transport of NCLS, and that the Arp2/3 complex is directly involved in the formation of actin tails in Huh7 cells.

3.4. Identification of Rac1/WAVE1/Arp2/3 Signaling Network that Is Involved in Long-Distance Transport of NCLS

The Arp2/3 complex resides, in inactive status, within the cell, and requires activation by NFPs, such as N-WASP or Wave proteins, that bind and activate the Arp2/3 complex via their conserved VCA-domain (Figure 4A) [29]. As it was recently demonstrated that the inhibition of *N*-WASP using Wiskostatin did not interfere with the long-distance transport of EBOV nucleocapsids [3], we focused on Wave1, an alternative activator of the Arp2/3 complex. Interestingly, siRNA directed against WASF1 (Wave1) faithfully recapitulated the phenotype of Arp3 siRNA treatment, leading to the reduced straightness of the NCLS (Figure 4B). This indicates that activation of the Arp2/3 complex downstream of Wave1 is involved in the directionality of long-distance trajectories. Wave1 activity itself can be controlled by the RhoGTPase Rac1 (Figure 4A–C) [37,38]. Again, the treatment of Huh7 cells with siRNA against Rac1 resulted in a decreased level of directed long-distance trajectories (Figure 4D,F,G). In contrast, the depletion of Cdc42, another member of the RhoGTPase family, did not affect the straightness of trajectories (Figure S1D), thereby further reinforcing the inference that Rac1 activity is involved in NCLS transport (Figure 4F,H). For further confirmation, we applied the Rac1 inhibitor NSC 23766, which also interfered with long-distance transport, confirming the relevance of Rac1 activity in the regulation of the long-distance transport of EBOV NCLS (Figure 4H).

Taken together, these findings indicate that EBOV nucleocapsids exploit the canonical Rac1/WAVE1/Arp2/3 signaling pathway for long-distance transport of NCLS inside the cell. This signaling pathway classically induces lamellipodial protrusions, in which Rac1 acts coordinately with other upstream signals to activate actin regulators. Future studies shall reveal whether other actin-regulatory mechanisms are also involved in the viral life cycle, and how alterations in long-distance transport affect virus propagation.

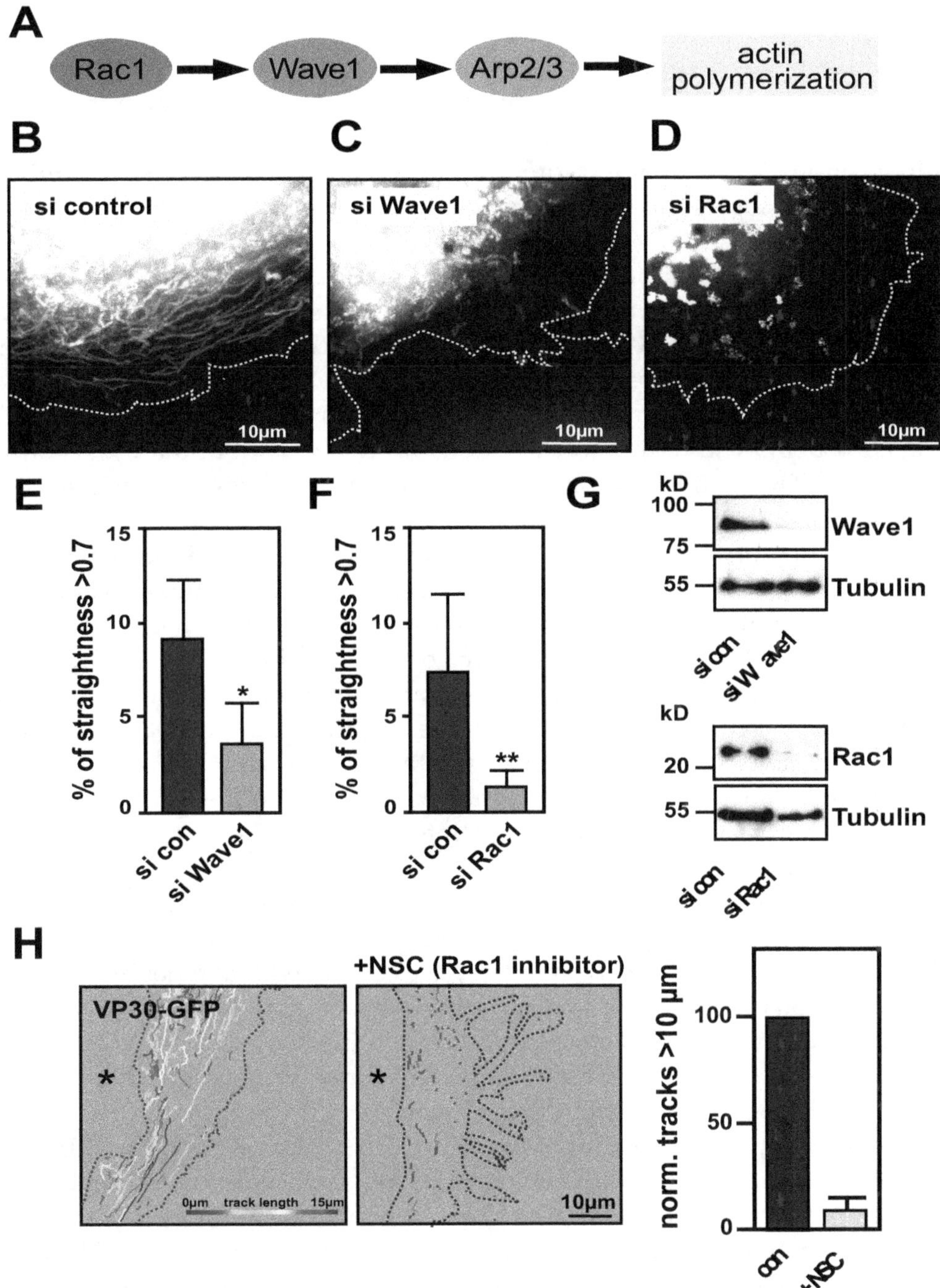

Figure 4. Identification of a Rac1/Wave1/Arp2/3 signaling. (**A**) Cartoon depicting canonical Arp2/3 signaling downstream of Rac1 [39]. (**B–D**) Images show maximum intensity projection of time-lapse images of cells recorded for 2 min; images were captured every 900 ms. Cells were transfected with control siRNA (**B**), siRNA against (**C**) Wave1 (*WASF1*) or (**D**) Rac1. (**E,F**) Analysis of live cell imaging comparing relative straightness of tracks > 10 µm and a straightness > 0.7 after either Wave1 siRNA (**E**) or Rac1 siRNA (**F**) transfection. Cells were transfected with siRNA against (**B**) Wave1 (*WASF1*) or (**C**) Rac1. Note that siRNA treatment against Wave1 and Rac1 results in a decrease in straight long-distance trajectories (taken from three independent experiments, at least 5 cells per experiments, ns = non-significant, ** $p > 0.001$, * $p > 0.01$). (**G**) Western Blots showing effective siRNA depletion of Wave1 and Rac1. (**H**) Incubation of Huh7 cells transfected with NCLS (VP30-GFP) was first recorded for 2 min, then they were incubated with 100 nM Rac1 inhibitor (NSC23766) for 1 h and then reimaged. The overview images reveal that NCS23766 also inhibits long-distance transport. Tracks are color-coded for mean speed (6 cells in $n = 3$, $p < 0.0079$, *t*-test).

4. Discussion

In this study, we have identified the Rac1/Wave1/Arp2/3 pathway as being involved in the actin-dependent transport of EBOV NCLS in human cell culture. Arp2/3 activity is also essential to inducing propulsive and polar-localized actin tails at the rear ends of NCLS, which can be robustly visualized in live cell imaging. Furthermore, the association of actin tails with EBOV NCLS and nucleocapsids can also be observed via super resolution in fixed samples.

Actin comet tails were initially characterized in *Listeria monocytogenes* in the late 1980s [40]. Since then, diverse types of intracellular pathogens have been identified as inducing actin polymerization at their surface. For instance, vaccinia virus protein A36 is able to recruit NCK and GRB2, which in turn recruit N-WASP to stimulate the Arp2/3 complex, or virus protein p78/83 mimics N-WASP and directly activates Arp2/3-induced actin polymerization [30,41]. Here, we showed that EBOV NCLS induce actin tails at one end, and actin tail formation is sensitive to the inhibition of Arp2/3 using CK666, indicating that the hijacking of the highly abundant and conserved Arp2/3 complex for induction of actin tails is a common mechanism in diverse types of pathogens without a common origin [30].

We further show that directed long-distance transport is regulated via Arp2/3 activity, yet the siRNA depletion of Arp3 in cells forming NCLS does not entirely block intracellular transport. Furthermore, the effects of siRNA treatment on track straightness appear more pronounced after Rac1 knockdown, when compared to the depletion of Arp3 or Wave1. These findings also suggest that other actin regulators downstream of Rac1 either compensate Arp2/3 activity, or synergistically regulate actin polymerization. One candidate could be the scaffolding protein IQGAP1 that interconnects multiple pathways of actin dynamics and interacts with Rac1. In cells infected with MARV, IQGAP1 was recruited to inclusion and to the read end of nucleocapsids, and the down-regulation of IQGAP1 resulted in the impaired release of MARV, suggesting a role for other major actin regulators in this process [14].

In addition, we also gained unprecedented evidence that Wave1 upstream of Arp2/3 is involved in regulating the long-distance transport of NCLS in Huh7 cells. Given that nucleocapsid transport in EBOV-infected cells does not depend on N-WASP [3], we concluded that other NFPs could be involved in the upstream regulation of Arp2/3. Further supporting this notion, N-WASP is typically involved in endocytotic events, where it regulates actin polymerization in a manner reminiscent of actin tails that propel pathogens through cells [42]. Consequently, enveloped viruses, such as vaccinia and EBOV, might profit from mimicking or recruiting N-WASP during their life cycle.

In recent years, additional NPFs, like WASH, WHAMM and JMY, have been described to promote actin tail formations at intracellular membranes like endosomes and autophagosomes [43–45]. In contrast, filoviral nucleocapsids travel in the cytosol without a membrane, and are highly structured protein complexes that probably utilize endocytosis-independent pathways for their transport, thereby likely avoiding membranous structures prior to their arrival at the plasma membrane. Wave1 and Rac1 signaling is considered primarily relevant to actin polymerization in lamellipodial cell protrusions [43], thus it is not surprising that long-distance NCLS transport is best observed in areas with a high activity of this pathway in Huh7 cells. Future studies using high resolution microscopy or electron microscopic approaches shall reveal whether and how actin regulators are actively recruited to EBOV nucleocapsids. It could be that at different stages of the nucleocapsid transport, from inclusion bodies to the budding sites, different actin tail-inducing machineries are exploited by the virus.

As described for other viruses, the long-distance transport of EBOV NCLS is accompanied by the induction of polar actin polymerization, likely resulting in the directionality of movement also observed in other actin tail-inducing pathogens and in in vitro reconstructions [41,46,47]. How this polar induction of actin polymerization is initiated is not understood in detail. One hypothesis derives from studies in *Listeria*, where it was shown that the surface protein ActA accumulates locally, likely during cell wall growth, which subsequently results in polar interactions with actin regulators [48]. In contrast, filoviral nucleocapsids are not enveloped when they leave inclusion bodies, and only encounter the viral proteins GP and VP40, which are transported independently, when they reach the plasma membrane to form infectious particles [49–52]. Thus, it remains unclear how this polar actin polymerization is induced in filoviral nucleocapsids, and which viral protein might be relevant for the actin polymerization. One explanation derives from a study investigating the structure of MARV capsids using cryo-electron microscopy, revealing that nucleocapsid assembly itself results in polar structures [10]. Here, it was demonstrated that MARV nucleocapsids are highly oriented towards the plasma membrane, with the pointed end of the nucleocapsid directed towards the plasma membrane prior to budding [11,53]. Thus, these findings support the notion that polar nucleocapsid assembly itself might result in specific protein conformations, thereby exposing the binding sites for cellular proteins that induce actin polymerization.

Taken together, our minimalistic transfection-based system reveals new opportunities to study the cellular transport of filoviral nucleocapsids, and to reliably quantify the dynamics of subviral structures. Future studies have to further characterize how filoviral proteins modulate highly spatial–temporal RhoGTPase signaling, and identify the direct interaction partners and structural changes that are required for long-distance transport during the EBOV life cycle.

Author Contributions: Conceptualization, K.G. and S.B.; methodology, K.G., Y.T., O.D., A.R.P.; formal analysis, K.G.; investigation, K.G., O.D., A.R.P.; writing—original draft preparation, K.G., S.B.; writing—review and editing, K.G., S.B., O.D.; visualization, K.G.; supervision, S.B.; funding acquisition, S.B. All authors have read and agreed to the published version of the manuscript.

Acknowledgments: We thank Astrid Herwig and Martina Weik for their technical support. We also wish to thank Katrin Roth (Marburg Imaging Facility) and Andreas Rausch (THM Giessen) for their support and helpful discussion.

References

1. Taylor, M.P.; Koyuncu, O.O.; Enquist, L.W. Subversion of the actin cytoskeleton during viral infection. *Nat. Rev. Microbiol.* **2011**, *9*, 427–439. [CrossRef]
2. Newsome, T.P.; Marzook, N.B. Viruses that ride on the coat-tails of actin nucleation. *Semin. Cell. Dev. Biol.* **2015**, *46*, 155–163. [CrossRef]
3. Schudt, G.; Dolnik, O.; Kolesnikova, L.; Biedenkopf, N.; Herwig, A.; Becker, S. Transport of Ebolavirus Nucleocapsids Is Dependent on Actin Polymerization: Live-Cell Imaging Analysis of Ebolavirus-Infected Cells. *J. Infect. Dis.* **2015**, *212*, 160–166. [CrossRef] [PubMed]
4. Schudt, G.; Kolesnikova, L.; Dolnik, O.; Sodeik, B.; Becker, S. Live-cell imaging of Marburg virus-infected cells uncovers actin-dependent transport of nucleocapsids over long distances. *Proc. Natl. Acad. Sci. USA* **2013**, *113*, 14402–14407. [CrossRef] [PubMed]
5. Takamatsu, Y.; Kolesnikova, L.; Becker, S. Ebola virus proteins NP, VP35, and VP24 are essential and sufficient to mediate nucleocapsid transport. *Proc. Natl. Acad. Sci. USA* **2018**, *115*, 1075–1080. [CrossRef]
6. Baseler, L.; Chertow, D.S.; Johnson, K.M.; Feldmann, H.; Morens, D.M. The Pathogenesis of Ebola Virus Disease. *Annu. Rev. Pathol.* **2017**, *12*, 387–418. [CrossRef]
7. Cantoni, D.; Rossman, J.S. Ebolaviruses: New roles for old proteins. *PLoS Negl. Trop. Dis.* **2018**, *12*, e0006349. [CrossRef]
8. Messaoudi, I.; Amarasinghe, G.K.; Basler, C.F. Filovirus pathogenesis and immune evasion: Insights from Ebola virus and Marburg virus. *Nat. Rev. Microbiol.* **2015**, *13*, 663–676. [CrossRef]
9. Hoenen, T.; Shabman, R.S.; Groseth, A.; Herwig, A.; Weber, M.; Schudt, G.; Dolnik, O.; Basler, C.F.; Becker, S.; Feldmann, H. Inclusion bodies are a site of ebolavirus replication. *J. Virol.* **2012**, *86*, 11779–11788. [CrossRef]
10. Wan, W.; Kolesnikova, L.; Clarke, M.; Koehler, A.; Noda, T.; Becker, S.; Briggs, J.A.G. Structure and assembly of the Ebola virus nucleocapsid. *Nature* **2017**, *551*, 394–397. [CrossRef] [PubMed]
11. Bharat, T.A.; Noda, T.; Riches, J.D.; Kraehling, V.; Kolesnikova, L.; Becker, S.; Kawaoka, Y.; Briggs, J.A. Structural dissection of Ebola virus and its assembly determinants using cryo-electron tomography. *Proc. Natl. Acad. Sci. USA* **2012** *109*, 4275–4280. [CrossRef]

12. Biedenkopf, N.; Hartlieb, B.; Hoenen, T.; Becker, S. Phosphorylation of Ebola virus VP30 influences the composition of the viral nucleocapsid complex: Impact on viral transcription and replication. *J. Biol. Chem.* **2013**, *288*, 11165–11174. [CrossRef] [PubMed]
13. Takamatsu, Y.; Kolesnikova, L.; Schauflinger, M.; Noda, T.; Becker, S. The Integrity of the YxxL Motif of Ebola Virus VP24 Is Important for the Transport of Nucleocapsid–Like Structures and for the Regulation of Viral RNA Synthesis. *J. Virol.* **2020**, *94*, e02170–e02219. [CrossRef] [PubMed]
14. Dolnik, O.; Kolesnikova, L.; Welsch, S.; Strecker, T.; Schudt, G.; Becker, S. Interaction with Tsg101 is necessary for the efficient transport and release of nucleocapsids in marburg virus-infected cells. *PLoS Pathog.* **2014**, *10*, e1004463. [CrossRef]
15. Dolnik, O.; Kolesnikova, L.; Stevermann, L.; Becker, S. Tsg101 is recruited by a late domain of the nucleocapsid protein to support budding of Marburg virus–like particles. *J. Virol.* **2010**, *84*, 7847–7856. [CrossRef]
16. Naghavi, M.H.; Walsh, D. Microtubule Regulation and Function during Virus Infection. *J. Virol.* **2017**, *91*, e00538–e00617. [CrossRef]
17. Pollard, T.D. Cell Motility and Cytokinesis: From Mysteries to Molecular Mechanisms in Five Decades. *Annu. Rev. Cell. Dev. Biol.* **2019**, *35*, 1–28. [CrossRef]
18. Murrell, M.; Oakes, P.W.; Lenz, M.; Gardel, M.L. Forcing cells into shape: The mechanics of actomyosin contractility. *Nat. Rev. Mol. Cell. Biol.* **2015**, *16*, 486–498. [CrossRef] [PubMed]
19. KCampellone, G.; Welch, M.D. A nucleator arms race: Cellular control of actin assembly. *Nat. Rev. Mol. Cell. Biol.* **2010**, *11*, 237–251. [CrossRef]
20. Pollard, T.D. Actin and Actin-Binding Proteins. *Cold Spring Harb. Perspect Biol.* **2016**, *18*, a018226. [CrossRef]
21. Lappalainen, P. Actin-binding proteins: The long road to understanding the dynamic landscape of cellular actin networks. *Mol. Biol. Cell.* **2016**, *27*, 2519–2522. [CrossRef] [PubMed]
22. Grikscheit, K.; Grosse, R. Formins at the Junction. *Trends Biochem. Sci.* **2016**, *41*, 148–159. [CrossRef] [PubMed]
23. Zech, T.; Calaminus, S.D.; Machesky, L.M. Actin on trafficking: Could actin guide directed receptor transport? *Cell. Adh. Migr.* **2012**, *6*, 476–481. [CrossRef] [PubMed]
24. Schaks, M.; Giannone, G.; Rottner, K. Actin dynamics in cell migration. *Essays Biochem.* **2019**, *63*, 483–495.
25. Schuh, M. An actin-dependent mechanism for long-range vesicle transport. *Nat. Cell. Biol.* **2011**, *13*, 1431–1436. [CrossRef]
26. Goley, E.D.; Welch, M.D. The ARP2/3 complex: An actin nucleator comes of age. *Nat. Rev. Mol. Cell. Biol.* **2006**, *7*, 713–726. [CrossRef]
27. Rodal, A.A.; Sokolova, O.; Robins, D.B.; Daugherty, K.M.; Hippenmeyer, S.; Riezman, H.; Grigorieff, N.; Goode, B.L. Conformational changes in the Arp2/3 complex leading to actin nucleation. *Nat. Struct. Mol. Biol.* **2005**, *12*, 26–31. [CrossRef]
28. Rodnick-Smith, M.; Luan, Q.; Liu, S.L.; Nolen, B.J. Role and structural mechanism of WASP-triggered conformational changes in branched actin filament nucleation by Arp2/3 complex. *Proc. Natl. Acad. Sci. USA* **2016**, *113*, E3834–E3843. [CrossRef]
29. Rotty, J.D.; Wu, C.; Bear, J.E. New insights into the regulation and cellular functions of the ARP2/3 complex. *Nat. Rev. Mol. Cell. Biol.* **2013**, *14*, 7–12. [CrossRef]
30. Welch, M.D.; Way, M. Arp2/3-mediated actin-based motility: A tail of pathogen abuse. *Cell. Host Microbe* **2013**, *14*, 242–255. [CrossRef]
31. Welch, M.D.; Rosenblatt, J.; Skoble, J.; Portnoy, D.A.; Mitchison, T.J. Interaction of human Arp2/3 complex and the Listeria monocytogenes ActA protein in actin filament nucleation. *Science* **1998**, *281*, 105–108. [CrossRef]
32. Jeng, R.L.; Goley, E.D.; D'Alessio, J.A.; Chaga, O.Y.; Svitkina, T.M.; Borisy, G.G.; Heinzen, R.A.; Welch, M.D. A Rickettsia WASP-like protein activates the Arp2/3 complex and mediates actin-based motility. *Cell. Microbiol.* **2004**, *6*, 761–769. [CrossRef]
33. Boujemaa–Paterski, R.; Gouin, E.; Hansen, G.; Samarin, S.; le Clainche, C.; Didry, D.; Dehoux, P.; Cossart, P.; Kocks, C.; Carlier, M.F.; et al. Listeria protein ActA mimics WASp family proteins: It activates filament barbed end branching by Arp2/3 complex. *Biochemistry* **2001**, *40*, 11390–11404. [CrossRef]
34. Hoenen, T.; Groseth, A.; Kolesnikova, L.; Theriault, S.; Ebihara, H.; Hartlieb, B.; Bamberg, S.; Feldmann, H.; Ströher, U.; Becker, S. Infection of naive target cells with virus-like particles: Implications for the function of ebola virus VP24. *J. Virol.* **2006**, *80*, 7260–7264. [CrossRef]

35. Biedenkopf, N.; Lier, C.; Becker, S. Dynamic Phosphorylation of VP30 Is Essential for Ebola Virus Life Cycle. *J. Virol.* **2016**, *90*, 4914–4925. [CrossRef]
36. Hetrick, B.; Han, M.S.; Helgeson, L.A.; Nolen, B.J. Small molecules CK-666 and CK-869 inhibit actin-related protein 2/3 complex by blocking an activating conformational change. *Chem. Biol.* **2013**, *20*, 701–712. [CrossRef]
37. Chen, B.; Chou, H.T.; Brautigam, C.A.; Xing, W.; Yang, S.; Henry, L.; Doolittle, L.K.; Walz, T.; Rosen, M.K. Rac1 GTPase activates the WAVE regulatory complex through two distinct binding sites. *Elife* **2017**, *6*, e29795. [CrossRef]
38. Eden, S.; Rohatgi, R.; Podtelejnikov, A.V.; Mann, M.; Kirschner, M.W. Mechanism of regulation of WAVE1-induced actin nucleation by Rac1 and Nck. *Nature* **2002**, *418*, 790–793. [CrossRef]
39. Pollitt, A.Y.; Insall, R.H. WASP and SCAR/WAVE proteins: The drivers of actin assembly. *J. Cell. Sci.* **2009**, *122*, 2575–2578. [CrossRef]
40. Tilney, L.G.; Portnoy, D.A. Actin filaments and the growth, movement, and spread of the intracellular bacterial parasite, Listeria monocytogenes. *J. Cell. Biol.* **1989**, *109*, 1597–1608. [CrossRef]
41. Mueller, J.; Pfanzelter, J.; Winkler, C.; Narita, A.; le Clainche, C.; Nemethova, M.; Carlier, M.F.; Maeda, Y.; Welch, M.D.; Ohkawa, T.C.; et al. Electron tomography and simulation of baculovirus actin comet tails support a tethered filament model of pathogen propulsion. *PLoS Biol.* **2014**, *12*, e1001765. [CrossRef]
42. Rottner, K.; Hänisch, J.; Campellone, K.G. WASH, WHAMM and JMY: Regulation of Arp2/3 complex and beyond. *Trends Cell. Biol.* **2010**, *20*, 650–661. [CrossRef] [PubMed]
43. Derivery, E.; Helfer, E.; Henriot, V.; Gautreau, A. Actin polymerization controls the organization of WASH domains at the surface of endosomes. *PLoS ONE* **2012**, *7*, e39774. [CrossRef] [PubMed]
44. Kast, D.J.; Dominguez, R. WHAMM links actin assembly via the Arp2/3 complex to autophagy. *Autophagy* **2015**, *11*, 1702–1704. [CrossRef]
45. Hu, X.; Mullins, R.D. LC3 and STRAP regulate actin filament assembly by JMY during autophagosome formation. *J. Cell Biol.* **2019**, *218*, 251–266. [CrossRef]
46. Shenoy, V.B.; Tambe, D.T.; Prasad, A.; Theriot, J.A. A kinematic description of the trajectories of Listeria monocytogenes propelled by actin comet tails. *Proc. Natl. Acad. Sci. USA* **2007**, *104*, 8229–8234. [CrossRef]
47. Dayel, M.J.; Akin, O.; Landeryou, M.; Risca, V.; Mogilner, A.; Mullins, R.D. In silico reconstitution of actin-based symmetry breaking and motility. *PLoS Biol.* **2009**, *7*, e1000201. [CrossRef]
48. Lacayo, C.I.; Soneral, P.A.; Zhu, J.; Tsuchida, M.A.; Footer, M.J.; Soo, F.S.; Lu, Y.; Xia, Y.; Mogilner, A.; Theriot, J.A. Choosing orientation: Influence of cargo geometry and ActA polarization on actin comet tails. *Mol. Biol. Cell.* **2012**, *23*, 614–629. [CrossRef]
49. Becker, S.; Klenk, H.D.; Mühlberger, E. Intracellular transport and processing of the Marburg virus surface protein in vertebrate and insect cells. *Virology* **1996**, *225*, 145–155. [CrossRef]
50. Mittler, E.; Schudt, G.; Halwe, S.; Rohde, C.; Becker, S. A Fluorescently Labeled Marburg Virus Glycoprotein as a New Tool to Study Viral Transport and Assembly. *J. Infect. Dis.* **2018**, *218*, S318–S326. [CrossRef]
51. Yamayoshi, S.; Noda, T.; Ebihara, H.; Goto, H.; Morikawa, Y.; Lukashevich, I.S.; Neumann, G.; Feldmann, H.; Kawaoka, Y. Ebola virus matrix protein VP40 uses the COPII transport system for its intracellular transport. *Cell. Host Microbe* **2008**, *3*, 168–177. [CrossRef]
52. Johnson, K.A.; Taghon, G.J.; Scott, J.L.; Stahelin, R.V. The Ebola Virus matrix protein, VP40, requires phosphatidylinositol 4,5-bisphosphate (PI(4,5)P2) for extensive oligomerization at the plasma membrane and viral egress. *Sci. Rep.* **2016**, *6*, 19125. [CrossRef]
53. Welsch, S.; Kolesnikova, L.; Krähling, V.; Riches, J.D.; Becker, S.; Briggs, J.A. Electron tomography reveals the steps in filovirus budding. *PLoS Pathog* **2010**, *6*, e1000875. [CrossRef]

11

Environmental Restrictions: A New Concept Governing HIV-1 Spread Emerging from Integrated Experimental-Computational Analysis of Tissue-Like 3D Cultures

Samy Sid Ahmed [1], Nils Bundgaard [2], Frederik Graw [2,*] and Oliver T. Fackler [1,*]

1 Department of Infectious Diseases, Integrative Virology, University Hospital Heidelberg, 69120 Heidelberg, Germany; Samy.SidAhmed@med.uni-heidelberg.de

2 BioQuant – Center for Quantitative Biology, Heidelberg University, 69120 Heidelberg, Germany; nils.bundgaard@bioquant.uni-heidelberg.de

* Correspondence: frederik.graw@bioquant.uni-heidelberg.de (F.G.); oliver.fackler@med.uni-heidelberg.de (O.T.F.)

Abstract: HIV-1 can use cell-free and cell-associated transmission modes to infect new target cells, but how the virus spreads in the infected host remains to be determined. We recently established 3D collagen cultures to study HIV-1 spread in tissue-like environments and applied iterative cycles of experimentation and computation to develop a first in silico model to describe the dynamics of HIV-1 spread in complex tissue. These analyses (i) revealed that 3D collagen environments restrict cell-free HIV-1 infection but promote cell-associated virus transmission and (ii) defined that cell densities in tissue dictate the efficacy of these transmission modes for virus spread. In this review, we discuss, in the context of the current literature, the implications of this study for our understanding of HIV-1 spread in vivo, which aspects of in vivo physiology this integrated experimental–computational analysis takes into account, and how it can be further improved experimentally and in silico.

Keywords: HIV-1 spread; cell-free infection; cell–cell transmission; 3D cultures; mathematical modeling; environmental restriction

1. Introduction

As obligate intracellular parasites, the replication of viruses depends on the infection of host cells that support the viral life cycle and the production of viral progeny. In order to establish virus replication in a new host, the virus has to efficiently spread following the initial infection at the portal of entry. The production of infectious progeny and infection of new target cells represents the central mechanism for virus spread. In principle, this can be achieved by the release of virus particles into the extracellular space, which can encounter and infect new target cells (cell-free infection) (Figure 1a). In addition, viruses can be transferred from infected donor cells to uninfected target cells via close physical contact between the cells (cell–cell transmission) (Figure 1b–d). Cell-associated modes of virus transmission include the short-distance transmission of cell-free virus at cell–cell contacts (Figure 1d), the transport of virus particles along or within cell protrusions connecting donor and target cells (Figure 1b,c), as well as cell–cell fusion [1,2], and are generally considered more efficient than cell-free infections. While cell-associated modes of virus transmission have been less explored than cell-free infection, evidence for the use of this transmission mode is steadily increasing and has been documented, e.g., for Vaccinia virus [3], Hepatitis C virus [4], Herpes Simplex virus [5], Epstein–Barr Virus [6] Dengue Virus [7], and the pathogenic human retroviruses Human Immunodeficiency Virus type 1 (HIV-1) and Human T-cell Lymphotropic Virus type 1 (HTLV-1) [8–12]. Most of these viruses are known to be able to spread by cell-free and cell-associated modes of transmission, but some viruses, such as HTLV-I, specialize in cell–cell transmission and appear to exclusively rely on this transfer mode, as cell-free infectious virus can seldom be isolated [12]. For viruses using both cell-free and cell-associated transmission, the relative contribution of each transmission mode to overall spread is difficult to assess, and the pathophysiological relevance of cell-associated transmission often remains unclear. It is therefore not surprising that traditional concepts in virology have focused on cell-free infections, which is still reflected in the majority of experimental studies conducted.

HIV-1 is an example of a virus for which the modes of transmission are particularly well studied. Initially assumed to spread exclusively via cell-free virus, early studies indicated that infected cells are a much better inoculum to drive virus spread in a new culture than cell-free virus [13]. The demonstration that constant agitation of infected $CD4^+$ T cells or physical separation of infected from uninfected cells by transwells disrupts the formation of cell–cell contacts as well as efficient virus spread then suggested that, in fact, cell-associated modes of transmission are essential for efficient HIV-1 spread in $CD4^+$ T-cell cultures [14,15]. A large series of imaging-based studies has now established that in addition to infection with cell-free virions, HIV-1 can efficiently spread via cell-cell contacts. Although probably relying on a slightly divergent mechanism, HIV-1 cell–cell transmission is observed between $CD4^+$ T cells, for the transfer from dendritic cells to $CD4^+$ T cells, between $CD4^+$ T cells and macrophages, and between myeloid cells [16]. The cell–cell contacts involved are referred to as virological synapses (VSs) between productively infected donor and target cells, or as infectious synapses when donor cells such as dendritic cells store virus for transmission without being productively infected [17–22]. This contact-dependent transmission mode has been found to be much more efficient than cell-free virus uptake, with an estimated 10-fold to 18,000-fold higher efficiency in mediating viral spread [14,23–25]. Despite this overwhelming evidence that HIV-1 efficiently uses cell-associated modes of transmission in experimental HIV-1 infection, the technical barriers to studying this aspect of HIV-1 biology in vivo limit the generation of evidence for the use of this transmission mode in the infected host. Only a few studies have reported visualization of cell–cell contacts reminiscent of a VS in vivo [26–28], or established the motility of cells loaded with infectious HIV-1 as essential for efficient HIV-1 spread in infected humanized mice [27,29]. By which transmission mode HIV-1 spreads in vivo, and how this might depend on the specific tissue environment, thus remains to be established.

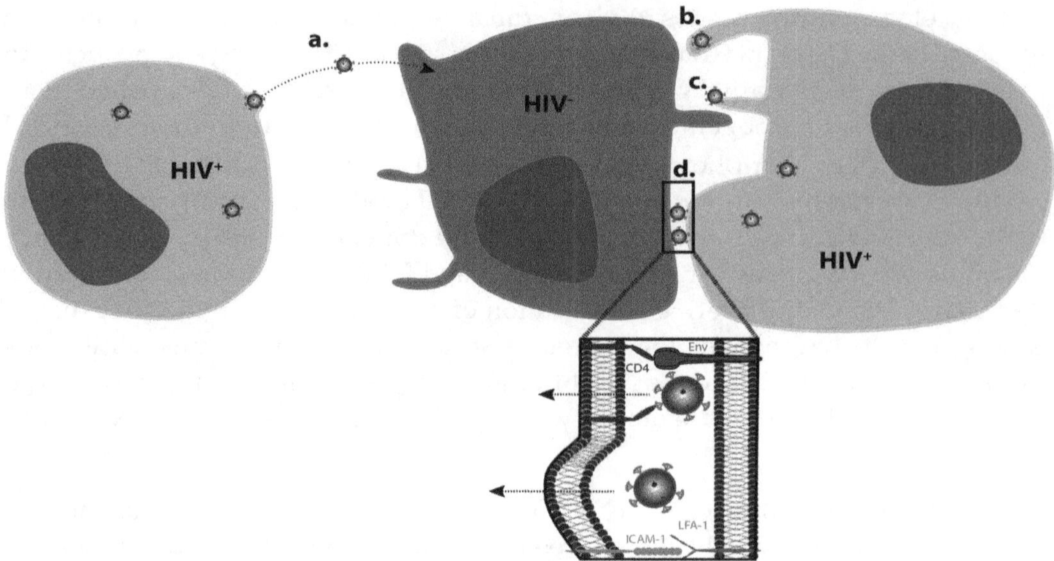

Figure 1. Transmission modes of HIV-1. Viral particles infect target cells via cell-free (**a**) or cell-associated (**b**-**d**) modes of transmission. (**a**) Viral particles bud at the surface of infected donor cells, mature, diffuse, and infect non-adjacent target cells. (**b**,**c**) Virions can bud at the tip (**b**) and surf along (**c**) filopodia to enter in adjacent target cells. In addition, infected and non-infected cells establish close contact, forming a virological synapse (**d**). Whether HIV-1 enters the target cell via fusion at the plasma membrane or following prior internalization [30,31] remains a matter of debate, and may depend on the nature of the target cell (reviewed in Reference [32]).

2. 3D Culture Systems

Questions such as the relevance of individual virus transmission modes for viral spread and disease progression would be best addressed in in vivo infection models, as these reflect the complex tissue organization and pathogen–host interactions of infected patients [33]. In the case of HIV-1 infection, various humanized mouse models that mirror a range of but not all pathological events of AIDS are available for such studies [34]. However, experimental parameters are very difficult to control in these complex in vivo systems, and the number of experiments that can be conducted is limited by logistic, financial, and/or ethical concerns. In turn, simple monotypic, two-dimensional (2D) cultures do not reflect the complex architecture and cell heterogeneity of HIV-1 target tissue and, even for suspension cells, such as $CD4^+$ T cells, quickly organize into a cell monolayer with dense cell packing. In vivo, the physical distance between donor and target cells often is the main barrier for efficient virus transmission, be it via diffusion of cell-free particles or the motility of the cell towards forming cell–cell contacts for virus transmission. Thus, densely packed 2D cultures are not suitable to study such processes.

As these constraints apply to all areas of biology, major efforts are being undertaken to establish and exploit experimental 3D systems that faithfully reflect specific aspects of tissue organization and physiology. One major area of development involves establishing organ or organoid cultures, either by culturing original tissues obtained from surgery or by reconstituting organ-like tissue ex vivo by differentiation from induced pluripotent stem cells [35,36]. In the case of HIV-1, such organotypic cell systems include cultures of human tonsil tissue as surrogates for lymph node tissues or mucosa explants as mimics of sites of virus transmission during primary HIV-1 infection [33,37]. While these models are of great value for advancing our understanding of pathological mechanisms, they do not offer tight control over critical experimental parameters such as cell density and composition, are subject to significant donor-to-donor variability, and their availability is limited. These limitations have precluded their use for defining the relative contribution of individual virus transmission modes and generated the need for the development of alternative experimental approaches. In, for example, immunology, significant efforts have been made to establish synthetic 3D systems to complement our portfolio of experimental systems with approaches that offer tight experimental control and direct access to visualization and genetic modification of individual cells. Particularly for lymphocytes and dendritic cells, 3D matrices made of collagen have emerged as the gold standard for such studies [38–40]. Collagen has the advantage of being a major constituent of extracellular matrix, supporting cell viability for up to several weeks, and allowing the easy removal of cells by collagenase digestion. Moreover, the good optical properties of the matrix provide access to live cell microscopy and the architecture of the matrix can be modified easily by adjusting polymerization conditions (collagen concentration, polymerization temperature, and medium used) [41–43]. The bovine and rat tail collagen classically used in 3D culture systems share high amino-acid similarities with the human type I collagen protein (91.1% identity for the alpha I chains and 87.4% for the alpha 2 chains), suggesting that they likely mirror the properties of human collagen. Consistent with this idea, $CD4^+$ cells embedded in bovine and rat collagen matrices adopt migrating behaviors reminiscent of in vivo situations [44]. The xenogeneic effects associated with culturing human cells in this type of culture thus seem to be minimal with respect to cell migration, although species differences in molecular regulation between different collagens cannot be entirely excluded. These aspects allow quantitative insight to be gained into, for example, $CD4^+$ T-cell function, such as their interaction and communication with DCs, or the ability to incorporate specific small RNAs into exosomes that match their behavior in vivo [45,46]. We therefore embarked on an attempt to reconstitute key aspects of HIV-1 spread in target tissue using 3D collagen cultures [47].

3. New Insights from Combining the Study of HIV-1 Spread in 3D Collagen Cultures with Computation

In our recent proof-of-concept study [47], 3D cultures used to study HIV-1 spread included exclusively lymphocytes, considering HIV-1-infected and uninfected $CD4^+$ T cells. Using a reporter virus expressing a sortable cell surface tag enabled the generation of pure HIV-1-infected donor cell populations that were mixed with uninfected target cells at a defined ratio prior to being embedded in 3D collagen. Cells were viable over several weeks in this system and cultures allowed the quantification of viral titers, quantification of expansion or depletion of specific cell populations, and the determination of cell motility and cell–cell contact formation/duration over time. Finally, the system allows for the use of collagen types with different density, with rat and bovine collagen resulting in denser or looser meshworks, respectively [48]. Recording the dynamics of HIV-1 spread and $CD4^+$ T-cell depletion in parallel suspension and 3D cultures immediately revealed that the accurate interpretation of this plethora of quantitative kinetic data required computational approaches. Moreover, addressing the potential relationship between cell motility and virus spread, as well as disentangling the relative contribution of cell-free vs. cell-associated virus transmission to the infection dynamics, was impossible to assess experimentally but required the development of customized computational models. Our analysis relied on several iterative cycles between experimental quantification and computation and resulted in an Integrative method to Study Pathogen spread by Experiment and Computation within Tissue-like 3D cultures(INSPECT-3D) workflow depicted in Figure 2.

Together, the development and first application of INSPECT-3D revealed that HIV-1 spread is delayed relative to suspension cultures when cells are placed in a dense matrix but, after an initial phase characterized by low virus replication, follows kinetics that are similar to those seen in suspension cultures. In contrast, low-density collagen matrices did not support a marked spread of HIV-1. Several findings allowed these surprising observations to be explained. First, 3D collagen, irrespective of its density, potently suppresses the infectivity of cell-free HIV-1 particles (approx. 20-fold). Moreover, virion diffusion rates are too slow to efficiently deliver virus particles to new target cells in the spacing of 3D collagen (i.e., diffusion of virions to the nineteen nearest neighboring cells requires ~22.6 h, in comparison to half-life HIV-1 particle infectivity of $t_{1/2}$ = 17.9 h). Together, these results revealed that 3D tissue-like environments potently restrict cell-free HIV-1 infection (see discussion of environmental restriction below). Analysis of the experimental data by mathematical modeling allowed us to estimate the relative contribution of cell-free and cell-associated HIV transmission under these various conditions. Hereby, consistent with the experimental observations, we found that HIV-1 spread in 3D collagen predominantly relies on cell–cell transmission (~78% (73–100%) and ~63% (55.8–100%) of all infections for loose and dense collagen, respectively). In contrast, in suspension cultures, there was no dominance of cell–cell transmission found and its contribution ranged widely between 0 and 100% (Table 1). Viral spread could thus be explained by the exclusive use of cell-free or cell–cell transmission. This revealed that in suspension cultures, no selection pressures for any of the transmission modes exist. This flexibility is consistent with the observation that physical separation of cells in suspension by agitation or transwells, which create conditions in which (i) cell-free infection depends on long-range diffusion of virus particles which compromises their infectivity and (ii) cell–cell contact is limited, markedly reduces virus spread [14,15]. Moreover, our analyses revealed that transmission dynamics substantially differ between suspension cultures and collagen environments.

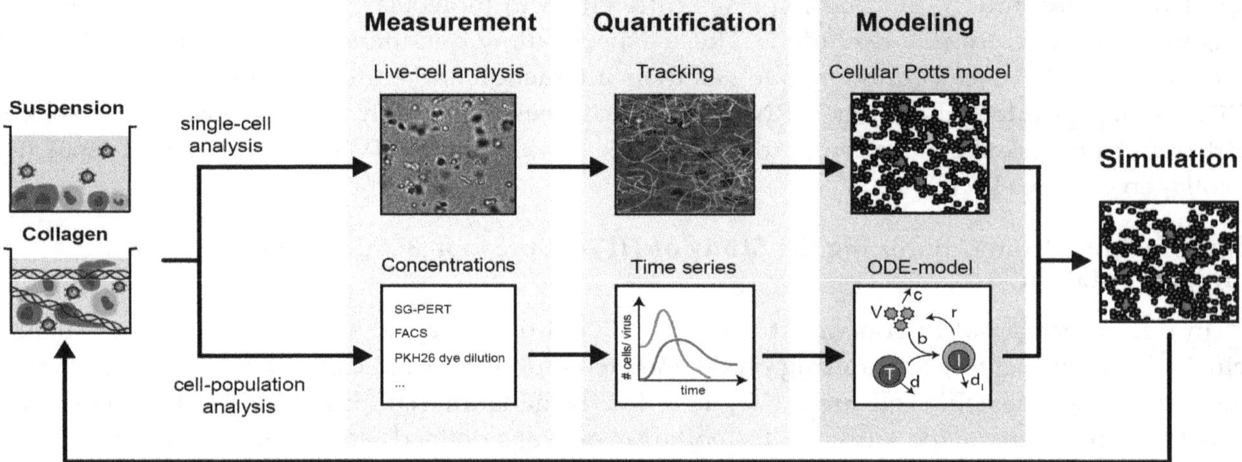

Figure 2. Schematic of the INSPECT-3D workflow that combined single-cell and cell-population measurements by mathematical modeling to reveal the dynamics of infection spread. Infected and non-infected Peripheral Blood Mononuclear Cells (PBMCs) were cultured in suspension or collagen cultures at a 1:20 ratio. In suspension cultures, cells rapidly settled at the bottom of the culture dish to form dense 2D monolayers. In collagen, cells were spatially separated from another, migrated along the 3D scaffold, and eventually made contact with other cells. Measurements from live-cell imaging of single cells and bulk dynamics allowed quantification of individual cell motilities and interactions, as well as long-term kinetics of viral load and cell concentrations. Appropriate mathematical model systems that either rely on spatially resolved cellular Potts models describing single-cell behavior or systems of ordinary differential equations (ODE) enabled us to disentangle processes of cell motility, cellular turnover, and viral transmission and replication. Theses analyses led to an integrative computational model for HIV-1 spread that provides a mechanistic and quantitative understanding of viral transmission within these multicellular systems. Recurrent experimental validation of model predictions by in silico manipulation of experimental conditions was used to refine the model system and reveal key processes governing the dynamics of HIV-1 spread.

Since 3D collagen suppressed cell-free HIV-1 infection and virus spread depended on cell-associated virus transmission, we investigated this scenario in more detail and found that in dense 3D collagen, contacts of extended duration between donor and target cells were more frequent in comparison to in loose collagen. These findings suggested that the dense 3D environment specifically promotes cell-associated HIV-1 transmission, while the low virus replication rates observed in loose collagen stem from the combination of suppression of cell-free infectivity by the 3D matrix and the lack of support for cell-associated virus transmission due to a lack of long-lasting cell–cell contacts. Importantly, the computational model was able to predict the extent to which cell densities had to be increased experimentally to overcome these barriers to HIV-1 spread in loose 3D collagen, suggesting that adopting a 2D monolayer type of configuration could override adverse effects of a 3D matrix on HIV-1 spread. These findings also imply that in a dense matrix, in which cells are not unphysiologically densely packed, cell motility is a key parameter governing the efficacy of HIV-1 spread, i.e., allowing transport of the infection to other areas as cell-free infection is impaired. Quantitative assessments of the relationship of HIV-1 spread in specific target organs will hence require the individual reconstitution of cell and matrix density as well as of the spacing between donor and target cells.

Correlating cell–cell contact duration and productive HIV-1 transmission also allowed us to predict the minimal donor–target cell contact time required for productive infection of new target cells, which was estimated to be in the range of ~25 min. Low-resolution inspection of productive donor–target cell contacts in dense 3D collagen revealed a strikingly close and intimate organization of these contacts, and it will be interesting to dissect how this morphology translates into elevated rates of virus transmission and how dense, but not loose, 3D environments induce these events.

4. How Well Do Experimental and Computational INSPECT-3D Data Match In Vivo?

Although ex vivo 3D collagen cultures allow an improved experimental assessment of physiologically relevant cellular behaviour, they still only provide a reduced representation of relevant tissue conditions and do not completely mirror specific target tissues. However, comparison of the kinetics inferred for viral transmission and cellular turnover dynamics within these culture systems can help to assess their relevance for understanding the spread of HIV-1 in vivo. Mathematical modeling has long been used as an essential tool to dissect the multifactorial complexity of viral transmission and replication dynamics (reviewed in References [49,50]). Standard models of viral dynamics describing the turnover of uninfected and infected cells and the progression of the viral load by systems of ordinary differential equations have been applied to various experimental and clinical data assessing the dynamics of HIV-1 viral spread [51–53]. In particular, these models allowed the quantification of rates describing the kinetics of viral production (ρ) or the death of infected cells (δ) based on patient data. Using HIV-1 plasma RNA levels as an indicator for viral spread in patients, several studies estimated an average production of ~10^{10} HIV-1 virions in total per day, with estimates of the death rate of infected cells between $\delta \approx 0.46-1.40\ \text{day}^{-1}$, corresponding to average half-lives of infected cells between $t_{1/2} = 0.5$ and 1.5 days ($t_{1/2} = \log 2/\delta$) [51–54]. Studies based on standard in vitro culture systems have produced estimates that are at the upper end of the half-lives determined for infected cells in vivo, with estimates of the corresponding death rates being in the order of $\delta = 0.5 +/- 0.1\ \text{day}^{-1}$ [55]. The estimates on the death rate of infected cells assessed via the INSPECT-3D workflow are within the same range, with a slight tendency towards higher death rates within 3D collagen environments compared to 2D suspension cultures ($\delta = 0.46$–$0.56\ \text{day}^{-1}$ in dense and loose collagen vs. $\delta = 0.40-0.44\ \text{day}^{-1}$ in suspension; ranges based on 95% confidence intervals of estimates, Table 1). Although the differences are minor, this might indicate that environmental factors, in combination with additional clearance mechanisms such as immune responses, could also contribute to the shorter half-lives of infected cells observed in vivo. While mathematical analyses of HIV-1 viral load kinetics in patients usually allow an appropriate quantification of viral and infected cell decay dynamics, as they work under the assumption that Highly Active Antiretroviral Therapy (HAART) effectively blocks novel infections [50,52], assessing the kinetics of viral transmission and spread in vivo is much more difficult. This requires an approximation of the available number of target cells, which might vary dependent on the time-point of infection and the different compartments of viral replication [56]. In vitro systems, with their associated complete knowledge of the initial experimental conditions, provide an advantage for assessing these dynamics, but have to account for the general lack of potential physiological relevance. As cell-free and cell-associated transmission modes are synergistic, unraveling their relative contributions to viral spread in vitro usually requires the impairment of at least one of them [4,55,57–59]. Iwami et al. [55] set out to assess the transmission rates for HIV-1 using shaken suspension cultures where cell–cell transmission was considered to be impaired [14]. Based on their data, they estimated that cell-associated viral transmission contributed to approximately 60% of viral infections. Without the assumption of efficiently blocking one transmission mode, we found similar contributions of cell-associated transmission modes to viral spread within 3D collagen (with only ~22% (0%, 27%) and ~37% (0%, 44.2%) of infections within loose and dense collagen, respectively, due to cell-free transmission). Thus, there was a clear difference in this contribution compared to spread within 2D suspension cultures. The simultaneous analysis of all three environments revealed a ~4–7-fold higher probability of cells becoming infected by cell–cell transmission within dense and loose collagen compared to suspension. The impaired contribution of cell-free transmission within 3D collagen compared to suspension cultures is further indicated by a ~50% reduction in viral production rates, and a measured relative infectivity of virions that was reduced to ~14% within collagen (β_f, Table 1). In addition to previous reports that blocking $CD4^+$ T-cell exit from lymph nodes restricted HIV-1 spread [27], these findings obtained for viral spread within 3D collagen environments support the argument that cell–cell transmission is the main contributor to HIV-1 spread in vivo. The importance of cell-contact-dependent transmission modes in tissue-like conditions is further indicated by the increased fraction of $CD4^+$ T cells predicted to be initially refractory to infection in collagen (loose ~50%; dense ~30%) compared to suspension (~17%; fractions at 2.5 days post infection). This could point towards cells that are inaccessible for contact-dependent transmission modes due to physical barriers, as they are blocked by collagen within certain areas of the culture.

Table 1. Estimates for HIV-1 infection dynamics obtained from the analysis of experimental infection in cell lines or primary cells and in vivo patient data. Obtained estimates vary over several orders of magnitude due to different mathematical models and data types used. Please note that viral production rates vary in terms of units due to different quantification methods used. Numbers marked with (*) indicate estimates obtained for Simian Immunodeficiency Virus (SIV) infection.

Parameter	Description	Unit	Cell Lines	Primary Cells [47]			In Vivo
				2D Suspension	3D Collagen		
					Loose	Dense	
ρ	Viral production rate	$\times 10^4$ day^{-1}	2.61 (1.55, 3.70) (RNA copies) [60]	1.02 (0.80, 1.38) (RT cell^{-1})	0.48 (0.37, 0.66) (RT cell^{-1})	0.43 (0.32, 0.61) (RT cell^{-1})	5.0 * (1.3, 12.0) (virion cell^{-1}) [61] 0.07–0.34 [56,62]
ρI	Total viral production rate	$\times 10^{10}$ virions day^{-1}					1.03 ± 1.17 [52,53]
δ_I	Death rate of infected cells	day^{-1}	0.50 ± 0.10 [55]	0.42 (0.40, 0.44)	0.48 (0.46, 0.50)	0.52 (0.51, 0.56)	0.48–1.36 ± 0.16 [53,63] 0.7 ± 0.25 [51] 1.0 ± 0.3 [54] 0.88 * ± 0.40 [61]
β_f	Cell-free infection rate	$\times 10^{-5}$ day^{-1}	0.42 ± 0.14 [55] (p24)	2.3 (0, 3.0) (RT)	0.14 × suspension	0.14 × suspension	
β_c	Cell–cell infection rate	$\times 10^{-5}$ (cell × day)$^{-1}$	0.11 ± 0.03 [55]	10^{-6} (0, 7.0)	4.3 (3.6, 5.4)	1.7 (1.4, 2.6)	
P_f	Predicted percentage of infections by cell-free transmission	%	57% ± 7% [55]	99.6% (0.0%, 100%)	22.0% (0.0%, 27.0%)	37% (0.0%, 44.2%)	

Combining live-cell microscopy data with a spatially detailed cellular Potts model in our INSPECT3D workflow also allowed us to estimate a minimal contact duration between infected and uninfected cells required for productive infection of ~25 min. Importantly, this matches very well the time required for the transfer of fluorescent viral material across the VS ex vivo [17,20], indicating that the events observed by these imaging approaches typically result in the productive infection of the target cell.

Based on the relative comparison of the viral kinetics between 2D suspension and 3D culture systems, our study indicated the need to account for physiological conditions when aiming to quantitatively understand viral replication and transmission dynamics in vivo. Arguably, the dense collagen conditions we used [47] might come close to mucosal tissue without reflecting the cellular composition of this tissue environment. Extending the existing culture systems by incorporating additional cell types and molecular factors will be needed to approximate specific tissue conditions. Other relevant physiological parameters that will have to be considered to be implemented to advance the experimental 3D model further include dynamic tissue perfusion, which might impact HIV-1 spread, e.g., by affecting the dynamics of cell-free infections or the frequency and/or stability of cell–cell interactions. It will be interesting to assess how these additional factors might impact the quantification of the viral processes obtained so far.

5. Requirements and Limitations of Current Computational Approaches to Simulating HIV-1 Spread in Tissue

The majority of previous studies combining mathematical models and experimental data have relied on model systems based on ordinary differential equations that use time-resolved measurements of cell or viral concentrations [50]. These analyses have increased our knowledge of various aspects of HIV-1 life-cycle kinetics and transmission dynamics, and are still the method of choice for gathering most types of experimental and clinical data. However, a more detailed analysis, and thus computational representation, of single-cell dynamics within tissues is needed to understand the dynamics of HIV-1 spread within multicellular systems.

Population dynamic models have their limitations when it comes to analyzing cell-based transmission modes, as their ability to describe single-cell transmission dynamics appropriately is, by definition, hampered. Several extensions have been developed that adjust standard models for viral dynamics to account for cell-contact requirements dependent on the tissue environment [64,65]. Nevertheless, the need to clearly quantify the kinetics of cell-associated viral transmission modes, and to determine the contributions of cell-free and cell-associated transmission, as well as local immunity to viral spread, provides novel challenges for computational analysis.

Individual-cell-based models that describe single-cell behavior have been used previously to simulate and analyze viral spread [66]. However, these model systems were mainly used for qualitative assessment of dynamics, e.g., for the spread of HIV-1 [67,68], as appropriate data and tools for model parameterization remained limited. Both limitations have now been overcome. Advanced imaging and visualization techniques used in vitro and in vivo [33,69,70] have increased our ability to observe and quantify cellular behavior and infection processes at a single-cell level. In addition, improved parameter inference tools and modeling systems have enabled us to quantitatively adapt individual-cell-based models to these novel types of data [71,72]. However, several challenges remain to appropriately analyze HIV-1 spread within different tissue environments using mathematical and computational models (Figure 3). A major challenge for computational approaches that analyze the spatiotemporal dynamics of HIV-1 spread is the appropriate representation of cell motility and the underlying tissue architecture, e.g., as represented by the collagen matrix within 3D ex vivo culture systems. Determining the influence of the environment on cell and infection dynamics requires a detailed description of these matrices and networks that structure the tissue or culture system. Previous attempts to describe the topology of the fibroblastic reticular network within lymph nodes in order to analyze T-cell motility and activation dynamics have shown the difficulty of such tasks [73–76]. High-resolution images of collagen matrices and fibroblastic reticular networks help to improve our understanding of the structures shaping and influencing tissue architecture and cell motility [48]. Nevertheless, the identification of appropriate quantities for network description, such as the pore size, fiber length, and connectivity, and how they relate to network density [42,43,77–79], is also required in order to provide a reliable computational representation of 3D tissue environments that accounts for the different resolutions of fibers and cells [73,76,80–82]. The challenges of describing cell motility based on live-cell microscopy data, and possible ways to refine these analyses, have been already discussed elsewhere [83]. Considering relevant physiological tissue conditions will also have to account for various cell types, such as T cells, dendritic cells, and macrophages, and requires a detailed characterization of their complex morphology and dynamics in 3D. In particular, this also applies to the appropriate representation of the diffusion dynamics of soluble factors, such as virions or chemokines, that are important for cell–cell communication, infection spread, or cell movement. Being able to measure gradients and concentrations of these factors, especially their release dynamics from single cells, would help to mimic their dynamics and to determine the influence of these factors on viral spread and the efficacy of the counteracting immune response [84].

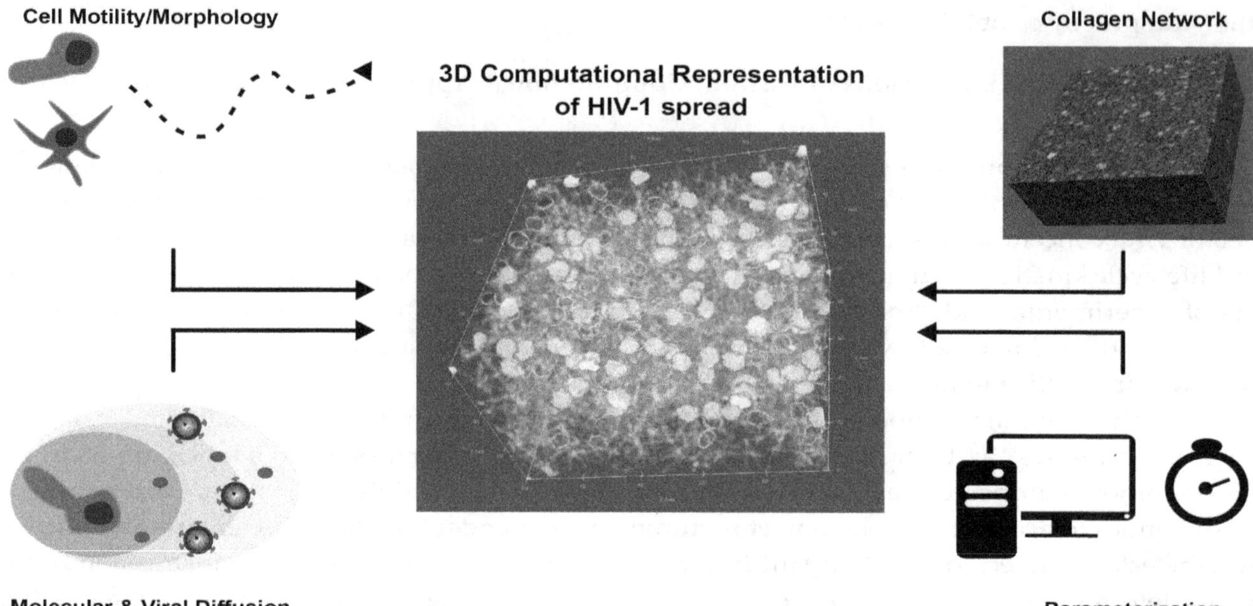

Figure 3. 3D computational representation of HIV-1 spread: Appropriate representations to infer the processes that govern the dynamics of infection require a detailed description and quantification of individual cell motility and morphology, the underlying tissue structure, and the diffusion dynamics of soluble components (chemokines, virions, etc.) within these environments. Data-driven parameterization and subsequent simulation of these 3D model systems requires computationally efficient methods. The sketch shows a computational representation of 100 HIV-1-infected (green) and 100 uninfected (red) cells within a 3D collagen matrix (blue fibers).

Finally, a major challenge for the computational modeling of 3D tissue conditions on a single-cell level is the increased level of detail required for its description, and especially its parameterization. The higher complexity of the model systems and the adaptation of their stochastic dynamics to heterogeneous experimental data requires the use of increased computational resources, including high-performance computing clusters and sophisticated parameter inference tools. Improved parallelization and advanced parameter estimation methods, such as approximated Bayesian computing [72], could help to reduce computational run time costs. In addition, simulation areas in terms of the size and number of cells considered have to be chosen with care in order to balance the need for sufficient tissue volumes that account for stochastic variability while simultaneously ensuring computational efficiency for the simulation of long-term infection dynamics. Thus, novel advances in imaging techniques and computational methods will improve our ability to conduct data-driven computational modeling of HIV-1 spread within multicellular systems, which will be essential to disentangling and understanding the processes that shape these dynamics within physiologically relevant conditions.

6. Environmental Restriction to Cell-Free HIV-1 Infection

A central finding of the studies on HIV-1 spread in 3D collagen is that residing within such a matrix reduces the infectivity of cell-free virus particles approx. 20-fold [47]. Since single-particle tracking has revealed that particles frequently undergo short and transient interactions with the collagen matrix, this negative impact on particle infectivity likely represents the consequence of physical stress imposed on virions in the context of these interactions. It is noteworthy that marked effects of the tissue environment on virus infectivity have been reported. One example is the interaction of viruses with bile acids, which can induce fusogenicity (e.g., norovirus [85]) or restrict infection (cytomegalovirus [86]; Hepatitis Delta virus [87]). Similarly, seminal fluids can impact virion infectivity via peptide-mediated enhancement of virus infectivity (e.g., retroviruses including HIV-1 [88], Ebola virus [89]) or by affecting particle aggregation [90,91]. Since the infectivity reduction in 3D collagen results from negative physical impact by the surrounding matrix and thus represents a direct effect of tissue architecture rather than the consequence of compounds released in the extracellular space, we propose the novel term "environmental restriction" for this phenomenon. Considering that this environmental restriction seems to represent a built-in antiviral activity of the 3D collagen matrix, this is conceptually similar to intrinsic immune barriers such as host cell restriction factors, and could thus be viewed as a novel element of innate immunity (Figure 4). In contrast to cell-autonomous mechanisms exerted by, for example, restriction factors, this activity would reflect a tissue-autonomous restriction. Since the restriction of virus spread by extracellular tissue architecture has only recently begun to become apparent, we can only speculate about the effector function mediating the restriction. Such environmental restrictions could include direct effects on the pathogen, such as the impairment of cell-free infectivity [92] (see below), but could also impact the host cells (e.g., via the promotion of longer-lasting cell contacts with architecture optimized for virus transmission). It can also be envisioned that, for example, transcriptional and epigenetic changes resulting from mechanosensing of the cells in the 3D environment would affect expression and activity of pro- and antiviral host cell factors. Assessing the effect of tissue-like 3D environments on the permissivity of various HIV-1 target cells for infection and their ability to sense the infection will be an important area of future research. Cell-associated HIV-1 transmission is currently considered to be beneficial for the virus, as it allows for more polarized and therefore efficient short-term transfer of particles to target cells, allows host cell restriction factors such as tetherin to be partially overcome, and avoids the recognition of certain neutralizing antibodies [25,93–95]. In addition, we propose that cell-associated HIV-1 transmission evolved from the need to bypass the environmental restriction against cell-free virion infectivity exerted by HIV-1 target tissue.

Following the initial description of the environmental restrictions on cell-free HIV-1 infectivity in 3D collagen cultures, it will be important to dissect the mechanism by which tissue-like 3D matrices impair the infectivity of cell-free virions. Intuitively, physical interactions with the matrix could enhance shedding of the Env glycoprotein. However, overall Env levels in virions were unaffected by interactions of virus particles with the 3D matrix. Alternatively, the 3D matrix could affect virion infectivity by alteration of their aggregation state [90]. However, single-particle tracking of HIV-1 particles in 3D matrices did not reveal virion aggregates in either suspension or 3D collagen, suggesting that HIV-1 particles do not form aggregates under these experimental conditions. Finally, the reduced particle infectivity in 3D collagen may also reflect adaptation of the producer cells to the 3D environment; however, the reduction of virion infectivity in 3D collagen was similar when particles were produced within the matrix or embedded in a cell-free matrix, indicating that the matrix exerts direct effects on HIV-1 virions. Future studies will be thus required to dissect whether the 3D matrix affects HIV-1 infection at the level of particle entry into target cells, e.g., by affecting the conformation of Env glycoproteins on virions, or at subsequent post entry steps (Figure 5). Assessing how conserved this effect may be among viruses other than HIV-1 will be helpful in dissecting the underlying mechanisms. Another key question is whether this barrier is indeed in place in HIV-1 target tissue in infected individuals. Analyzing this question is complicated by the fact that HIV-1 particles derived from patient samples represent a pool of particles of heterogeneous and undefined origin, and that potential effects of environmental tissue restrictions could not easily be distinguished from effects due to soluble factors present in bodily fluids. Organotypic cultures will likely be instrumental for addressing this question.

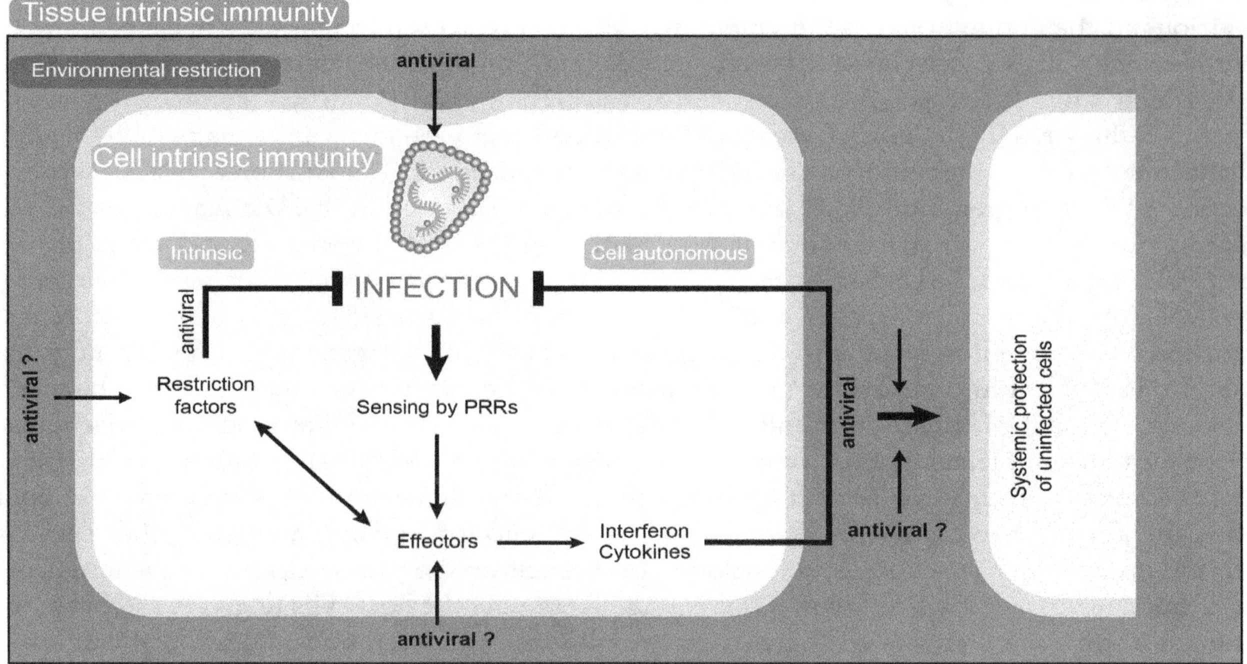

Figure 4. Schematic representation of the environmental restriction concept. Human cells bear cell-intrinsic mechanisms to limit virus spread, such as intrinsic immunity by restriction factors and cell-autonomous immunity by recognition of viral components by pattern-recognition receptors (PRRs). Both pathways can lead to the induction of antiviral effectors such as interferons and cytokines, which can impair virus replication in both the infected and bystander cells. In addition, we propose that the tissue environment can provide tissue-intrinsic immunity by exerting environmental restrictions on virus replication. Such activities could be exerted via direct effects on virus particles, but we also hypothesize that tissue interactions impact the expression and activity of cell-autonomous immune mechanisms.

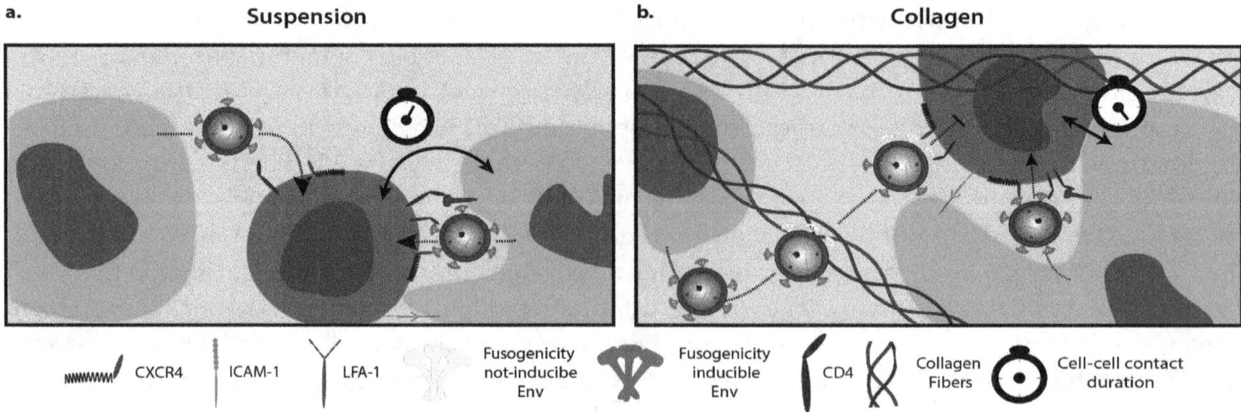

Figure 5. HIV-1 spread in suspension and 3D collagen cultures. (**a**) In suspension cultures, virions can be transmitted to uninfected target cells via both cell-free and cell-associated modes of transmission. (**b**) In collagen cultures, cell-associated transmission is the main driving force for HIV-1 spread. This reflects on one hand that the infectivity of cell-free particles is impaired, which coincides with frequent transient interactions with collagen fibres. These interactions could impair the ability of Env proteins to adopt a fusogenic conformation in response to interaction with the HIV-1 receptor/coreceptor complex on target cells, and thus inhibit virus entry. Moreover, cell-associated HIV-1 transmission may benefit from alterations in the architecture of contacts between donor and target cells, which are more stable and involve larger areas of membrane compared to cells in suspension.

7. Conclusions and Future Perspectives

The efficacy of virus spread in vivo is governed by a complex array of parameters that can only be disentangled by the combined application of synthetic and organoid 3D models and computational analyses. Our proof-of-concept study on HIV-1 spread in 3D collagen cultures revealed a significant impact of this tissue-like environment on modes and efficacy of HIV-1 spread, and suggests the existence of tissue-intrinsic mechanisms that suppress viral replication (environmental restriction). With the constant development of physiological ex vivo 3D culture systems and improved computational analyses, this approach is likely to continue to reveal more relevant details about the mechanisms of virus spread in vivo. An important aspect of such future efforts will be to decipher the contribution of antibody-mediated neutralization of HIV-1 during cell–cell spread (reviewed in Reference [96]) by extending the mathematical model and experimental validations within the INSPECT-3D workflow. Moreover, exosomes that resemble retroviral particles can also play a role in viral spread by affecting susceptibility to infection or the life-span of bystander cells [97–99]. However, the impact of tissue environments on these activities is unexplored. Finally, the INSPECT-3D workflow can directly be used to investigate the spread of other T-cell-tropic viruses, such as HTLV-1, for which the concept of cell-associated transmission via a VS was first established [10], and for which infection is thought to almost exclusively depend on cell–cell transmission, even in suspension cultures [100,101].

Author Contributions: O.T.F. and F.G. conceptualized the review. S.S.A., F.G. and O.T.F. wrote the manuscript. N.B. provided important intellectual content and contributed to the figures. All authors have read and agreed to the published version of the manuscript.

References

1. Zhong, P.; Agosto, L.M.; Munro, J.B.; Mothes, W. Cell-to-cell transmission of viruses. *Curr. Opin. Virol.* **2013**, *3*, 44–50. [CrossRef] [PubMed]
2. Sattentau, Q. Avoiding the void: Cell-to-cell spread of human viruses. *Nat. Rev. Microbiol.* **2008**, *6*, 815–826. [CrossRef] [PubMed]
3. Roberts, K.L.; Smith, G.L. Vaccinia virus morphogenesis and dissemination. *Trends Microbiol.* **2008**, *16*, 472–479. [CrossRef] [PubMed]
4. Graw, F.; Martin, D.N.; Perelson, A.S.; Uprichard, S.L.; Dahari, H. Quantification of Hepatitis C Virus Cell-to-Cell Spread Using a Stochastic Modeling Approach. *J. Virol.* **2015**, *89*, 6551–6561. [CrossRef] [PubMed]
5. Polcicova, K.; Goldsmith, K.; Rainish, B.L.; Wisner, T.W.; Johnson, D.C. The extracellular domain of herpes simplex virus gE is indispensable for efficient cell-to-cell spread: Evidence for gE/gI receptors. *J. Virol.* **2005**, *79*, 11990–12001. [CrossRef] [PubMed]
6. Imai, S.; Nishikawa, J.; Takada, K. Cell-to-cell contact as an efficient mode of Epstein-Barr virus infection of diverse human epithelial cells. *J. Virol.* **1998**, *72*, 4371–4378. [CrossRef] [PubMed]
7. Yang, C.F.; Tu, C.H.; Lo, Y.P.; Cheng, C.C.; Chen, W.J. Involvement of Tetraspanin C189 in Cell-to-Cell Spreading of the Dengue Virus in C6/36 Cells. *PLoS Negl. Trop. Dis.* **2015**, *9*, e0003885. [CrossRef]
8. Jolly, C.; Kashefi, K.; Hollinshead, M.; Sattentau, Q.J. HIV-1 cell to cell transfer across an Env-induced, actin-dependent synapse. *J. Exp. Med.* **2004**, *199*, 283–293. [CrossRef]
9. Bracq, L.; Xie, M.; Benichou, S.; Bouchet, J. Mechanisms for Cell-to-Cell Transmission of HIV-1. *Front. Immunol.* **2018**, *9*, 260. [CrossRef]
10. Igakura, T.; Stinchcombe, J.C.; Goon, P.K.; Taylor, G.P.; Weber, J.N.; Griffiths, G.M.; Tanaka, Y.; Osame, M.; Bangham, C.R. Spread of HTLV-I between lymphocytes by virus-induced polarization of the cytoskeleton. *Science* **2003**, *299*, 1713–1716. [CrossRef]

11. Pais-Correia, A.M.; Sachse, M.; Guadagnini, S.; Robbiati, V.; Lasserre, R.; Gessain, A.; Gout, O.; Alcover, A.; Thoulouze, M.I. Biofilm-like extracellular viral assemblies mediate HTLV-1 cell-to-cell transmission at virological synapses. *Nat. Med.* **2010**, *16*, 83–89. [CrossRef] [PubMed]
12. Gross, C.; Thoma-Kress, A.K. Molecular Mechanisms of HTLV-1 Cell-to-Cell Transmission. *Viruses* **2016**, *8*, 74. [CrossRef] [PubMed]
13. Dimitrov, D.S.; Willey, R.L.; Sato, H.; Chang, L.J.; Blumenthal, R.; Martin, M.A. Quantitation of human immunodeficiency virus type 1 infection kinetics. *J. Virol.* **1993**, *67*, 2182–2190. [CrossRef] [PubMed]
14. Sourisseau, M.; Sol-Foulon, N.; Porrot, F.; Blanchet, F.; Schwartz, O. Inefficient human immunodeficiency virus replication in mobile lymphocytes. *J. Virol.* **2007**, *81*, 1000–1012. [CrossRef]
15. Orlandi, C.; Flinko, R.; Lewis, G.K. A new cell line for high throughput HIV-specific antibody-dependent cellular cytotoxicity (ADCC) and cell-to-cell virus transmission studies. *J. Immunol. Methods* **2016**, *433*, 51–58. [CrossRef]
16. Xie, M.; Leroy, H.; Mascarau, R.; Woottum, M.; Dupont, M.; Ciccone, C.; Schmitt, A.; Raynaud-Messina, B.; Verollet, C.; Bouchet, J.; et al. Cell-to-Cell Spreading of HIV-1 in Myeloid Target Cells Escapes SAMHD1 Restriction. *Mbio* **2019**, *10*. [CrossRef]
17. Hubner, W.; McNerney, G.P.; Chen, P.; Dale, B.M.; Gordon, R.E.; Chuang, F.Y.; Li, X.D.; Asmuth, D.M.; Huser, T.; Chen, B.K. Quantitative 3D video microscopy of HIV transfer across T cell virological synapses. *Science* **2009**, *323*, 1743–1747. [CrossRef]
18. Sowinski, S.; Jolly, C.; Berninghausen, O.; Purbhoo, M.A.; Chauveau, A.; Kohler, K.; Oddos, S.; Eissmann, P.; Brodsky, F.M.; Hopkins, C.; et al. Membrane nanotubes physically connect T cells over long distances presenting a novel route for HIV-1 transmission. *Nat. Cell Biol.* **2008**, *10*, 211–219. [CrossRef]
19. Felts, R.L.; Narayan, K.; Estes, J.D.; Shi, D.; Trubey, C.M.; Fu, J.; Hartnell, L.M.; Ruthel, G.T.; Schneider, D.K.; Nagashima, K.; et al. 3D visualization of HIV transfer at the virological synapse between dendritic cells and T cells. *Proc. Natl. Acad. Sci. USA* **2010**. [CrossRef]
20. Real, F.; Sennepin, A.; Ganor, Y.; Schmitt, A.; Bomsel, M. Live Imaging of HIV-1 Transfer across T Cell Virological Synapse to Epithelial Cells that Promotes Stromal Macrophage Infection. *Cell Rep.* **2018**, *23*, 1794–1805. [CrossRef]
21. McDonald, D.; Wu, L.; Bohks, S.M.; KewalRamani, V.N.; Unutmaz, D.; Hope, T.J. Recruitment of HIV and its receptors to dendritic cell-T cell junctions. *Science* **2003**, *300*, 1295–1297. [CrossRef] [PubMed]
22. Earl, L.A.; Lifson, J.D.; Subramaniam, S. Catching HIV 'in the act' with 3D electron microscopy. *Trends Microbiol.* **2013**, *21*, 397–404. [CrossRef] [PubMed]
23. Chen, P.; Hubner, W.; Spinelli, M.A.; Chen, B.K. Predominant mode of human immunodeficiency virus transfer between T cells is mediated by sustained Env-dependent neutralization-resistant virological synapses. *J. Virol.* **2007**, *81*, 12582–12595. [CrossRef] [PubMed]
24. Martin, N.; Welsch, S.; Jolly, C.; Briggs, J.A.; Vaux, D.; Sattentau, Q.J. Virological synapse-mediated spread of human immunodeficiency virus type 1 between T cells is sensitive to entry inhibition. *J. Virol.* **2010**, *84*, 3516–3527. [CrossRef]
25. Kolodkin-Gal, D.; Hulot, S.L.; Korioth-Schmitz, B.; Gombos, R.B.; Zheng, Y.; Owuor, J.; Lifton, M.A.; Ayeni, C.; Najarian, R.M.; Yeh, W.W.; et al. Efficiency of cell-free and cell-associated virus in mucosal transmission of human immunodeficiency virus type 1 and simian immunodeficiency virus. *J. Virol.* **2013**, *87*, 13589–13597. [CrossRef]
26. Sewald, X.; Gonzalez, D.G.; Haberman, A.M.; Mothes, W. In vivo imaging of virological synapses. *Nat. Commun.* **2012**, *3*, 1320. [CrossRef]
27. Murooka, T.T.; Deruaz, M.; Marangoni, F.; Vrbanac, V.D.; Seung, E.; von Andrian, U.H.; Tager, A.M.; Luster, A.D.; Mempel, T.R. HIV-infected T cells are migratory vehicles for viral dissemination. *Nature* **2012**, *490*, 283–287. [CrossRef]
28. Ladinsky, M.S.; Khamaikawin, W.; Jung, Y.; Lin, S.; Lam, J.; An, D.S.; Bjorkman, P.J.; Kieffer, C. Mechanisms of virus dissemination in bone marrow of HIV-1-infected humanized BLT mice. *ELife* **2019**, *8*. [CrossRef]
29. Sewald, X.; Ladinsky, M.S.; Uchil, P.D.; Beloor, J.; Pi, R.; Herrmann, C.; Motamedi, N.; Murooka, T.T.; Brehm, M.A.; Greiner, D.L.; et al. Retroviruses use CD169-mediated trans-infection of permissive lymphocytes to establish infection. *Science* **2015**, *350*, 563–567. [CrossRef]
30. Miyauchi, K.; Kim, Y.; Latinovic, O.; Morozov, V.; Melikyan, G.B. HIV enters cells via endocytosis and dynamin-dependent fusion with endosomes. *Cell* **2009**, *137*, 433–444. [CrossRef]

31. Dale, B.M.; McNerney, G.P.; Thompson, D.L.; Hubner, W.; de Los Reyes, K.; Chuang, F.Y.; Huser, T.; Chen, B.K. Cell-to-cell transfer of HIV-1 via virological synapses leads to endosomal virion maturation that activates viral membrane fusion. *Cell Host Microbe* **2011**, *10*, 551–562. [CrossRef] [PubMed]
32. Melikyan, G.B. HIV entry: A game of hide-and-fuse? *Curr. Opin. Virol.* **2014**, *4*, 1–7. [CrossRef] [PubMed]
33. Fackler, O.T.; Murooka, T.T.; Imle, A.; Mempel, T.R. Adding new dimensions: Towards an integrative understanding of HIV-1 spread. *Nat. Rev. Microbiol.* **2014**, *12*, 563–574. [CrossRef] [PubMed]
34. Marsden, M.D.; Zack, J.A. Humanized Mouse Models for Human Immunodeficiency Virus Infection. *Annu. Rev. Virol.* **2017**, *4*, 393–412. [CrossRef] [PubMed]
35. Rossi, G.; Manfrin, A.; Lutolf, M.P. Progress and potential in organoid research. *Nat. Rev. Genet.* **2018**, *19*, 671–687. [CrossRef] [PubMed]
36. Bar-Ephraim, Y.E.; Kretzschmar, K.; Clevers, H. Organoids in immunological research. *Nat. Rev. Immunol.* **2019**. [CrossRef]
37. Merbah, M.; Introini, A.; Fitzgerald, W.; Grivel, J.C.; Lisco, A.; Vanpouille, C.; Margolis, L. Cervico-vaginal tissue ex vivo as a model to study early events in HIV-1 infection. *Am. J. Reprod. Immunol.* **2011**, *65*, 268–278. [CrossRef]
38. Sixt, M.; Lammermann, T. In vitro analysis of chemotactic leukocyte migration in 3D environments. *Methods Mol. Biol.* **2011**, *769*, 149–165. [CrossRef]
39. Wolf, K.; Alexander, S.; Schacht, V.; Coussens, L.M.; von Andrian, U.H.; van Rheenen, J.; Deryugina, E.; Friedl, P. Collagen-based cell migration models in vitro and in vivo. *Semin. Cell Dev. Biol.* **2009**, *20*, 931–941. [CrossRef]
40. Lammermann, T.; Bader, B.L.; Monkley, S.J.; Worbs, T.; Wedlich-Soldner, R.; Hirsch, K.; Keller, M.; Forster, R.; Critchley, D.R.; Fassler, R.; et al. Rapid leukocyte migration by integrin-independent flowing and squeezing. *Nature* **2008**, *453*, 51–55. [CrossRef]
41. Miron-Mendoza, M.; Seemann, J.; Grinnell, F. The differential regulation of cell motile activity through matrix stiffness and porosity in three dimensional collagen matrices. *Biomaterials* **2010**, *31*, 6425–6435. [CrossRef] [PubMed]
42. Olivares, V.; Condor, M.; Del Amo, C.; Asin, J.; Borau, C.; Garcia-Aznar, J.M. Image-based Characterization of 3D Collagen Networks and the Effect of Embedded Cells. *Microsc. Microanal.* **2019**, *25*, 971–981. [CrossRef] [PubMed]
43. Jansen, K.A.; Licup, A.J.; Sharma, A.; Rens, R.; MacKintosh, F.C.; Koenderink, G.H. The Role of Network Architecture in Collagen Mechanics. *Biophys. J.* **2018**, *114*, 2665–2678. [CrossRef] [PubMed]
44. Friedl, P.; Brocker, E.B. The biology of cell locomotion within three-dimensional extracellular matrix. *Cell. Mol. Life Sci.* **2000**, *57*, 41–64. [CrossRef]
45. Thippabhotla, S.; Zhong, C.; He, M. 3D cell culture stimulates the secretion of in vivo like extracellular vesicles. *Sci. Rep.* **2019**, *9*, 13012. [CrossRef]
46. Abu-Shah, E.; Demetriou, P.; Balint, S.; Mayya, V.; Kutuzov, M.A.; Dushek, O.; Dustin, M.L. A tissue-like platform for studying engineered quiescent human T-cells' interactions with dendritic cells. *ELife* **2019**, *8*. [CrossRef]
47. Imle, A.; Kumberger, P.; Schnellbacher, N.D.; Fehr, J.; Carrillo-Bustamante, P.; Ales, J.; Schmidt, P.; Ritter, C.; Godinez, W.J.; Muller, B.; et al. Experimental and computational analyses reveal that environmental restrictions shape HIV-1 spread in 3D cultures. *Nat. Commun.* **2019**, *10*, 2144. [CrossRef]
48. Wolf, K.; Te Lindert, M.; Krause, M.; Alexander, S.; Te Riet, J.; Willis, A.L.; Hoffman, R.M.; Figdor, C.G.; Weiss, S.J.; Friedl, P. Physical limits of cell migration: Control by ECM space and nuclear deformation and tuning by proteolysis and traction force. *J. Cell Biol.* **2013**, *201*, 1069–1084. [CrossRef]
49. Graw, F.; Perelson, A.S. Modeling Viral Spread. *Annu. Rev. Virol.* **2016**, *3*, 555–572. [CrossRef]
50. Perelson, A.S. Modelling viral and immune system dynamics. *Nat. Rev. Immunol.* **2002**, *2*, 28–36. [CrossRef]
51. Perelson, A.S.; Neumann, A.U.; Markowitz, M.; Leonard, J.M.; Ho, D.D. HIV-1 dynamics in vivo: Virion clearance rate, infected cell life-span, and viral generation time. *Science* **1996**, *271*, 1582–1586. [CrossRef] [PubMed]
52. Ramratnam, B.; Bonhoeffer, S.; Binley, J.; Hurley, A.; Zhang, L.; Mittler, J.E.; Markowitz, M.; Moore, J.P.; Perelson, A.S.; Ho, D.D. Rapid production and clearance of HIV-1 and hepatitis C virus assessed by large volume plasma apheresis. *Lancet* **1999**, *354*, 1782–1785. [CrossRef]

53. Perelson, A.S.; Essunger, P.; Cao, Y.; Vesanen, M.; Hurley, A.; Saksela, K.; Markowitz, M.; Ho, D.D. Decay characteristics of HIV-1-infected compartments during combination therapy. *Nature* **1997**, *387*, 188–191. [CrossRef] [PubMed]
54. Markowitz, M.; Louie, M.; Hurley, A.; Sun, E.; Di Mascio, M.; Perelson, A.S.; Ho, D.D. A novel antiviral intervention results in more accurate assessment of human immunodeficiency virus type 1 replication dynamics and T-cell decay in vivo. *J. Virol.* **2003**, *77*, 5037–5038. [CrossRef] [PubMed]
55. Iwami, S.; Takeuchi, J.S.; Nakaoka, S.; Mammano, F.; Clavel, F.; Inaba, H.; Kobayashi, T.; Misawa, N.; Aihara, K.; Koyanagi, Y.; et al. Cell-to-cell infection by HIV contributes over half of virus infection. *ELife* **2015**, *4*. [CrossRef]
56. De Boer, R.J.; Ribeiro, R.M.; Perelson, A.S. Current estimates for HIV-1 production imply rapid viral clearance in lymphoid tissues. *PLoS Comput. Biol.* **2010**, *6*, e1000906. [CrossRef]
57. Timpe, J.M.; Stamataki, Z.; Jennings, A.; Hu, K.; Farquhar, M.J.; Harris, H.J.; Schwarz, A.; Desombere, I.; Roels, G.L.; Balfe, P.; et al. Hepatitis C virus cell-cell transmission in hepatoma cells in the presence of neutralizing antibodies. *Hepatology* **2008**, *47*, 17–24. [CrossRef]
58. Barretto, N.; Sainz, B., Jr.; Hussain, S.; Uprichard, S.L. Determining the involvement and therapeutic implications of host cellular factors in hepatitis C virus cell-to-cell spread. *J. Virol.* **2014**, *88*, 5050–5061. [CrossRef]
59. Barretto, N.; Uprichard, S.L. Hepatitis C virus Cell-to-cell Spread Assay. *Bio Protoc.* **2014**, *4*. [CrossRef]
60. Iwami, S.; Holder, B.P.; Beauchemin, C.A.; Morita, S.; Tada, T.; Sato, K.; Igarashi, T.; Miura, T. Quantification system for the viral dynamics of a highly pathogenic simian/human immunodeficiency virus based on an in vitro experiment and a mathematical model. *Retrovirology* **2012**, *9*, 18. [CrossRef]
61. Chen, H.Y.; Di Mascio, M.; Perelson, A.S.; Ho, D.D.; Zhang, L. Determination of virus burst size in vivo using a single-cycle SIV in rhesus macaques. *Proc. Natl. Acad. Sci. USA* **2007**, *104*, 19079–19084. [CrossRef] [PubMed]
62. Reilly, C.; Wietgrefe, S.; Sedgewick, G.; Haase, A. Determination of simian immunodeficiency virus production by infected activated and resting cells. *AIDS (London, England)* **2007**, *21*, 163–168. [CrossRef] [PubMed]
63. Althaus, C.L.; De Vos, A.S.; De Boer, R.J. Reassessing the human immunodeficiency virus type 1 life cycle through age-structured modeling: Life span of infected cells, viral generation time, and basic reproductive number, R0. *J. Virol.* **2009**, *83*, 7659–7667. [CrossRef] [PubMed]
64. Kumberger, P.; Durso-Cain, K.; Uprichard, S.L.; Dahari, H.; Graw, F. Accounting for Space-Quantification of Cell-To-Cell Transmission Kinetics Using Virus Dynamics Models. *Viruses* **2018**, *10*, 200. [CrossRef] [PubMed]
65. Pilyugin, S.S.; Antia, R. Modeling immune responses with handling time. *Bull. Math. Biol.* **2000**, *62*, 869–890. [CrossRef] [PubMed]
66. Bauer, A.L.; Beauchemin, C.A.; Perelson, A.S. Agent-based modeling of host-pathogen systems: The successes and challenges. *Infor. Sci.* **2009**, *179*, 1379–1389. [CrossRef]
67. Strain, M.C.; Richman, D.D.; Wong, J.K.; Levine, H. Spatiotemporal dynamics of HIV propagation. *J. Theor. Biol.* **2002**, *218*, 85–96. [CrossRef]
68. dos Santos, R.M.Z.; Coutinho, S. Dynamics of HIV Infection: A cellular automata approach. *Phys. Rev. Lett.* **2001**, *87*, 168102. [CrossRef]
69. Coombes, J.L.; Robey, E.A. Dynamic imaging of host-pathogen interactions in vivo. *Nat. Rev. Immunol.* **2010**, *10*, 353–364. [CrossRef]
70. Germain, R.N.; Miller, M.J.; Dustin, M.L.; Nussenzweig, M.C. Dynamic imaging of the immune system: Progress, pitfalls and promise. *Nat. Rev. Immunol.* **2006**, *6*, 497–507. [CrossRef]
71. Starruss, J.; de Back, W.; Brusch, L.; Deutsch, A. Morpheus: A user-friendly modeling environment for multiscale and multicellular systems biology. *Bioinformatics* **2014**, *30*, 1331–1332. [CrossRef] [PubMed]
72. Klinger, E.; Rickert, D.; Hasenauer, J. pyABC: Distributed, likelihood-free inference. *Bioinformatics* **2018**, *34*, 3591–3593. [CrossRef] [PubMed]
73. Novkovic, M.; Onder, L.; Cheng, H.W.; Bocharov, G.; Ludewig, B. Integrative Computational Modeling of the Lymph Node Stromal Cell Landscape. *Front. Immunol.* **2018**, *9*, 2428. [CrossRef] [PubMed]
74. Novkovic, M.; Onder, L.; Bocharov, G.; Ludewig, B. Graph Theory-Based Analysis of the Lymph Node Fibroblastic Reticular Cell Network. *Meth. Mol. Biol.* **2017**, *1591*, 43–57. [CrossRef]

75. Novkovic, M.; Onder, L.; Cupovic, J.; Abe, J.; Bomze, D.; Cremasco, V.; Scandella, E.; Stein, J.V.; Bocharov, G.; Turley, S.J.; et al. Topological Small-World Organization of the Fibroblastic Reticular Cell Network Determines Lymph Node Functionality. *PLoS Biol.* **2016**, *14*, e1002515. [CrossRef]
76. Novkovic, M.; Onder, L.; Bocharov, G.; Ludewig, B. Topological Structure and Robustness of the Lymph Node Conduit System. *Cell Rep.* **2020**, *30*, 893–904. [CrossRef]
77. Licup, A.J.; Munster, S.; Sharma, A.; Sheinman, M.; Jawerth, L.M.; Fabry, B.; Weitz, D.A.; MacKintosh, F.C. Stress controls the mechanics of collagen networks. *Proc. Natl. Acad. Sci. USA* **2015**, *112*, 9573–9578. [CrossRef]
78. Lindstrom, S.B.; Vader, D.A.; Kulachenko, A.; Weitz, D.A. Biopolymer network geometries: Characterization, regeneration, and elastic properties. *Phys. Rev. E.* **2010**, *82*, 051905. [CrossRef]
79. Stein, A.M.; Vader, D.A.; Jawerth, L.M.; Weitz, D.A.; Sander, L.M. An algorithm for extracting the network geometry of three-dimensional collagen gels. *J. Microsc.* **2008**, *232*, 463–475. [CrossRef]
80. Harjanto, D.; Zaman, M.H. Modeling extracellular matrix reorganization in 3D environments. *PLoS ONE* **2013**, *8*, e52509. [CrossRef]
81. Graw, F.; Balagopal, A.; Kandathil, A.J.; Ray, S.C.; Thomas, D.L.; Ribeiro, R.M.; Perelson, A.S. Inferring viral dynamics in chronically HCV infected patients from the spatial distribution of infected hepatocytes. *PLoS Computat. Biol.* **2014**, *10*, e1003934. [CrossRef]
82. Beltman, J.B.; Henrickson, S.E.; von Andrian, U.H.; de Boer, R.J.; Maree, A.F. Towards estimating the true duration of dendritic cell interactions with T cells. *J. Immunol. Methods* **2009**, *347*, 54–69. [CrossRef] [PubMed]
83. Beltman, J.B.; Maree, A.F.; de Boer, R.J. Analysing immune cell migration. *Nat. Rev. Immunol.* **2009**, *9*, 789–798. [CrossRef] [PubMed]
84. Howat, T.J.; Barreca, C.; O'Hare, P.; Gog, J.R.; Grenfell, B.T. Modelling dynamics of the type I interferon response to in vitro viral infection. *J. R. Soc. Interface* **2006**, *3*, 699–709. [CrossRef] [PubMed]
85. Nelson, C.A.; Wilen, C.B.; Dai, Y.N.; Orchard, R.C.; Kim, A.S.; Stegeman, R.A.; Hsieh, L.L.; Smith, T.J.; Virgin, H.W.; Fremont, D.H. Structural basis for murine norovirus engagement of bile acids and the CD300lf receptor. *Proc. Natl. Acad. Sci. USA* **2018**, *115*, E9201–E9210. [CrossRef]
86. Schupp, A.K.; Trilling, M.; Rattay, S.; Le-Trilling, V.T.K.; Haselow, K.; Stindt, J.; Zimmermann, A.; Haussinger, D.; Hengel, H.; Graf, D. Bile Acids Act as Soluble Host Restriction Factors Limiting Cytomegalovirus Replication in Hepatocytes. *J. Virol.* **2016**, *90*, 6686–6698. [CrossRef] [PubMed]
87. Veloso Alves Pereira, I.; Buchmann, B.; Sandmann, L.; Sprinzl, K.; Schlaphoff, V.; Dohner, K.; Vondran, F.; Sarrazin, C.; Manns, M.P.; Pinto Marques Souza de Oliveira, C.; et al. Primary biliary acids inhibit hepatitis D virus (HDV) entry into human hepatoma cells expressing the sodium-taurocholate cotransporting polypeptide (NTCP). *PLoS ONE* **2015**, *10*, e0117152. [CrossRef] [PubMed]
88. Munch, J.; Rucker, E.; Standker, L.; Adermann, K.; Goffinet, C.; Schindler, M.; Wildum, S.; Chinnadurai, R.; Rajan, D.; Specht, A.; et al. Semen-derived amyloid fibrils drastically enhance HIV infection. *Cell* **2007**, *131*, 1059–1071. [CrossRef]
89. Bart, S.M.; Cohen, C.; Dye, J.M.; Shorter, J.; Bates, P. Enhancement of Ebola virus infection by seminal amyloid fibrils. *Proc. Natl. Acad. Sci. USA* **2018**, *115*, 7410–7415. [CrossRef]
90. Sanjuan, R. Collective Infectious Units in Viruses. *Trends Microbiol.* **2017**, *25*, 402–412. [CrossRef]
91. Andreu-Moreno, I.; Sanjuan, R. Collective Infection of Cells by Viral Aggregates Promotes Early Viral Proliferation and Reveals a Cellular-L0evel Allee Effect. *Curr. Biol.* **2018**, *28*, 3212–3219. [CrossRef] [PubMed]
92. Zotova, A.; Atemasova, A.; Pichugin, A.; Filatov, A.; Mazurov, D. Distinct Requirements for HIV-1 Accessory Proteins during Cell Coculture and Cell-Free Infection. *Viruses* **2019**, *11*, 390. [CrossRef] [PubMed]
93. Malbec, M.; Porrot, F.; Rua, R.; Horwitz, J.; Klein, F.; Halper-Stromberg, A.; Scheid, J.F.; Eden, C.; Mouquet, H.; Nussenzweig, M.C.; et al. Broadly neutralizing antibodies that inhibit HIV-1 cell to cell transmission. *J. Exp. Med.* **2013**, *210*, 2813–2821. [CrossRef] [PubMed]
94. Zhong, P.; Agosto, L.M.; Ilinskaya, A.; Dorjbal, B.; Truong, R.; Derse, D.; Uchil, P.D.; Heidecker, G.; Mothes, W. Cell-to-Cell Transmission Can Overcome Multiple Donor and Target Cell Barriers Imposed on Cell-Free HIV. *PLoS ONE* **2013**, *8*, e53138. [CrossRef]
95. Jolly, C.; Booth, N.J.; Neil, S.J. Cell-cell spread of human immunodeficiency virus type 1 overcomes tetherin/BST-2-mediated restriction in T cells. *J. Virol.* **2010**, *84*, 12185–12199. [CrossRef]
96. Dufloo, J.; Bruel, T.; Schwartz, O. HIV-1 cell-to-cell transmission and broadly neutralizing antibodies. *Retrovirology* **2018**, *15*, 51. [CrossRef]

97. Chen, P.; Chen, B.K.; Mosoian, A.; Hays, T.; Ross, M.J.; Klotman, P.E.; Klotman, M.E. Virological synapses allow HIV-1 uptake and gene expression in renal tubular epithelial cells. *J. Am. Soc. Nephrol.* **2011**, *22*, 496–507. [CrossRef]
98. Lee, J.H.; Schierer, S.; Blume, K.; Dindorf, J.; Wittki, S.; Xiang, W.; Ostalecki, C.; Koliha, N.; Wild, S.; Schuler, G.; et al. HIV-Nef and ADAM17-Containing Plasma Extracellular Vesicles Induce and Correlate with Immune Pathogenesis in Chronic HIV Infection. *EBioMedicine* **2016**, *6*, 103–113. [CrossRef]
99. Ostalecki, C.; Wittki, S.; Lee, J.H.; Geist, M.M.; Tibroni, N.; Harrer, T.; Schuler, G.; Fackler, O.T.; Baur, A.S. HIV Nef- and Notch1-dependent Endocytosis of ADAM17 Induces Vesicular TNF Secretion in Chronic HIV Infection. *EBioMedicine* **2016**, *13*, 294–304. [CrossRef]
100. Mazurov, D.; Ilinskaya, A.; Heidecker, G.; Lloyd, P.; Derse, D. Quantitative comparison of HTLV-1 and HIV-1 cell-to-cell infection with new replication dependent vectors. *PLoS Pathog.* **2010**, *6*, e1000788. [CrossRef]
101. Shunaeva, A.; Potashnikova, D.; Pichugin, A.; Mishina, A.; Filatov, A.; Nikolaitchik, O.; Hu, W.S.; Mazurov, D. Improvement of HIV-1 and Human T Cell Lymphotropic Virus Type 1 Replication-Dependent Vectors via Optimization of Reporter Gene Reconstitution and Modification with Intronic Short Hairpin RNA. *J. Virol.* **2015**, *89*, 10591–10601. [CrossRef] [PubMed]

12

The Cellular Protein CAD is Recruited into Ebola Virus Inclusion Bodies by the Nucleoprotein NP to Facilitate Genome Replication and Transcription

Janine Brandt, Lisa Wendt, Bianca S. Bodmer, Thomas C. Mettenleiter and Thomas Hoenen *

Institute of Molecular Virology and Cell Biology, Friedrich-Loeffler-Institut, 17493 Greifswald-Insel Riems, Germany; janine.brandt@fli.de (J.B.); lisa.wendt@fli.de (L.W.); bianca.bodmer@fli.de (B.S.B.); ThomasC.Mettenleiter@fli.de (T.C.M.)

* Correspondence: thomas.hoenen@fli.de

Abstract: Ebola virus (EBOV) is a zoonotic pathogen causing severe hemorrhagic fevers in humans and non-human primates with high case fatality rates. In recent years, the number and extent of outbreaks has increased, highlighting the importance of better understanding the molecular aspects of EBOV infection and host cell interactions to control this virus more efficiently. Many viruses, including EBOV, have been shown to recruit host proteins for different viral processes. Based on a genome-wide siRNA screen, we recently identified the cellular host factor carbamoyl-phosphate synthetase 2, aspartate transcarbamylase, and dihydroorotase (CAD) as being involved in EBOV RNA synthesis. However, mechanistic details of how this host factor plays a role in the EBOV life cycle remain elusive. In this study, we analyzed the functional and molecular interactions between EBOV and CAD. To this end, we used siRNA knockdowns in combination with various reverse genetics-based life cycle modelling systems and additionally performed co-immunoprecipitation and co-immunofluorescence assays to investigate the influence of CAD on individual aspects of the EBOV life cycle and to characterize the interactions of CAD with viral proteins. Following this approach, we could demonstrate that CAD directly interacts with the EBOV nucleoprotein NP, and that NP is sufficient to recruit CAD into inclusion bodies dependent on the glutaminase (GLN) domain of CAD. Further, siRNA knockdown experiments indicated that CAD is important for both viral genome replication and transcription, while substrate rescue experiments showed that the function of CAD in pyrimidine synthesis is indeed required for those processes. Together, this suggests that NP recruits CAD into inclusion bodies via its GLN domain in order to provide pyrimidines for EBOV genome replication and transcription. These results define a novel mechanism by which EBOV hijacks host cell pathways in order to facilitate genome replication and transcription and provide a further basis for the development of host-directed broad-spectrum antivirals.

Keywords: Ebola virus; filovirus; inclusion bodies; CAD; pyrimidine synthesis

---

1. Introduction

Ebola virus (EBOV) is a zoonotic pathogen belonging to the genus *Ebolavirus* within the order *Filoviridae*, and is the causative agent of severe hemorrhagic fevers in humans and non-human primates with high case fatality rates [1,2]. Increasing numbers of EBOV outbreaks in Africa highlight the importance of understanding the molecular mechanisms of the EBOV life cycle and virus-host cell interactions better in order to develop new countermeasures against this virus. EBOV possesses a non-segmented single-stranded RNA genome of negative polarity that forms a helical nucleocapsid in the center of virions together with the ribonucleoprotein (RNP) complex proteins. During assembly of the nucleocapsid, the RNA genome is tightly coated with the viral nucleoprotein (NP), which protects it from degradation and recognition by the cellular immune response [3]. During EBOV infection, NP-associated RNA genomes serve as templates for mRNA transcription and genome replication [4]. For viral replication, NP interacts with the polymerase cofactor VP35, which acts as a linker between NP and the RNA-dependent RNA polymerase L [5]. NP, VP35, and L are sufficient to facilitate EBOV genome replication, while for viral transcription the transcriptional activator VP30 is additionally required [6,7]. EBOV replication and transcription takes place in cytoplasmic inclusion bodies, which represent a characteristic feature of EBOV infections in cells [8,9]. Their formation can be driven by the expression of NP alone [5,10,11]. Due to the limited number of viral genes, successful genome replication and transcription is highly dependent on host cell factors, which play an important role during the EBOV life cycle. For instance, the host factor STAU1 has been shown to interact with multiple EBOV RNP components, and to redistribute into NP-induced or virus-induced inclusion bodies, suggesting that STAU1 plays a crucial role during viral RNA synthesis by facilitating the interaction between the viral genome and RNP proteins [12]. EBOV has also been shown to recruit SMYD3 into inclusion bodies, which modulates NP-VP30 interaction and enhances mRNA transcription [13]. Similarly, RBBP6 was found to influence EBOV replication by disrupting the interaction between NP and VP30 [14]. Importin-α7 was described as being required for the efficient formation of inclusion bodies [15]. Furthermore, several cellular kinases and phosphatases are known to localize in inclusion bodies to support EBOV replication and transcription [16–18]. Finally, we previously showed that EBOV NP recruits the nuclear RNA export factor 1 (NXF1) into inclusion bodies to facilitate viral mRNA export from these structures into the cytoplasm [19]. Despite this recent progress in our understanding of the interplay between host factors and EBOV, there remains a considerable need to identify and, more importantly, characterize further host factors required for EBOV replication to identify novel targets for antiviral drug development.

We previously performed a genome-wide siRNA screen using a minigenome system to identify potential host-directed targets [20]. In this system, a minigenome, i.e., a truncated version of the EBOV genome lacking all viral open reading frames (ORF) and consisting of a reporter gene (e.g., a luciferase or green fluorescent protein) flanked by the viral non-coding terminal leader and trailer regions, is expressed from a plasmid in mammalian cells together with the plasmids encoding the viral RNP proteins [6]. For initial transcription of the minigenome RNAs from the minigenome-encoding plasmids most existing EBOV minigenome systems use a T7 RNA polymerase (T7) promoter, and therefore require expression of T7 polymerase, which is usually provided via a T7-expressing plasmid that is cotransfected with the plasmids encoding the RNP proteins [6,21]. However, recently, an EBOV minigenome system using the cellular RNA polymerase II (Pol-II) for initial minigenome RNA transcription has also been established and shown to be more efficient at least in some cell types [22]. After initial transcription and encapsidation by RNP proteins, minigenome RNAs are recognized as authentic templates by the viral polymerase due to their leader and trailer regions, and are replicated and transcribed into mRNAs, which results in expression of the reporter protein. Thus, minigenome assays allow us to study viral genome replication and transcription, as well as viral protein expression, outside of maximum containment laboratories, simplifying the identification of host factors involved in these processes. By using this system, we recently identified the trifunctional protein carbamoyl-phosphate synthetase 2, aspartate transcarbamylase, and dihydroorotase (CAD) as being important for the EBOV life cycle [20].

CAD is an important component of the pyrimidine pathway that catalyzes the first three steps during the de novo biosynthesis of pyrimidine nucleotides using its four distinct enzymatic domains [23–25]. The first domain, glutaminase (GLN), initiates the pathway by catalyzing the hydrolysis of glutamine. This is followed by the synthesis of carbamoyl phosphate facilitated by the carbamoyl-phosphate synthetase (CPS). Carbamoyl phosphate is in turn the substrate for the aspartate transcarbamylase (ATC), which catalyzes the reaction of aspartate with carbamoyl phosphate to carbamoyl aspartate [26,27]. Finally, carbamoyl aspartate is converted to dihydroorotate by dihydroorotase (DHO) [28]. In response to cell growth and proliferation, CAD activity is upregulated by phosphorylation through MAP kinases at position Thr-456, while in resting cells Thr-456 is dephosphorylated [29]. Furthermore, CAD is known to primarily localize in the cytoplasm of resting cells, but in response to cell growth and Thr-456 phosphorylation a small fraction is translocated into nuclear compartments, suggesting a cellular function of CAD in the nucleus [30,31]. However, little is known about the role of CAD during virus infection, and particularly the role of CAD in the EBOV life cycle still needed to be further analyzed. Therefore, we wanted to characterize the interaction of CAD with EBOV on both a biochemical and functional level. Based on our results, we suggest that CAD is important for both genome replication and transcription due to its function in pyrimidine synthesis and that it is recruited into NP-induced and virus-induced inclusion bodies to facilitate the de novo biosynthesis of pyrimidine nucleotides.

2. Materials and Methods

2.1. Cell Lines

Human embryonic kidney cells (HEK 293T, Collection of Cell Lines in Veterinary Medicine CCLV-RIE 1018), African green monkey kidney cells (Vero E6, kindly provided by Stephan Becker, Philipps University Marburg), and human hepatocellular carcinoma cells (Huh7, kindly provided by Stephan Becker, Philipps University Marburg) were cultured in Dulbecco's modified Eagle's medium (DMEM; Thermo Fisher Scientific, Darmstadt, Germany) supplemented with 10% fetal calf serum (FCS), 100 U/mL penicillin, 100 µg/mL streptomycin (PS; Thermo Fisher Scientific), and 1× GlutaMAX (Thermo Fisher Scientific). All cells were incubated at 37°C and 5% CO_2.

2.2. Plasmids and Cloning

Minigenome assay components, including expression plasmids coding for the EBOV RNP proteins, T7 polymerase, firefly luciferase, and a classical T7-driven monocistronic minigenome (pT7-1cis-EBOV-vRNA-nLuc) have been previously described [20,32]. A NanoLuc luciferase-expressing T7-driven replication-deficient minigenome was cloned from a classical minigenome expressing NanoLuc luciferase as a reporter by deletion of 55 nucleotides (nt) in the antigenomic replication promoter as previously described [32]. Based on this, a Pol-II-driven replication-deficient minigenome was generated by PCR to amplify a linear version of the replication-deficient minigenome flanked by hammerhead and hepatitis delta virus ribozymes using primers #4571 (5′-AGC TTA CGT GAC TAC TTC CTT CGG ATG CCC AGG TCG GAC CGC G-3′) and #4572 (5′-GAC CGG TAG AAA ACT GAT GAG TCC GTG AGG ACG AAA CGG AGT CTA GAC TCC GTC TTT TCC AGG AAT CCT TTT TGC AAC GTT TAT TCT G-3′). The linearized construct was subsequently inserted into pCAGGS. The CAD gene was cloned from 293T cells into pCAGGS, and deletion mutants and domains of CAD were then generated using PCR-based approaches. All constructs were first cloned into pCAGGS, followed by subcloning into a pCAGGS plasmid encoding an N-terminal FLAG/HA-tag (DYKDDDDKLDGGYPYDVPDYA) immediately upstream of a BsmBI cloning site, allowing a seamless insertion of the open reading frame of interest. The expression plasmid for N-terminally myc-tagged VP35 was constructed by cloning a myc-tag (EQKLISEEDL) immediately before the VP35 ORF. Detailed cloning strategies are available on request.

2.3. Antibodies

The anti-FLAG (clone M2) antibody used for immunofluorescence analyses (IFA), co-immuno precipitation (coIP), and Western blot analyses was purchased from Sigma-Aldrich (Munich, Germany) [F1804], and the anti-c-myc antibody used for IFA analysis was obtained from Thermo Fisher Scientific [A-21281]. Primary antibodies against NP (rabbit anti-EBOV NP polyclonal antibody), GAPDH (mouse anti-GAPDH clone 0411), and CAD (rabbit anti-CAD clone EP710Y) were ordered from IBT Bioservices (San Jose, USA; anti-NP [0301-012]), Santa Cruz (Heidelberg, Germany; anti-GAPDH [sc47724]), or Abcam (Cambridge, UK; anti-CAD [ab40800]). Secondary antibodies used for IFA analysis against mouse (Alexa Fluor 488 anti-mouse [A-11029]), rabbit (Alexa Fluor 568 anti-rabbit [A-11036]), and chicken IgY (Alexa Fluor 647 anti-chicken [A-21449]) were obtained from Thermo Fisher Scientific. For Western blotting, secondary antibodies against mouse (IRDye 680RD anti-mouse [926-68070]) and rabbit IgG (IRDye 800CW anti-rabbit [926-68071]) were purchased from Li-COR (Bad Homburg, Germany), while anti-mouse IgG (Kappa light chain) Alexa Fluor 680 [115-625-174] used for coIP analyses was ordered from Dianova (Hamburg, Germany).

2.4. Viruses

Zaire ebolavirus rec/COD/1976/Mayinga-rgEBOV (GenBank accession number KF827427.1), which is identical in sequence to the EBOV Mayinga isolate with the exception of four silent mutations as genetic markers [33], was used for all infection experiments. rgEBOV was propagated in VeroE6 cells and virus titers were determined by 50% tissue culture infectious dose (TCID50) assay. All work with the infectious virus was performed under BSL-4 conditions at the Friedrich-Loeffler-Institut (Federal Research Institute of Animal Health, Greifswald Insel-Riems, Germany) following approved standard operating procedures.

2.5. Chemical Compounds

100mM uridine or cytidine (both Sigma-Aldrich) stock solutions were prepared in dimethyl sulfoxide (DMSO) and further diluted in cell culture medium. Diluted pyrimidines or DMSO corresponding to 1% of the supernatant volume in 12-well plates was added to the cells at the time of transfection and after medium changes. All concentrations indicated in the figures are final concentrations.

2.6. siRNA Knockdown with EBOV Minigenomes and Pyrimidine Complementation

For siRNA knockdown of endogenous CAD, 293T cells were reverse transfected (i.e., transfected in suspension and subsequently seeded into plates) with 12 pmol pre-designed silencer select siRNAs (CAD-siRNA#1: s2320 [5′-GAG GGU CUC UUC UUA AGU A-3′]; CAD-siRNA#2: 117891 [5′-GCU AGC UGA GAA AAA CUU U-3′]; Negative Control siRNA #2; all Thermo Fisher Scientific) or a self-designed EBOV-anti-L siRNA [5′-UUU AUA UAC AGC UUC GUA CUU-3′] ordered from Eurofins Genomics (Ebersberg, Germany). Transfection was performed in 12-well plates using Lipofectamine RNAiMax (Thermo Fisher Scientific) following the manufacturer's instructions. 48 h post-siRNA transfection, the cells were transfected using Transit-LT1 (Mirus Bio LLC, Madison, USA) with all minigenome assay components, i.e., pCAGGS-based expression plasmids for NP (62.5 ng), VP35 (62.5 ng), VP30 (37.5 ng), L (500 ng), codon-optimized T7-polymerase (125 ng), firefly luciferase (as a control, 125 ng), and the T7-driven monocistronic minigenome (pT7-EBOV-1cis-vRNA-nLuc; 125 ng). For analyses of vRNA and mRNA levels the control firefly luciferase was replaced with GFP (200 ng), and for the replication-deficient minigenome assay a Pol-II-driven replication-deficient minigenome (pCAGGS-EBOV-1cis-vRNA-nLuc-RdM) was used. Transfections were performed using Transit LT1 as previously described [32]. All samples were harvested 48 h post-transfection for either determination of reporter activity or RNA isolation (see below). For measuring the luciferase activity, cells were lysed for 10 min in 1x Lysis Juice (PJK, Kleinblittersdorf, Germany) at room temperature and lysates were cleared of cell debris by centrifugation for 3 min at 10,000× g. Then, 40 µL of the cleared lysates were added to either 40 µL of Beetle Juice (PJK) or NanoGlo Luciferase Assay Reagent (Promega, Madison, USA) in opaque 96-well plates and luminescence was measured using a Glomax Multi (Promega) microplate reader. NanoLuc luciferase activities were normalized to firefly luciferase activities.

2.7. RNA Isolation and RT-qPCR

RNA isolation from minigenome cell lysates was performed following the manufacturer's instructions using a NucleoSpin RNA kit (Machery-Nagel, Düren, Germany). After RNA purification, all samples were treated with DNase (TURBO DNA-free kit; Thermo Fisher Scientific) following the manufacturer's instructions to avoid plasmid contamination. For cDNA generation, RNA samples were incubated with an oligo(dT)-primer for mRNA quantification or with a strand-specific primer (5′-AGT GTG AGC TTC TAA AGC AAC C-3′) for vRNA quantification using the RevertAid Reverse Transcriptase (Thermo Fisher Scientific) following the manufacturer's instructions. The subsequent qPCR was performed using a PowerUp SYBR Green Master Mix (Thermo Fisher Scientific) with 1 μL of cDNA and primers targeting either the reporter gene (5′-TTC AGA ATC TCG GG GTG TCC-3′, 5′-CGT AAC CCC GTC GAT TAC CA-3′), or GFP as a control (5′-CTT GTA CAG CTC GTC CAT GC-3′, 5′-CGA CAA CCA CTA CCT GAG CAC-3′). Values for vRNA and mRNA levels were normalized to control GFP mRNA levels.

2.8. Immunofluorescence Analysis

Huh7 cells, which are more suitable for IFA than 293T cells, were seeded on coverslips in 12-well plates and transfected 24 h later with 500 ng pCAGGS-EBOV-NP and 500 ng pCAGGS-FLAG-HA-CAD (or CAD mutants) and, in selected experiments, additionally with 500 ng pCAGGS-myc-VP35 as indicated. For a mock control, cells were transfected with pCAGGS. Transfection was performed using polyethylenimine (Sigma-Aldrich) following the manufacturer's instructions. 48 h post-transfection, cells were fixed using 4% paraformaldehyde (Roth, Karlsruhe, Germany) in DMEM for 20 min and then treated with 1 M glycine (in phosphate-buffered saline^{++} (PBS with 0.9M Ca^{2+} and 0.5M Mg^{2+})) for 10 min. Then, cells were permeabilized with 0.1% Triton X-100 in PBS for another 10 min and incubated with 10% fetal calf serum (FCS) in PBS for 45 min. Primary antibodies (rabbit anti-EBOV-NP 1:500; mouse anti-FLAG 1:2500; chicken anti-myc 1:1200) were diluted in PBS with 10% FCS and cells were incubated for 1 h at room temperature with the prepared antibody solutions. Secondary antibodies (Alexa Fluor 488 anti-mouse 1:1200; Alexa Fluor 568 anti-rabbit 1:500; Alexa Fluor 647 anti-chicken 1:1200) were prepared as described for the primary antibodies. After 45 min of staining, cells were washed with PBS and water before mounting with ProLong Diamond Antifade mountant with 4′,6-diamidino-2-phenylindole (DAPI) (Thermo Fisher Scientific). Slides were analyzed by confocal laser scanning microscopy using a Leica SP5.

2.9. Infection of Transfected Huh7 Cells

To investigate the localization of CAD during EBOV infection, Huh7 cells were seeded in 8-well chambered slides (ibidi, Martinsried, Germany) and transfected as described above (immunofluorescence analysis) with 500 ng pCAGGS-FLAG/HA-CAD. At 48 h post-transfection, the transfected cells were infected with EBOV at an MOI of 1, and the samples were fixed 16 h post-infection in 10% formalin twice overnight prior to removal from the BSL4 facility and immunofluorescence analysis.

2.10. Co-Immunoprecipitation of Viral Proteins

CoIPs were performed as previously described [19]. Briefly, 293T cells were seeded in 6-well plates and transfected with expression plasmids encoding FLAG/HA-tagged CAD and EBOV-NP using Transit LT-1 (Mirus Bio LLC) following the manufacturer's instructions. The medium was changed after 24 h and the cells were harvested 48 h post-transfection. For coIP, cells were lysed in 1 mL coIP lysis buffer (1% NP-40; 50 mM Tris pH 7.4; 167 mM NaCl in water) with protease inhibitors (cOmplete; Roche, Mannheim, Germany). To investigate a possible RNA dependency of the interaction between CAD and NP, 100 μg/mL RNase A (Machery-Nagel) were added to the samples. Subsequently, the samples were incubated rotating at 15 RPM for 2 h at 4 °C. Then, 150 μL of the cleared lysates were taken as an input control (representing a sixth of the complete pre-immune lysate and 20% of the sample used for immunoprecipitation) and subjected to acetone precipitation. The remaining 750 μL of cell lysate were mixed with the prepared bead-antibody solution (Dynabeads Protein G, Thermo Fisher Scientific; 1 μL anti-FLAG M2 antibody per 10 μL beads). Immunoprecipitation was performed for 10 min, as recommended by the manufacturer, at room temperature and rotation at 15 RPM. Then, samples were transferred to new tubes and boiled for 10 min at 99 °C. Input and coIP samples were analyzed by SDS-PAGE and Western blotting.

2.11. Western Blotting

For validation of CAD knockdown efficiency and analyses of coIP input and lysates, samples were subjected to SDS-PAGE and Western blotting as previously described [34]. FLAG-tagged CAD was detected using a monoclonal anti-FLAG antibody (1:2000), while NP, wild type CAD, and GAPDH were detected using anti-NP (1:1000), anti-CAD (1:250), and anti-GAPDH (1:1000) antibodies. As secondary antibodies, 680RD-coupled goat-anti-mouse, goat-anti-mouse-Alexa Fluor 680, and 800CW-coupled goat-anti-rabbit antibodies (1:14000) were used. Fluorescent signals were detected and quantified using an Odyssey CLx infrared imaging system (Li-Cor Biosciences). For knockdown quantification, CAD signals were normalized to GAPDH signals.

2.12. Statistical Analyses

One-way ANOVA with a Dunnett's multiple comparisons test was performed using the GraphPad Prism 8.1.0 software.

3. Results

3.1. CAD Knockdown Affects Both EBOV Genome Replication and Transcription

Using a genome-wide siRNA screen, we previously identified CAD to be important for EBOV RNA synthesis and/or viral protein expression [20]. However, since only the effect of CAD knockdown on the sum of these processes had been tested, we now analyzed the role of CAD on individual aspects of the EBOV life cycle. As a first step, we assessed the efficiency of endogenous CAD knockdown using two different siRNAs via quantitative Western blotting, which revealed a 60% to 80% reduction in endogenous CAD expression levels for the two siRNAs (Figure 1A,B).

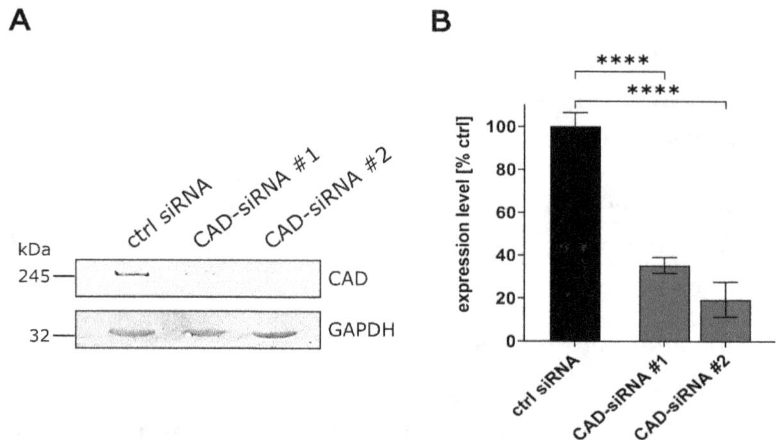

Figure 1. Quantification of CAD knockdown. (**A**) Analysis of CAD knockdown. 293T cells were transfected with siRNAs targeting CAD (CAD-siRNA), or a negative control (ctrl siRNA). The cells were harvested 48 h post-transfection and the lysates were subjected to SDS-PAGE and Western blotting. (**B**) Quantification of CAD knockdown. The Western blot signals for CAD knockdown (as shown in Figure 1A) were measured and normalized to the GAPDH signals. The negative control (ctrl siRNA) was set to 100% and the efficiency of CAD knockdown was calculated (**** $p \leq 0.0001$).

Next, we performed a classical minigenome assay (Figure 2A) in connection with an siRNA knockdown of CAD. As previously shown, knockdown of CAD led to a 40 to 53-fold reduction in reporter activity, verifying an influence of CAD on EBOV viral RNA synthesis and protein expression (Figure 2B) [20]. In order to identify whether CAD knockdown affects transcription and/or protein expression independent of replication, we next used a replication-deficient minigenome system [32]. In contrast to a replication-competent minigenome, the replication-deficient minigenome lacks 55 nt in the antigenomic replication promoter leading to a block of minigenome vRNA replication, while minigenome transcription still takes place [32]. However, when using this system, which is based on T7-driven initial transcription of minigenomes, we observed a very low dynamic range between our controls, which made it difficult to evaluate a possible influence of CAD knockdown (Figure S1). Therefore, in order to increase the dynamic range of this system, we generated a Pol-II-driven replication-deficient minigenome that resulted in a ~10-fold higher dynamic range (Figure S1). Using this system, CAD knockdown resulted in a clear reduction in reporter activity, indicating that CAD is important for EBOV transcription and/or protein expression independent of viral genome replication (Figure 2C).

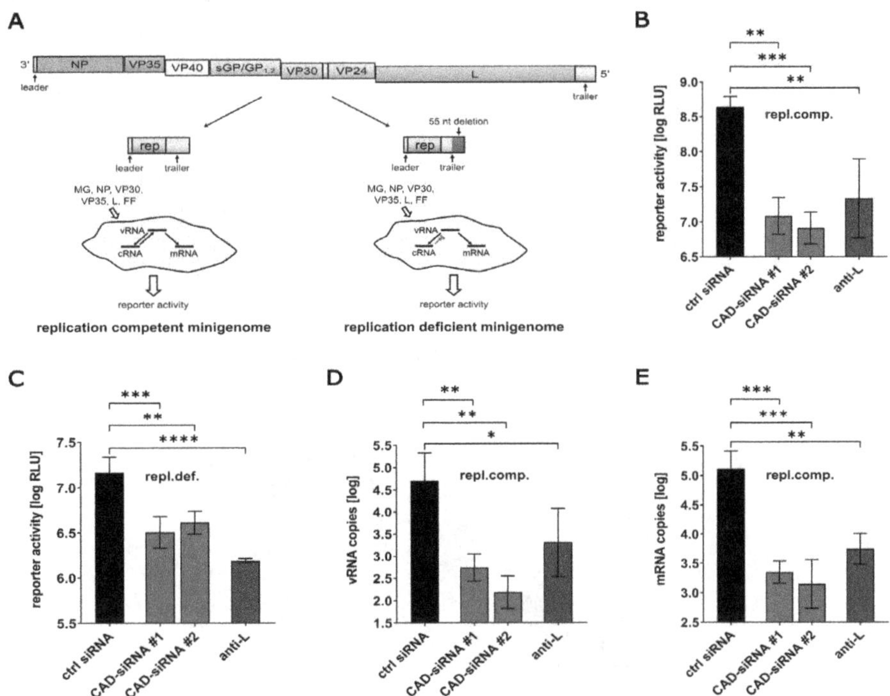

Figure 2. Influence of CAD knockdown on the Ebola virus life cycle. (**A**) Replication-competent and -deficient minigenome systems. The full-length genome structure of EBOV, as well as replication-competent and -deficient minigenomes derived from this full-length genome, are shown. Abbrevations: MG: minigenome, rep: reporter; FF: Firefly luciferase. Figure modified from [35] under CC BY 4.0 license. (**B**) Influence of CAD knockdown on EBOV RNA synthesis. 293T cells were transfected with siRNAs targeting either CAD (CAD-siRNA), EBOV-L (anti-L), or a negative control (ctrl siRNA). 48 h post-transfection, cells were transfected with all the components required for a replication-competent minigenome assay (repl.comp.). Another 48 h later, cells were harvested and the reporter activity was measured. (**C**) Analysis of CAD knockdown on EBOV transcription and gene expression. 293T cells were transfected with siRNAs targeting either CAD (CAD-siRNA), EBOV-L (anti-L), or a negative control (ctrl siRNA). 48 h post-transfection, cells were transfected with all the components required for a replication-deficient minigenome assay (repl.def.). Another 48 h later, cells were harvested and the reporter activity was measured. (**D**) Impact of CAD knockdown on EBOV replication. Cells were treated as described in 2B. After cell harvesting, RNA was extracted from the cell lysates and RT-qPCR for vRNA was performed. (**E**) Influence of CAD knockdown on EBOV mRNA levels. Cells were treated as described in 2B. After cell harvesting, RNA was extracted from cell lysates and RT-qPCR for mRNA was performed. The means and standard deviations of 3 independent experiments are shown for each panel. Asterisks indicate p-values from a one-way ANOVA (* $p \leq 0.05$; ** $p \leq 0.01$; *** $p \leq 0.001$; **** $p \leq 0.0001$; ns: $p > 0.05$).

To further dissect the influences of CAD on viral genome replication, mRNA transcription, and later steps of viral protein expression, we performed classical minigenome assays in the context of an siRNA knockdown of CAD and measured vRNA and mRNA levels in cell lysates using RT-qPCR. For this, we used either an oligo-dT primer for reverse transcription of mRNAs, or a strand-specific primer for reverse transcription of vRNA, followed by qPCR against the reporter gene. CAD siRNA-treated cells showed a strong reduction in both vRNA and mRNA levels in comparison to the control cells, demonstrating that CAD is important for both EBOV transcription and viral genome replication (Figure 2D,E).

3.2. The Effect of CAD Knockdown Can Be Compensated for by Exogenous Pyrimidines

As CAD is an important component for pyrimidine synthesis [23], we wanted to investigate the effect of providing exogenous pyrimidines on EBOV transcription and replication during siRNA knockdown of CAD. To this end, we performed an siRNA-mediated knockdown of CAD with EBOV minigenomes and treated the cells with 1 mM of either uridine or cytidine. Complementation of uridine resulted in reporter activities similar to the positive controls, indicating that the effect of CAD knockdown on EBOV genome replication and transcription is due to a lack of pyrimidines (Figure 3). When providing cytidine, a similar rescue effect was seen, albeit less pronounced, possibly because cytidine is not metabolized into uridine, whereas exogenous uridine can be metabolized into cytidine during natural pyrimidine synthesis.

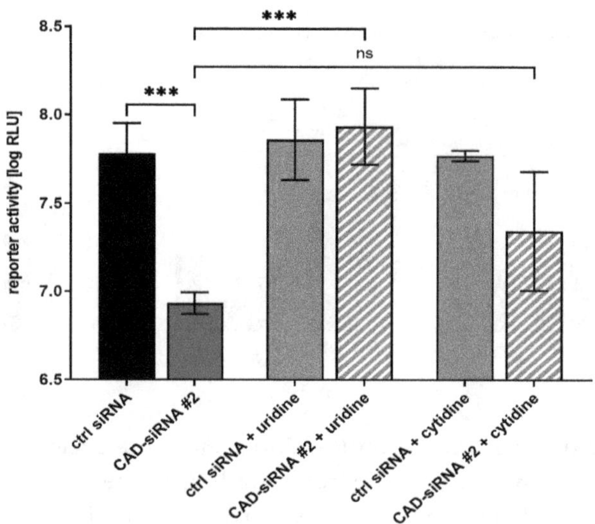

Figure 3. Supplementation of pyrimidines compensates for the effect of CAD knockdown. 293T cells were transfected with siRNAs targeting CAD (CAD-siRNA) or a negative control (ctrl siRNA). 48 h post-transfection, the cells were transfected with all the components required for a replication-competent minigenome assay and treated with 1 mM pyrimidines, either uridine or cytidine. Another 48 h later, the cells were harvested and the reporter activity was measured. The means and standard deviations of 3 independent experiments are shown. Asterisks indicate p-values from a one-way ANOVA (*** $0.0001 < p \leq 0.001$; ns: $p > 0.05$).

3.3. CAD Colocalizes with NP-Induced Inclusion Bodies

Similar to other negative-sense RNA viruses, EBOV and in particular its nucleoprotein NP is known to induce the formation of cytoplasmic inclusion bodies, which are sites of viral genome replication and transcription [8,9]. Since we had shown that CAD is important for EBOV replication and transcription, we wanted to investigate whether the presence of inclusion bodies has an influence on the intracellular distribution of CAD, and in particular whether recruitment of CAD into NP-induced inclusion bodies can be detected. As previously reported, expression of only NP resulted in the formation of inclusion bodies, predominantly in the perinuclear region [5,10,11], while sole expression of CAD led to an even distribution throughout the cytoplasm, with small amounts of CAD present in the nucleus [30] (Figure 4A). During coexpression of NP and CAD we observed relocalization of CAD into NP-induced inclusion bodies (with clear accumulation in inclusion bodies in 70% of the cells, clear exclusion in 0%, and an unclear phenotype in 30%). When we additionally coexpressed VP35, which is involved in nucleocapsid formation during EBOV infection, together with NP [36], we observed a similar relocalization (Figure 4B). To confirm these results, we also performed experiments with infectious EBOV and stained the samples for NP as an inclusion body marker and CAD (Figure 5). Colocalization of CAD and inclusion bodies was still detectable, albeit not as apparent as under conditions of recombinant overexpression of NP and VP35. Taken together, these results suggest that CAD is recruited into viral inclusion bodies to provide sufficient amounts of pyrimidines for EBOV genome replication and transcription.

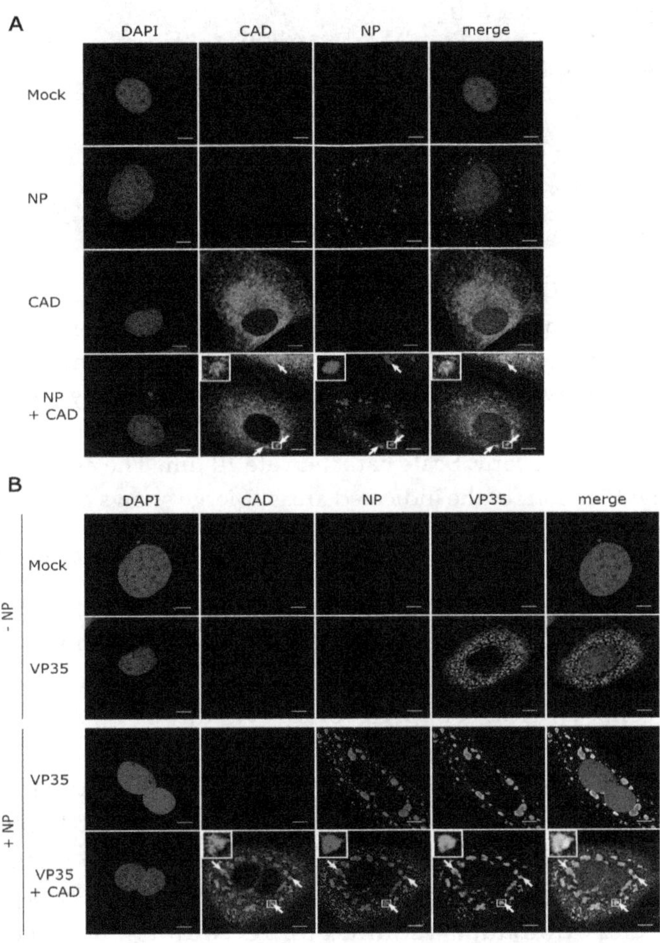

Figure 4. Recruitment of CAD into NP-induced inclusion bodies. (**A**) Colocalization between CAD and NP-induced inclusion bodies. Huh7 cells were transfected with plasmids encoding FLAG/HA-CAD and EBOV-NP as indicated. 48 h post-transfection, the cells were fixed with 4% paraformaldehyde and permeabilized with 0.1% Triton X-100. FLAG-tagged CAD (shown in green) was detected using an anti-FLAG antibody and NP (shown in red) was stained with anti-EBOV NP antibodies. (**B**) Recruitment of CAD into inclusion bodies occurs in the presence of VP35. Huh7 cells were transfected with plasmids encoding FLAG/HA-CAD, EBOV-NP, and myc-EBOV-VP35 as indicated. 48 h post-transfection, the cells were fixed with 4% PFA and permeabilized with 0.1% Triton X-100. FLAG-tagged CAD (shown in green) was detected using an anti-FLAG antibody, NP (shown in red) was stained with anti-EBOV NP antibodies, and myc-tagged VP35 (shown in turquoise) with an anti-myc antibody. The nuclei were stained with DAPI (shown in blue), and the cells were visualized by confocal laser scanning microscopy. The scale bars indicate 10 µm. The arrows point out colocalization, and the insets show magnifications of the indicated areas. Merge shows an overlay of all three channels.

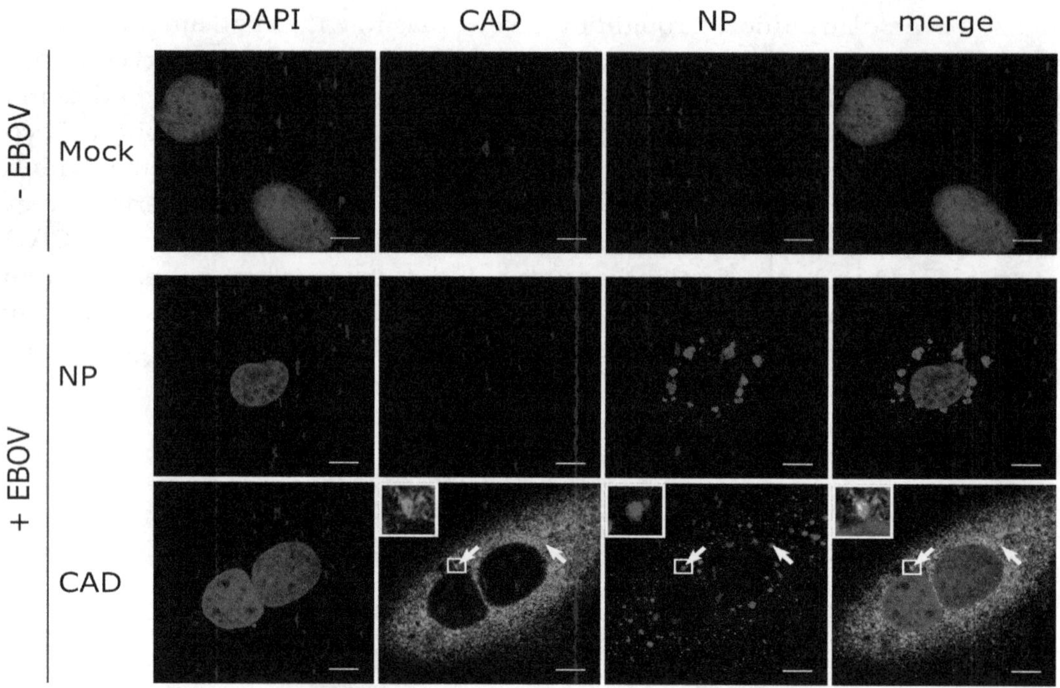

Figure 5. CAD localizes in EBOV inclusion bodies. Huh7 cells were transfected with a plasmid encoding FLAG/HA-CAD. 48 h post-transfection, the cells were infected with rgEBOV at an MOI of 1. After incubation for 16 h, the cells were fixed with 10% formalin and permeabilized with Triton X-100. CAD (shown in green) was detected with an anti-FLAG antibody and NP (shown in red) with an anti-NP antibody. The nuclei were stained with DAPI (shown in blue), and the cells were visualized by confocal laser scanning microscopy. Scale bars indicate 10 μm. The arrows point out colocalization, and the insets show magnifications of the indicated areas. Merge shows an overlay of all three channels.

3.4. The GLN Domain of CAD Is Required for its Accumulation in Inclusion Bodies

To assess the contribution of individual domains of CAD in its recruitment into NP-induced inclusion bodies, we focused on the GLN and the CPS domains. When we expressed deletion mutants lacking these domains, they showed a similar intracellular distribution compared to wild-type CAD when expressed alone in cells. During coexpression of NP and CAD-ΔCPS, we observed recruitment of this mutant into NP-driven inclusion bodies (with clear accumulation in inclusion bodies in 50% of the cells, clear exclusion in 0%, and an unclear phenotype in 50%), indicating that the CPS domain of CAD is not required for its accumulation in inclusion bodies (Figure 6). In stark contrast, when NP was expressed together with CAD-ΔGLN, colocalization with inclusion bodies was abolished (with clear accumulation in inclusion bodies in 0% of the cells, clear exclusion in 68%, and unclear phenotype in 32%), suggesting that the GLN domain is required for recruitment and accumulation in NP-induced inclusion bodies.

3.5. CAD Interacts with NP in an RNA-Independent Manner

As NP recruits CAD into EBOV inclusion bodies, we next assessed whether CAD interacts with NP. To this end, we performed coIP assays using FLAG-CAD expressed in the presence of NP by precipitating CAD with an anti-FLAG antibody and then detecting NP by Western blotting. We could readily co-precipitate NP with CAD, indicating that CAD is able to interact with NP (Figure 6). Because NP is an RNA-binding protein [37], we also tested whether this interaction between CAD and NP is RNA-dependent by treating the samples prior to coIP with RNase A. Under these conditions, we were still able to co-precipitate NP with CAD, demonstrating that the interaction between CAD and NP is not dependent on the presence of RNA (Figure 7).

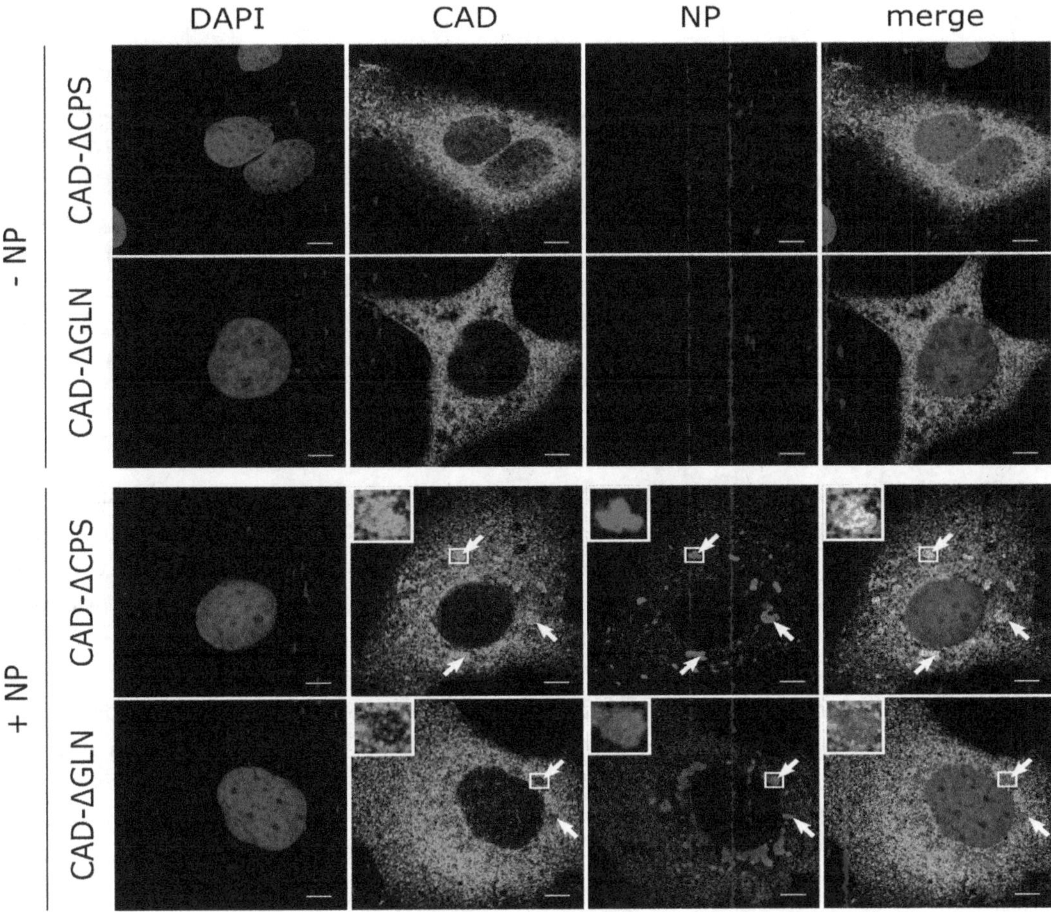

Figure 6. Recruitment of CAD deletion mutants into inclusion bodies. Huh7 cells overexpressing FLAG/HA-CAD-ΔGLN, FLAG/HA-CAD-ΔCPS and EBOV-NP, as indicated, were fixed with 4% PFA and permeabilized with 0.1% Triton X-100 48 h post-transfection. FLAG-tagged CAD (shown in green) was detected using an anti-FLAG antibody and NP (shown in red) was stained with EBOV anti-NP antibodies. The nuclei were stained with DAPI (shown in blue), and the cells were visualized by confocal laser scanning microscopy. Scale bars indicate 10 μm. The arrows point out inclusion bodies, and the insets show magnifications of the indicated areas. Merge shows an overlay of all three channels.

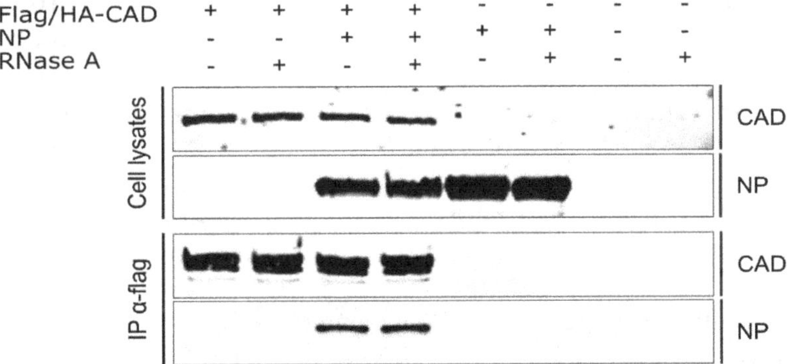

Figure 7. Interaction of CAD with NP. 293T cells were transfected with plasmids encoding FLAG/HA-CAD and EBOV-NP. 48 h post-transfection, the cells were lysed and treated with RNase A (100 μg/mL) or remained untreated. FLAG/HA-CAD was precipitated using anti-FLAG antibodies, and input and precipitates were analyzed via SDS-PAGE and Western blotting using anti-FLAG and anti-NP antibodies. In the CAD IP sample, several bands for CAD are visible, possibly due to posttranslational modifications that are not visible in the lysates because of the overall lower CAD signals in those samples.

4. Discussion

In this work, we identified CAD, an essential component of the de novo pyrimidine synthesis pathway, to be important for both EBOV genome replication and transcription, and demonstrated that the function of CAD in pyrimidine synthesis is responsible for this effect. Knockdown of CAD was also shown to affect replication and transcription of other viruses, e.g., hepatitis C viruses [38]. Furthermore, inhibitors of CAD, e.g., the antinucleoside N-phosphonacetyl-L-aspartate (PALA), which transiently inhibits the aspartate transcarbamylase activity of CAD, were effective in vitro against various viruses, including vaccinia virus and arenaviruses [39,40]. The fact that these compounds exhibit antiviral activity against a broad range of viruses qualifies CAD as a promising indirect antiviral target. However, whether PALA shows antiviral efficiency against EBOV remains to be investigated. Further, whether targeting viral RNA synthesis by inhibition of CAD will be synergistic with other inhibitors of EBOV RNA synthesis, such as remdesivir [41], will have to be addressed in future studies.

Our results are consistent with the fact that several pyrimidine synthesis inhibitors are effective against EBOV in vitro, underlining the importance of the pyrimidine pathway for these viruses [20,42]. Examples are the FDA-approved drug leflunomide and its active metabolite teriflunomide, as well as SW835, a racemic version of GSK983, which has been described to exhibit a broad-spectrum antiviral activity [20,42,43]. These compounds all impair de novo pyrimidine biosynthesis through inhibition of dihydroorotate dehydrogenase (DHODH), an enzyme downstream of CAD in the pyrimidine pathway. Interestingly, treatment with these inhibitors seems to have similar inhibitory effects on EBOV minigenome assays compared to the effect we observe for CAD knockdown, although CAD activity is not directly affected [20,42]. Provision of pyrimidines or upstream metabolites, e.g., orotic acid, reversed antiviral activity of all pyrimidine pathway inhibitors in EBOV minigenome assays, which is consistent with our observation that supplementation with pyrimidines restores reporter activity after CAD knockdown. Interestingly, inhibition of DHODH by using SW835 not only showed pyrimidine depletion, but also stimulated ISG (interferon-stimulated gene) expression, which contributes to the innate immune response [42]. However, currently, the mechanism behind this stimulation of the innate immune response by DHODH inhibitors remains incompletely understood and needs to be further analyzed. Further supporting the importance of CAD for the EBOV lifecycle is the fact that the de novo pyrimidine synthesis activity of CAD is a prerequisite for cell division, which has been suggested to be necessary for productive infection of cells with EBOV [44].

We were further able to show that CAD is recruited to EBOV inclusion bodies, which represent the site of EBOV replication and transcription [8,9]. Since we observed CAD recruitment into NP-induced inclusion bodies during expression of NP alone and detected an interaction of CAD with NP using CoIP studies, we suggest that this recruitment is mediated via an interaction of CAD with NP. So far, knowledge regarding direct interactions between CAD and the proteins of other viruses is limited, but Angeletti et al., showed that CAD recruits the preterminal protein (pTP) of adenoviruses to the site of adenovirus replication in the nuclear matrix via direct interaction. This interaction is believed to be required for anchorage of the adenovirus replication complex at the nuclear matrix in close proximity of the cellular factors required to segregate replicated and genomic viral DNA [45,46].

In the context of its cellular function, CAD has been shown to localize primarily in the cytoplasm, although small amounts can also be detected in the nucleus of dividing cells. Redistribution of CAD into nuclear compartments during cell growth and proliferation is believed to be in response to phosphorylation by MAP kinases at position Thr-456, which results in upregulation of the enzymatic activity of CAD [30]. Since NP is known to recruit a number of factors, including kinases and phosphatases, into inclusion bodies [16–18], it is possible that recruited CAD is activated in inclusion bodies in order to provide pyrimidines for EBOV replication and transcription. However, CAD lacking the CPS domain, which contains Thr-456, was still recruited into NP-induced inclusion bodies, excluding selective recruitment of Thr-456-phosphorylated and thus activated CAD into inclusion bodies.

Overall, we have shown that CAD is recruited into NP-induced and virus-induced inclusion bodies to provide sufficient amounts of pyrimidines for EBOV genome replication and transcription. Furthermore, we demonstrated that the GLN domain of CAD is required for recruitment into inclusion bodies. These findings increase our understanding of EBOV and its host cell interactions, and provide a basis for future identification of molecular targets for the development of novel therapies against this virus.

Author Contributions: Conceptualization, J.B., L.W., T.H.; investigation, J.B., L.W., B.S.B., T.H.; supervision, T.C.M., T.H.; visualization, J.B., T.H.; funding acquisition, T.H.; writing—original draft preparation, J.B., T.H. All authors have read and agreed to the published version of the manuscript.

Acknowledgments: The authors would like to thank Logan Banadyga (Public Health Agency of Canada) for his help in establishing the coIP procedure and Stephan Becker (Philipps University Marburg) for providing cell lines. We further thank Luca Zaeck (Friedrich-Loeffler-Institut) for technical assistance with the confocal microscope, as well as Allison Groseth and Sandra Diederich (Friedrich-Loeffler-Institut) for technical assistance with the BSL4 work.

References

1. Feldmann, H.; Geisbert, T.W. Ebola haemorrhagic fever. *Lancet* **2011**, *377*, 849–862. [CrossRef]
2. Burk, R.; Bollinger, L.; Johnson, J.C.; Wada, J.; Radoshitzky, S.R.; Palacios, G.; Bavari, S.; Jahrling, P.B.; Kuhn, J.H. Neglected filoviruses. *FEMS Microbiol. Rev.* **2016**, *40*, 494–519. [CrossRef]
3. Bharat, T.A.; Noda, T.; Riches, J.D.; Kraehling, V.; Kolesnikova, L.; Becker, S.; Kawaoka, Y.; Briggs, J.A. Structural dissection of Ebola virus and its assembly determinants using cryo-electron tomography. *Proc. Natl. Acad. Sci. USA* **2012**, *109*, 4275–4280. [CrossRef]
4. Ruigrok, R.W.; Crepin, T.; Kolakofsky, D. Nucleoproteins and nucleocapsids of negative-strand RNA viruses. *Curr. Opin. Microbiol.* **2011**, *14*, 504–510. [CrossRef]
5. Becker, S.; Rinne, C.; Hofsass, U.; Klenk, H.D.; Muhlberger, E. Interactions of Marburg virus nucleocapsid proteins. *Virology* **1998**, *249*, 406–417. [CrossRef]
6. Muhlberger, E.; Weik, M.; Volchkov, V.E.; Klenk, H.D.; Becker, S. Comparison of the transcription and replication strategies of marburg virus and Ebola virus by using artificial replication systems. *J. Virol.* **1999**, *73*, 2333–2342. [CrossRef]
7. Weik, M.; Modrof, J.; Klenk, H.D.; Becker, S.; Muhlberger, E. Ebola virus VP30-mediated transcription is regulated by RNA secondary structure formation. *J. Virol.* **2002**, *76*, 8532–8539. [CrossRef]
8. Hoenen, T.; Shabman, R.S.; Groseth, A.; Herwig, A.; Weber, M.; Schudt, G.; Dolnik, O.; Basler, C.F.; Becker, S.; Feldmann, H. Inclusion bodies are a site of ebolavirus replication. *J. Virol.* **2012**, *86*, 11779–11788. [CrossRef]
9. Lier, C.; Becker, S.; Biedenkopf, N. Dynamic phosphorylation of Ebola virus VP30 in NP-induced inclusion bodies. *Virology* **2017**, *512*, 39–47. [CrossRef]
10. Boehmann, Y.; Enterlein, S.; Randolf, A.; Muhlberger, E. A reconstituted replication and transcription system for Ebola virus Reston and comparison with Ebola virus Zaire. *Virology* **2005**, *332*, 406–417. [CrossRef]
11. Groseth, A.; Charton, J.E.; Sauerborn, M.; Feldmann, F.; Jones, S.M.; Hoenen, T.; Feldmann, H. The Ebola virus ribonucleoprotein complex: A novel VP30-L interaction identified. *Virus Res.* **2009**, *140*, 8–14. [CrossRef] [PubMed]
12. Fang, J.; Pietzsch, C.; Ramanathan, P.; Santos, R.I.; Ilinykh, P.A.; Garcia-Blanco, M.A.; Bukreyev, A.; Bradrick, S.S. Staufen1 Interacts with Multiple Components of the Ebola Virus Ribonucleoprotein and Enhances Viral RNA Synthesis. *mBio* **2018**, *9*. [CrossRef] [PubMed]

13. Chen, J.; He, Z.; Yuan, Y.; Huang, F.; Luo, B.; Zhang, J.; Pan, T.; Zhang, H.; Zhang, J. Host factor SMYD3 is recruited by Ebola virus nucleoprotein to facilitate viral mRNA transcription. *Emerg. Microbes Infect.* **2019**, *8*, 1347–1360. [CrossRef] [PubMed]
14. Batra, J.; Hultquist, J.F.; Liu, D.; Shtanko, O.; Von Dollen, J.; Satkamp, L.; Jang, G.M.; Luthra, P.; Schwarz, T.M.; Small, G.I.; et al. Protein Interaction Mapping Identifies RBBP6 as a Negative Regulator of Ebola Virus Replication. *Cell* **2018**, *175*, 1917–1930. [CrossRef]
15. Gabriel, G.; Feldmann, F.; Reimer, R.; Thiele, S.; Fischer, M.; Hartmann, E.; Bader, M.; Ebihara, H.; Hoenen, T.; Feldmann, H. Importin-alpha7 Is Involved in the Formation of Ebola Virus Inclusion Bodies but Is Not Essential for Pathogenicity in Mice. *J. Infect. Dis.* **2015**, *212* Suppl 2, S316–S321. [CrossRef]
16. Morwitzer, M.J.; Tritsch, S.R.; Cazares, L.H.; Ward, M.D.; Nuss, J.E.; Bavari, S.; Reid, S.P. Identification of RUVBL1 and RUVBL2 as Novel Cellular Interactors of the Ebola Virus Nucleoprotein. *Viruses* **2019**, *11*, 372. [CrossRef]
17. Kruse, T.; Biedenkopf, N.; Hertz, E.P.T.; Dietzel, E.; Stalmann, G.; Lopez-Mendez, B.; Davey, N.E.; Nilsson, J.; Becker, S. The Ebola Virus Nucleoprotein Recruits the Host PP2A-B56 Phosphatase to Activate Transcriptional Support Activity of VP30. *Mol. Cell* **2018**, *69*, 136–145. [CrossRef]
18. Takamatsu, Y.; Krahling, V.; Kolesnikova, L.; Halwe, S.; Lier, C.; Baumeister, S.; Noda, T.; Biedenkopf, N.; Becker, S. Serine-Arginine Protein Kinase 1 Regulates Ebola Virus Transcription. *mBio* **2020**, *11*. [CrossRef]
19. Wendt, L.; Brandt, J.; Bodmer, B.S.; Reiche, S.; Schmidt, M.L.; Traeger, S.; Hoenen, T. The Ebola Virus Nucleoprotein Recruits the Nuclear RNA Export Factor NXF1 into Inclusion Bodies to Facilitate Viral Protein Expression. *Cells* **2020**, *9*, 187. [CrossRef]
20. Martin, S.; Chiramel, A.I.; Schmidt, M.L.; Chen, Y.C.; Whitt, N.; Watt, A.; Dunham, E.C.; Shifflett, K.; Traeger, S.; Leske, A.; et al. A genome-wide siRNA screen identifies a druggable host pathway essential for the Ebola virus life cycle. *Genome Med.* **2018**, *10*, 58. [CrossRef]
21. Uebelhoer, L.S.; Albarino, C.G.; McMullan, L.K.; Chakrabarti, A.K.; Vincent, J.P.; Nichol, S.T.; Towner, J.S. High-throughput, luciferase-based reverse genetics systems for identifying inhibitors of Marburg and Ebola viruses. *Antiviral Res.* **2014**, *106*, 86–94. [CrossRef] [PubMed]
22. Nelson, E.V.; Pacheco, J.R.; Hume, A.J.; Cressey, T.N.; Deflube, L.R.; Ruedas, J.B.; Connor, J.H.; Ebihara, H.; Muhlberger, E. An RNA polymerase II-driven Ebola virus minigenome system as an advanced tool for antiviral drug screening. *Antiviral Res.* **2017**, *146*, 21–27. [CrossRef] [PubMed]
23. Coleman, P.F.; Suttle, D.P.; Stark, G.R. Purification from hamster cells of the multifunctional protein that initiates de novo synthesis of pyrimidine nucleotides. *J. Biol. Chem.* **1977**, *252*, 6379–6385. [PubMed]
24. Jones, M.E. Pyrimidine nucleotide biosynthesis in animals: Genes, enzymes, and regulation of UMP biosynthesis. *Annu. Rev. Biochem.* **1980**, *49*, 253–279. [CrossRef]
25. Lee, L.; Kelly, R.E.; Pastra-Landis, S.C.; Evans, D.R. Oligomeric structure of the multifunctional protein CAD that initiates pyrimidine biosynthesis in mammalian cells. *Proc. Natl. Acad. Sci. USA* **1985**, *82*, 6802–6806. [CrossRef]
26. Christopherson, R.I.; Jones, M.E. The overall synthesis of L-5,6-dihydroorotate by multienzymatic protein pyr1-3 from hamster cells. Kinetic studies, substrate channeling, and the effects of inhibitors. *J. Biol. Chem.* **1980**, *255*, 11381–11395.
27. Irvine, H.S.; Shaw, S.M.; Paton, A.; Carrey, E.A. A reciprocal allosteric mechanism for efficient transfer of labile intermediates between active sites in CAD, the mammalian pyrimidine-biosynthetic multienzyme polypeptide. *Eur. J. Biochem.* **1997**, *247*, 1063–1073. [CrossRef]
28. Evans, D.R.; Guy, H.I. Mammalian pyrimidine biosynthesis: Fresh insights into an ancient pathway. *J. Biol. Chem.* **2004**, *279*, 33035–33038. [CrossRef]
29. Sigoillot, F.D.; Berkowski, J.A.; Sigoillot, S.M.; Kotsis, D.H.; Guy, H.I. Cell cycle-dependent regulation of pyrimidine biosynthesis. *J. Biol. Chem.* **2003**, *278*, 3403–3409. [CrossRef]
30. Sigoillot, F.D.; Kotsis, D.H.; Serre, V.; Sigoillot, S.M.; Evans, D.R.; Guy, H.I. Nuclear localization and mitogen-activated protein kinase phosphorylation of the multifunctional protein CAD. *J. Biol. Chem.* **2005**, *280*, 25611–25620. [CrossRef]
31. Chaparian, M.G.; Evans, D.R. Intracellular location of the multidomain protein CAD in mammalian cells. *FASEB J.* **1988**, *2*, 2982–2989. [CrossRef] [PubMed]
32. Hoenen, T.; Jung, S.; Herwig, A.; Groseth, A.; Becker, S. Both matrix proteins of Ebola virus contribute to the regulation of viral genome replication and transcription. *Virology* **2010**, *403*, 56–66. [CrossRef] [PubMed]

33. Shabman, R.S.; Hoenen, T.; Groseth, A.; Jabado, O.; Binning, J.M.; Amarasinghe, G.K.; Feldmann, H.; Basler, C.F. An upstream open reading frame modulates ebola virus polymerase translation and virus replication. *PLoS Pathog.* **2013**, *9*, e1003147. [CrossRef] [PubMed]
34. Kamper, L.; Zierke, L.; Schmidt, M.L.; Muller, A.; Wendt, L.; Brandt, J.; Hartmann, E.; Braun, S.; Holzerland, J.; Groseth, A.; et al. Assessment of the function and intergenus-compatibility of Ebola and Lloviu virus proteins. *J. Gen. Virol.* **2019**, *100*, 760–772. [CrossRef] [PubMed]
35. Schmidt, M.L.; Hoenen, T. Characterization of the catalytic center of the Ebola virus L polymerase. *PLoS Negl. Trop. Dis.* **2017**, *11*, e0005996. [CrossRef] [PubMed]
36. Huang, Y.; Xu, L.; Sun, Y.; Nabel, G.J. The assembly of Ebola virus nucleocapsid requires virion-associated proteins 35 and 24 and posttranslational modification of nucleoprotein. *Mol. Cell.* **2002**, *10*, 307–316. [CrossRef]
37. Noda, T.; Hagiwara, K.; Sagara, H.; Kawaoka, Y. Characterization of the Ebola virus nucleoprotein-RNA complex. *J. Gen. Virol.* **2010**, *91*, 1478–1483. [CrossRef]
38. Borawski, J.; Troke, P.; Puyang, X.; Gibaja, V.; Zhao, S.; Mickanin, C.; Leighton-Davies, J.; Wilson, C.J.; Myer, V.; Cornellataracido, I.; et al. Class III phosphatidylinositol 4-kinase alpha and beta are novel host factor regulators of hepatitis C virus replication. *J. Virol.* **2009**, *83*, 10058–10074. [CrossRef]
39. Ortiz-Riano, E.; Ngo, N.; Devito, S.; Eggink, D.; Munger, J.; Shaw, M.L.; de la Torre, J.C.; Martinez-Sobrido, L. Inhibition of arenavirus by A3, a pyrimidine biosynthesis inhibitor. *J. Virol.* **2014**, *88*, 878–889. [CrossRef]
40. Katsafanas, G.C.; Grem, J.L.; Blough, H.A.; Moss, B. Inhibition of vaccinia virus replication by N-(phosphonoacetyl)-L-aspartate: Differential effects on viral gene expression result from a reduced pyrimidine nucleotide pool. *Virology* **1997**, *236*, 177–187. [CrossRef]
41. Warren, T.K.; Jordan, R.; Lo, M.K.; Ray, A.S.; Mackman, R.L.; Soloveva, V.; Siegel, D.; Perron, M.; Bannister, R.; Hui, H.C.; et al. Therapeutic efficacy of the small molecule GS-5734 against Ebola virus in rhesus monkeys. *Nature* **2016**, *531*, 381–385. [CrossRef] [PubMed]
42. Luthra, P.; Naidoo, J.; Pietzsch, C.A.; De, S.; Khadka, S.; Anantpadma, M.; Williams, C.G.; Edwards, M.R.; Davey, R.A.; Bukreyev, A.; et al. Inhibiting pyrimidine biosynthesis impairs Ebola virus replication through depletion of nucleoside pools and activation of innate immune responses. *Antiviral Res.* **2018**, *158*, 288–302. [CrossRef] [PubMed]
43. Deans, R.M.; Morgens, D.W.; Okesli, A.; Pillay, S.; Horlbeck, M.A.; Kampmann, M.; Gilbert, L.A.; Li, A.; Mateo, R.; Smith, M.; et al. Parallel shRNA and CRISPR-Cas9 screens enable antiviral drug target identification. *Nat. Chem. Biol.* **2016**, *12*, 361–366. [CrossRef] [PubMed]
44. Kota, K.P.; Benko, J.G.; Mudhasani, R.; Retterer, C.; Tran, J.P.; Bavari, S.; Panchal, R.G. High content image based analysis identifies cell cycle inhibitors as regulators of Ebola virus infection. *Viruses* **2012**, *4*, 1865–1877. [CrossRef] [PubMed]
45. Angeletti, P.C.; Engler, J.A. Adenovirus preterminal protein binds to the CAD enzyme at active sites of viral DNA replication on the nuclear matrix. *J. Virol.* **1998**, *72*, 2896–2904. [CrossRef] [PubMed]
46. Fredman, J.N.; Engler, J.A. Adenovirus precursor to terminal protein interacts with the nuclear matrix in vivo and in vitro. *J. Virol.* **1993**, *67*, 3384–3395. [CrossRef]

Permissions

All chapters in this book were first published by MDPI; hereby published with permission under the Creative Commons Attribution License or equivalent. Every chapter published in this book has been scrutinized by our experts. Their significance has been extensively debated. The topics covered herein carry significant findings which will fuel the growth of the discipline. They may even be implemented as practical applications or may be referred to as a beginning point for another development.

The contributors of this book come from diverse backgrounds, making this book a truly international effort. This book will bring forth new frontiers with its revolutionizing research information and detailed analysis of the nascent developments around the world.

We would like to thank all the contributing authors for lending their expertise to make the book truly unique. They have played a crucial role in the development of this book. Without their invaluable contributions this book wouldn't have been possible. They have made vital efforts to compile up to date information on the varied aspects of this subject to make this book a valuable addition to the collection of many professionals and students.

This book was conceptualized with the vision of imparting up-to-date information and advanced data in this field. To ensure the same, a matchless editorial board was set up. Every individual on the board went through rigorous rounds of assessment to prove their worth. After which they invested a large part of their time researching and compiling the most relevant data for our readers.

The editorial board has been involved in producing this book since its inception. They have spent rigorous hours researching and exploring the diverse topics which have resulted in the successful publishing of this book. They have passed on their knowledge of decades through this book. To expedite this challenging task, the publisher supported the team at every step. A small team of assistant editors was also appointed to further simplify the editing procedure and attain best results for the readers.

Apart from the editorial board, the designing team has also invested a significant amount of their time in understanding the subject and creating the most relevant covers. They scrutinized every image to scout for the most suitable representation of the subject and create an appropriate cover for the book.

The publishing team has been an ardent support to the editorial, designing and production team. Their endless efforts to recruit the best for this project, has resulted in the accomplishment of this book. They are a veteran in the field of academics and their pool of knowledge is as vast as their experience in printing. Their expertise and guidance has proved useful at every step. Their uncompromising quality standards have made this book an exceptional effort. Their encouragement from time to time has been an inspiration for everyone.

The publisher and the editorial board hope that this book will prove to be a valuable piece of knowledge for researchers, students, practitioners and scholars across the globe.

List of Contributors

Mayumi K. Holly and Jason G. Smith
Department of Microbiology, University of Washington, 1705 NE Pacific St., Seattle, WA 98195, USA

Ashley A. Stegelmeier, Jacob P. van Vloten, Robert C. Mould, Elaine M. Klafuric, Jessica A. Minott, Sarah K. Wootton, Byram W. Bridle and Khalil Karimi
Department of Pathobiology, Ontario Veterinary College, University of Guelph, Guelph, ON N1G 2W1, Canada

Madlin Potratz, Luca Zaeck, Michael Christen, Antonia Klein, Conrad M. Freuling, Thomas Müller and Stefan Finke
Friedrich-Loeffler-Institut (FLI), Federal Research Institute for Animal Health, Institute of Molecular Virology and Cell Biology, 17493 Greifswald-Insel Riems, Germany

Verena te Kamp
Thescon GmbH, 48653 Coesfeld, Germany

Tobias Nolden
ViraTherapeutics GmbH, 6020 Innsbruck, Austria

Sara Goglia, Chiara Vicentini, Enrico Tombetti, Micaela Garziano and Mara Biasin
Department of Biomedical and Clinical Sciences-L. Sacco, University of Milan, 20157 Milan, Italy

Irma Saulle and Claudia Vanetti
Department of Biomedical and Clinical Sciences-L. Sacco, University of Milan, 20157 Milan, Italy
Department of Pathophysiology and Transplantation, University of Milan, 20122 Milan, Italy

Mario Clerici
Department of Pathophysiology and Transplantation, University of Milan, 20122 Milan, Italy
Don C. Gnocchi Foundation ONLUS, IRCCS, 20148 Milan, Italy

Gergely Tekes
Institute of Virology, Justus Liebig University Giessen, 35390 Giessen, Germany

Rosina Ehmann
Bundeswehr Institute of Microbiology, 80937 Munich, Germany

Magda Wąchalska, Małgorzata Graul, Aleksandra W. Babnis, Emmanuel J. H. J. Wiertz, Krystyna Bieńkowska-Szewczyk and Andrea D. Lipińska
Laboratory of Virus Molecular Biology, Intercollegiate Faculty of Biotechnology, University of Gdańsk, Abrahama 58, 80-307 Gdańsk, Poland

Patrique Praest and Rutger D. Luteijn
Department of Medical Microbiology, University Medical Center Utrecht, Heidelberglaan 100, 3584CX Utrecht, The Netherlands

Hong-My Nguyen, Kirsten Guz-Montgomery and Dipongkor Saha
Department of Immunotherapeutics and Biotechnology, Texas Tech University Health Sciences Center School of Pharmacy, Abilene, TX 79601, USA

Mélissa K. Mariani, Audray Fortin, Elise Caron, Mario Kalamujic and Espérance Mukawera
CRCHUM—Centre Hospitalier de l'Université de Montréal, Montréal, QC H2X 0A9, Canada

Diana I. Hotea, Dacquin M. Kasumba and Nathalie Grandvaux
CRCHUM—Centre Hospitalier de l'Université de Montréal, Montréal, QC H2X 0A9, Canada
Department of Biochemistry and Molecular Medicine, Faculty of Medicine, Université de Montréal, Montréal, QC H3T 1J4, Canada

Pouria Dasmeh and Adrian W. R. Serohijos
Department of Biochemistry and Molecular Medicine, Faculty of Medicine, Université de Montréal, Montréal, QC H3T 1J4, Canada
Centre Robert Cedergren en Bioinformatique et Génomique, Université de Montréal, Montréal, QC H3T 1J4, Canada

Alexander N. Harrison
CRCHUM—Centre Hospitalier de l'Université de Montréal, Montréal, QC H2X 0A9, Canada
Department of Microbiology and Immunology, McGill University, Montréal, QC H3A 2B4, Canada

Sandra L. Cervantes-Ortiz
CRCHUM—Centre Hospitalier de l'Université de Montréal, Montréal, QC H2X 0A9, Canada
Department of Microbiology, Infectiology and Immunology, Faculty of Medicine, Université de Montréal, Montréal, QC H3T 1J4, Canada

Nicole C. Bilz, Janik Böhnke, Denise Hübner, Uwe G. Liebert and Claudia Claus
Institute of Virology, University of Leipzig, 04103 Leipzig, Germany

Edith Willscher and Hans Binder
Interdisciplinary Center for Bioinformatics, University of Leipzig, 04107 Leipzig, Germany

Megan L. Stanifer
Schaller Research Group at CellNetworks, Department of Infectious Diseases, Virology, Heidelberg University Hospital, 69120 Heidelberg, Germany

Steeve Boulant
Schaller Research Group at CellNetworks, Department of Infectious Diseases, Virology, Heidelberg University Hospital, 69120 Heidelberg, Germany
Research Group "Cellular Polarity and Viral Infection" (F140), German Cancer Research Center (DKFZ), 69120 Heidelberg, Germany

Olga Dolnik
Institute of Virology, Philipps University Marburg, Hans-Meerwein-Str. 2, 35043 Marburg, Germany

Katharina Grikscheit and Stephan Becker
Institute of Virology, Philipps University Marburg, Hans-Meerwein-Str. 2, 35043 Marburg, Germany
German Center for Infection Research (DZIF), Partner Site: Giessen-Marburg-Langen, Hans-Meerwein-Str. 2, 35043 Marburg, Germany

Yuki Takamatsu
Institute of Virology, Philipps University Marburg, Hans-Meerwein-Str. 2, 35043 Marburg, Germany
Department of Virology I, National Institute of Infectious Diseases, Tokyo 208-0011, Japan

Ana Raquel Pereira
Oxford Nano Imager, Linacre House, Banbury Rd, Oxford OX2 8TA, UK

Samy Sid Ahmed and Oliver T. Fackler
Department of Infectious Diseases, Integrative Virology, University Hospital Heidelberg, 69120 Heidelberg, Germany

Nils Bundgaard and Frederik Graw
BioQuant - Center for Quantitative Biology, Heidelberg University, 69120 Heidelberg, Germany

Janine Brandt, Lisa Wendt, Bianca S. Bodmer, Thomas C. Mettenleiter and Thomas Hoenen
Institute of Molecular Virology and Cell Biology, Friedrich-Loeffler-Institut, 17493 Greifswald-Insel Riems, Germany

Index